AF880248

Springer Proceedings in Earth and Environmental Sciences

Series Editors

Natalia S. Bezaeva, The Moscow Area, Russia

Heloisa Helena Gomes Coe, Niterói, Rio de Janeiro, Brazil

Muhammad Farrakh Nawaz, Institute of Environmental Studies, University of Karachi, Karachi, Pakistan

The series Springer Proceedings in Earth and Environmental Sciences publishes proceedings from scholarly meetings and workshops on all topics related to Environmental and Earth Sciences and related sciences. This series constitutes a comprehensive up-to-date source of reference on a field or subfield of relevance in Earth and Environmental Sciences. In addition to an overall evaluation of the interest, scientific quality, and timeliness of each proposal at the hands of the publisher, individual contributions are all refereed to the high quality standards of leading journals in the field. Thus, this series provides the research community with well-edited, authoritative reports on developments in the most exciting areas of environmental sciences, earth sciences and related fields.

Fernando Luís Almeida
José Carlos Morais • José Duarte Santos
Editors

Multidimensional Sustainability: Transitions and Convergences

Proceedings of ISPGAYA 2022

Springer

Editors
Fernando Luís Almeida
Instituto Superior Politécnico Gaya
Vila Nova de Gaia, Portugal

José Carlos Morais
Instituto Superior Politécnico Gaya
Vila Nova de Gaia, Portugal

José Duarte Santos
Instituto Superior de Contabilidade e Administração do Porto
São Mamede, Infesta, Portugal

ISSN 2524-342X ISSN 2524-3438 (electronic)
Springer Proceedings in Earth and Environmental Sciences
ISBN 978-3-031-24891-7 ISBN 978-3-031-24892-4 (eBook)
https://doi.org/10.1007/978-3-031-24892-4

© The Editor(s) (if applicable) and The Author(s), under exclusive license to Springer Nature Switzerland AG 2023
This work is subject to copyright. All rights are solely and exclusively licensed by the Publisher, whether the whole or part of the material is concerned, specifically the rights of translation, reprinting, reuse of illustrations, recitation, broadcasting, reproduction on microfilms or in any other physical way, and transmission or information storage and retrieval, electronic adaptation, computer software, or by similar or dissimilar methodology now known or hereafter developed.
The use of general descriptive names, registered names, trademarks, service marks, etc. in this publication does not imply, even in the absence of a specific statement, that such names are exempt from the relevant protective laws and regulations and therefore free for general use.
The publisher, the authors, and the editors are safe to assume that the advice and information in this book are believed to be true and accurate at the date of publication. Neither the publisher nor the authors or the editors give a warranty, expressed or implied, with respect to the material contained herein or for any errors or omissions that may have been made. The publisher remains neutral with regard to jurisdictional claims in published maps and institutional affiliations.

This Springer imprint is published by the registered company Springer Nature Switzerland AG
The registered company address is: Gewerbestrasse 11, 6330 Cham, Switzerland

Paper in this product is recyclable.

Preface

We are becoming increasingly aware that the resources made available by the planet are finite and that the planet has a limited regeneration capacity. The UN's Agenda 2020 has the merit of outlining a new vision of the planet and humanity as we know them, defining concrete objectives and indicators for their fulfilment, and it has the even greater merit of arguments that working in partnership, in a network logic, is the path to the feasibility of this agenda and the viability of acceptable scenarios for the planet in the short-, medium-, and long-term future.

The path to sustainability crosses the social, cultural, economic, political, and environmental domains.

The congress which ISPGAYA organized in Vila Nova de Gaia, Portugal, on 29 and 30 September 2022 is based on the theme of sustainability, meeting the current academic, social, and political agenda.

In the same event are gathered several different areas: accountancy and fiscal organization; electronics and industrial automation; energy efficiency; information technology and cybersecurity; management and administration; marketing; mechanics; sustainability and innovation; tourism and leisure.

Vila Nova de Gaia, Portugal

Fernando Luís Almeida
José Carlos Morais
José Duarte Santos

Introduction

Nowadays, the community of teachers, students, researchers, professionals and citizens in general is gathered around the theme of sustainability. We are becoming increasingly aware that the resources made available by the planet are finite and that the planet has a limited regeneration capacity. The creation of wealth in the most diverse forms and the transformation of raw materials available on the increasingly globalised market entail the consideration of development scenarios, in their social, economic, technological and environmental aspects in an increasingly smart sense. The need for green goes hand in hand with sustainable development objectives outlined by the UN, foreshadowing scenarios for our planet that will necessarily lead to the issue of climate change.

The environmental issue is perhaps the most visible in an agreement that should combine the characteristics of being international, inter-institutional, multicultural and interdisciplinary. If the UN's Agenda 2020 has the merit of outlining a new vision of the planet and humanity as we know them, defining concrete objectives and indicators for their fulfilment, it has the even greater merit of arguing that working in partnership, in a network logic, is the path to the feasibility of this agenda and the viability of acceptable scenarios for the planet in the short-, medium- and long-term future.

New forms of energy generation and use are at stake, circumventing the absolute dependence on solid fuels for human activity, making way for environmentally friendly technology. But the changes required are more complex than the mere application of new technology. The structural changes that have been identified refer to fundamental changes in the world status quo and are mainly social changes, prefiguring individual and collective action in various fields and a new vision on the political programming of local, regional and global development. A new planet is what is basically under discussion, and the changes may take longer than one might imagine, but we are on our way.

Issues of redistribution of wealth between social groups or financing the development of countries with emerging economies are not easy to address these days, and indeed never have been in human history. But the notion of the association between indicators of social wellbeing and quality of life of populations and the wellbeing and health of the planet has never been so explicit as it is now. Waste production and its treatment, recycling, the issue of waste (in several domains), poverty, education or state action, social and environmental policies and the attitude of organisations in terms of social responsibility are seen by public opinion as inseparable matters.

These days are also a reminder that old rifts between economic blocs are disruptive in terms of international efforts towards sustainability. The theme of war between countries and economic and political blocs is very present, but it also invites to an urgent consideration of the use of new energy resources and a redesigning of old dependencies and financial logics associated with the use of fossil fuels in such sensitive issues as domestic heating.

Basically, humanity is invited to rethink itself in terms of the consumption that characterises it, what to consume, by whom and under what conditions.

This path to sustainability crosses the social, cultural, economic, political and environmental domains. The development of new environmentally friendly technology only makes sense when accompanied by new modes of production and consumption, new configurations of productive design in organisations and of marketing. The objectives of production and consumption can no longer be restricted to profit and pure financial logic. The economic domain also necessarily has social dimensions.

The congress which ISPGAYA organised in Vila Nova de Gaia, Portugal, on 29 and 30 September 2022 is based on the theme of sustainability, meeting the current academic, social and political agenda. This event brings together the different scientific areas of this higher education institution and seeks to highlight the set of interdisciplinary and multilevel efforts to change the existing reality, supported by the affinity of ideas and attitudes, knowledge and practices.

Contributions from different national and international institutions that integrate the technological and scientific system were made, asserting the concertation of strategies at local, regional and global levels, and the interconnection between synergistic contributions of the living forces to mobilize towards the consolidation of the planetary sustainability over the next few decades.

In the event are gathered areas such as: accountancy and fiscal organisation; electronics and industrial automation; energy efficiency; information technology and cybersecurity; management and administration; marketing; mechanics; sustainability and innovation; tourism and leisure.

It only makes sense to approach sustainability with a multidimensional approach, taking into account the transitions and convergences that have taken place, affirming it as the theme of reference for the academy and for organisations in general, in essence, for human action.

Keywords: Entrepreneurship; Smart cities; Technology innovation; Sustainability; Green technologies

An Analytical Synopsis of the Contents of Each Part

Accountancy and Fiscal Organization

A vast wave of change is making itself felt in an increasingly noticeable way in the sense that organizations, public and private, show to the outside, but also to their internal audience, sustainability concerns, and their indicators are seen by the set of employees closer or further away as indicators of security or guarantee of continuity in the medium and long term of these same organisations. The sustainability reporting of organisations is gaining ground, being appreciated by the insurance industry, credit entities, reflected directly in stock market prices and also in the reputation of companies. The public administration (local and national) is also in this collaborative logic, working in partnership and assuming the values of sustainability indicators of central importance with the local populations. There are several tools that can be used to achieve sustainability goals, being mandatory the references to ISO standards (international standard organization) that resulted, directly or indirectly, from the United Nations's Green Agenda. These subjects are addressed in Chaps. 1, and 2.

Electronics and Industrial Automation

This part starts with Chap. 3 and links induction heating, energy efficiency, eddy current, microprocessor and power electronics. Electromagnetic field (EMF) imaging, magnetic field sensors integration, and electrical machines improvement are the issue of Chap. 4. Next, sustainability is the issue of the Chap. 5. In the same line, sustainability and energy saving are addressed in Chap. 6.

Energy Efficiency

The part starts with Chap. 7. Sustainability and energy consumption are the subject of Chap. 8, aiming to analyse the hydrogen ecosystem in technical, economic and geopolitical aspects facing the international efforts in sustainable decarbonisation solutions and in the search for energetic independence.

Information Technology and Cybersecurity

Chapter 9 approaches varying attitudes towards eBooks among individuals, specifically between male and female. Gamification, cybersecurity and lecturing are the subject of Chap. 10. In the Chap. 11, innovation is addressed as a key driver to meet the challenges of sustainability. Sustainability, active learning, soft skills, digital skills are addressed in Chap. 12. Sustainable education, sustainability, learning environments and university education are the subject of the Chap. 13, following the methodology of integrative review of the literature.

Management and Administration

Entrepreneurship, sustainable entrepreneurship, higher education and social business are the keywords of Chap. 14. Personality, sustainability, equity crowdfunding, asymmetric Information and venture finance are the issue in the Chap. 15. Private agricultural insurance is studied in Chap. 16. Chapter 17 examines how sustainability is incorporated into decision making today and what might lead to strengthening the successful integration of sustainability into business activities. Next, in Chap. 18, ISO 37120 and sustainable cities, sustainable development goals, sustainable development and urban planning are discussed.

Marketing

Marketing has not been alien to the developments that can be placed in the area of sustainability, integrating a global movement that is called the green wave. The strategic vision on the design and production of products and services, with attention to factors such as pollution control, product reuse and recycling, is a current of thought that has been installed in the marketing area. This area considers as determinant the marketing processes that starts internally, in the organization's closest employees, counting on a contagion effect that will have positive impacts on the perspectives of stakeholders such as customers and society in general. This part is composed by Chap. 19, addressing issues such as sustainability, corporate social responsibility, brand image, brand value and structural equations.

Mechanics

Chapter 21 links topics such as robotics, AGV, ArUco, Indoor navigation, OpenCV, Raspberry Pi, Arduino. Chapter 20 addresses alternatives to experimental and expensive tests by performing virtual testing and stress analysis through the use of finite element method.

Sustainability and Innovation

House building, traditional lightning, intelligent lightning, plastic and copper reduction and sustainability are addressed in Chap. 22. Innovation, Society 5.0 and sustainability are approached in Chap. 23. Next, Chap. 24 brings the issue of mathematics, statistics and modelling. In Chap. 25, sustainability is addressed based on a program were socio-emotional learning, executive function, wellbeing and teachers are interconnected. Current times ask, also, for global questionings such as those addressed in Chap. 26. This part includes, also, the Chap. 27 addressing recycling in cork industry.

Tourism and Leisure

Chapter 28 brings some insights on qualitative approach to studies in tourism. Museums and tourism are addressed in Chap. 29. The importance of fairs to sustainability is addressed in Chap. 30. In Chap. 31 is presented an embryonic approach embodied in an exploratory study that includes a brief survey of cases of sustainable use of the railroad heritage, and the study points to different degrees of relationship that museums establish with the community, involving it in their activities, which can be a reason for attracting tourists. Social responsibility and ethics are linked with education on "Approaches to Sustainable Tourism Through Education". In Chap. 32 is given an managerial advice, suggesting that companies in the tourism sector should act essentially in the promotion of programs that involve leisure, holidays, culture, nature and gastronomy to obtain a greater impact, thus increasing the demand. The Covid-19 pandemic impacts, in the order of the day, are addressed in Chap. 33.

Contents

About the Editors

Fernando Luís Almeida Holds a PhD in Engineering and Computer Science from the Faculty of Engineering of the University of Porto (FEUP). He holds a Masters in Innovation and Technological Entrepreneurship and a degree in Computer Engineering and Computing by FEUP. In the last 10 years of activity, he has performed professional functions in the field of university polytechnic education, business, and applied research in the areas of innovation, entrepreneurship, and information technologies.He is an Integrated Researcher at INESC TEC's Center for Innovation, Technology and Entrepreneurship. He is a founding member of the US-based International Association of Innovation Professionals (IAOIP).

Coordinating Professor at ISPGAYA, he is Visiting Researcher in Budapesti Corvinus Egyetem, Hungary.

José Carlos Morais Holds a PhD in Sociology from Faculdade de Letras of University of Porto since 2010. He is the Director of the *Journal of Psicologia, Educação e Cultura*. He is a Researcher at CEOS.PP (Center for Organizational and Social Studies of the Polytechnic of Porto). From 2014 to the present, he has held the presidency of the ISPGAYA Technical-Scientific Council. He teaches courses in undergraduate and masters cycles in the area of social sciences and humanities.

Since 2017, he has regularly participated in international collaboration programs – ERASMUS+.

José Duarte Santos holds a PhD in Management and Business Directing Management from the University of Vigo. He is a specialist in Marketing and Advertising and has a degree in Business Sciences with specialization in Marketing from Instituto Superior de Linguas e Administração. He is a Master in Marketing from Portucalense University. Since 1999, has been a professor in higher education in Portugal in the fields of management and marketing. He is currently a marketing professor at the Accounting and Business School of the Polytechnic of Porto (ISCAP/P.PORTO).

He is also a senior researcher at the CEOS.PP – Centre for Organizational and Social Studies of the Polytechnic of Porto, Portugal. His main research areas are Strategic Marketing and Digital Marketing Strategies.

Part I

Accountancy & Fiscal Organization

1 Sustainability Reporting in the Portuguese Municipalities: An Empirical Analysis of the 25 Largest Municipalities

Fernando Pinto, Cristina Couto, Dulce Gomes, Joaquim Oliveira, Ronaldo Silva, and Domingos Oliveira

1.1 Introduction

In view of the growing global concern about environmental damage and its implications for everyday life, it means that the theme of sustainability is of a strong importance both at the business level and at the public level. It is because of these implications and the very evolution of society with regard to social demands and community life that are central to the theme of sustainability. Our huge productive economy leads us to an ever-increasing consumerism, leading to products being consumed, destroyed, replaced and disposed of at an accelerated rate. For this reason, it is urgent to adopt a development model based on economic growth that integrates social progress and the protection of the planet.

The *sustainability report is* an expression that was created in the preparation of the Rio +20 Conference.[1] Therefore, this issue is relatively recent and is not yet a common practice between entities (private or public). The essence of the management report is that communication on the economy, the social and the environment is crucial for governments and entities (private or public) to assess their contribution to sustainable development.

The sustainability report appears as a complement to traditional financial reporting and today the largest entities (public or private) voluntarily issue (or should) their sustainability reports, although it is not a common practice among all entities as it should be and many of these reported are not audited, although this affects the credibility of their information.

[1] The United Nations Conference on Sustainable Development – or Rio +20 – took place in Rio de Janeiro, Brazil, on June 20 and 22, 2012.

F. Pinto (✉) · C. Couto · D. Gomes · J. Oliveira · R. Silva · D. Oliveira
ISPGAYA, Vila Nova de Gaia, Portugal
e-mail: fmpinto@ispgaya.pt

© The Author(s), under exclusive license to Springer Nature Switzerland AG 2023

F. L. Almeida et al. (eds.), *Multidimensional Sustainability: Transitions and Convergences*, Springer Proceedings in Earth and Environmental Sciences,
https://doi.org/10.1007/978-3-031-24892-4_1

In addition to the importance of its content, the sustainability report allows benchmarking of entities that reveal the extent of the gap between them in terms of sustainability organizations and practices.

Together with financial information, the sustainability report is essential to understand and assess the contribution of organizations to sustainable development (economic, social and environmental).

According to Hahn and Kühnen (2013), the focus of the sustainability report had a historical evolution: in the 1970s, traditional financial reporting was sometimes complemented by social reports; in the 1980s, the focus shifted to environmental issues.

There are several frameworks for guiding the preparation of SR and for disseminating sustainability performance: The Global Reporting Initiative (GRI) is the most widely used framework for this purpose; International Organization Standards of Standardization (ISO) namely ISO 14001 – Environmental Management Systems and ISO 37120 – Sustainable cities and communities, which is of particular importance for this study, UN Sustainable Development objectives in accordance with the 2030 Agenda, sector-specific reports, company-specific reports, etc.

Integrated reports should also be analysed in the future, integrating finances and SR into the same report. These frameworks are not mutually exclusive, and it is customary to find sustainability reports using and reporting according to various frameworks.

The biggest benefits of sustainability reporting are: transparency of sustainability data is the most important, but also: organizational governance, stakeholder engagement and data comparability.

Referring to weaknesses or critical sustainability reports that need to be resolved, some are pointed out: only a few sustainability reporters, the principle of materiality should be observed in the sustainability report (only material information should be reported), lack of comparability, many sustainability reports are not guaranteed or audited (which affects their credibility) and there is no consensus on who should ensure SR and, finally, costs (SR can represent a burden for smaller entities).

On the future of sustainability reporting, 5 pathways or trends are considered: consolidation of sustainability reports, integrated sustainability reports, specific or sectoral sustainability reports, to make sustainability mandatory by governments and, less optimistic, the sustainability report will disappear.

In any case, and given the history of financial reporting, it seems that more standardization and regulation will be needed if we consider that sustainability is really important.

Confirming the importance of this issue for accounting, the international accounting standardization body IFRS announced on 3 November 2021 the creation of a new standardization board: the International Sustainability Council to respond to companies' demand for reporting on environmental, social and governance issues.[2]

To study sustainability reports in Portuguese municipalities is to understand which path is being taken by elected officials and managers at the local level. Yes,

[2] Known as ESG matters.

to understand first of all whether the information provided in these reports is about sustainability and, if there is a check whether the way in which the reports are perceived, is in line with accountability, transparency and the disclosure of non-financial information.

It is common understanding that elected officials and managers have an obligation to manage resources efficiently, effectively, and economically. In other words, they have an obligation that in the public sphere resources belonging to all are being managed and that therefore citizens and other stakeholders should have access to information at the level of sustainability. In the context of municipalities and in the light of the agency's theory, it is expected that sustainability reports will be used either by mayors or by citizens or other stakeholders as a means of monitoring political action (Maher, 1979; Zimmerman, 1977.)

We are faced with a relationship between the Mayor (principal) and the citizens and other stakeholders (agent) which will necessarily lead to accountability. This means that in view of the mayor's objectives, citizens will check whether their objectives are being met and how they are being met in the light of sustainability. We will have control and monitoring of procedures and activities here from the outset. This control and monitoring is undoubtedly increasingly evident due to the importance and relevance that the theme of sustainability has today.

The theory of legitimacy has been used as a theoretical reference with regard to studies on sustainability (Greco et al., 2015), in the sense that this theory has underpinning the idea of social contract, that is, mayors have the fundamental role of understanding how stakeholders and other external users see municipalities. It is essential to provide information and that it is available to all users, both internal and external, to retain a positive image of the municipalities. It is essential that local government correctly uses legitimation strategies.

In view of the above, our research problem is: is the inclusion of sustainable development objectives associated with socio-economic indicators such as education and GDP?

Thus, considering sustainability, the present study aims to:

- to know which of the largest Portuguese citizens hold sustainability reports;
- identify what the regulations are followed in the preparation of the sustainability report;
- identify the objectives for sustainable development included in the projects and best practices of the municipalities evaluated;
- to evaluate the association between the frequency with which the objectives for sustainable development in projects and good practices are included and the level of education of the population and GDP of the municipalities.

In order to achieve the proposed objectives, the following research questions arise:

- Q1: Which municipalities considered to be large-scale publish sustainability reports?

- Q2: Is ISO 37120 considered in the preparation of sustainability reports?
- Q3: What are the objectives of sustainable development taken into account in the projects and good practices developed by large municipalities?
- Q4: Are projects and good practices developed in large municipalities associated with socio-economic indicators such as education and GDP?

This study contributes to the literature on sustainability reports, namely, to seek to identify what information is disclosed and whether there is an association between the adoption of the SDGs and two socio-economic variables: GDP and education.

The research methodology used is qualitative and quantitative and the data were collected through the analysis of the websites of the larger municipalities. For the processing of the collected data, descriptive and statistical analysis was used. The results show that many are large municipalities that prepare and publish sustainability reports on their website. They also show that there is a statistically significant relationship between most SDGs and the two variables: GDP per capita and level of education.

In addition to this introduction, the study is organized as follows: in Sect. 1.1, we present the literature review. In Sect. 1.2, we present the theoretical framework, as well as the problem of investigation, the questions and the hypotheses of investigation. Section 1.3 presents the research methodology, in particular the method of data collection and the method of data analysis. In Sect. 1.4, the results are displayed. Finally, in Sect. 1.5, the conclusions, limitations and clues to future research are presented.

1.2 Literature Review

1.2.1 Sustainability: Definition and Objectives

Our huge productive economy leads us to be depleted, and products to be consumed, destroyed, replaced and disposed of at an accelerated pace, impacting ecosystems and the biosphere. On the other hand, there are high levels of social inequalities and poverty, and lack of corporate management ethics.

There is an urgent need to adopt a model of development, based on economic growth that integrates social progress and the protection of the planet, capable of ensuring to future generations the opportunities and quality of life that the current generations (still) enjoy.

Literally "sustainability", which derives from Latin "sustentare" can be understood as the ability to sustain a system, referring to the intertemporal character[3] of the maintenance of its attributes.

As stated by Haris Alibasic in "Sustainability And Resilience Planning For Local Governments", the term sustainability is often described as a positive effect on social, economic, environmental and governmental issues within an organizational structure. The main focuses on sustainability programs and resilience programs are

[3] Meaning of sustaining, supporting, conserving in good condition, maintaining, resisting.

the aptitude of communities and organizations in adapting to environmental, social and economic change. Sustainability and resilience topics are under incessant reviews and inspections.

It is often regarded as an additional cost to society. However, with increased pressure on rising energy costs, climate change policies and less financial influence, sustainability (in the short and long term) and resilience are increasingly seen as part of the solution.

There are several ideas, concepts, paradigms and theories used to build sustainability, sustainability development grids and the corresponding economic, social and environmental bases, both for organizations and for society.

This concept of sustainability has increasingly attracted the attention of academics and professionals worldwide over the past two decades. Sustainability deals with a balanced integration between the performances of three pillars: social, environmental and economic, as well as human life with society, the environment and the economy, aiming at the benefit of current and future generations (Ranjbari et al., 2021).

> "Sustainability is now an ambiguous catchword that results in a series of parallel practical approaches, often superficial and/or contradictory, not always coincident with the originally proposed objectives". (Guerra, 2011)

According to the UNITED NATIONS (United Nations) in 1987, "Sustainability is meeting the needs of the present without compromising the ability of future generations to meet their own needs".

Relate directly to economic and material development, using natural resources intelligently, without harming the environment, so that they remain in the future, ensuring sustainable development, this being "the development *that meets the needs of the present without compromising the capacity of future generations to meet their own needs, ensuring the balance between economic growth, care for the environment and social well-being*" (Brundtland, 1987).

The definition of "Sustainable Development" can be found in the famous report "Our Common Future", authored *by the World Commission on Environment and Development* (WCED), established by the UN and led by former Norwegian Prime Minister Gro Brundtland. According to WCED, sustainable development is *one that meets the needs of the present without compromising the ability of future generations to meet their own needs.*

It is universally understood that this concept of sustainable development, which promotes human well-being and at the same time the preservation of the natural environment, especially when assessing government issues, has proved elusive (Adger, 2009) superficial (Redclift, 2005) and inconsequential (Carter, 2001; Lélé, 2018).

Alibasic, in "Sustainability And Resilience Planning For Local Governments" advocates the definition by the UN on the fundamental principles of sustainability and sustainable development as the rational handling of resources in the present, by organizations and individuals, without compromising the needs of future generations.

Elkington in 1997 created the triple baseline definition (**TLB**) to measure and assess the social and environmental impact of sustainability, in addition to economic benefit.

Initially, this vision focused on the private sector; however, a broader view of the principles is also applied to the public sector. Thus, with the evolution of sustainability, his static view of the three pillars was forced to evolve, giving rise to the quadruple baseline theory **(QLB), which** approaches the subject from a governance perspective.

The sustainability report is a document integrating information on the performance of entities, on issues and indicators that reflect significant (positive and negative) impacts, which can substantially influence stakeholder assessments and decisions at economic, **social and environmental level** l. It integrates financial and non-financial indicators, thus reporting to stakeholders on how they develop their activities and manage their sustainability risks.

Among the various reporting/communication tools, such as report and accounts, environmental report; corporate citizenship report; integrated report; newsletter/site/intranet/, the sustainability report and the ESG report are important.

The sustainability report is a document that addresses several stakeholders (customers, investors, employees, partners, regulators, among others) and reports the economic, social and environmental performance of entities in material aspects.

On the other hand, the ESG Report, with a more technical language, addresses investors and reports the performance ESG: Environmental (Environmental) – Social (Social) – Governance.

The report on sustainability can be developed by GRI[4] (Global Reporting Initiative) standards, which is the most widely used model in the world. GRI, whose Sustainability Reporting Standards are developed with real contributions from various stakeholders and rooted in the public interest, is an essential tool to ensure transparency, and helps entities and governments around the world to understand and communicate their impact on critical sustainability issues such as climate change, human rights, governance and social welfare.

The latest milestones in the implementation of measures for sustainable development are presented (Fig. 1.1).

2014:

Directive 2014/95/EU of the European Parliament and the Council of 22 October 2014 (amending Directive 2013/34/EU on the disclosure of non-financial information and financial information on diversity by certain large companies and groups), of a mandatory nature for public interest undertakings with more than 500 employees, motto "Together for new growth". It aims to increase the transparency of social

[4]Neither GRI 101, 102 th 103 are standards applicable to all organisations; The GRI series 200, 300 and 400 are specific standards related to the 3 ESG pillars (Economic, Environmental and Social). Organizations select and use only those that relate to their material topics.

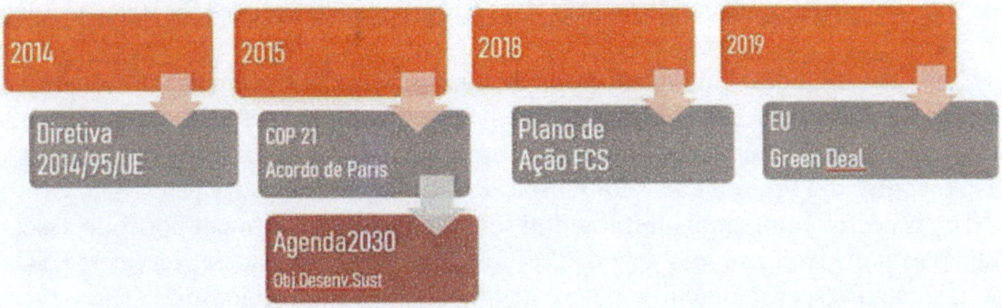

Fig. 1.1 Agenda for sustainable development

and environmental information provided by companies in the various sectors in a comparable way in all Member States;[5]

This Directive entered into force in 2018 in Portugal, transposed by DL 89/2017, leading to the analysis of companies' performance and identification of sustainability risks in society, strengthening consumer and investor confidence.

2015:

PARIS AGREEMENT – It is a treaty that commits 195 countries under the UN Framework Convention on Climate Change (UNFCCC) to reduce GHG emissions from 2020, and keep global warming below 2 °C. The EU aims to be the first economy without climate impact by 2050.

This treaty, signed on 12 December 2015, entered into force 30 days after its ratification in 55 countries representing 55% of greenhouse gas emissions. In less than a year, the goal had been reached.

AGENDA 2030 SUSTAINABLE DEVELOPMENT GOALS: it was adopted by the United Nations General Assembly in September 2015, agreed by the 193 Member States. Of a universal and inclusive nature. Based on the experience of the 8 Millennium Development Goals between 2000 and 2015, the 17 sustainable development goals (SDGs) and 169 targets were defined by 2030, measured through indicators.

It combines social, economic and environmental concerns by promoting the collective effort and shared responsibility of all countries, with the aim of promoting prosperity and well-being for all, protecting the environment and combating climate change.

THE EUROPEAN COMMISSION's 2018 ACTION PLAN to FINANCE SUSTAINABLE GROWTH: This Plan aims to create financial strategies to support the EU's climate and sustainable development agenda by financing the green transition using private capital, achieving sustainable finance and sustainable and responsible investments (including non-financial considerations such as environmental and social, in decision-making, in order to better manage the company's risks and

[5] Source: https://eur-lex.europa.eu/legal-content/PT/TXT/PDF/?uri=CELEX:32014L0095&from=PT

generate long-term sustainable returns "United Nations – Principles for Responsible Investment".

This plan includes actions covering the main agents of the financial system in matters of standardization of procedures (classification system (taxonomies)), responsibilities (sustainability and disclosure), advice (on sustainable investment), sustainability of requirements and transparency in information.

The need for comparable information for responsible investment portfolio asset managers has given rise to a new market for sustainability financial services: ESG ratings, rankings, sustainability indices and (socially) responsible funds. These services are provided to managers with the analytical information needed to assess ESG performance, risks and opportunities of assets under management. Thus, "Sustainable Investment" is defined as Investment in an economic activity that:[6]

E – contributes to an environmental objective
S – contributes to a social objective, provided that such investments do not significantly prejudice any of these objectives
G – provided that the companies receiving the investment use good governance practices

2019:
GREEN DEAL – EUROPEAN GREEN PACT: Package of measures and legislative proposals to make EU policies on climate, energy, transport and taxation, which will enable businesses and citizens to benefit from a sustainable ecological transition, with the intention of achieving a reduction by 2030 of net GHG emissions by at least 55% (compared to 1990). Supported by investments in green technologies, sustainable solutions and new businesses, it contributes to improving the well-being and health of citizens and future generations. It could be a new EU growth strategy.

1.2.2 The United Nations: Objectives for Sustainable Development (SDGs)

As already mentioned, the sustainable development goals (SDGs) are a global agenda adopted during the United Nations Summit on Sustainable Development in September 2015 signed by more than 190 countries to be achieved by 2030.

There are 17 Sustainable Development Goals, and 169 targets that define global priorities and aspirations for 2030 and require global action by governments, businesses and civil society to eradicate poverty and create a life with dignity and opportunities for all, within the limits of the planet.

[6] DJSI Dow Jones Sustainability Index: Released by S&P Dow Jones Indices and RobecoSAM9 in 1999. ESG factors are covered by more than 120 questions in each industry grouped into about 20 to 25 criteria; FTSE4Good: The company evaluation methodology results from an analysis of the 3 pillars of ESG sustainability: environment, social and governance; sustainalytics: ESG information and research analysis provider that supports investor decision-making.

Fig. 1.2 Sustainable development goals as adopted by all UN member states. (United Nations Department of Economic and Social Affairs, 2016)

The analysis of the evolution of sustainability must take into account certain particularities. There are SDGs that interconnect and complement each other (e.g. SDG 2 (eradicating hunger) and 3 (quality health)) so their indicators should be interpreted according to the goal/objective, and the same indicator may have adverse impacts depending on the SDGs in which they are included.[7]

The UN database has an "o.f.Related indicators" field linking SDS indicators with other existing indicators. It should be noted that the SDGs with more interconnections with other indicators are SDGs 1 (eradicating poverty) and 11 (sustainable cities and communities) (Fig. 1.2).[8]

At national level, 6 SDGs have been established as priorities and strategic:[9]

Quality Education – SDG 4
Gender Equality – SDG 5
Industry, Innovation and Infrastructure – SDG 9
Reduction of Inequalities – SDG 10
Climatic Action – SDG 13
and Marine Life Protection – SDG 14

[7] By way of illustration, GDP growth has a positive impact on the economy (SDGs 8) but may have negative impacts on environmental SDGs; or the number of road and air passengers, the growth of which would be considered favorable for SDG9 (infrastructure), but whose impact would be considered negative if inserted in an environmental SDG.

[8] Publication ODS 2022; INE.

[9] It should be said that no reasons for the choice of these SDDs as priorities have been presented, nor has there been any intervention by the Assembly of the Republic in the selection and validation of these priorities.

The 2030 Agenda is extremely comprehensive and challenging, making real advances require deep involvement of diverse organizations representing civil society, local governments, the private sector and academia.

The achievement of the sustainable development goals will be achieved through cooperation, as well as the action of dynamic regulated participatory *initiatives (through deliberative tools such as (i) the 21 Local Agendas (A21L), (ii) participatory budgets (POs) and (iii) other more traditional mechanisms such as municipal environmental plans*), among all individual and collective subjects. The agenda should monitor the commitments established, as well as the actions of the necessary resources and the intervention of the various interdependent actors of society.

As appropriately mentioned, compliance with the 2030 Agenda for Sustainable Development is dependent on the active role of various actors, with local and regional governments, more specifically local authorities, agents of high relevance, because they are a level of governance closer to the population, with democratic legitimacy, and relate, directly or indirectly, to the various goals contained in all the SDDs. Thus, they play an important role in the quality of the agenda and sustainability (Lafferty, 2001a) and are in a privileged position to promote the sustainability report (Williams et al., 2010).

Regarding the institutional model, the general coordination of implementation of the 2030 Agenda was attributed to the Ministry of Foreign Affairs, in conjunction with the Ministry of Planning and Infrastructure, and the coordination of each of the SDDS was attributed to a Ministry responsible for its implementation, monitoring and review.

Other entities, such as the National Institute of Statistics (INE), which monitors and implements the 2030 Agenda at the statistical level, and the Agency for Development and Cohesion are also involved.

The National Institute of Statistics – INE provides a portal on the subject (permanently updated) and publishes, since 2018, annual reports with progress in the main indicators.[10] The monitored indicators are those defined by the United Nations for each SDS.

The SDGs follow the Millennium Goals (MDGs), extending the challenges that must be addressed to eradicate poverty and cover a wide range of interrelated topics, dividing into the dimensions of sustainable development that are set below:

- Social: related to human needs, health, education, improvement of quality of life and justice.
- Environmental: it deals with the preservation and conservation of the environment, with actions ranging from the reversal of deforestation, protection of forests and biodiversity, combating desertification, sustainable use of oceans and marine resources to the adoption of effective measures against climate change.
- Economic: addresses the use and depletion of natural resources, waste production, energy consumption, among others.

[10] Portal INE: bit.ly/2mDiPOG

- Institutional: it concerns the capabilities to implement the SDGs.
- To achieve the 17 Objectives (SDGs), different local, regional and national initiatives are carried out according to the logic of the 5Ps:
  - People (promoting a world where all individuals enjoy healthy environments and live with dignity and equality) – SGD 1 to 6
  - Prosperity (fostering the prosperity of individuals as well as nature) _ SGD 7 to 10
  - Planet (protect the planet through sustainable production and consumption, using resources responsibly) – SGDs 11 to 15
  - Peace (building peaceful, just and inclusive societies) – SGD16
  - Partnerships (achieving the objectives through a spirit of solidarity, with the participation of various entities), which together will enable the eradication of poverty and the achievement of sustainable development through health, education, environmental protection, peace and justice – SDGs 17

1.2.3 ISO 37120

The International Organization for Standardization (ISO) is an international standardization and standardization entity. It aims to approve international regulations in all technical areas, which are regulated by uniform rules, procedures and processes. They are thus tools with relevant impact on to achieve each of the SDDs, as presented below.

In 2014, the International Organization for Standardization (ISO) developed the first ISO 37120 to normalize the sustainable development of cities: "ISO 37120:2014 Sustainable development of communities First Editions – Indicators for city services and quality of life", allowing to measure and measure (quantitatively and qualitatively) results through their indicators. In this way, it is possible to compare the practices and measure the performance of the services and the quality of life in the city. NP ISO 37120:2017, Sustainable community development – Indicators for urban services and quality of life; constitutes the Portuguese version of ISO 37120:2014 – "Sustainable development of communities — Indicators for city services and quality of life".

The adoption of ISO 37120 can benefit local governance (including local authorities) in providing effective services, infrastructure and employment, with international targets as a reference. They help make improvement and progress comparisons, leading to the quality of life of the local population and structured sustainability planning.

According to WCCD (World Council on City Data (n.d), there was a clear improvement in the more than 60 cities that achieved iso 37120 certification. The organization also states that for the first time cities were able to communicate with each other, using globally standardized comparable data, allowing them to gain insights into other cities and learning from each other like never before.

ISO 37120 establishes a uniform criterion in the measurement and comparison of thematically grouped indicators in 17 sections that are presented below, which compare cities and communities among themselves.

5. Economy
6. Education
7. Energy
8. Environment
9. Finance
10. Fires and Emergency Response
11. Government
12. Health
13. Leisure
14. Security
15. Social Protection
16. Waste
17. Telecommunications and Innovation
18. Transportation
19. Urban Planning
20. Wastewater
21. Supply and Sanitation

ISO 37120 was revised in 2018 and was accompanied by two other complementary standards on indicators for smart cities (ISO 37122) and resilient cities (ISO 37123).

To meet the principles of urban sustainability, specifically SD1, the process of reviewing this ISO was initiated, share both of objectives that lead to the extinction of poverty, improving the health of populations and providing sustainable growth of cities.

The following is a table that relates the indicators of SO1 with ISA 37120. It should be noted, however, that the SDS 11 indicators are generic, while those of ISO 37120 are specific. In addition, the differentiation in the name of key indicators and support indicators.

It should be noted that there is a relationship between ISO 37120 and the SDGs as you can see in Table 1.1.

1.2.4 The Sustainability Report in Local Government

The public sector, as in the private sector, has a duty to account to the most diverse stakeholders,[11] having essentially social obligations, in addition to legal and economic obligations (Nagy & Robb, 2008; Dagger, 1997), so the maximum

[11] Strategic audience.

Table 1.1 Relationship between ISO and SGD standards

SGD	Standards ISO						
	9001 - Quality management	14001 - Environmental Management	26000 - Performance evaluation	50001 - Implementation, maintenance, review and improvement of energy management systems	13485 - Quality management of medical devices	16000-40 - indoor air quality management	20000-1 - Service Management
SGD 1	X	X	X				
SGD 2		X	X				
SGD 3		X	X		X	X	
SGD 4		X	X				
SGD 5			X				
SGD 6		X	X				
SGD 7		X	X	X			
SGD 8		X	X				
SGD 9	X	X	X				X
SGD 10			X		X		
SGD 11			X	X		X	
SGD 12	X	X	X	X		X	
SGD 13		X	X	X		X	
SGD 14	X	X	X			X	
SGD 15		X	X				
SGD 16			X				
SGD 17							

Source: Authors

transparency is expected in the reporting of its sustainability activities (GRI-Global Reporting Iniciative, 2005).

In a generalist way when we look at the concept of sustainability, it is urgent to associate with the idea of managing resources efficiently, effectively and efficiently so that the continuity and satisfaction of the needs of future generations is not called into question. However, the enormous productive economy leads us to an increasing consumerism, leading to the depleted natural resources, and that products are consumed, destroyed, replaced and discarded at an accelerated rate, impacting ecosystems and the biosphere. On the other hand, there are high levels of social inequalities and poverty and lack of corporate management ethics.

All of this above is a very high importance both in the private sector, but essentially in the public sector. From the outset, in the public sector, we are managing money and resources that belong to everyone. Thus, given that the public sector must be guided by transparency and accountability, it arises from among the various practices and tools of accountability: sustainability reports. It should also be noted that when we look at this theme of sustainability, three important dimensions should be taken into account: the environmental dimension, the social dimension and the economic dimension.

It is verified that in many cases there is no serious "commitment" to the sustainability report, and this disclosure is included in the annual reports and accounts of the public sector. According to Chambers et al. (2003), "the greater the extent of the report, the more committed the organization is to this same report".

It is certain that in recent years there has been a greater concern with sustainability and the definition of *accountability and transparency strategies and* mechanisms with regard to the theme of sustainability. This concern was verified through a review of the literature in which we found that there are studies both in the private sector and also in the public sector. Furthermore we found that the mainly voluntary basis of sustainability reports and the various labels used to cite this type of reports: corporate social responsibility report, sustainable development report, sustainability report, etc. and more recently integrated reports (integrating traditional financial reports and sustainability reports into a single report, thus providing a holistic picture of the entity). The only internal determinant that is consistently found to have a positive effect on SR is the size of the company; researchers also provide a significant influence of industry membership and tend to confirm a positive effect of listed companies, capital intensity and government ownership in SR (Hahn & Kühnen, 2013)".

In analysing the relationship between the sustainability report and the financial performance, Aggarwal (2013) concluded that there is a very positive relationship between the two which entails several advantages, namely greater involvement of various stakeholders, channels of communication of poems for customers, innovative products, highly motivated employees and their retention and compliance with regulations. In addition to these aspects, Jianu er al. (2015) concluded that the client as one of the elements in the company's sustainability report. Consumer feedback validates the company's measures that address environmental and social issues.

However, Kolk (2003) in looking at his study on the different types of sustainability reports presented in a group of multinational companies in Fortune Globe 250 concluded that there are different types of sustainability reports carried out and with very generalist information. He further concluded that only a few of them obtained independent verification and assurance of the information provided. From his study, Kolk (2003) found that in 2004, 64% of respondents, compared to 45% in 2001 and 35% in 1998, provided different types of sustainability reports. The author also observed an increase in integrated reports (20% of the sample against a company in 2001).

Keeping the business perspective, Dias (2009) studied empirical data on the practices and degree of dissemination of sustainability reports according to GRI standard in a sample of 49 Portuguese listed companies at Euronext Lisbon 2005. It concluded that only 45% of these companies report this type of information, although the sustainability communication in Portugal at the time was voluntary (currently the non-financial declaration is mandatory in Portugal for large companies that are companies of public interest, according to law no. 89/2017, of July 28).

However, this concern, on the one hand, and importance given to the business sector to sustainability reports is unison with the public sector. Alibašić (2018) addresses a few themes: what is sustainability and resilience in communities and entities, what challenges cities will face in the future, how can local governments maintain sustainability and resilience planning and how can resilience be incorporated into organizations? The main global is thus the planning of sustainability and resilience, a step ahead of the sustainability report. Local governments should take an appropriate holistic and systematic approach. The relevance of cities is really important because they account for more than 75% of the world's energy consumption and emit more than 75–80% of all greenhouse gas emissions. Thus, to build sustainability and resilience, it is crucial to adopt an appropriate planning approach. The author proposes the quadruple baseline (QBL) approach to resilience planning, including four pillars: social, economic, environmental and, like the new, governance. Resilience is seen as an extension and expansion of sustainability, with a particular focus on the threats of climate change. The author mentions the most relevant components for successful sustainability and resilience planning: leadership, monitoring, reporting mechanism (and here we have a link to the sustainability report), stakeholder engagement, commitment, etc.

Galera et al. (2014) analysed the dissemination of sustainability information by Local Anglo-Saxon and Nordic governments on their websites; its findings are that local governments are strongly committed to providing transparent sustainability information using their websites as communication tools with their stakeholders, despite the lack of a legal obligation to do so. They also concluded that the level of development of a country and/or its quality of governance does not imply greater transparency on the part of local governments with regard to sustainability; the level of debt and deficit do not seem to be factors affecting the involvement of local administrations in sustainability reporting. The results of the study support the conclusion that there is scope to improve the dissemination of sustainability by local authorities, including (a) designing a sustainability report model (comprehensive, single document, formally presented, (b) identifying key items, (c) identifying stakeholders addressed and (d) studying the information needs of these stakeholders, in order to set information priorities.

The report of sustainability in local government mirrors the image of the local government before its target public, determining the strategies of the Municipalities, as well as the decisions regarding concerns about environmental, social and economic issues.

In Portugal, there is an obligation, by the Labour Code, to communicate annually about the company's social activity regarding the staff and other related information (conclusion and termination of fixed-term employment contracts; additional work; continuing vocational training; safety and health services at work; the Social Balance Sheet) (GEP, 2011) (Table 1.2).

In view of the conclusions of empirical studies on sustainability, we found that there are studies that show that both the private sector and the public sector have a concern with the theme of sustainability, and that the private business fabric has more this need to "sell" the product as being the most sustainable. This means that

Table 1.2 Empirical studies on sustainability

Authors	Research objective	Sample	Methodology research	Conclusions
Guerra (2011)	Assess initiatives regulated and coordinated by local government on sustainable development	Municipalities	Questionnaires	Concrete and continued actions are being taken to pursue the principles of sustainability and participatory democracy; However, there are also many a21L ghost initiatives, with no real existence, without a national policy that allows the framing or evaluation of this process on the ground, the aim of which is to safeguard the benefits that a local Agenda 21 can represent on the national scene: prestige, symbolic capital, and Community funds
Eccles et al. (2014)	Effect and implications of regulations on sustainability disclosure in the dissemination and evaluation practices of companies in China, Denmark, Malaysia and South Africa.	Companies	Documentary analysis	Companies subject to this "obligation": (i) increase their disclosure; (ii) add efforts to increase the comparison and credibility of the information they disseminate; (iii) in addressing sustainability they reach a wider group of stakeholders beyond shareholders, and with typically non-financial objectives (environmental, social and governance). Although this regulation may bring costs for some companies, this article suggests that in the end the balance sheet is positive

(continued)

Table 1.2 (continued)

Authors	Research objective	Sample	Methodology research	Conclusions
Niemann and Hoppe (2018)	Assess the effects of sustainability reporting involvement	Municipalities	Evaluation of actual results	1 – Various types of registrations are valuable to regional governments as a weapon of learning, acting and registration; 2 – costs can be very limited; 3 – have organizational benefits, but if the objectives are not met can arise the "fatigue of the register" and with this its suspension or radical change
Martins (2011)	Realize: (1) how sustainability communication is processed on the web pages of the various municipalities studied; (2) differences in the form of communication on web pages (3) Importance of communication.	Websites	Website Analysis	It is concluded that sustainability communication is of great importance, since all the cities considered in the sample disclose this theme, however, only 36.84% of the sample communicates information related to sustainability on the homepage
Dias (2009)	Check the trends, practices and degrees of dissemination of the sustainability assessment, according to the GRI (Global Reporting Initiative) model.	Companies	Website Analysis	45% of the companies that make up the sample voluntarily disclosed information related to corporate sustainability. Few (or none) present numerical indicators, indicating that this disclosure has a purely commercial purpose rather than constituting a commitment to the responsibility of social

(continued)

Table 1.2 (continued)

Authors	Research objective	Sample	Methodology research	Conclusions
Galera et al. (2014)	Assess initiatives regulated and coordinated by local government on sustainable development.	Municipalities	Questionnaires	Concrete and continued actions are being taken to pursue the principles of sustainability and participatory democracy; However, there are also many a21L ghost initiatives, with no real existence, without a national policy that allows the framing or evaluation of this process on the ground, the aim of which is to safeguard the benefits that a local Agenda 21 can represent on the national scene: prestige, symbolic capital and community funds
Williams et al. (2010)	Effect and implications of regulations on sustainability disclosure in the dissemination and evaluation practices of companies in China, Denmark, Malaysia and South Africa.	Companies	Documentary analysis	Companies subject to this "obligation": (i) increase their disclosure; (ii) add efforts to increase the comparison and credibility of the information they disseminate; (iii) In addressing sustainability they reach a wider group of stakeholders beyond shareholders, and with typically non-financial objectives (environmental, social and governance). Although this regulation may bring costs for some companies, this article suggests that in the end the balance sheet is positive

(continued)

Table 1.2 (continued)

Authors	Research objective	Sample	Methodology research	Conclusions
War (2011)	Assess initiatives regulated and coordinated by local government on sustainable development.	Local government	Obtained 107 valid answers (almost 35% of the total Portuguese municipalities, include the whole territories equitably).	Census of participatory processes of local sustainability taking place in the country and model of analysis of the participatory sustainability processes that are taking place in the municipalities
Barut et al. (2016)	Assess the importance of sustainability reports, including the relationship between local government objectives and related biodiversity benchmarks	Local governments	Document analysis (sites)	Lack of disclosure to stakeholders in order to assess the responsibility for the management of common resources The information in the sustainability reports is underdeveloped
	Know the main reasons for the preparation of sustainability reports and what influence stakeholders can have on their own preparation	Local governments	Interview	The reports are prepared for the sake of legitimacy and accountability The involvement of stakeholders contributes to legitimizing strategies in order to find a solution between political programs and the objectives/concerns/ interests of stakeholders
Navarro-Galera et al. (2017)	Assess the level of association between transparency of local government sustainability and international guidelines	Local governments	Document analysis (sites)	The transparency of the sustainability of the various local governments under study reaches a level of 58.05% of association with international guidelines. This level of association shows, on the one hand, that local government websites emerge as drivers of sustainability transparency and, on the other hand, to strengthen the dissemination of information on sustainability among local governments in Europe

there is a greater sensitivity on the part of the private business fabric to the dissemination of information about sustainability in the sense that the client will be more attentive to these issues when their "option to buy", as well as their interest in environmental and sustainability issues have been increasing. However, in both sectors, it should be noted that the social context and the need to be transparent and accountable are increasingly demanding elected officials and managers to disseminate information about sustainability. It should also be noted that, as Navarro-Galera et al. refer. (2017) the various international bodies highlight and recommend that local governments take good governance practices that meet the favouring of public sustainability policies. This recommendation aims at requiring transparency on economic, social and environmental issues.

1.3 Methodology

1.3.1 Research Protocol

First, it was defined that the research to be carried out would be in the larger municipalities. For information on whether or not a sustainable development report existed, an email was sent to each of the 25 municipalities that were the target of this study. Of all the emails sent, we only got a response from seven municipalities. To point out that of these seven municipalities only three responded to the email about the requested information. The other municipalities only accused the receipt and registration of the same or sent the email to another department, not getting any kind of response.

After submitting this research, we proposed to verify that the reports dubbed "sustainability reports" meet the precept in ISO 37210 and meet the 17 Objectives SDGs by 2030. Then we did a search through the LOCAL ODS platform.[12] This platform is created by the various municipalities of Portugal. It allows monitoring the projects or good practices that each municipality is implementing; what SDG objectives are working on; allows monitoring the performance of the various indicators of the 17 SDG objectives; and also to understand if they are in the alignment to be achieved by 2030 or if they are far from this intended mark or even if it is not even being implemented.

Finally, we checked the Pearson coefficient for the relationship between the SDGs and also verified the association between the SDGs and two socio-economic variables: GDP per capita and level of education.

1.3.2 Method of Data Collection and Analysis

The problem and its research issues lead to the combination of different methodologies and the use of different forms of data analysis. Therefore, the present study is descriptive and exploratory.

[12] https://odslocal.pt/

This study uses qualitative research, since it allows us to understand in a generalist way the situations, experiences, meanings of actions through their analysis, understanding and description. It is denoted that with regard to qualitative research, the form of study deals with the collection, analysis, description and interpretation of the data. In turn, quantitative research aims to find relationships between variables and to make descriptions taking into account statistical data obtained (Bogdan & Biklen, 1994).

It is important to note that both methodologies have positive and negative aspects compared to the study developed. From the outset, the quantitative methodology has as its main advantage the fact that conclusions can be drawn and knowledge that is useful in most situations, which imputes to it a superior external validity. However, in the reverse of the coin, and as (Pérez Serrano, 1998) points out uncertainty as to what is measured.

The documentary analysis that constitutes an important technique and necessary in the context of any type of research, and this importance is increased in the field of accounting (Sparrow & Correia, 1995; Sousa & Baptista, 2014; Vieira et al., 2009). For analysis of the information about sustainability available on the website: existence of sustainability reports; information about the vision, mission, and value in the scope of sustainability and normative guiding of the elaboration we use the index of disclosure.

After this documental analysis, we resorted to quantitative and statistical analysis of information. For this, we calculate descriptive statistics, more specifically counts, medians (a central trend measure used for variables with a non-normal distribution) and the interquartile interval (dispersion measure for variables with non-normal distribution that results from the difference between the first quartile and the third quartile). To evaluate the correlation between the SDS, a correlation matrix was constructed, and Spearman's correlation coefficient was calculated. In addition, to evaluate the association between the number of sustainable development objectives and socio-economic variables, *Poisson regression models* (appropriate for counting data) were used. The significance level was the p-value of <0.05. Statistical analysis was performed in the statistical analysis software R version 4.1.3.

1.4 Results

1.4.1 Sample Characterization

The target population considered in this study comprises 25 municipalities considered large, taking into account the data of the 2021 Census, and Portugal consists of 308 municipalities (Table 1.3).

1.4.2 Results and Their Interpretation

When analysing the sites of the 25 largest Portuguese municipalities, we found that only three municipalities (Braga, Vila Nova de Famalicão and Porto, located in the Northern region of Portugal) have available sustainability reports.

Table 1.3 Municipalities targeted for study

Position	Municipality	Region	Population	GDP per capita	Level of education
1	Lisboa	Lisbon Metropolitan Area	544.851	205.62	36.78
2	Sintra	Lisbon Metropolitan Area	385.954	93.61	16.78
3	Vila Nova de Gaia	North	304.149	100.55	18.92
4	Porto	North	231.962	154.02	32.06
5	Cascais	Lisbon Metropolitan Area	214.134	117.95	28.18
6	Loures	Lisbon Metropolitan Area	201.646	100.99	17.61
7	Braga	North	193.333	108.78	23.77
8	Almada	Lisbon Metropolitan Area	177.400	109.16	21
9	Matosinhos	North	172.669	130.63	21.91
10	Oeiras	Lisbon Metropolitan Area	171.802	153.13	33.08
11	Amadora	Lisbon Metropolitan Area	171.719	100.07	18.69
12	Seixal	Lisbon Metropolitan Area	166.693	91.16	16.93
13	Gondomar	North	164.255	84.89	14.64
14	Guimarães	North	156.852	91.3	13.32
15	Odivelas	Lisbon Metropolitan Area	148.156	88.78	19.91
16	Coimbra	Center	140.796	126.28	30.67
17	Vila Franca de Xira	Lisbon Metropolitan Area	137.659	97.35	16.45
18	Santa Maria da Feira	North	136.720	85.76	13.44
19	Maia	North	134.959	110.64	6.64
20	Vila Nova de Famalicão	North	133.590	89.86	13.56
21	Leiria	Center	128.640	103.21	18.69
22	Setúbal	Lisbon Metropolitan Area	123.684	107.95	18.01
23	Barcelos	North	116.777	80.12	11.21
24	Funchal	Wood	105.919	115.71	19.18
25	Viseu	Center	99.693	95.7	21.22

Source: Authors

Regarding the "vision, mission, and strategy" that municipalities present in the information they provide and, essentially, in the context of sustainability, we found that 13 of the 25 municipalities show concern and interest on this topic. In relation to the regulations adopted for the realization of sustainability reports, we verified that they refer to GRI and ISO 37120. Porto and Vila Nova de Famalicão refer to GRI and ISO 37120; Braga does not mention what regulations are used in the preparation of sustainability reports. Although most municipalities do not submit sustainability reports, they refer to the SDGs and present projects and good practices under these SDGs.

We also tried to analyse in the light of ISO 37120 what sub-items were included in the sustainability reports. We found that only one report (from the municipality of Porto) takes into account all sub-items related to finance issues; fires and emergency response; government; health; leisure; waste; telecommunications and innovation; transport; urban planning; wastewater; supply and sanitation. In this same report, with regard to energy, environment, social protection and safety, all sub-items are not covered, but a good part of them is still considered (73.93%).

In the two other reports (from Braga and Vila Nova de Famalicão) we found that they do not have information about all the sub-items related to finance, social protection, urban planning, telecommunications, and innovation.

By analyzing the local SDG site (municipal platform of sustainable development goals) we found that of the 25 municipalities targeted for study eight municipalities do not present any project or good practices under the objectives for sustainable development. We also found that the SDGs that municipalities have more projects and good practices are SDGs 1 – Eradication of poverty (9.4%); SDGs 3 – Health and quality (9.3%); SDGs 4 – Quality education (12.4%) and SDGs 11 – sustainable cities and communities (10.5%). On the other hand, those SDGs that have fewer projects and good practices in the municipalities under analysis are: SDGs 6 – Clean water and sanitation and SDGs 7 – renewable energies.

We also found that the three municipalities that have the most projects and good practices considering the number of SDGs covered are: Vila Nova de Gaia (18.0%), Cascais (17.9%) and Lisbon (13.2%) (See Figs. 1.3 and 1.4) (Table 1.4).

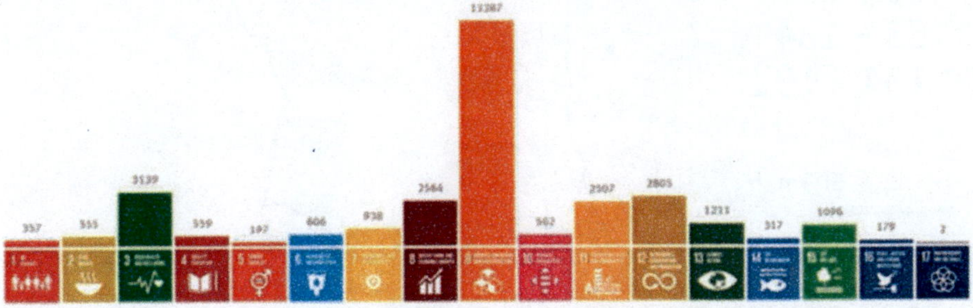

Fig. 1.3 Impact tools to achieve with each of the SDGs. (Source: https://www.iso.org/sdgs.html)

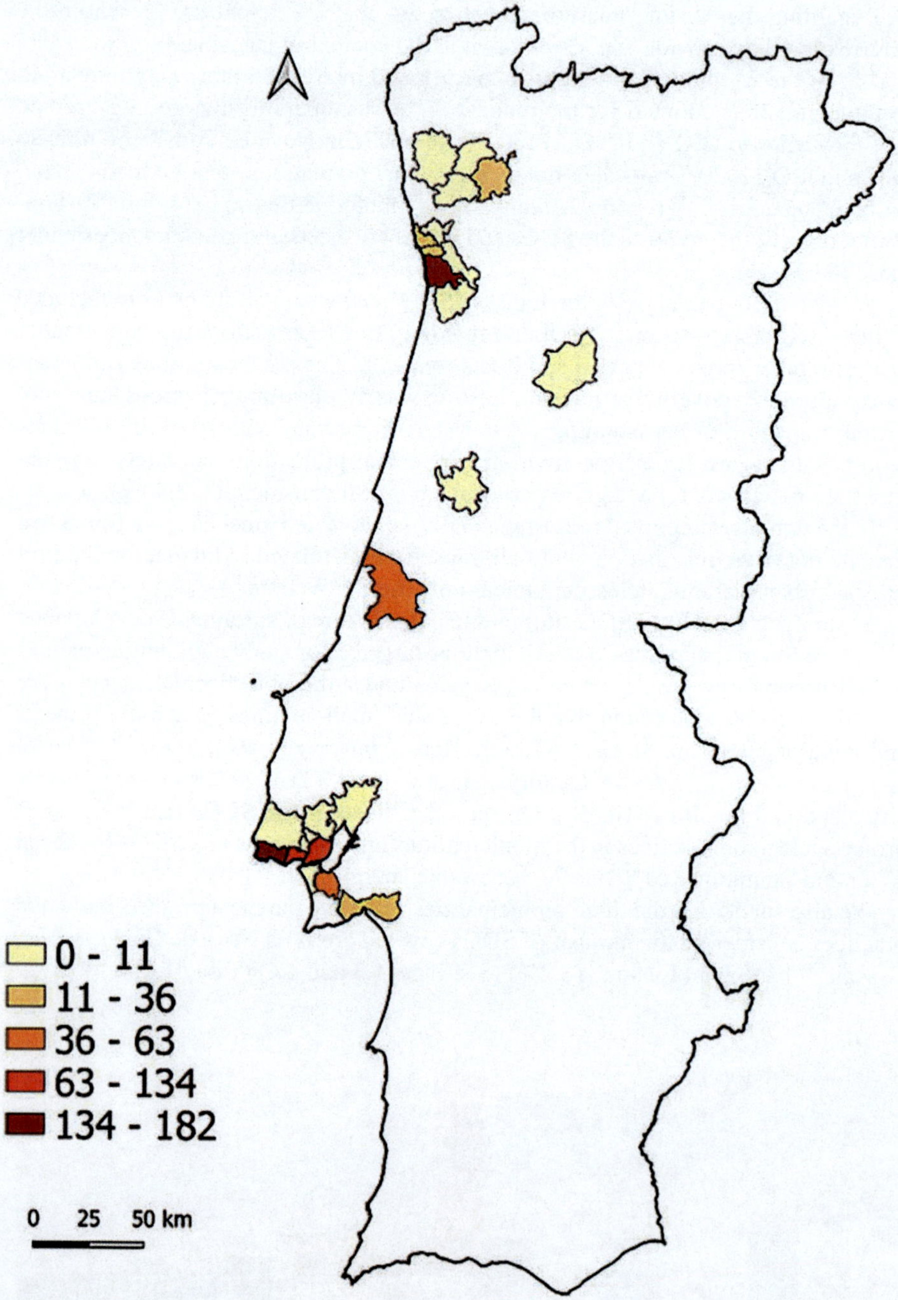

Fig. 1.4 SDGs adopted by municipalities. (Source: Authors)

Table 1.4 SDGs adopted by municipalities

Municipality	SDG1	%	SDG2	%	SDG3	%	SDG4	%	SDG5	%	SDG6	%	SDG7	%	SDG8	%	SDG9	%
Lisboa	11	**8.2**	14	**10.4**	6	**4.5**	13	**9.7**	7	**5.2**	5	**3.7**	0	**0.0**	4	**3.0**	1	**0.7**
Sintra	0	**0.0**	0	**0.0**	0	**0.0**	0	**0.0**	0	**0.0**	1	**20.0**	0	**0.0**	0	**0.0**	0	**0.0**
Vila Nova de Gaia	25	**13.7**	12	**6.6**	20	**11.0**	29	**15.9**	16	**8.8**	2	**1.1**	0	**0.0**	7	**3.8**	1	**0.5**
Porto	6	**18.2**	5	**15.2**	4	**12.1**	5	**15.2**	5	**15.2**	1	**3.0**	0	**0.0**	1	**3.0**	0	**0.0**
Cascais	25	**13.8**	14	**7.7**	17	**9.4**	20	**11.0**	11	**6.1**	2	**1.1**	1	**0.6**	6	**3.3**	3	**1.7**
Loures	0	**0.0**	0	**0.0**	0	**0.0**	0	**0.0**	0	**0.0**	0	**0.0**	0	**0.0**	0	**0.0**	0	**0.0**
Braga	0	**0.0**	0	**0.0**	0	**0.0**	0	**0.0**	0	**0.0**	0	**0.0**	0	**0.0**	0	**0.0**	0	**0.0**
Almada	2	**18.2**	2	**18.2**	1	**9.1**	2	**18.2**	1	**9.1**	0	**0.0**	0	**0.0**	0	**0.0**	0	**0.0**
Matosinhos	2	**6.5**	0	**0.0**	4	**12.9**	5	**16.1**	3	**9.7**	0	**0.0**	1	**3.2**	1	**3.2**	1	**3.2**
Oeiras	7	**6.3**	3	**2.7**	13	**11.7**	11	**9.9**	1	**0.9**	1	**0.9**	0	**0.0**	6	**5.4**	2	**1.8**
Amadora	0	**0.0**	0	**0.0**	0	**0.0**	0	**0.0**	0	**0.0**	0	**0.0**	0	**0.0**	0	**0.0**	0	**0.0**
Seixal	5	**10.0**	3	**6.0**	7	**14.0**	6	**12.0**	5	**10.0**	0	**0.0**	1	**2.0**	1	**2.0**	0	**0.0**
Gondomar	0	**0.0**	0	**0.0**	0	**0.0**	1	**11.1**	0	**0.0**	0	**0.0**	0	**0.0**	1	**11.1**	1	**11.1**
Guimarães	0	**0.0**	0	**0.0**	4	**11.8**	6	**17.6**	3	**8.8**	3	**8.8**	3	**8.8**	3	**8.8**	3	**8.8**
Odivelas	0	**0.0**	0	**0.0**	0	**0.0**	0	**0.0**	0	**0.0**	0	**0.0**	0	**0.0**	0	**0.0**	0	**0.0**
Coimbra	2	**22.2**	2	**22.2**	1	**11.1**	2	**22.2**	2	**22.2**	0	**0.0**	0	**0.0**	0	**0.0**	0	**0.0**
Vila Franca de Xira	0	**0.0**	0	**0.0**	0	**0.0**	0	**0.0**	0	**0.0**	0	**0.0**	0	**0.0**	0	**0.0**	0	**0.0**
Santa Maria da Feira	0	**0.0**	0	**0.0**	0	**0.0**	0	**0.0**	0	**0.0**	0	**0.0**	0	**0.0**	0	**0.0**	0	**0.0**
Mayan	0	**0.0**	0	**0.0**	0	**0.0**	0	**0.0**	0	**0.0**	0	**0.0**	0	**0.0**	0	**0.0**	0	**0.0**

(continued)

Table 1.4

Municipality	SDG1	%	SDG2	%	SDG3	%	SDG4	%	SDG5	%	SDG6	%	SDG7	%	SDG8	%	SDG9	%
Vila Nova de Famalicão	0	**0.0**	0	**0.0**	0	**0.0**	0	**0.0**	0	**0.0**	0	**0.0**	0	**0.0**	0	**0.0**	0	**0.0**
Leiria	3	**4.8**	5	**7.9**	3	**4.8**	8	**12.7**	3	**4.8**	0	**0.0**	3	**4.8**	2	**3.2**	2	**3.2**
Setúbal	0	**0.0**	0	**0.0**	6	**16.7**	6	**16.7**	0	**0.0**	1	**2.8**	0	**0.0**	0	**0.0**	0	**0.0**
Barcelos	0	**0.0**	0	**0.0**	0	**0.0**	0	**0.0**	0	**0.0**	0	**0.0**	0	**0.0**	0	**0.0**	0	**0.0**
Funchal	7	**6.0**	4	**3.4**	7	**6.0**	12	**10.3**	5	**4.3**	4	**3.4**	3	**2.6**	4	**3.4**	5	**4.3**
Viseu	0	**0.0**	1	**20.0**	1	**20.0**	0	**0.0**	0	**0.0**	1	**20.0**	0	**0.0**	0	**0.0**	0	**0.0**
TOTAL	**95**	**9.4**	**65**	**6.4**	**94**	**9.3**	**126**	**12.4**	**62**	**6.1**	**21**	**2.1**	**12**	**1.2**	**36**	**3.6**	**19**	**1.9**

Table 1.5 Median, range, minimum and maximum – ODS

SDG	Median	Interquartile Range	Minimal	Maximum
Total	1	50	0	182
1	0	5	0	25
2	0	3	0	14
3	1	6	0	20
4	2	6	0	29
5	0	3	0	16
6	0	1	0	5
7	0	0	0	3
8	0	2	0	7
9	0	1	0	5
10	0	1	0	25
11	1	5	0	22
12	1	3	0	18
13	1	4	0	19
14	0	1	0	6
15	1	4	0	12
16	0	1	0	14
17	1	3	0	15

Source: Authors

In analysing the median, we found that the SDS with the highest median is SDG 4 – Quality Education, followed by SDG 3 – Health and Well-being, SDG 11 – Sustainable cities and communities, SDG 12 – Sustainable production and consumption, SDG 13 – Climate action, SDS 15 – Protecting terrestrial life; and SDG 17 – Partnership for implementing the objectives. This demonstrates that projects and good practices tend to focus on these themes as you can see in Table 1.5.

Below is the correction matrix between the different SDGs. It is verified that most correlations are statistically significant, positive and moderate to strong (with rho above 0.5), i.e. municipalities that have a high number of SDGs in a given thematic data also tend to have a high number of SDGs in other themes. It should be noted, however, that some of the correlations observed are weaker (rho <0.5) and are not statistically significant ($p > 0.05$), such as the correlation between SDG 2 – Eradicating hunger and SDG 7 – Ensuring access to reliable, sustainable and modern energy sources for all, SDG 1 – Eradicating poverty and 7 – Ensuring access to reliable energy sources, sustainable and modern for all, SDG 6 – Ensuring the availability and sustainable management of drinking water and sanitation for all and 7 – Ensuring access to reliable, sustainable and modern energy sources for all and SDG 14 and 6 – Ensuring the availability and sustainable management of drinking water and sanitation for all and SDG 15 and 7 – Ensuring access to reliable energy sources, sustainable and modern for all (Table 1.6).

Table 1.6 Spearman correlation between SGDs (in Portuguese ODS)

Coeficiente de correlação de Spearman entre ODS

	ODS1	ODS2	ODS3	ODS4	ODS5	ODS6	ODS7	ODS8	ODS9	ODS10	ODS11	ODS12	ODS13	ODS14	ODS15	ODS16	ODS17	TOTAL
ODS1	1.00	0.93	0.85	0.87	0.92	0.50	0.38	0.81	0.59	0.89	0.77	0.68	0.59	0.60	0.58	0.81	0.86	0.85
ODS2	0.93	1.00	0.78	0.79	0.85	0.54	0.30	0.71	0.49	0.76	0.65	0.73	0.51	0.51	0.55	0.66	0.78	0.81
ODS3	0.85	0.78	1.00	0.94	0.86	0.69	0.50	0.81	0.64	0.88	0.85	0.78	0.72	0.62	0.68	0.83	0.90	0.94
ODS4	0.87	0.79	0.94	1.00	0.89	0.65	0.54	0.89	0.77	0.87	0.90	0.83	0.81	0.62	0.71	0.85	0.94	0.97
ODS5	0.92	0.85	0.86	0.89	1.00	0.56	0.57	0.84	0.65	0.90	0.76	0.68	0.58	0.64	0.46	0.83	0.84	0.86
ODS6	0.50	0.54	0.69	0.65	0.56	1.00	0.24	0.64	0.57	0.54	0.61	0.76	0.57	0.36	0.57	0.51	0.59	0.71
ODS7	0.38	0.30	0.50	0.54	0.57	0.24	1.00	0.56	0.67	0.60	0.55	0.44	0.50	0.53	0.29	0.65	0.57	0.51
ODS8	0.81	0.71	0.81	0.89	0.84	0.64	0.56	1.00	0.88	0.89	0.91	0.80	0.74	0.70	0.65	0.85	0.91	0.87
ODS9	0.59	0.49	0.64	0.77	0.65	0.57	0.67	0.88	1.00	0.73	0.82	0.68	0.72	0.55	0.60	0.80	0.76	0.73
ODS10	0.89	0.76	0.88	0.87	0.90	0.54	0.60	0.89	0.73	1.00	0.83	0.67	0.63	0.73	0.54	0.95	0.92	0.86
ODS11	0.77	0.65	0.85	0.90	0.76	0.61	0.55	0.91	0.82	0.83	1.00	0.82	0.91	0.62	0.82	0.81	0.91	0.90
ODS12	0.68	0.73	0.78	0.83	0.68	0.76	0.44	0.80	0.68	0.67	0.82	1.00	0.83	0.52	0.82	0.64	0.84	0.88
ODS13	0.59	0.51	0.72	0.81	0.58	0.57	0.50	0.74	0.72	0.63	0.91	0.83	1.00	0.54	0.93	0.68	0.78	0.84
ODS14	0.60	0.51	0.62	0.62	0.64	0.36	0.53	0.70	0.55	0.73	0.62	0.52	0.54	1.00	0.45	0.63	0.65	0.64
ODS15	0.58	0.55	0.68	0.71	0.46	0.57	0.29	0.65	0.60	0.54	0.82	0.82	0.93	0.45	1.00	0.58	0.70	0.78
ODS16	0.81	0.66	0.83	0.85	0.83	0.51	0.65	0.85	0.80	0.95	0.81	0.64	0.68	0.63	0.58	1.00	0.85	0.82
ODS17	0.86	0.78	0.90	0.94	0.84	0.59	0.57	0.91	0.76	0.92	0.91	0.84	0.78	0.65	0.70	0.85	1.00	0.93
TOTAL	0.85	0.81	0.94	0.97	0.86	0.71	0.51	0.87	0.73	0.86	0.90	0.88	0.84	0.64	0.78	0.82	0.93	1.00

Colour scale: 1, 0.8, 0.6, 0.4, 0.2, 0, -0.2, -0.4, -0.6, -0.8, -1

Source: Authors

The association between the SDGs and the two socio-economic variables: GDP per capita and level of education is presented below. From the analysis we conclude that there is a statistically significant association between the majority of the SDDs and the two socio-economic variables. This means that by analysing all the SDBs, we found that for each unit increase in the value of GDP the number of SDBs included in projects/good practice swells by 1.016%. A similar situation occurs regarding the level of education.

There is no longer a significant association between the two socio-economic variables and the SDS 7 – Ensuring access to reliable, sustainable and modern energy sources for all, SDS 9 – Building resilient infrastructure, promoting inclusive and sustainable industrialization and fostering innovation and SDS 16 – Peace, justice and effective institutions (Table 1.7).

Table 1.7 Association (relative risk) between number of SDGs and socio-economic variables

	GDP per capita	Level of education
Total	1.016 ($p < 0{,}001$)	1.084 ($p < 0{,}001$)
SDG 1	1.015 ($p < 0{,}001$)	1.097 ($p < 0{,}001$)
SDG 2	1.020 ($p < 0.001$)	1.119 ($p < 0{,}001$)
SDG 3	1.012 ($p < 0{,}001$)	1.072 ($p < 0{,}001$)
SDG 4	1.013 ($p < 0.001$)	1.068 ($p < 0{,}001$)
SDG 5	1.013 ($p < 0.001$)	1.070 ($p < 0{,}001$)
SDG 6	1.020 ($p < 0.001$)	1.080 ($p > 0{,}01$)
SDG 7	0.994 (n.s.)	0.963 (n.s.)
SDG 8	1.015 ($p < 0{,}001$)	1.081 ($p < 0{,}001$)
SDG 9	1.009 (n.s.)	1.034 (n.s.)
SDG 10	1.008 ($p > 0{,}05$)	1.067 ($p < 0{,}001$)
SDG 11	1.022 ($p < 0{,}001$)	1.116 ($p < 0.001$)
SDG 12	1.022 ($p < 0{,}001$)	1.095 ($p < 0{,}001$)
SDG 13	1.023 ($p < 0.001$)	1.105 ($p < 0.001$)
SDG 14	1.008 (n.s.)	1.067 ($p > 0{,}05$)
SDG 15	1.021 ($p < 0{,}001$)	1.103 ($p < 0.001$)
SDG 16	1.004 (n.s)	1.025 (n.s.)
SDG 17	1.013 ($p < 0.001$)	1.078 ($p < 0{,}001$)

Source: Own Development

1.5 Conclusions

The present study had the following objectives: to know which of the largest Portuguese municipalities have sustainability reports; to identify what regulation scans followed in the preparation of the sustainability report; to identify the objectives for sustainable development included in the projects and in the good practices of the municipalities evaluated and to evaluate the association between the frequency with which the objectives for sustainable development in projects and good practices and the level of education of the population and GDP of the municipalities are included.

When analysing which of the 25 largest municipalities had available on their website's sustainability reports, we found that the three municipalities had a Sustainability Report, or at least one report that addressed the three fundamental strands that was based on our study (GRI, ISO 37120 and SDGs). These municipalities were the Municipality of Porto, Braga and Vila Nova de Famalicão (located in the Northern Region). This goes against the results of Barut et al. (2016), which emphasized the underdevelopment of the report on sustainability by local government. In turn, we conclude that, as Greco et al. states. (2015), information about sustainability is a form of legitimacy for municipalities, and the involvement of stakeholders becomes fundamental in this legitimation.

In terms of the relevant information provided by the reports the Porto and Braga in fact stood out from that of Vila Nova de Famalicão. This was because they presented both the SDGs and the GRI that were following concretely, while the Vila Nova de Famalicão only addressed the SDGs. The latter claimed to follow the GRI but did not show which ones specifically.

The Municipalities of Porto and Braga present numerous indicators related to the economic, environmental and social factors that affect their municipality. Regarding Vila Nova de Famalicão, it presents several indicators, however, in terms of relevance does not rise to the levels of the others studied.

In terms of design, once again those of Porto and Braga presented a high quality with tables that allowed us to understand the evolution of indicators over the years, as well as the colours chosen were well distributed which provides the reader with a more fun reading.

We would like to give a detail that places the Sustainability Report of the Municipality of Porto above the others, which is based on the interconnection that it places in each SDA. This interconnection allows the reader to access the SO website and understand the meaning of it and its contribution to the sustainability of society.

Then, in analysing these reports, we concluded that the other municipalities, despite not presenting sustainability reports, their concern with this theme was noted in the sense that the majority presented projects and good practices addressed the SDDs promoted by the 2030 Agenda. We also found that the three municipalities that have the most projects and good practices taking into account the number of SDGs covered are: Vila Nova de Gaia (18.0%), Cascais (17.9%) and Lisbon (13.2%).

We conclude that eight municipalities do not present any project or good practices under the objectives for sustainable development. We also found that the SDGs that municipalities have more projects and good practices are SDGs 1 – Eradication of poverty (9.4%); SDGs 3 – Health and quality (9.3%); SDGs 4 – Quality education (12.4%) and SDGs 11 – sustainable cities and communities (10.5%). On the other hand, those SDGs that have fewer projects and good practices in the municipalities under analysis are: SDGs 6 – Clean water and sanitation and SDGs 7 – renewable energies.

We found, in analysing the correlation between the SDG that a significant correlation is made between most SDDs, with the exception of SDG 2 – Eradicating hunger and SDG 7 – Ensuring access to reliable, sustainable and modern energy sources for all, SDG 1 – Eradicating poverty and 7 – Ensuring access to reliable energy sources, sustainable and modern for all, SDG 6 – Ensuring the availability and sustainable management of drinking water and sanitation for all and 7 – Ensuring access to reliable, sustainable and modern energy sources for all and SDG 14 and 6 – Ensuring the availability and sustainable management of drinking water and sanitation for all and SDG 15 and 7 – Ensuring access to reliable energy sources, sustainable and modern for all.

Finally, when assessing the association between the frequency with which the SDGs are included in projects and good practices and the level of education of the population and GDP of the municipalities, we found that there is an association between the SDGs and the two socioeconomic variables mentioned above.

We point out as a limitation to the present study the fact that we only made the analysis to the largest municipalities, which does not allow us to conclude in a generalist way what occurs in the 308 municipalities.

This study contributes to the literature on sustainability essentially in two aspects. The first is that sustainability is effectively, although it is a theme that is on the agenda, there are still practical terms to go. In other words, more and more public sector officials should have information to be disclosed through sustainability reports in accordance with current standards. On the other hand, we have found that the dissemination of information about sustainability becomes an important responsibility tool for both internal and external users.

As clues to future research, we suggest investigating alternatives for the sustainability communication of municipalities, as well as their framework and standards. We also suggest the analysis of the association with other indicators such as budget and population density. And finally, the analysis of all Portuguese municipalities to be possible to compare with local governments at European level.

References

Adger, W. N. (2009). *Governing sustainability*. Cambridge University Press.

Aggarwal, P. (2013). Sustainability reporting and its impact on corporate financial performance: A literature review. *Indian Journal of Commerce and Management Studies, 4*(3), 51.

Alibašić, H. (2018). *Sustainability and resilience planning for local governments*. Springer.

Barut, M., Raar, J., & Azim, M. I. (2016). Biodiversity and local government: A reporting and accountability perspective. *Managerial Auditing Journal, 31*(2), 197–227. https://doi.org/10.1108/MAJ-08-2014-1082

Bogdan, R., & Biklen, S. (1994). *Qualitative research in education: An introduction to theory and methods*. Porto Publishing.

Brundtland, G. H. (1987). Brundtland report. Our common future. *Comissão Mundial, 4*(1), 17–25.

Dagger, R. (1997). *Civic virtues: Rights, citizenship, and republican liberalism*. Oxford University Press.

Dias, A. C. G. (2009). The report of corporate sustainability: Practices in Portugal. *Portuguese Journal of Accounting and Management, 8*, 111–150.

Eccles, R. G., Ioannou, I., & Serafeim, G. (2014). The impact of corporate sustainability on organizational processes and performance. *Management Science, 60*(11), 2835–2857. https://doi.org/10.1287/mnsc.2014.1984

Galera, A. N., de los Ríos Berjillos, A., Lozano, M. R., & Valencia, P. T. (2014). Transparency of sustainability information in local governments: English-speaking and Nordic cross-country analysis. *Journal of Cleaner Production, 64*, 495–504.

GEP. (2011). *Estudos Ministério da Economia, da Inovação e do Desenvolvimento-Gabinete de Planeamento, Estratégia, Avaliação e Relações Internacionais-Ministério das Finanças.*

Greco, G., Sciulli, N., & D'Onza, G. (2015). The influence of stakeholder engagement on sustainability reporting: Evidence from Italian local councils. *Public Management Review, 17*(4), 465–488. https://doi.org/10.1080/14719037.2013.798024

Gri, G. R. I. (2005). *Sector supplement for public agencies. Global Reporting Initiative GRI.*

Guerra, J. (2011). *Municípios, participação e sustentabilidade: dinâmicas locais de imperativos globais.*. https://repositorio.ul.pt/handle/10451/4755

Hahn, R., & Kühnen, M. (2013). Determinants of sustainability reporting: A review of results, trends, theory, and opportunities in an expanding field of research. *Journal of Cleaner Production, 59*, 5–21. https://doi.org/10.1016/j.jclepro.2013.07.005

Jianu, I., Ţurlea, C., & Guşatu, I. (2015). The reporting and sustainable business marketing. *Sustainability, 8*(1), 23. https://doi.org/10.3390/su8010023

Kolk, A. (2003). Trends in sustainability reporting by the Fortune Global 250. *Business Strategy and the Environment, 12*(5), 279–291. https://doi.org/10.1002/bse.370

LéLé, S. M. (2018). 18 sustainable development: A critical review. In *Green Planet Blues* (pp. 195–207). Routledge.

Maher, M. W. (1979). An analysis of reporting practices in American cities.

Martins, M. (2011). *Comunicação da Sustentabilidade nas páginas web dos municípios da Europa dos 15*. Faculdade de Economia, Universidade do Porto.

Nagy, J., & Robb, A. (2008). Can universities be good corporate citizens? *Critical Perspectives on Accounting, 19*(8), 1414–1430. https://doi.org/10.1016/j.cpa.2007.10.001

Navarro-Galera, A., Ruiz-Lozano, M., Tirado-Valencia, P., & de los Rios-Berjillos, A. (2017). Promoting sustainability transparency in European local governments: An empirical analysis based on administrative cultures. *Sustainability, 9*(3), 20. https://doi.org/10.3390/su9030432

Niemann, L., & Hoppe, T. (2018). Sustainability reporting by local governments: a magic tool? Lessons on use and usefulness from European pioneers. *Public Management Review, 20*(1), 201–223. https://doi.org/10.1080/14719037.2017.1293149

Pérez Serrano, G. (1998). Qualitative research challenges and questions.

Ranjbari, M., Saidani, M., Shams Esfandabadi, Z., Peng, W., Lam, S. S., Aghbashlo, M., Quatraro, F., & Tabatabaei, M. (2021). Two decades of research on waste management in the circular economy: Insights from bibliometric, text mining, and content analyses. *Journal of Cleaner Production, 314*(128009), 128009. https://doi.org/10.1016/j.jclepro.2021.128009

Redclift, M. (2005). Sustainable development (1987–2005): an oxymoron comes of age. *Sustainable Development, 13*(4), 212–227. https://doi.org/10.1002/sd.281

Sousa, M. J., & Baptista, C. S. (2014). *How to do research, dissertations, theses and reports: According to Bologna*. Pactor.

Sparrow, L., & Correia, E. (1995). Methods and techniques of social research. 1995. Port: Areal. PATRICIO, A, 1–18.

United Nations Department of Economic and Social Affairs. (2016). Measure what we treasure: sustainable data for sustainable development. In *The Millennium Development Goals Report 2015* (pp. 10–13). UN.

Vieira, R., Major, M. J., & Sea bass, R. (2009). Qualitative research in accounting. Accounting and management control. *Theory, Methodology and Practice, 1*, 131–163.

Williams, B., Wilmshurst, T., & Clift, R. (2010). The role of accountants in sustainability reporting-a local government study. In *Asia Pacific Interdisciplinary Research in Accounting Conference (APIRA)* (pp. 1–18).

World Council on City Data. (n.d.). *World Council on City Data*. Retrieved 10 November 2023, from https://www.dataforcities.org

Zimmerman, J. L. (1977). The municipal accounting maze: An analysis of political incentives. *Journal of Accounting Research, 15*, 107–144.

2 Understanding Sustainable Innovations in Malaysia's Public Procurement Initiatives

Nur Faiza Ishak and Vinesh Thiruchelvam

2.1 Introduction

Public policy is outlined according to its specific requirement, target group and value consequences. According to Walker (2000), the goal is to provide an overview that can clarify the problem, define the targets and provide guidance on how to achieve them. As a start, Malaysia begins to develop sustainability efforts through the environmental dimensions in the year 2016 and to move towards other dimensions as of now. When the United Nations (UN) first launched the sustainable development goals (SDGs) Agenda in 2016, the 11th Malaysia Plan was reorganized, SDGs targets and indicators are closely associated with the 11th Malaysia Plan (11MP) from 2016 to 2020, where Malaysia committed to pursuing green growth and adopting sustainable consumption and production concepts (EPU, 2015). Currently, the 12th Malaysia Plan (12MP) enlisted three main themes related to sustainability, innovation, procurement and green economic growth; generating income through green practice and encouraging SME participation, reinforcing inclusivity through increasing Bumiputera resiliency and encouraging green growth, and finally, increase sustainability towards the low-carbon city and green economy growth (Economic Planning Unit, 2021b).

Predominantly, public procurement can be a lever to deliver broader government objectives, such as boosting more innovation in supply markets, using public money

N. F. Ishak (✉)
Integrated Sustainability and Urban Creativity Research Centre, Asia Pacific University of Technology and Innovation, Kuala Lumpur, Malaysia
e-mail: tp062102@mail.apu.edu.my

V. Thiruchelvam
Asia Pacific University of Technology and Innovation, Kuala Lumpur, Malaysia

© The Author(s), under exclusive license to Springer Nature Switzerland AG 2023

F. L. Almeida et al. (eds.), *Multidimensional Sustainability: Transitions and Convergences*, Springer Proceedings in Earth and Environmental Sciences,
https://doi.org/10.1007/978-3-031-24892-4_2

to encourage social and environmental goals and supporting the local industry (Nijboer et al., 2017). Furthermore, the Economic Planning Unit (2017) reported that since 2013, the government's spending on green procurement has achieved up to RM429 million. By implementing sustainable procurement as the biggest single buyer, the country as a whole will benefit economically without damaging the environment and harming the society. Tremendous studies focus on either only the environmental dimension or the sustainable dimension without integrating the innovation aspect.

From a narrower perspective, Eikelboom et al. (2018) and Rolfstam (2012) defined sustainable innovation through public procurement in a much simpler way as purchasing activities which do not compromise the needs of future generations being carried out by public agencies that lead to innovation. The practice applies not only to products, services and technologies but also to new business and organizational models. Sustainable innovations through public procurement can be developed in various options, such as an enhanced product, a synergized product or a product-service system that will benefit the environment, economy and social elements of sustainability (Alhola & Nissinen, 2018). Furthermore, sustainable innovations through public procurement can be implemented at all levels to assist the country in achieving goals for the sustainability agenda.

Through a more comprehensive picture, this study allows the industry to recognize the main objective of overall policies is to achieve green economic growth regardless of sector. Another aim of this study also allows us to organize individual policies into a strategic framework to understand the relationship and prove that they can contribute to each other's goals. Furthermore, towards reviewing the core policies related to sustainability, innovations and public procurement in Malaysia, this study will evaluate the policy perspective in achieving the nation's SDGs Goals 2030.

This study makes several contributions to the body of knowledge. First, this study is beneficial for the administrator executive amongst the civil servant to associate with actual on-the-ground practices. Since the policies underlined are of different ministries and stakeholders, the final framework can help the policy enabler oversee the whole perspective of sustainable innovations in the public procurement context. Besides, the key prominent goal of each stand-alone policy is explained further in this study.

The rest of this paper is structured as follows. Section 2.2 describes previous research on sustainable innovation implementation in public procurement and addresses the missing context of the existing research. Section 2.3 provides a brief introduction to all sustainable innovations in public procurement-related policies in Malaysia. Section 2.4 describes the policy framework proposed in this study, and Sect. 2.5 presents the study's conclusions and recommendations.

2.2 Existing Public Procurement Action in Achieving Sustainable Innovations

Recent discussions of sustainability and innovations in public procurement mainly focuses on various implementation globally. A study by Bucea-Manea-ţoniş et al. (2021) in Romanian agriculture proposed that the sustainable incorporation of innovative products and services will be able to address the barriers encountered, such as the ability to produce constant supply, reduce the cost and longer shelf life of the product. This study is similar to Morley (2020) who also agrees that the act of purchasing by public organizations in the UK has the ability to stimulate a broader sustainability practice through the knowledge spill over in the industry. Furthermore, Zipperer (2019) also suggested that it is appropriate to further invest in sustainable innovation solutions to reduce the price in the long run in Germany's context. However, following the incorporation of sustainable innovation initiatives, a further study by Todeschini et al. (2020) found that only a small number of consumers decided to purchase the fashion product in Brazil due to no further price cuts contributed.

Another study by Ghisetti (2017) in the EU revealed that 44% of the respondents who have won at least one tender stated that they had included environmental innovation in their procurement contract. Among the organizations are well-established firms that are innovators themselves. However, research in Germany's public procurement context between 2006 and 2016 only identified 2.19% of the tender that had included GPP innovation (Zipperer, 2019).

Since sustainable innovation practice in public procurement has a minimal overview from the previous researcher, it is significant to focus on the ground rule of sustainable public procurement to encourage innovation implementation. Besides, policy commitment will be able to be the benchmark for the decision-making process by public and private organizations on a broader scale. Hence, it is wise to start with understanding Malaysia's policy outline that can be utilized to implement sustainable innovations through the public procurement process.

To date, there have been decent studies on implementing sustainable procurement in Malaysia (Ishak & Thiruchelvam, 2022; Vejaratnam et al., 2020). However, most recent studies only focused on the construction sectors (Bidin et al., 2020; Jaaffar & Kaman, 2020) and the manufacturing industry (Chan et al., 2016; Yusr et al., 2020). Nevertheless, an early study by Adham and Siwar (2012c) reviewed Malaysia's current policy related to the economics and overall public procurement directions. His study was then supported by other subsequent findings by Musa et al. (2013) which indicated four (4) key factors that drive organizations towards implementation success; one of them is the familiarity of policies framework. Bidin et al. (2020) and Liu et al. (2019) mentioned similar findings related to the importance of solid policy coherences too. Hence, conducting a document review of Malaysia's latest policy will address the policy continuity to ensure that Malaysia is on the right path towards successfully implementing sustainable innovations through the public procurement process.

2.3 Towards Sustainable Innovations in Malaysia Public Procurement

When people discuss sustainability, the most crucial thing is to achieve the needs without compromising the environment, social and economic aspects. Even though many have related sustainability to the environmental dimension, holistically, sustainability encompasses a broader target in terms of benefitting social welfare and boosting the local economy. Sustainability terms are becoming increasingly popular and are widely known as one of the drivers of innovation (Spindler, 2013). On top of that, public procurement can also drive the innovation cycle through R&D management, commercialization and project and risk management (Obwegeser & Müller, 2018).

In Malaysia, sustainability and innovation in public procurement are in line with the national sustainability objectives through various commitments and policies. The spill over from sustainable public procurement implementation has driven the market for innovation in related products or services (Haugbølle & Raffnsøe, 2019; Rainville, 2016). Furthermore, the contribution of sustainable innovations through public procurement will be able to deliver "the most economically advantageous tender" and at the same time, influence the supply chain market in prioritizing sustainable innovations market. Besides, Uyarra (2013) agreed that the inclusivity of sustainable innovations through the procurement process might improve the service delivery of the public sector itself.

This study employed a qualitative study using document analysis research design in order to address the main objective. The document analysis method mainly focused on three (3) steps; namely skimming, reading and interpretation (Bowen, 2009). The document analysis method is based on secondary data obtained from the official website of Malaysia's Government and its related ministries, filtered with only the latest policy, which will be analysed. Moreover, the document analysis method helps understand the historical flow, legislation requirement, current policy constraint and the expected output and outcome of each policy. Besides, according to Mackieson et al. (2019), official secondary data from the government are considered a good data source of official provenance with high-quality content.

There are a few selection criteria when choosing the related official documents. The policy must be current and active, ongoing implementation and related to either sustainability or innovations with the procurement process included. Besides, the policies must be applicable to all key stakeholders, either the ministries, state Government offices or even the private organizations in all sectors. After carefully exploring Malaysia's government's official websites, seven (7) main policies related to sustainable innovations in public procurement were selected. The complete flow for retrieving secondary data for this study is illustrated in Fig. 2.1.

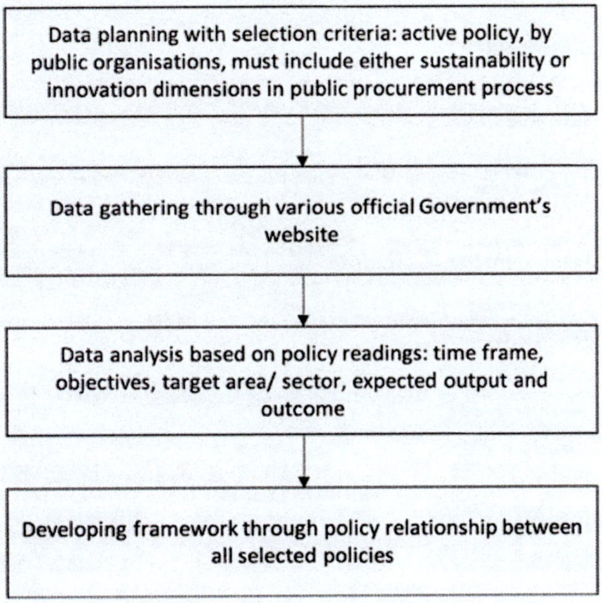

Fig. 2.1 Data gathering flow for this study

2.3.1 12th Malaysia Plan

The Malaysia Plan was first introduced in the year 1970 to address economic and social issues within a span of five (5) years period. This includes welfare, employment opportunities, new economic activities, as well as health and social well-being. The Five-Year Development Plans set out the macro-economic growth targets and the size and allocation of the public sector development programs. At the same time, it states the indicative role envisaged for the private sector. After more than 50 years, in 2021, Malaysia's Government launched Twelfth Plan 2021–2025 with the goal of "*Keluarga Malaysia* – Prosperous, Inclusive, Sustainable". The 12th Malaysia Plan (12MP) is anchored on three (3) themes, four (4) policy enablers, nine (9) focus areas and 14 game changers (Jabatan Perdana Menteri, 2021).

According to Economic Planning Unit (2021a), the 12MP aims to concentrate on existing problems whilst simultaneously reviving Malaysia's socio-economic development. Unlike other Malaysia Plans, the 12MP addresses several economic challenges, an ever-ready workforce and good well-being of the whole community due to the terrible pandemic aftermath. This measure also includes sustainability and prosperity in the long term by improving society's well-being. Ultimately, 12MP addresses the issues faced in implementing 11MP and targets to deliver the principle of togetherness and a resilient economy. The 12MP will further continue to embed each strategy aligned to the SDGs Agenda. Hence, we can consider the 12MP as the "umbrella" surrounding the following policies in the following subsections to make Malaysia a sustainable growing nation for future advancements. Figure 2.2 shows the overall structure of the 12MP according to the theme, policy enabler, game changer and strategy.

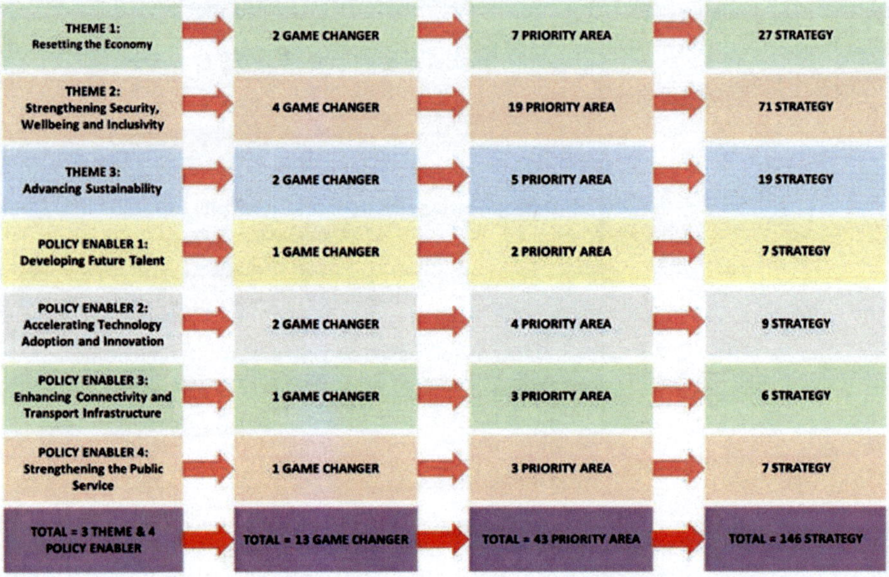

Fig. 2.2 12th Malaysia Plan structure

2.3.2 The National Science, Technology and Innovation Policy (NSTIP)

Another significant agenda of innovation is through various policy initiatives since 2010 such as Innovating Malaysia Framework (2011), Public Sector Initiatives (2012) and the National Science, Technology and Innovation Policy (2015) (Ramli et al., 2017). Specifically, Malaysia's government has enlisted the National Science, Technology, Innovation and Economic Policy 2021–2030 (NSTIP) to cater to new innovative developments and support the nation's economy. To achieve the mission of a sustainable, inclusive and scientifically enriched society towards a high-tech nation, six (6) strategic thrusts were introduced as shown in Fig. 2.3.

One of the programs embedded in NSTIP is the National Technology and Innovation Sandbox. This program encourages organizations to spur innovation and technology from local suppliers where they can participate in public procurement with no restrictions. When innovation and sustainability are being implemented across the country as a policy instrument, it can stimulate the local economy, promote labour rights and encourage the development of new technologies.

On top of that, Malaysia's GGP initiatives in the long-term action plan have also included innovative solutions in public procurement which is known as public procurement promoting innovation (PPPI). Furthermore, sustainable innovation solutions will create higher value-added products at competitive prices in the long term. Likewise, the market will be encouraged to supply innovative products or solutions instead of the conventional supply available. Through this effort, there will be smart

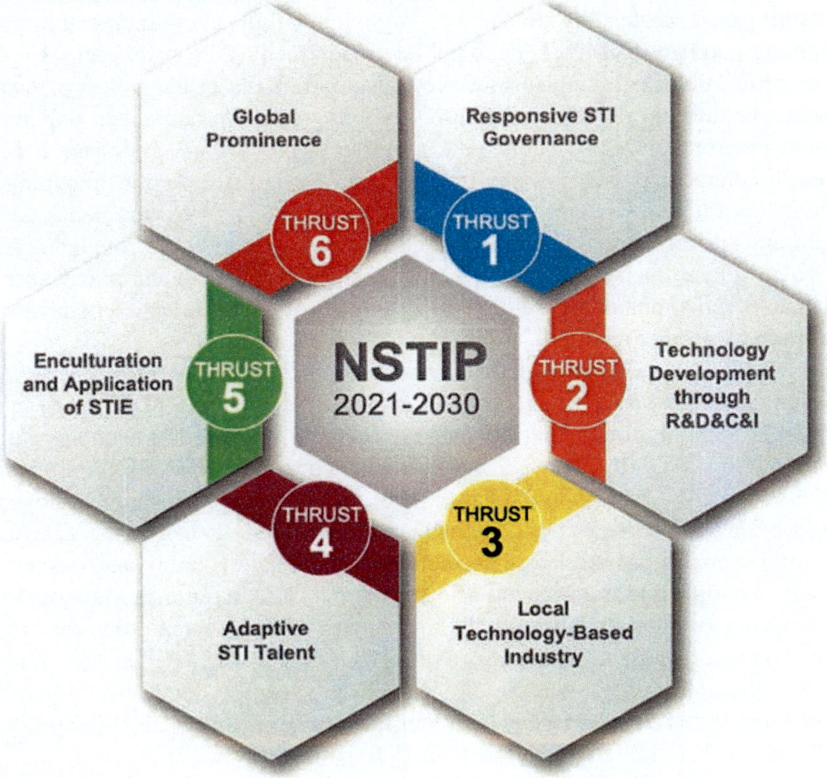

Fig. 2.3 Six (6) strategic thrusts of NSTIP. (Source: Ministry of Science Technology and Innovation, 2021)

funding for sustainable innovation research that will benefit both the government and the industry. According to Ramli et al. (2017), the presence of innovation policy will encourage more innovation-related activities in fast-growing countries. Besides, with the government being the single biggest spender in the country, innovations through public procurement will promote sustainability solutions to spur more demand (Knutsson & Thomasson, 2014). Through this action, the rise of sustainable innovative products and services through public procurement demand can push the industry player to provide more innovative products and solutions.

2.3.3 Government Green Procurement (GGP)

Various countries have begun to set up national policies and capacity building to promote sustainable and innovative solutions through the act of public purchasing (Alhola & Nissinen, 2018). Kahlenborn et al. (2014) defined green procurement as "procurement that is consistent with the principles of sustainable development, such as ensuring a strong, healthy and just society, living within environmental limits and promoting good governance." Green public procurement (GPP) is expected to play an essential role in adopting sustainable consumption and production framework. Besides, the GPP initiative is parallel to promoting public procurement policies to develop environmentally friendly goods and services (Adham & Siwar, 2012a). Green procurement is also known as Environmental Responsible Public Procurement, Sustainable Public Procurement, Environmental Product Procurement, Green Purchasing and Eco-Procurement (Michelsen & de Boer, 2009). Through an effective policy, governments can also demonstrate the effectiveness and practicality of sustainable procurement by setting an example for other private organizations (European Commission, 2020).

Like GPP, Malaysia through Kahlenborn et al. (2014) has released its own Government Green Procurement (GGP) to improve the efficiency of public procurement by utilizing public market influence to transform Malaysia's economy into a green economy. GGP in Malaysia uses the same concept as the GPP term that has been used globally. Adham and Siwar (2012b) defined GGP as the purchase of supplies, services and works by public organizations considering environmental criteria to minimize environmental harm, accelerate the national economy and foster sustainable development. Malaysia has also considered GGP in the 11th Malaysia Plan (11MP) from 2016 to 2020 through its fifth pillar which is to achieve at least 20% of GGP on selected green products and services. Additionally, GGP has been listed as one of the pathways to increase the demand for green products and services by encouraging industries to meet green requirements. To achieve this, Malaysia has developed GGP Short-Term Action Plan (STAP) in the year 2014–2015 and the Long-Term Action Plan (LTAP) by targeting at least 20% of procurement to be green by the year 2020, 50% green by 2025 and by 2030, 100% of procurement to be green (Economic Planning Unit, 2016).

Even though most of the developed countries acknowledge the importance of green practices, developing countries such as Malaysia and other southeast Asian countries are just at their starting point (Bohari et al., 2017). Amongst all, GGP benefited in terms of supporting local innovation, the economy and the innovative solution market. The inclusivity of environmental, social and economic dimensions can be seen in Malaysia's GGP initiative through its Long-Term Action Plan 2016–2025, which will indirectly encourage innovation through local vendors. Through GGP 3.0 Policy Guidelines too, the product list has been extended to 40 products and services compared to only six (6) when the policy was first introduced (EPU, 2020). Entering 2021, GGP 3.0 Policy Guidelines' key highlight is expanding from only the environmental dimension to social and innovation aspects in public procurement.

2.3.4 The National Sustainable Consumption and Production Blueprint

According to UNDP (2018), Malaysia's leading cause of environmental degradation and resource depletion is due to the current consumption and production trends. It is crucial for Malaysia to economically benefit the people without harming the environment and affecting the economy. A study by Adham et al. (2013) shows that since 2010, Malaysia has considered several policy objectives such as the New Economic Model and 10th Malaysia Plan (2011–2015) that integrated life cycle objectives with fiscal and economic incentives. Apart from the above-mentioned policies, Economic Planning Unit (EPU) in the year 2016 introduced The National Sustainable Consumption and Production (SCP) Blueprint 2016–2030 with its own tagline "more and better with less" (EPU, 2016). Associating SCP with existing policies will complement all national objectives, development plans and achieve global goals by promoting economic growth without compromising the environment or jeopardizing the needs of future generations.

The National SCP Blueprint enlisted ten (10) pathways through a regulatory framework, economic growth, social inclusivity and environment protection. The purpose is to improve the standard of living and achieve a better quality of life for its people for the following 15 years of implementation until 2030. It will also address the most relevant sectors for green growth, which account for a higher percentage of Malaysia's Gross Domestic Product (GDP). Among the ten pathways paved, the first pathway is "Leading by example through Government Green Procurement (GGP)," where the overall implementation target would be "20% green procurement until 2020, 50% until 2025 and 100% until 2030" (EPU, 2016).

2.3.5 Malaysia's New Economic Model

In addition, the New Economic Model (NEM) is a framework to drive the nation towards a high-nation status by embracing three goals: high income, inclusiveness and sustainability. Apart from that, NEM also proposed that economic sustainability and environmental protection must be the key component to enhancing public sector efficiency. Through NEM too, the latest public procurement guidelines emphasized the preferential margin and quota for local products and services of Bumiputera companies (Adham & Siwar, 2012a; Ministry of Finance, 2020). Preferences favouring the indigenous Malays or Bumiputera existed in colonial Malaya through Article 153 of the new Constitution as part of a package of criteria aimed at securing multi-ethnic support (McCrudden & Gross, 2006). Bumiputera priority is similar to the indigenous program that was introduced in Indonesia (Sarulla Operations Limited, 2015), Australia (Australian government, 2015), Nigeria (Akenroye, 2013) and Canada (Mah, 2014). Furthermore, it is crucial to consider the inclusivity criteria of local products and Bumiputera's priorities do not defeat the purpose of achieving sustainable procurement and supporting economic growth.

In addition to that, the government also mandated tenderers be awarded a specific value of tenders to participate and contribute to the Industrial Collaboration Program (ICP). ICP is a proposition consisting of activities that will add value to public procurement considering cost-effectiveness. The ICP implementation can strategize procurement through global supply chain participation by the local industry players. According to Abdullah and Safari (2018), the ICP program is also known as a countertrade and offset programs such as a buy-back scheme, barter trade or direct offset program. The industry players shall contribute to the government in terms of technology transfer, intellectual property rights, capacity building, human resources, global supply chain or global R&D with local institutions as well as reduce the outflow of local currency (Technology Depository Agency: TDA, 2021). A study by Abdullah and Safari (2018) also revealed that Malaysia had piloted 34 projects valued at up to RM45 billion and a 400% return investment.

2.3.6 Malaysia's Public Procurement Guidelines

In public organizations, the procurement regime is usually associated with a critical decision-making process by procurement officers and top management who are obliged to regulations. Potentially, public procurement exercises can use the purchasing power to help the organization achieve a better sustainability goal which includes expanding an environmentally friendly market, encouraging a good social value for the society and driving the local economy (Cravero, 2017; Zeppel, 2014). Through the broader reach too, public procurement can stimulate innovations in developing a new product, service or even process model in an organization to ensure the best value for money.

In the current year, Malaysia's public sector mainly consists of the organizations under the Federal Government with 27 Federal Ministries commanded by respective Ministers and administrative Secretary General (Ministry of Finance, 2021). Each ministry has various departments and agencies responsible for effectively accommodating specific tasks. In Malaysia, public procurement is primarily regulated by the Ministry of Finance (MOF) through Financial Procedure Act 1957 and the Government Contract Act 1949. The procurement methods, procedures to be followed and the guiding principles are laid down in the Treasury Instructions (*Arahan Perbendaharaan*), supplemented by Treasury Circulars in 1-*Pekeliling Perbendaharaan* (1PP). The government stipulates that any procurement shall follow five fundamental principles: public accountability, transparency, value for money, open and fair competition and fair dealings (Ministry of Finance, 2013). A summary of a typical Malaysia's public procurement procedures is shown in Fig. 2.4.

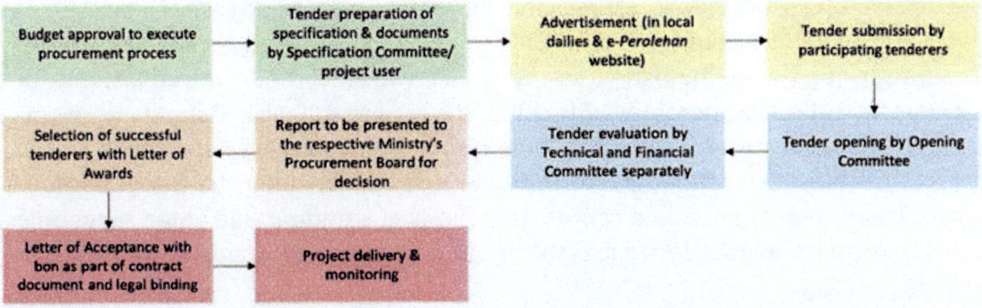

Fig. 2.4 Malaysia's public procurement procedures according to Treasury Circulars

2.3.7 Impact-Driven Public Social Procurement Programme (PPISK)

Compared to the environmental dimension in GPP that has been exercised for quite some time, socially responsible criteria in public procurement (SRPP) are new with limited global policies. As one of the pioneers, the European Commission intends to facilitate the uptake of socially responsible criteria in public procurement and to promote their use across the EU (European Commission, 2021a). According to European Commission (2021b), SRPP is grounded in the 2014 EU harmonized public procurement rules based on the Treaty principles and social aspects. Public authorities can engage in socially responsible public procurement by buying ethical products and services and using public tenders to create job opportunities, decent work, social and professional inclusion and better conditions for disabled and disadvantaged people. On the other hand, SRPP in Australia focuses on the Indigenous Procurement Policy (2020), Equal Opportunity for Women in the Workplace Policy and the Victorian Social Procurement Framework (2017) (Barraket et al., 2021; Troje, 2021). Similarly, the Canadian government has recently adopted the Community Employment Benefit Policy to improve employment opportunities for under-represented groups in significant infrastructure projects (Social Enterprise Ecosystem Project, 2015).

Despite the main policy on GGP, Malaysia has currently initiated a Public Social Procurement Program or known as *Program Perolehan Impak Sosial Kerajaan* (PPISK) in the year 2021, which aims to leverage the purchasing power in positive social outcomes for the country. The program offers priority to organizations and enterprises that can propose products or services that are readily available that comply with social criteria. As a start, PPISK onboarded 11 products and services that can offer social benefits such as fair opportunities, empowering local skills, improving the likelihoods of women, building safe and sustainable livelihoods to the community, offering job employment for people with disabilities or supporting aboriginal villagers (MaGIC, 2021a). The objective of PPISK is mainly to encourage social entrepreneurs and innovations to offer measurable social benefits through the procured product groups.

Furthermore, PPISK was established through three initiatives: (1) priority is given to registered enterprises/organizations that pledge to offer social inclusions, (2) no restriction of participation in any tender values and (3) waived criteria for *ePerolehan* (e-Procurement) (MaGIC, 2021b). As a result, the Malaysian government has pledged a total of RM20 million in social procurement options that can be utilized. On the other hand, the effort of PPISK is expected to offer fair opportunities for businesses to grow in terms of social-impact solutions and at the same time to increase more awareness regarding the importance of sustainable growth and equitable society.

2.3.8 Malaysia's Policy Timeline

After a fundamental strategic policy review, it is evident that all existing policies that address sustainable innovations in public procurement are grounded in the 12th Malaysia Plan. Apparently, even though the ultimate goal of the 12th Malaysia Plan is to achieve green economic growth, but the strategic planning does not solely focus on the environmental or economic dimension. Instead, any economic development should not compromise the environment, social considerations and innovative solutions.

The review of this study also shows Malaysia's strategic policy on sustainable innovations in public procurement is quite fragmented. Each dimension of sustainable innovation and public procurement has its own policy that was outlined independently. However, these policies underline that implementing sustainable innovations in public procurement is essential in the long term. Hence, Fig. 2.5 illustrates the timeline of ongoing active policy for sustainable innovations and public procurement in Malaysia's development plan.

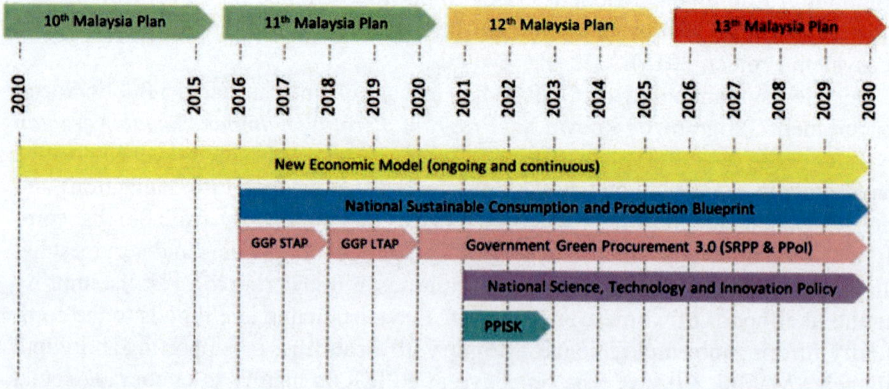

Fig. 2.5 Timeline of sustainable innovations in public procurement policies in Malaysia

2.4 Overall Policy Framework

This study focuses on how sustainable innovations in public procurement address Malaysia's strategic policy to achieve the nation's objective. Trying to address any of the three sustainability pillars in isolation will often result in a trade-off and cause the whole framework to collapse (Aila & Ototo, 2018). Therefore, a balancing approach between all three aspects of sustainability in public procurement will be able to drive the opportunities and see the impact it has at the end of the process. Based on the strategic policy review too, the overall policy framework is grounded in the umbrella of the 12MP which holistically covers each aspect of the environment, social, economic and innovation in public procurement. Each element is bound by at least one active policy as the guidelines that can be referred to during the decision-making process. Furthermore, the whole framework shows the right balance between the environment, society and economy to ensure long-term innovative achievement. Figure 2.6 shows the compilation of policies that connect this study as a framework.

Through this fundamental policy review, we can summarize that Malaysia's government is doing well at sketching structured strategic planning. Overall, when visualizing the policy framework holistically, necessary resources are required to make it happen. For example, a strong capacity building is crucial to ensure that the implementation of these policies will be materialized within the allocated time frame. Besides, enabling economic resources and encouraging the industry to work hand-in-hand can effectively expedite Malaysia's green economic growth.

The idea of collaborating on both sustainability and innovation as a catalyst is nothing new. Even though Malaysia is still at the infancy level of implementing sustainable innovation criteria in the public procurement process, but the action plan shows a promising path. Moreover, the existing regulations and legal acts in the

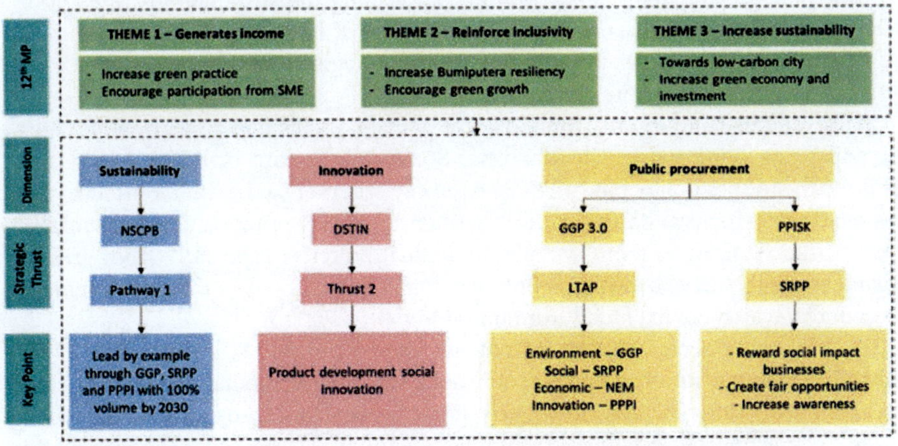

Fig. 2.6 Sustainable innovations of public procurement policy framework in Malaysia

social and economic dimensions will ease the process of implementation too. For example, Malaysia's Employment Act clearly indicates the employee's rights such as minimum wages, annual leave entitlement, gender equality, age limit and working hours (Employment Act 1955, 2012). Similarly, a study by Salmoral et al. (2020) shows that robust enforcement and governance should be initiated if the nation's key objective is to be achieved soon.

The formulation of sustainability and innovation criteria in the public procurement process requires the involvement of all key stakeholders. Even though the policies are formulated by the government, those policies are treated as the nation's guideline for other organizations and industry players to follow through. Mainly, these policies were considered the main parameter before any other subpolicy could be used as their additional standards.

2.5 Conclusions

As the global world evolves rapidly, exclusive focus on only a specific target is no longer practical. This perspective brings to light the need for a more practical application of the sustainability concept in public procurement, which seeks to balance the environmental, social and economic pillars in consumption decision-making. The findings of this study managed to connect the invisible bridge and set the policy directions towards the same goal: green economic growth. It also proves that these policies complement each other towards the national fulfilment of sustainability. Rest assured, implementing these policies can result in quality assurance and transparency in the public procurement process. More importantly, this study contributes to the overall understanding of sustainable innovations in public procurement policies and extends the literature by integrating existing policies to develop a new framework for public procurers. Moreover, this study proves that on top of regulating new policies, the main concern of accomplishing the sustainability agenda of SDGs is being addressed accordingly. Nevertheless, this study can be used deliberately as a path towards synchronizing various strategic initiatives and policies by systematically looking at the policies' objectives.

However, this study is only limited to the active policies that are currently ongoing and being implemented in Malaysia. Some policies are being enhanced and replaced the former policy but still embed the key criteria listed earlier. Furthermore, this study only focuses on the policies written by public organizations and ministries. At the same time, there are also individual policies, especially from private organizations that have been practising sustainable innovations in their procurement procedures, which are excluded from this study.

Even though a decent number of policies exist to guide Malaysia towards sustainability and innovations, it does not mean that there are no challenges in practising them. To further benefit all related stakeholders, this study can be extended towards the actual performance and implementation of sustainable innovations in the public procurement process. Besides, it is also possible to compare policy analysis between developed and developing countries to assess the actual performance of sustainable innovations in public procurement. Having said that, future studies on a comprehensive view of sustainable innovations in public procurement implementation would be an excellent benchmark for the overall impact of the organization's actual performance.

References

Abdullah, A., & Safari, Z. (2018). Industry Collaboration Program (ICP): Empowering technology development for national economic growth. *Journal of Advanced Manufacturing Technology, 12*, 159–172.

Adham, K. N., & Siwar, C. (2012a). An empirical investigation of government green procurement (GGP) practices in Malaysia. *OIDA International Journal of Sustainable Development, 4*(4), 77–87.

Adham, K. N., & Siwar, C. (2012b). The establishment of government green procurement practices in Malaysia: The way forward. *International Journal of Sustainable Development, 05*, 41–50.

Adham, K. N., & Siwar, C. (2012c). Transformation of government procurement in Malaysia: Directions and initiatives. In *5th international public procurement conference*, April 2016, pp. 1621–1641. http://www.ippa.org/IPPC5/Proceedings/Part7/PAPER7-1.pdf

Adham, K. N., Merle, K., & Weihs, G. (2013). Sustainable consumption & production in Malaysia: A baseline study on government policies, institutions and initiatives. *Economic Planning Unit* (Issue January). https://doi.org/10.1007/978-1-4020-8939-8_68

Aila, F. O., & Ototo, R. N. (2018). Sustainable procurement concept: Does it all add up? *International Journal of Development and Sustainability, 7*(2), 448–457.

Akenroye, T. O. (2013). An appraisal of the use of social criteria in public procurement in Nigeria. *Journal of Public Procurement, 13*(3), 364–397.

Alhola, K., & Nissinen, A. (2018). Integrating cleantech into innovative public procurement process – Evidence and success factors. *Journal of Public Procurement, 18*(4), 336–354. https://doi.org/10.1108/JOPP-11-2018-020

Australian Government. (2015). Commonwealth Indigenous Procurement Policy. https://www.niaa.gov.au/sites/default/files/publications/indigenous-procurementpolicy.pdf

Barraket, J., Hiruy, K., Walshe, J., & Moussa, B. (2021). *The state of social procurement in Australia and New Zealand 2021*. IPA Personnel Services and Chartered Institute of Procurement and Supply. https://www.cips.org/australia-new-zealand/the-state-of-social-procurement/

Bidin, Z. A., Mohamad Bohari, A. A., Amat Rais, S. L., Mohamad Saferi, M., & Olanipekun, A. (2020). Challenges and drivers of green procurement among construction practitioners in Malaysia. *International Journal of Service Management and Sustainability, 5*(1), 149–176. http://ir.uitm.edu.my/id/eprint/32537/

Bohari, A. A. M., Skitmore, M., Xia, B., & Teo, M. (2017). Green oriented procurement for building projects: Preliminary findings from Malaysia. *Journal of Cleaner Production, 148*, 690–700. https://doi.org/10.1016/j.jclepro.2017.01.141

Bowen, G. A. (2009). Document analysis as a qualitative research method. *Qualitative Research Journal, 9*(2), 27–40. https://doi.org/10.3316/QRJ0902027

Bucea-Manea-ţoniş, R., Martins, O. M. D., Ilic, D., Belous, M., Bucea-Manea-ţoniş, R., Braicu, C., & Simion, V. E. (2021). Green and sustainable public procurement – An instrument for nudging consumer behavior. A case study on Romanian green public agriculture across different sectors of activity. *Sustainability (Switzerland), 13*(1), 1–25. https://doi.org/10.3390/su13010012

Chan, S. W., Abdullah, M. E., Zaman, I., & Fong, P. P. (2016). Identification of sustainable procurement in manufacturing industry. *Journal of Engineering and Applied Sciences, 11*(10), 2141–2145.

Cravero, C. (2017). Socially responsible public procurement and set-asides: A comparative analysis of the US, Canada and the EU. *Arctic Review on Law and Politics, 8*, 174–192. https://doi.org/10.23865/arctic.v8.739

Economic Planning Unit (2016). *SDG RoadMap For Malaysia Phase 1:2016–2020*. https://www.ekonomi.gov.my/sites/default/files/2021-05/SDG_Roadmap_Phase_I_2016-2020.pdf

Economic Planning Unit. (2017). *Malaysia Sustainable Development Goals voluntary national review 2017*. Economic Planning Unit.

Economic Planning Unit. (2021a). *Twelfth Malaysia plan: A prosperous, inclusive, sustainable Malaysia*. EPU. https://rmke12.epu.gov.my/bm

Economic Planning Unit. (2021b). *Twelfth Malaysia plan 2021–2025*. Economic Planning Unit, Prime Minister's Department. https://rmke12.epu.gov.my/bm%0A, https://rmke12.epu.gov.my/about-us

Eikelboom, M. E., Gelderman, C., & Semeijn, J. (2018). Sustainable innovation in public procurement: The decisive role of the individual. *Journal of Public Procurement, 18*(3), 190–201. https://doi.org/10.1108/JOPP-09-2018-012

Employment Act 1955, Business Law Review. (2012). https://doi.org/10.54648/bula1982153

EPU. (2015). *Eleventh Malaysia plan 2016–2020*. https://doi.org/10.1109/CLUSTR.2004.1392655

EPU. (2016). *Final Draft The National SCP Blueprint 2016–2030 The Pathways for Sustainable Consumption and Production (SCP) in Malaysia*. http://www.switch-asia.eu/fileadmin/user_upload/PSCs/SCP-BP_Final_Draft_Jan2016.pdf

EPU. (2020). *Garis Panduan Perolehan Hijau Kerajaan (GGP) 3.0*. EPU. https://doi.org/10.22146/bb.60958

European Commission. (2020). Innovation procurement. https://commission.europa.eu/funding-tenders/tools-public-buyers/innovation-procurement_en

European Commission. (2021a). *Social procurement*. https://ec.europa.eu/info/policies/public-procurement/tools-public-buyers/social-procurement_en

European Commission. (2021b). Socially responsible public procurement. https://www.socialenterprisebsr.net/wp-content/uploads/2021/02/WeBuySocialEUFAQ_final.pdf

Ghisetti, C. (2017). Demand-pull and environmental innovations: Estimating the effects of innovative public procurement. *Technological Forecasting and Social Change*, 178–187. https://doi.org/10.1016/j.techfore.2017.07.020

Haugbølle, K., & Raffnsøe, L. M. (2019). Rethinking life cycle cost drivers for sustainable office buildings in Denmark. *Facilities, 37*(9), 624–638. https://doi.org/10.1108/F-01-2018-0003

Ishak, N. F., & Thiruchelvam, V. (2022). Sustainable public procurement in Malaysia: A literature review. *International Journal of Early Childhood Special Education (INT-JECS), 14*(03), 7251–7268. https://doi.org/10.9756/INT-JECSE/V14I3

Jaaffar, A. H., & Kaman, Z. K. (2020). Green supply chain management practices and environmental performance: A study of employee's practices in Malaysia chemical related industry. *Journal of Environmental Treatment Techniques, 8*(1), 125–131.

Jabatan Perdana Menteri. (2021). *Twelfth Malaysia plan 2021–2025 speech*. https://rmke12.epu.gov.my/storage/mediastatementandspeech/2021092820_official_translation_speech_of_yab_prime_minister_tabling_of_the_twelfth_malaysia_plan_2021_2025.pdf

Kahlenborn, W., Mansor, N., & Adham, K. N. (2014). Government Green Procurement (GGP) guidelines for government procurers. Economic Planning Unit, Malaysia.

Knutsson, H., & Thomasson, A. (2014). Innovation in the Public Procurement Process: A study of the creation of innovation-friendly public procurement. *Public Management Review, 16*(2), 242–255. https://doi.org/10.1080/14719037.2013.806574

Liu, J., Shi, B., Xue, J., & Wang, Q. (2019). Improving the green public procurement performance of Chinese local governments: From the perspective of officials' knowledge. *Journal of Purchasing and Supply Management*. https://doi.org/10.1016/j.pursup.2018.05.002

Mackieson, P., Shlonsky, A., & Connolly, M. (2019). Increasing rigor and reducing bias in qualitative research: A document analysis of parliamentary debates using applied thematic analysis. *Qualitative Social Work, 18*(6), 965–980. https://doi.org/10.1177/1473325018786996

MaGIC. (2021a). *Buy for impact*. https://www.buyforimpact.co/

MaGIC. (2021b). *Program Perolehan Impak Sosial Kerajaan (PPISK)*. https://www.mymagic.my/ppisk/en

Mah, E. (2014). An evaluation of Canada's procurement policies for Aboriginal business. *Manitoba Policy Perspectives, 1*(1), 65–81.

McCrudden, C., & Gross, S. G. (2006). WTO government procurement rules and the local dynamics of procurement policies: A Malaysian case study. *European Journal of International Law, 17*(1), 151–185. https://doi.org/10.1093/ejil/chi162

Michelsen, O., & de Boer, L. (2009). Green procurement in Norway; a survey of practices at the municipal and county level. *Journal of Environmental Management, 91*(1), 160–167. https://doi.org/10.1016/j.jenvman.2009.08.001

Ministry of Finance Malaysia. (2013). *Malaysia's government procurement regime*. Government Gazette. http://www.treasury.gov.my/pdf/lain-lain/msia_regime.pdf

Ministry of Finance Malaysia. (2020). *Punca kuasa, prinsip dan dasar perolehan kerajaan*. Pekeliling Perbendaharaan Malaysia. http://www.treasury.gov.my/pekeliling/topik/1.PK1Combine - 1.pdf

Ministry of Finance Malaysia. (2021). Anggaran Perbelanjaan Persekutuan 2021. https://www.mof.gov.my/portal/arkib/perbelanjaan/exp2021.html

Ministry of Science Technology and Innovation Malaysia. (2021). National Science, Technology and Innovation Policy 2021–2030. https://www.mosti.gov.my/wp-content/uploads/2022/03/National-Science-Technology-and-Innovation-Policy-2021-2030.pdf

Morley, A. (2020). Procuring for change: An exploration of the innovation potential of sustainable food procurement. *Journal of Cleaner Production, 279*. https://doi.org/10.1016/j.jclepro.2020.123410

Musa, N. D., Buniamin, S., Johari, N. H., Ahmad, N., Rauf, F. H. A., & Rashid, A. A. (2013). Key indicators towards the implementation of green government procurement in Malaysia. *World Applied Sciences Journal, 28*(13), 127–135. https://doi.org/10.5829/idosi.wasj.2013.28.efmo.27020

Nijboer, K., Senden, S., & Telgen, J. (2017). Cross-country learning in public procurement: An exploratory study. *Journal of Public Procurement, 17*(4), 449–482.

Obwegeser, N., & Müller, S. D. (2018). Innovation and public procurement: Terminology, concepts, and applications. *Technovation, 74–75*(March 2018), 1–17. https://doi.org/10.1016/j.technovation.2018.02.015

Rainville, A. (2016). From whence the knowledge came: Heterogeneity of innovation procurement across Europe. *Journal of Public Procurement, 16*(4), 463–504. https://doi.org/10.1108/jopp-16-04-2016-b003

Ramli, R. I., Abu-hassan, N., Arifin, A. S., & Jasmi, A. N. (2017). Implementation of policy initiatives to foster public sector innovation in Malaysia: The need for measurement. *Journal of Science, Technology and Innovation Policy, 3*(1), 23–29. https://doi.org/10.1136/bjsports-2015-095179

Rolfstam, M. (2012). An institutional approach to research on public procurement of innovation. *Innovation: The European Journal of Social Science Research, 25*(3), 303–321. https://doi.org/10.1080/13511610.2012.717475

Salmoral, G., Zegarra, E., Vázquez-Rowe, I., González, F., del Castillo, L., Saravia, G. R., Graves, A., Rey, D., & Knox, J. W. (2020). Water-related challenges in nexus governance for sustainable development: Insights from the city of Arequipa, Peru. *Science of the Total Environment, 747*, 141114. https://doi.org/10.1016/j.scitotenv.2020.141114

Sarulla Operations Limited. (2015). *Indigenous peoples plan* (Issue February).

Social Enterprise Ecosystem Project. (2015). *A guide to social procurement*. Buy Social Canada CCC Ltd. 2018.

Spindler, E. A. (2013). The history of sustainability the origins and effects of a popular concept. In *Sustainability in tourism* (pp. 9–31). Springer Gabler. https://doi.org/10.1007/978-3-8349-7043-5_1

Technology Depository Agency: TDA. (2021). *Policy and guidelines on Industrial Collaboration Programme (ICP) in Government Procurement.* https://www.tda.my/wpcontent/uploads/2022/06/ICP-Policy-and-Guidelines-Second-Edition.pdf

Todeschini, B. V., Cortimiglia, M. N., & de Medeiros, J. F. (2020). Collaboration practices in the fashion industry: Environmentally sustainable innovations in the value chain. *Environmental Science and Policy, 106*(April), 1–11. https://doi.org/10.1016/j.envsci.2020.01.003

Troje, D. (2021). Policy in practice: Social procurement policies in the Swedish construction sector. *Sustainability (Switzerland), 13*(14). https://doi.org/10.3390/su13147621

UNDP. (2018). SDG localization in Asean: Experiences in shaping policy and implementation pathways. https://www.undp.org/sites/g/files/zskgke326/files/publications/RBAP-DG-2019-SDG-Localization-in-ASEAN.pdf

Uyarra, E. (2013). Review of measures in support of public procurement of innovation. In *NESTA Compendium of Evidence on Innovation Policy Intervention* (Issue November). http://www.innovation-policy.net/compendium/

Vejaratnam, N., Mohamad, Z. F., & Chenayah, S. (2020). A systematic review of barriers impeding the implementation of government green procurement. *Journal of Public Procurement, 20*(4), 451–471. https://doi.org/10.1108/JOPP-02-2020-0013

Walker, W. E. (2000). Policy analysis: A systematic approach to supporting policymaking in the public sector. *Journal of Multi-Criteria Decision Analysis, 9*(1–3), 11–27. https://doi.org/10.1002/1099-1360(200001/05)9:1/3<11::AID-MCDA264>3.0.CO;2-3

Yusr, M. M., Salimon, M. G., Mokhtar, S. S. M., Abaid, W. M. A. W., Shaari, H., Perumal, S., & Saoula, O. (2020). Green innovation performance! How to be achieved? A study applied on Malaysian manufacturing sector. *Sustainable Futures, 2*(September). https://doi.org/10.1016/j.sftr.2020.100040

Zeppel, H. (2014). *Green procurement by local government: A review of sustainability criteria.* https://eprints.usq.edu.au/25109/1/GPLocalGov-ACSBDWP8(Zeppel, 2014).pdf

Zipperer, V. (2019). Green public procurement and the innovation activities of firms. *SSRN Electronic Journal.* https://doi.org/10.2139/ssrn.3456520

Part II

Electronics & Industrial Automation

Induction Heating System for Industrial Bearings or Common Appliances

3

Alexandre Diogo Leites da Silva,
José Alberto Campos Neves, and Nelson Neves

3.1 Introduction

The conventional methods for heating steel bearings, to facilitate their placement on a machine shaft, such as engines and alternators, are open flame, oil bath heating, resistive plates and resistive furnace heaters.

According to SKF (2022), the first solution brings risks to the user and can cause structural damage to the bearing. The second solution is environmentally unadvisable, besides contaminating the bearing with oil and possibly compromising its future performance. The third solution is characterized by excessive thermal energy loss. The fourth solution, although more efficient than the previous one, is more expensive, due to the cost of the furnace. In addition, the clogging of the furnace to accommodate large bearings makes it difficult to heat them near the electric machine shaft, increasing risks of improper bearing assembly.

Currently, induction heating is one of the key metal industry applications, based on three main effects: electromagnetic induction, skin effect and heat transfer. The proposed system uses an electronic power converter for coil current control, and consequently, the temperature, an AVR microcontroller plus a set of sensors to measure the necessary variables of the induction heating process. Such system reduces the risk of incorrect use and improves energy efficiency (Mohan et al., 2003) when compared to conventional heating devices. This work aims to develop high-energy efficiency induction heater system, without environmental risks, that minimizes

A. D. L. da Silva (✉) · N. Neves
ISPGAYA, Vila Nova de Gaia, Portugal
e-mail: adiogo@ispgaya.pt

J. A. C. Neves
ISPGAYA, Vila Nova de Gaia, Portugal

ISVOUGA, Santa Maria da Feira, Portugal

© The Author(s), under exclusive license to Springer Nature Switzerland AG 2023

F. L. Almeida et al. (eds.), *Multidimensional Sustainability: Transitions and Convergences*, Springer Proceedings in Earth and Environmental Sciences,
https://doi.org/10.1007/978-3-031-24892-4_3

problems of inadequate assembly, and fulfilling the requirements of the heating process of steel alloys bearings. This induction-based heater guarantees the appropriate process temperature as the temperature differential between the inner and outer rings, within manufacturer specified limits.

This chapter is divided into five parts. In the first part, presents an introduction to the subject. The second part is a review of the induction heating methods. The third part describes the applied method, and in the fourth, results are presented and evaluated. Finally, a conclusion is drawn up indicating practical implications, limitations and other issues as well as future lines of development.

3.2 Literature Review

In induction heating, the temperature rise in the workpiece is caused by its electrical conduction characteristic. This phenomenon is called Foucault currents, also known as Eddy currents, caused by electromagnetic induction (Umans et al., 2014) that is frequency dependent. When compared to conventional heating systems, as is the case of oil bath heating, open flame heating and resistive furnace heaters, induction heating is safe, clean, quick and efficient (Mohan et al., 2003), allowing a defined section of the workpiece to be heated accurately. This is the heating technology of choice in many industrial, domestic (Khatroth et al., 2021) and medical applications due to its advantages regarding efficiency (Lucia et al., 2013), heating speed, safety, cleanness and accurate control. Low frequency and high frequency power converters are, both, used in induction heaters (Mohan et al., 2003; Shen et al., 2006).

The magnitude of the induced currents in workpiece decreases exponentially with the distance x from the surface, by the equation $I_{(x)} = I_0 e^{-x/\delta}$ where I_0 is the current at the surface and and δ is the penetration depth at which the current is reduced to I_0 times a factor $1/e$ (Mohan et al., 2003). The penetration depth is inversely proportional to the square root of the frequency f (Hz) and the magnetic permeability μ (H/m) of the workpiece and is proportional to the square root of the material resistivity ρ (Ω.m), as shown by Eq. 3.1 (Mohan et al., 2003; Rudnev et al., 2017).

$$\delta = \sqrt{\frac{\rho}{\pi . f . \mu_r \mu_0}} \tag{3.1}$$

Consequently, the workpiece apparent cross section in alternated current (S_{ac}) is smaller than the real cross section in direct current (S_{real}). The resulting effect is greater electric resistance $R_{ac} = R_{dc} \frac{S_{real}}{S_{ac}}$ that heats more by Joule effect. Figure 3.1 depicts the apparent cross section of a circular conductor carrying an alternating current.

Circulating currents are caused in the workpiece by currents in the induction coil. The induction load can be represented by an equivalent resistance R in series with the coil reactance $j\omega L$. The equivalent resistance R_{ac} is the electric resistance of the workpiece (Mohan et al., 2003), which is dependent of the penetration depth and temperature.

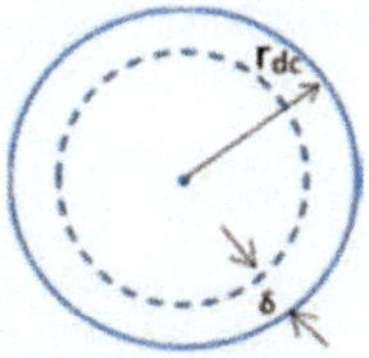

Fig. 3.1 Apparent and real cross section

Using a power converter, such an AC/AC working at constant frequency (Séguier et al., 2015), the knowledge of the current and voltage phase difference in the circuit, is essential to find the minimum value of the power switches firing angle, e.g. thyristors. In an RL load, the phase difference between voltage and current is $\varphi = \text{arctg}\left(\frac{\omega L}{R}\right)$.

The firing angle of the power switches Ψ must have the following conditions: $\varphi < \Psi < \pi$ for one thyristor and $\pi + \varphi < \pi + \Psi < 2\pi$ for the other one. Eddy currents generate, by Joule effect, a thermal power per mass unit (W/Kg), calculated by the equation: $P_e = K_e.B_{\max}^n.f^2$ (Umans et al., 2014). In this equation, B represents the density of the magnetic flux. Using a coil to produce the desired magnetic flux, its density can be calculated by applying the Ampère's Law to a solenoid (Villate, 1999) $B = \mu \frac{N}{l} I$ where I is the current in the solenoid. Thus, the thermal power developed in the workpiece and consequently its temperature depend on the control of that current.

The temperature in the workpiece can be calculated as follows (Perrot, 2010):

$$T = \left(\frac{P}{hA} + T_0\right) - \frac{P}{hA} e^{-\left(\frac{hA}{mC_p}t\right)} \tag{3.2}$$

where h is the heat transfer coefficient of air (W/m^{2}K), A is the dissipation area, m is the workpiece mass, C_p is its specific heat,T_0 is the room temperature and P is the electric power supplied. For time t long enough, the temperature T tends to the steady state value: $T = \left(\frac{P}{hA} + T_0\right)$.

The electric power needed to raise temperature to a desired value after a specified time, its calculated rearranging Eq. 3.2:

$$P = \frac{hA\left(T - T_0\right)}{\left(1 - e^{-\frac{hA}{mCp}t}\right)} \tag{3.3}$$

3.3 Methodology

The proposed solution consists in taking advantage of the temperature rise that occurs in materials, such as steel, which is caused by hysteresis losses and induced Eddy currents (Umans et al., 2014), when the workpiece is exposed to a

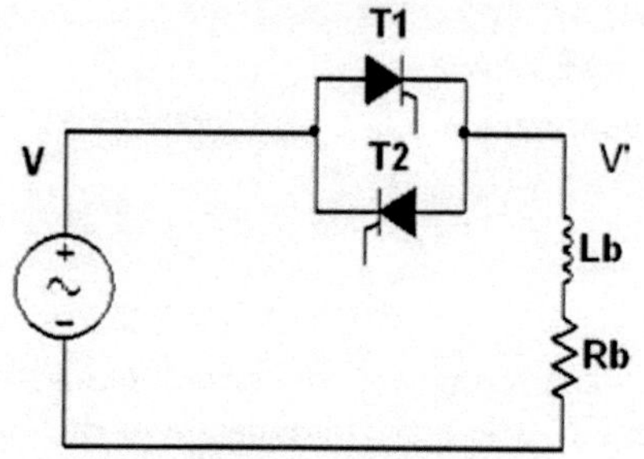

Fig. 3.2 AC/AC power converter

time-varying magnetic field. The thermal power by mass unit (W/Kg) caused by hysteresis losses is proportional to the hysteretic cycle area and is calculated by Steinmetz's empirical formula: $P_h = K_h.B^n_{\max}.f$. Eddy current losses, on the other hand, occur by Joule effect and result from the induced current in the ferromagnetic material. The thermal power by mass unit (W/Kg) caused by Eddy currents is calculated from: $P_e = K_e.B^n_{\max}.f^2$. Thus, the thermal power, by mass unit, that will raise the temperature of the workpiece is $P_t = P_h + P_e$, increasing with the amplitude of the magnetic field.

By controlling the magnitude of the magnetic induction field B, using current control in an induction coil is controlled the thermal power developed and thus the temperature. In order to keep the temperature under specific control in the heating phase, a single-phase AC/AC power electronic converter at constant frequency is used, with a microprocessor control circuit, with a feedback loop. This feedback loop of the control system uses a current sensor and an electric voltage sensor, to get the phase angle between voltage and current, along with two more sensors to measure temperatures in the bearing's inner and outer rings. Figure 3.2 shows the power converter, excluding the thyristor drive circuits, which are electrically isolated by HF transformers. In the scheme, L_b and R_b represent the inductance and ohmic resistance of the coil, respectively.

3.3.1 Calculating the Ohmic Resistance of the Induction Coil

Due to the dimensions of the ferromagnetic core of the prototype under study, the coil is limited to a maximum length l = 35 cm. Using copper conductor wire of 25 mm² cross section (r = 2.82 mm), coiled onto the core of straight section $A = l_1 * l_1 = 5$ cm * 5 cm, with a pitch between turns $p = 7$ mm, the calculated number of turns is

$$N = \frac{l}{p} = \frac{0.35}{0.007} = 50\,\text{turns}.$$

The length of the copper conductor wire obtained is

$L_{\text{cond}} = N * 4 * l_1 = 10$ m. Its ohmic resistance is, according to the physics:

$$R_b = \rho \frac{L_{\text{cond}}}{S} \tag{3.4}$$

Using Eq. 3.4, the resistivity defined for hard copper, $\rho = 0.0179\ \Omega\text{mm}^2/\text{m}$:

$$R_b = 0.0179 * \frac{10}{\pi(2.82)^2} = 7.16\,\text{m}\Omega$$

3.3.2 Calculating the Inductance of the Induction Coil, by Approximation to a Solenoid

Figure 3.3 schematically represents a solenoid. The magnetic flux density inside a solenoid of length l, with N coils ($n = N/l$ coils per meter), crossed by an electric current I_c, is obtained, by application of Ampère's Law (Villate, 1999):

$$\oint_c \vec{B}.d\vec{l} = B\int_b^a dh = B.h = \mu_0 I_{int} \Leftrightarrow Bh = \mu_0.h.\frac{N}{l}.I_c \Leftrightarrow B = \mu_0\frac{N}{l}I_c$$

where μ_0 represents the magnetic permeability of an air core ($\mu_0 = 4\pi.10^{-7}$ H/m).

The magnetic flux through each coil is $\phi = B.A = \mu_0\frac{N}{l}i.A$, where A is the straight section of the solenoid. From the general expression for the inductance of a coil $L = N\frac{\phi}{i}$, the inductance is

$$L = \mu_0\frac{N^2}{l}A \tag{3.5}$$

For a core made of packed silicon steel sheets of magnetic permeability $\mu_r = 1000$, the inductance calculated with Eq. 3.5 is

$$L = \mu_r\mu_0\frac{N^2}{l}A = 1000 * 4\pi * 10^{-7} * \frac{50^2}{0.35} * 0.05 * 0.05 = 22.4\,\text{mH}$$

The induction coil will, therefore, have an impedance:

$$Z = \sqrt{R^2 + X^2} = 7\arg(-90°)\Omega, \text{for } f = 50\,\text{Hz}.$$

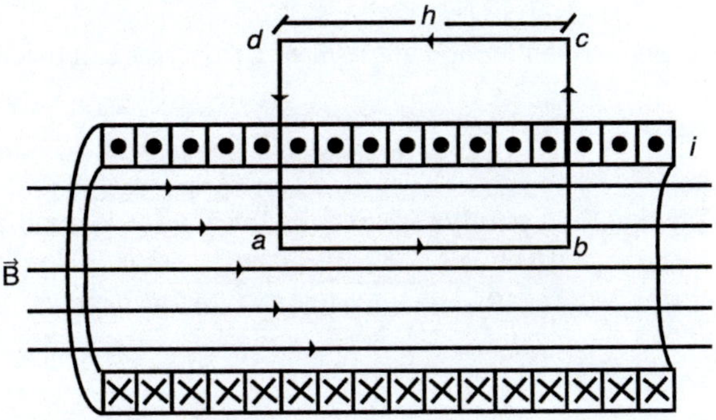

Fig. 3.3 Representation of a solenoid. (Source: Villate, 1999)

So, under a voltage of 230 V_{rms}, it would have a current intensity $I_c = 33$ A (7.5 KVA). Therefore, thyristors in the converter will need to have a stipulated current $I_{\text{rms}} > \frac{33}{\sqrt{2}} = 23\text{A}$.

3.3.3 AC/AC Converter with RL Load

Knowing the phase difference between the current and voltage in the circuit is essential to calculate the minimum value of the thyristor firing angle (Séguier et al., 2015). In *RL* loads, there is a phase difference between the voltage and current $\varphi = \text{arctg}\left(\frac{\omega L}{R}\right)$. The firing angle of the thyristors Ψ must have the following conditions: $\varphi < \Psi < \pi$ for the thyristor T_1 and $\pi + \varphi < \pi + \Psi < 2\pi$ for thyristor T_2. Firing one of the thyristors, for instance T_1 with $\Psi < \varphi$, leads to one of the undesirable operating scenarios depending on the type of command signal. Using a pulse at the thyristor gate, it is obtained a rectified current in half-wave, so the coil would be crossed by direct current of very high value, limited only by the ohmic resistances of the coil and the conductors connecting to the electrical installation. In case of using a pulse train of width $\pi - \Psi$ radians, the current will be sinusoidal with an effective value limited to $I = V/Z$, but without the possibility of control. As will be shown later, this mode of operation is used at system *start-up* and during two cycles of the grid voltage, to calculate, at runtime, the initial phase angle of the load circuit. In some circumstances, it is possible to run the heater at full power with sinusoidal current.

Figure 3.4 depicts the current and voltage waveforms in the induction coil using a suitable thyristor firing angle. It should also be noted that the control range of the converter will be ($\pi - \phi$) radians. However, this is not a real scenario, because, as already mentioned, the coil inductance is in series with the workpiece electric resistance, which is directly proportional to the temperature and inversely proportional to the penetration depth of the Eddy currents. This means that the control range of the power converter is a little greater than 90°.

3.3.4 Command Characteristic with RL Load

The end-of-conduction angle θ_1 of a thyristor can be calculated as follows:

$$\sin(\theta_1 - \varphi) = \sin(\Psi - \varphi)e^{-\frac{\theta_1 - \Psi}{Q}} \quad \text{where } Q = \frac{\omega L}{R}.$$

Note that for a purely resistive load, we get $\theta_1 = \pi$ radians.

The effective value of the voltage at the load comes: $V' = V.\sqrt{\frac{\theta_1 - \Psi}{\pi} - \frac{1}{2}.\frac{\sin 2\theta_1 - \sin 2\Psi}{\pi}}$

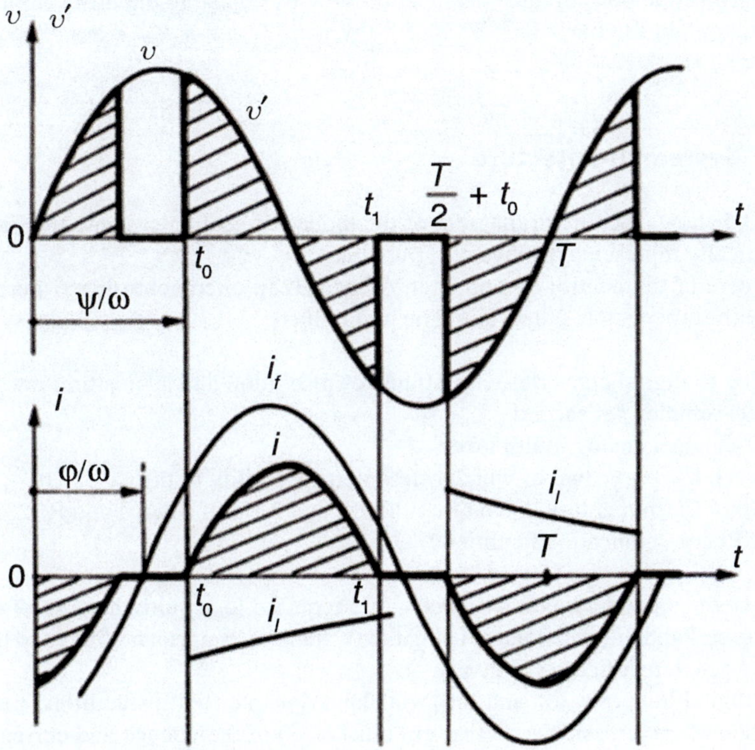

Fig. 3.4 Load voltage and current. (Source: Séguier et al., 2015)

whose Fourier series development leads to V_1', V_3', V_5', V'_7, etc. The effective value of the first harmonics of the current in the coil can obtained by

$$I_1 = \frac{V_1'}{\sqrt{R^2 + (\omega L)^2}} = \frac{V_1'}{R\sqrt{1+Q^2}}, I_3 = \frac{V_3'}{R\sqrt{1+9Q^2}},$$

$$I_5 = \frac{V_5'}{R\sqrt{1+25Q^2}}$$

For predominantly inductive loads, i.e. with $\varphi > 45°$, as is the case of our load, the effective value of the 3rd harmonic is higher than the effective value of the fundamental component of the current, for very high firing angles, i.e. for a low power operation scenario (Séguier et al., 2015).

As already explained, the thermal power generated by mass unit in the workpiece – due to hysteresis and Eddy currents – increases with the density of the magnetic flux in the coil $\left(B = \mu_0 \mu_r \frac{N}{l} I_c \right)$, which in turn is directly proportional to the current. Thus, the power adjustment can be done by adjusting the firing angle Ψ of the thyristors, in the range $\varphi < \Psi < \pi$ for thyristor T_1 and $\pi + \varphi < \pi + \Psi < 2\pi$ for thyristor T_2.

3.3.5 System Architecture

Figure 3.5 shows the circuit diagram of the induction heating system, and Fig. 3.6 schematically represents its physical structure.

The core of the control system is an ATmega328p microcontroller (Microchip, 2022) with some features that should be highlighted:

- Analog to digital converter with 10 bits of resolution and acquisition rate up to 10,000 samples per second
- Eight-channel analog multiplexer
- Fourteen I/O ports, two of which offer interrupt facility to the CPU
- 32 Kb of Flash memory, 1 Kb of EEPROM and 2 Kb of SRAM and RS232, I^2C and SPI communication facilities

Two of the analogic inputs, A_0 and A_1, are required to acquire the current in the induction coil and the grid voltage using the G_1 trans-resistance amplifier and the G_2 precision active rectifier, respectively.

Two digital I/O ports, Int_1 and Int_2, which provide interrupting facilities, are connected to the zero crossing detection circuits (ZCC) of the voltage and current signals, enabling the run-time measurement of the phase angle φ between grid voltage and current in the induction coil. In addition, the ZCC associated with the grid voltage makes it possible to find, in runtime, the delay to be generated for the firing angle of the power converter thyristors.

Temperature measurement sensors on the inner and outer bearing rings, T_{C1} and T_{C2}, respectively, connect to other analogic inputs A_2 and A_3 of the microcontroller,

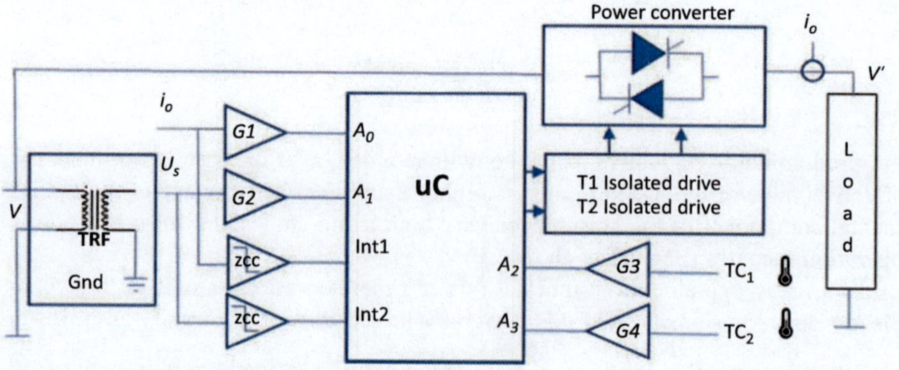

Fig. 3.5 Circuit diagram

through the cold junction temperature-compensated voltage amplifiers, G_3 and G_4. The operator's interface is user friendly, using an alphanumeric LCD display with I²C interface and a small keyboard.

3.3.6 Thermal Behaviour

Figure 3.6 depicts the physical structure of the induction heater.

Assuming that there are no losses, the energy required to raise the temperature from T_0 to T of a workpiece of mass m and specific heat C_p is $W = m.\ C_p.\ (T - T_0)$.

And the needed power is

$$P = m.C_p \frac{dT}{dt}. \tag{3.6}$$

Therefore, the temperature increases linearly with time: $T = T_0 + \frac{P}{m.C_p} t$

However, because there are convection and conduction losses:

$$P_{\text{loss}} = hA\left(T - T_0\right) \tag{3.7}$$

Equation 3.6 must be written to:

$$P - P_{\text{loss}} = m.C_p \frac{dT}{dt}$$

Fig. 3.6 Physical structure

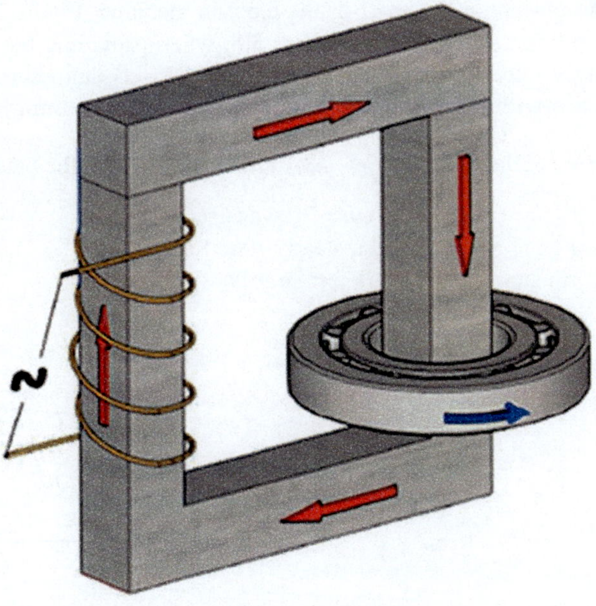

Solving this differential equation, it is obtained the equation to calculate the temperature of the workpiece (Perrot, 2010):

$$T = T_0 + \frac{P}{hA}\left(1 - e^{-\frac{hA}{mC_p}t}\right) \tag{3.8}$$

In Eq. 3.8, h is the air heat transfer coefficient (W/m^{2}K), A is the workpiece thermal contact area, T is its temperature and T_0 is the room temperature. For the workpiece with the shape depicted in Fig. 3.7, the thermal contact area is

$$A = 2A_1 + A_2$$

where:

$$A_1 = \pi\left[\left(D/2\right)^2 - \left(d/2\right)^2\right]$$

and:

$$A_2 = \pi D b$$

It is shown by Eq. 3.8 that in steady state, the workpiece temperature will be: $T = T_0 + \frac{P}{hA}$

There is a wide range of bearings, with great variety in characteristics such as mass and diameter, from 0,009 kg–5 mm to 1200 kg–800 mm. Taking as reference SKF's recommendation (SKF, 2022) and the good practices of electric machine manufacturers – e.g. Asea Brown Boveri (Radvan, 2014), the temperature in the bearings should be raised from 20 to 110 °C in approximately 20 min, before they are placed on the shaft of the electric machine. Figure 3.8 shows the temperature evolution, calculated using Eq. 3.8, when providing the power equal to the workpiece's convection and conduction losses – as defined in Eq. 3.7 – to heat a workpiece with the following characteristics: D = 140 mm, d = 80 mm, b = 26 mm; m = 1650 g. For steel $C_p = 0.466\left(\frac{J}{°C.g}\right)$. Since the heat transfer coefficient of air is $5 < h < 30$ (W/m^{2}K), h = 30 was used.

For T_0 = 20 °C and a steady state temperature T = 110 °C the calculated power loss, using Eq. 3.7 is: P_{loss} = 86.9 W.

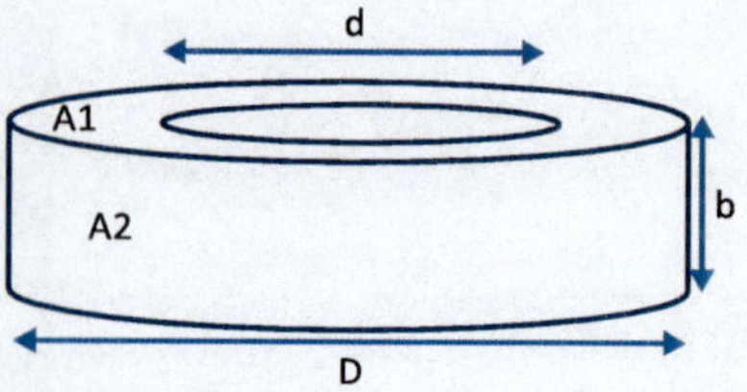

Fig. 3.7 Workpiece thermal contact area

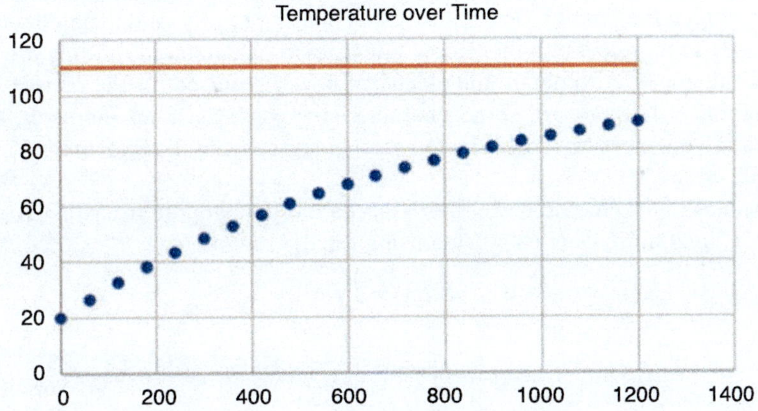

Fig. 3.8 Temperature (°C) over time (seconds) with $P = P_{loss}$

Table 3.1 Heating power for several workpieces

WP	*m*(Kg)	*D*(mm)	*d*(mm)	*b*(mm)	*P*(W)	Ptot(W)
1	1,65	140	80	26	311,4	318,0
2	3,40	160	90	40	614,0	623,9
3	3,70	180	100	34	672,0	683,3
4	4,05	170	80	39	730,6	742,3
5	5,15	200	110	38	923,6	937,7
6	6,50	215	120	42	1155,5	1171,8
7	6,80	200	110	53	1201,8	1217,9
8	7,50	220	150	56	1317,4	1333,9
9	8,30	230	130	46	1460,7	1479,4
10	8,45	190	90	64	1473,9	1491,0
11	8,50	310	200	34	1538,2	1563,4
12	10,7	280	200	60	1879,3	1902,8
13	11,3	320	240	60	2000,8	2028,0
14	12,5	215	100	73	2158,6	2180,7

There is also power dissipated by radiation:

$$P_{\text{rad}} = \sigma \varepsilon A T^4. \tag{3.9}$$

In Eq. 3.9, σ is the Stefan–Boltzmann constant, ε is the workpiece emissivity, A is the radiant surface and T its temperature in Kelvin. For the given data, and using Eq. 3.9, the power dissipated by radiation is $P_{\text{rad}} = 5,\ 67 \times 10^{-8} \times 0,\ 17 \times 0,\ 0322 \times (110 + 273{,}15)^4 = 6,\ 7$ W.

Total power losses are $P_{\text{loss}} + P_{\text{rad}} = 86.9 + 6.7 = 93.6$ W. However, to reach a temperature of 110 °C in 20 min, as recommended, the system must provide 318 W (see Table 3.1), instead of 93.6 W.

In the system under study, the user selects from a list stored in software the family to which the bearing belongs to. Each bearing family is associated with its mass m and some relevant dimensions, allowing to calculate, in runtime, the thyristors firing angle Ψ, to obtain the electrical power required for the correct heating. In Table 3.1, there are some characteristics such as inner diameter, outer diameter and the mass of some selected steel bearings from the SKF manufacturer (SKF, 2022).

The values of the electric power necessary to rise temperature from $T_0 = 20$ °C to $T = 110$ °C in 20 min, were calculated using Eq. 3.8 rewritten as

$$P = \frac{hA\left(T - T_0\right)}{\left(1 - e^{-\frac{hA}{mC_p}t}\right)} \tag{3.10}$$

In Eq. 3.10, it was used for the specific heat capacity of steel $C_p = 0.466\left(\frac{J}{°C.g}\right)$, for the heat transfer coefficient of air h = 30 W/m^{2}K and for the heating time t = 1200 s. In the right column of Table 3.1, it was added the power lost by radiation calculated using Eq. 3.9.

Figure 3.9 shows the temperature evolution over time for workpieces nr.1, nr.7 and nr.14, when supplying the calculated power shown in Table 3.1, using Eq. 3.10, to rise the temperature from $T_0 = 20$ °C to $T = 110$ °C in 20 min. These three pieces were chosen because they correspond to the minimum, intermediate and maximum values of electrical power to be supplied. The temperature profile is the similar for all other workpieces.

With the proposed 3.6 KVA system, we are left with the possibility of using it in a wide range of bearings.

Eddy currents are induced and circulates essentially on the inner surface of the ring closest to the core, since that is where the magnetic field is stronger, and it is

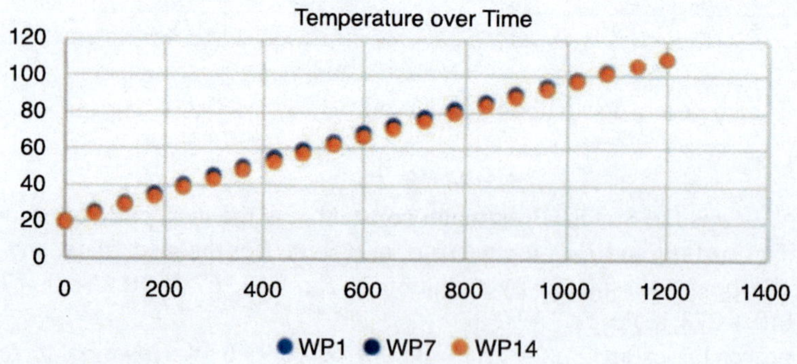

Fig. 3.9 Temperature (°C) over time (seconds) for WP1, WP7 and WP14

also on this surface that the skin effect occurs. The penetration depth of the eddy current comes (Rudnev et al., 2017):

$$\delta_{(T)} = \sqrt{\frac{\rho_{(T)}}{\pi . f . \mu_r \mu_0}} \tag{3.11}$$

The thermal power generated in the workpiece due the Eddy currents can be calculated by the formula (Perrot, 2010):

$$P_o = \frac{\rho}{b}\frac{d}{2\delta}\pi I_w^2 \tag{3.12}$$

where I_w is the induced current in the workpiece, d is the inner bearing diameter and b is its width.

Using Eq. 3.11 on Eq. 3.12:

$$P_o = \frac{\rho}{2b}\frac{d}{\sqrt{\frac{\rho_{(T)}}{\pi . f . \mu_r \mu_0}}}\pi I_w^2 \tag{3.13}$$

The effective eddy current resistance of the workpiece is (from Joule's Law):

$$R_{w(T)} = \frac{\rho}{2b}\frac{d\pi}{\sqrt{\frac{\rho_{(T)}}{\pi . f . \mu_r \mu_0}}} \tag{3.14}$$

where the workpiece electrical resistivity $\rho_{(T)} = \rho_{(20)}[1 + \alpha(T - 20)]$ and α is its the thermal resistivity coefficient. Then

$$R_{w(T)} = \rho_{(20)}\left[1+\alpha(T-20)\right]\frac{d\pi}{2b.\sqrt{\frac{\rho_{(20)}\left[1+\alpha(T-20)\right]}{\pi . f . \mu_r \mu_0}}} \tag{3.15}$$

The current in the workpiece is $I_w = N.\ I_c$ where $N.\ I_c$ is the magnetomotive force (mmf) of the system coil. Thus, the equation to calculate the power developed in the workpiece due to the Joule effect comes:

$$P_o = R_{w(T)}.I_w^2 = R_{w(T)}.\frac{N^2}{1}I_c^2 = \left(R_{w(T)}.N^2\right)I_c^2 \tag{3.16}$$

In this system, the coil, the magnetic core and the workpiece behave like a transformer. Therefore, in Eq. 3.16, $R_{w(T)}.\ N^2$ is the workpiece electric resistance referred to the transformer primary.

It is on this inner surface of the bearing that the temperature rises faster and propagates to the outer ring through the spheres or cylinders. This justifies the need to monitor the temperature difference between the two bearing rings, to avoid structural damage to the spheres/cylinders that would occur due to different thermal dilation of the bearing rings.

3.3.7 Control Algorithm

The proposed algorithm is presented in simplified form in the flowchart of Fig. 3.10. After reading some operator's data, such as the temperature *setpoint*, the maximum temperature difference between the inner and the outer bearing's ring and some other physical bearing characteristics (from bearing family), the initial thyristors firing angle is setup in order to minimize the energy consumption. The *trade-off* power vs heating time is optimized, also ensuring that no structural damage is caused to the bearing, which would occur as result of an excessive temperature difference between its inner and outer rings. The phase load angle is updated at each cycle of the mains. The firing angle is updated at each half cycle of the grid voltage (10 ms, for 50 Hz), using a CPU interrupts routine, so the system's response time is small enough when compared to the thermal time constant of the bearing ($\frac{hA}{mC_p}$ in Eq. 3.8). Such power control method is known as *phase control* (Séguier et al., 2015). In addition, the demagnetization is also assured at the final stage of the heating process, to avoid the aggregation of residues or metallic dust during transportation and placement of the bearing on the machine shaft.

3.4 Experimental Results

All signals are generated in an electronic circuit simulation environment for validation of the proposed control system. Figure 3.11 shows the synchronism pulses generated at zero crossing of the grid voltage (ZCC), the firing pulses for the thyristors T_1 and T_2 with a firing angle $\Psi \approx 95°$ and the voltage at the induction coil.

Figure 3.12 shows the waveforms of the voltage and current in the induction coil for a firing angle $\Psi < \varphi$. As already explained, this is an operating scenario of maximum power at the load but without control. This operation mode, with $\Psi < \phi$, is used at system start up and for a time interval corresponding to only two complete cycles of the grid voltage ($t = 2 \times 20$ ms, for $f = 50$ Hz) to obtain the phase angle φ of the induction coil and thus calculate, at runtime, the minimum possible value to be used for the thyristor firing angle Ψ. However, in some circumstances, it is possible to run the heater at full power with sinusoidal current.

In Figs. 3.13 and 3.14 it is represented, in addition to the firing pulses of the thyristors, the voltage and current in the induction coil for a firing angle $\Psi = 135°$ and $\Psi = 95°$, respectively, showing how current increases for lower firing angles.

In all figures, virtual oscilloscope vertical scales were configured with 20 V/div for load voltage measurements and 0.1 V/div for load current measurements. Current is measured as the voltage drop in a shunt resistor of 0.01 Ohm, resulting in 10 A/div vertical scale. For synchronism pulses, it was selected 1 V/div for vertical scale.

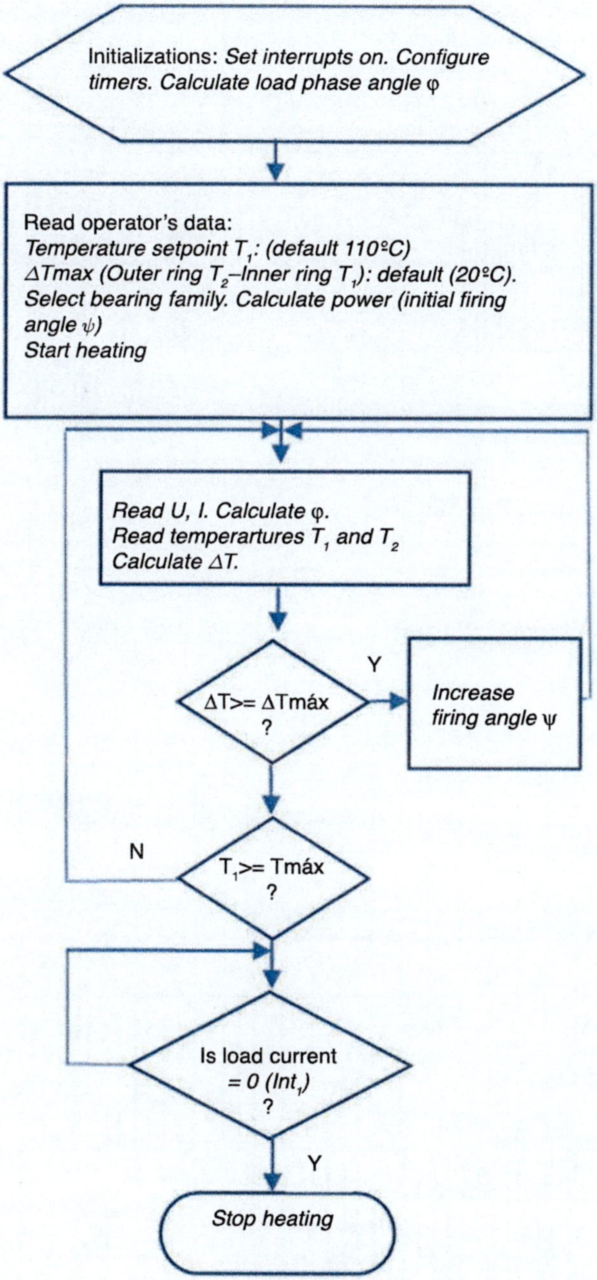

Fig. 3.10 Control algorithm

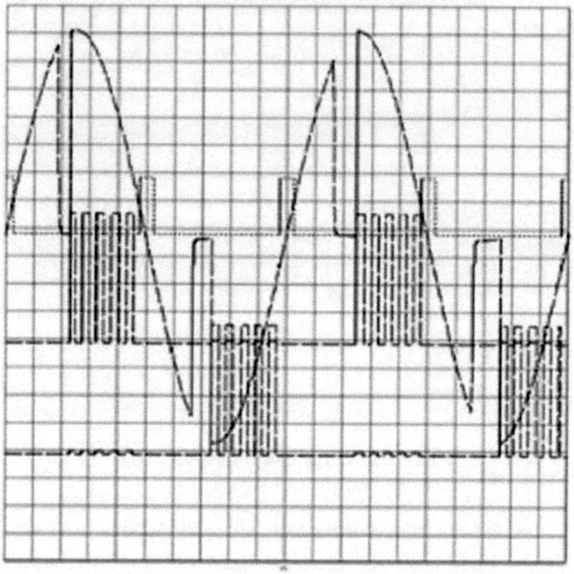

Fig. 3.11 VSync, VtriggerT1, VtriggerT2, V'

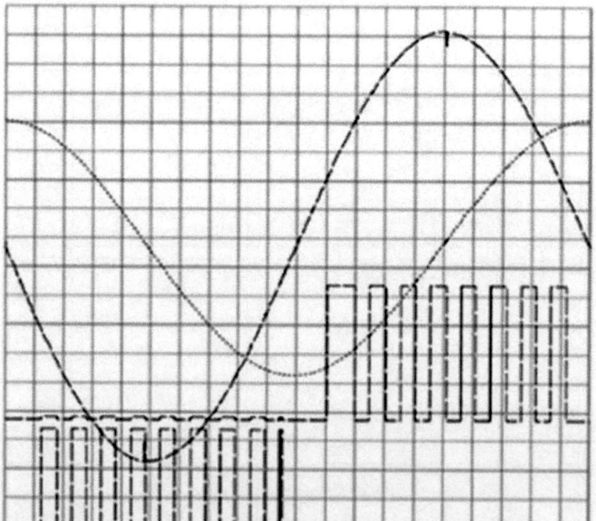

Fig. 3.12 VtriggerT1, VtriggerT2, V' and Io for $\Psi < \phi$. Time base: 1 ms/div

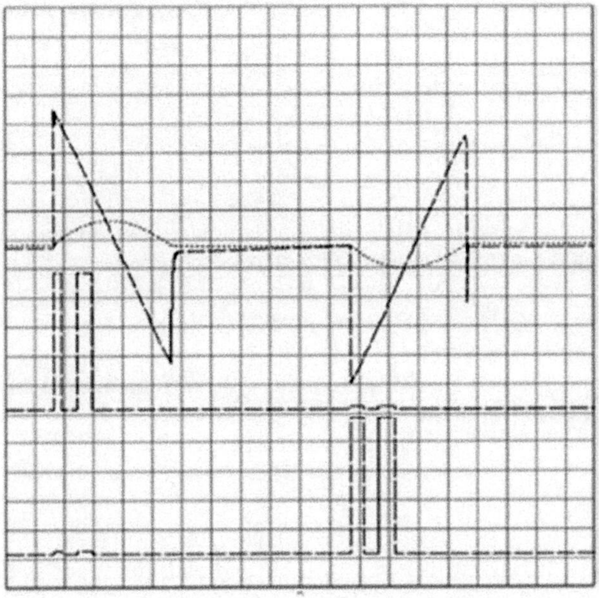

Fig. 3.13 VtriggerT1, VtriggerT2, V', Io, $\Psi = 135°$. Time base: 2 ms/div

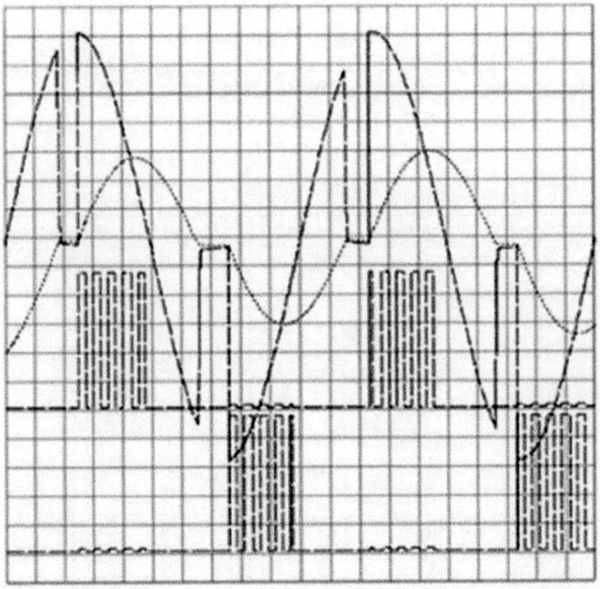

Fig. 3.14 VtriggerT1, VtriggerT2, V', Io, $\Psi = 95°$. Time base: 1 ms/div

As already stated in Eq. 3.14, the effective Eddy current resistance of the workpiece is

$$R_{w(T)} = \rho_{(T)} \frac{d\pi}{2b.\sqrt{\dfrac{\rho_{(T)}}{\pi.f.\mu_r\mu_0}}}$$

As the current in the workpiece is $I_w = N.\ I_c$, the power developed in the workpiece due the Joule effect is, according to the Eq. 3.16:

$$P_o = \left(R_{w(T)}.N^2 \right) I_c^2$$

For each workpiece listed in Table 3.2, the coil current needed to develop the required thermal power in the workpiece is

$$I_c = \sqrt{\frac{P_o}{R_{w(T)}.N^2}} \tag{3.17}$$

It was used for the mains frequency $f = 50$ Hz and considered for steel: $\mu_r = 7500$, $\rho_{(20)} = 1.6 \times 10^{-7}\Omega$. m and $\alpha = 6.5 \times 10^{-3}\ K^{-1}$.

In Table 3.2, P_o was calculated using Eq. 3.10, R_w was calculated using Eq. 3.14 and the coil current I_c was calculated using Eq. 3.17.

The payback of a new and more efficient heating system compared to a conventional system can be calculated as follows:

$$\text{Payback}\left(\text{years}\right) = \frac{\text{Investment difference}\left(\text{EUR}\right)}{\text{Annual savings}\left(\text{EUR / year}\right)} \tag{3.18}$$

Table 3.2 Coil current vs. workpiece power

WP	m(Kg)	D(mm)	d(mm)	b(mm)	P_o(W)	$R_w.N^2(\Omega)$	I_C(A)
1	1,65	140	80	26	318,0	7,40	6,6
2	3,40	160	90	40	623,9	5,41	10,7
3	3,70	180	100	34	683,3	7,08	9,8
4	4,05	170	80	39	742,3	4,94	12,3
5	5,15	200	110	38	937,7	6,97	11,6
6	6,50	215	120	42	1171,8	6,87	13,1
7	6,80	200	110	53	1217,9	4,99	15,6
8	7,50	220	150	56	1333,9	6,45	14,4
9	8,30	230	130	46	1479,4	6,80	14,7
10	8,45	190	90	64	1491,0	3,38	21,0
11	8,50	310	200	34	1563,4	14,15	10,5
12	10,7	280	200	60	1902,8	8,02	15,4
13	11,3	320	240	60	2028,0	9,62	14,5
14	12,5	215	100	73	2180,7	3,30	25,7

Annual savings in Eq. 3.18 can be calculated using the following formula:

$$\text{Annual savings} = \left(\frac{1}{\eta_{\text{std}}} - \frac{1}{\eta_{\text{new}}} \right) \times Pn \times T \times EC \tag{3.19}$$

where η_{std} and η_{new} are the energy efficiency of conventional heating and induction heating systems, respectively. P_n is the electrical power, in kW, T is the annual operation time, in hours and EC the energy cost in Eur/KWh. The energy efficiency of the heating system is

$$\eta_{\text{new}} = \frac{P_o}{P_o + P_i} \tag{3.20}$$

where P_o is the power in the workpiece and P_i is the power dissipated in the induction coil. According to (Callebaut, 2007), it is expected an energy efficiency in the range between 90% and 97%. The coil power loss P_i is minimized adopting the rules described in the Sect. 3.3, Methodology to build de coil, such as using low resistivity copper, low pitch between turns and a geometric configuration close to that of a solenoid.

As explained before, the thermal power generated in the workpiece by Eddy currents can be calculated by Eq. 3.12: Using Eq. 3.1 in

Eq. 3.12: $P_o = \frac{\rho}{b} \pi I_w^2 \frac{d}{2} \sqrt{\frac{\pi.f.\mu_r \mu_0}{\rho}} = \frac{\rho}{b} \pi I_w^2 \frac{d}{2} \sqrt{\frac{\pi.f.\mu_r.4\pi.10^{-7}}{\rho}}$

$$P_o = \frac{d}{b} \pi^2 \sqrt{10^{-7} \rho \mu_r f} \, I_w^2 = \frac{d}{b} \pi^2 \sqrt{10^{-7} \rho \mu_r f} \, (N.I_c)^2 \tag{3.21}$$

where d is the diameter of the inner ring, b is its width, μ_r is the relative magnetic permeability, ρ is the workpiece resistivity (at 110 °C) and f is the frequency of induced current I_w. Equation 3.21 tells that to increase the thermal power in the workpiece, it is better to increase coil current I_c than increasing frequency. In turn, increasing the frequency would increase the reactance of the coil $j\omega L$, which, according to the Ohm's Law, would decrease the current.

To calculate the system's energy efficiency, it is used Eq. 3.20. In that equation, the power loss in the coil is calculated using Joule's Law: $P_i = R_b.I_c^2$, where R_b is the coil electric resistance, and I_c it is the coil current. For the system under study, coil resistance was calculated using Eq. 3.4: $R_b = 7.16$ mΩ.

For workpiece WP1 listed in Table 3.2 (d = 80 mm, b = 26 mm), P_o is 318 W, I_c = 6.6 A and $P_i = R_b.I_c^2 = 0.3$ W. The system's energy efficiency is

$$\eta_{\text{new}} = \frac{318}{318 + 0.3} = 99.9\%$$

For workpiece WP14 listed in Table 3.2 ($d = 100$ mm, $b = 73$ mm), P_o is 2180,7 W, $I_c = 25.7$ A and $P_i = R_b.I_c^2 = 4{,}7$ W. The system's energy efficiency is

$$\eta_{\text{new}} = \frac{2180.7}{2180.7 + 4.7} = 99.8\%$$

In the system under study, it was expected to achieve a relevant energy efficiency for the minimum and the maximum electric power, when compared to conventional systems as those described in (SKF, 2022). That being said, there is still the need to verify in the real environment as it was not taken into account the magnetic losses in the system core.

3.5 Conclusions

Induction heating is based on the three following effects: electromagnetic induction, skin effect and heat transfer. Despite the undesirable skin effect in many electrotechnical applications, here, on the contrary, it is used to generate heat with high efficiency (Khatroth et al., 2021; Lucia et al., 2013; Mohan et al., 2003). The increase in electric resistance caused by the workpiece apparent cross section reduction due to skin effect allows getting more thermal power generated by Joule effect.

The proposed solution, it is a system safe and easy to use for its operators, does not pose risks to the environment, ensures low energy consumption to heat steel alloy bearings without compromising their mechanical structural characteristics.

A low frequency such as the utility frequency is used in the proposed system. That is enough for heating a workpiece or even to melt it (Mohan et al., 2003) should that be the case. The phase control used in the AC/AC power converter leads us to have to correct the power factor using an input capacitive filter. This is important to do, especially in low power operating scenario, where the third harmonic amplitude is predominant and the reactive power exceeds the active power (Séguier et al., 2015). In future developments, the focus will be also in high frequency induction heating, using resonant power converters, to achieve a better *power factor* but also to heat materials other than metals (e.g. ceramic workpieces).

The usage of the proposed system is not restricted to the described application. In fact, there is a wide spectrum of applications where energy consumption and user safety are mandatory in heating systems, as in the case of cooking devices (Khatroth et al., 2021) and other household and industrial appliances (Shen et al., 2006; Takau, 2015).

It is also pursued the objective of developing a fully functional prototype of an induction heater to support the training of our students and engineers.

References

Callebaut, J. (2007). Laborelec, «Chauffage par induction». *Leonardo ENERGY. Guide Power Quality* (7). https://www.econologie.com/file/technologie_energie/Chauffage_par_induction.pdf

Khatroth, S., Porpandiselvi, S., & Vishwanathan, N. (2021). Three-load cyclic controlled single-stage AC-AC resonant converter for induction cooking applications. In *2021 IEEE 2nd international conference on Smart Technologies for Power, Energy and Control (STPEC)* (pp. 1–6). IEEE. https://doi.org/10.1109/STPEC52385.2021.9718631

Lucia, O., Maussion, P., Dede, E. J., & Burdío, J. M. (2013). Induction heating technology and its applications: Past developments, current technology, and future challenges. *IEEE Transactions on Industrial Electronics, 61*(5), 2509–2520. https://doi.org/10.1109/TIE.2013.2281162

Microchip. (2022). *Atmega328*. https://www.microchip.com/en-us/product/atmega328p

Mohan, N., Undeland, T. M., & Robbins, W. P. (2003). *Power electronics: Converters, applications, and design*. Wiley.

Perrot, O. (2010). Cours D'Électrothermie. *IUT de Saint-Omer dunkerque département Génie Thermique et l'énergie, (2010–2011)*. https://sitelec.org/download_page.php?filename=cours/electrothermie.pdf

Radvan, L. (2014). *Fast thyristors. When burning for induction heating solutions*. https://search.abb.com/library/Download.aspx?DocumentID=PEE_82014&LanguageCode=en&DocumentPartId=&Action=Launch

Rudnev, V., Loveless, D., & Cook, R. L. (2017). *Handbook of induction heating*. CRC Press. https://doi.org/10.1201/9781315117485

Séguier, G., Delarue, P., & Labrique, F. (2015). *Electronique de puissance* (10e éd.)*: Structures, commandes, applications*. Dunod.

Shen, H., Yao, Z. Q., Shi, Y. J., & Hu, J. (2006). Study on temperature field induced in high frequency induction heating. *Acta Metallurgica Sinica (English Letters), 19*(3), 190–196. https://doi.org/10.1016/S1006-7191(06)60043-4

SKF. (2022). *Aquecedores por indução*. https://www.skf.com/pt/products/maintenance-products/bearing-heaters/heaters-for-mounting/induction-heaters

Takau, L. (2015). Improved modelling of induction and transduction heaters. PhD thesis in Electrical and Computer Engineering at the University of Canterbury. https://doi.org/10.26021/2756

Umans, S. D., Fitzgerald, A. E., & Kingsley, C. (2014). *Máquinas elétricas* (7ªed.). Bookman.

Villate, J. E. (1999). *Eletromagnetismo*. McGraw-Hill.

4 Magnetic Field Imaging Prototype

José Alberto Campos Neves
and Alexandre Diogo Leites da Silva

4.1 Introduction

This chapter is divided into six parts. The first part presents an introduction to the subject. The second part is a review of the sensor technologies available. The third part describes the prototype hardware. In the fourth part, the basic software, the image formation and colouring techniques, the interpolation and filtering methods are described. Results are presented in the fifth part. Finally, a conclusion is drawn up comprising practical implications, limitations and other issues as well as future lines of development.

Magnetic field is a basic element used in most of the modern machines and electronic devices. Technological innovation allows devices improvement and better understanding of magnetic fields, especially magnetic sensors devices.

Measuring the magnetic field intensity and direction at a point using sensors is simple, but mapping this information in space would improve machines and devices behaviour, particularly when many of the equipment around are electromagnetic based.

The aim of this chapter is to present a prototype capable of visualizing constant and variable magnetic fields for research and academic purposes and for industrial uses. In a world strongly dependent on EMF, improving magnetic field visualization, and the possibility to see them, even if in a limited form, is a meaningful challenge and positive contribution for industry and research in this domain.

J. A. C. Neves (✉)
ISPGAYA, Vila Nova de Gaia, Portugal

ISVOUGA, Santa Maria da Feira, Portugal

A. D. L. da Silva
ISPGAYA, Vila Nova de Gaia, Portugal

© The Author(s), under exclusive license to Springer Nature Switzerland AG 2023

F. L. Almeida et al. (eds.), *Multidimensional Sustainability: Transitions and Convergences*, Springer Proceedings in Earth and Environmental Sciences,
https://doi.org/10.1007/978-3-031-24892-4_4

4.2 Literature Review

Common inductive or magnetic sensors on the market are widely applied in industrial production, as control variables in numerous processes. In general, magnetic fields are auxiliaries of position systems, current measurement, displacement tools or many others. They essentially use Hall-effect, magnetic resistance sensors, transistors magnetic, magneto-impedance, optical-magnetic or some more other sensor devices in industrial processes. Due to operation simplicity, cost, reliability and availability, Hall-effect sensors, as well as magneto-resistive and magnetic impedance sensors, are the most frequent in industry.

The Hall sensor is based on the Hall principle: a semiconductor device crossed by a current I, immersed in a magnetic field B, produces a voltage V proportional to the product of the magnetic field B, the current I and the sine of the angle between the surface sensor and the magnetic field vector, as shown in Fig. 4.1.

Traditional industrial inductive sensors are based on electromagnetic induction principle. They have a circuit oscillator with two coils, the first powered by AC high-frequency signal, and the second where tension change occurs due to Foucault currents. These sensors are used as a position detector of metallic objects (Fig. 4.2).

In case of magnetoresistive sensors (Resistive Magnet Sensor – manufacturer, n.d.; Zhukov et al., 2020; Mohri et al. 2015; Sensor Magneto Resistivo – fabricante n.d.), in the presence of a magnetic field, a variation in electrical resistance occurs in a semiconductor element, based on the magnetoresistive effect (Resistivo, n.d.).

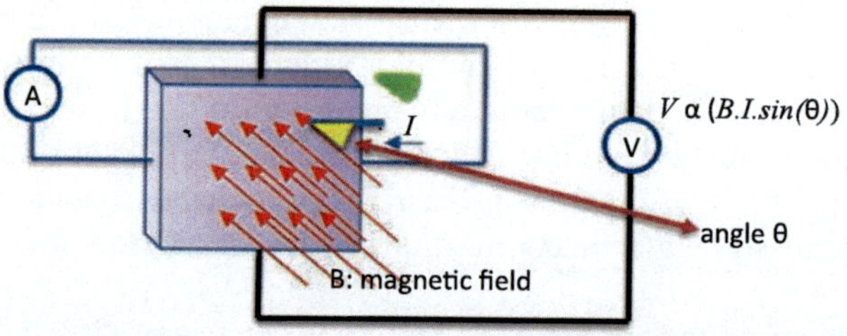

Fig. 4.1 Hall-effect principle

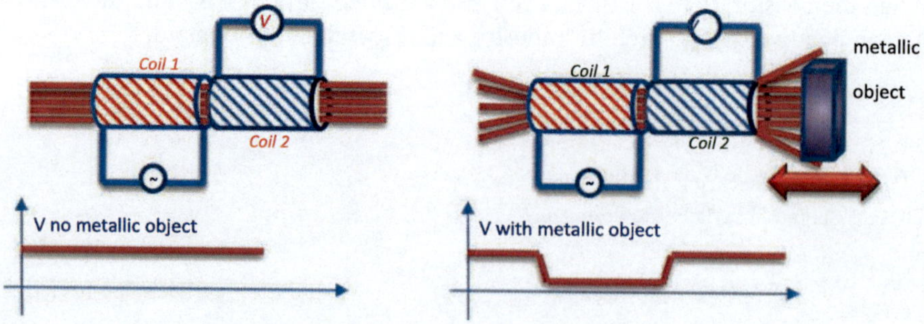

Fig. 4.2 Basic principle of inductive sensor

This variation takes into account the intensity of the field and its direction. These sensor elements are simple to find and have a wide range of applications in positioning systems and in electromechanical devices.

Magnetoimpedance sensors (Shi & Uchiyama, 2020; Uchiyama et al., 2012; Magnetic Impedance Sensing, n.d.; Magneto Impedance Sensor – JP, n.d.) and its operation are based on non-linear behaviour of the impedance variation of an element in the presence of a magnetic field. It provides a fast response where the impedance value changes in a nonlinear way for small variations of magnetic field.

Magneto-optical sensors are crystal elements placed at the end of an optical fiber, having a high Verdet constant, indexed to the variation in the polarization plan of a light beam passing through the fiber. The magnetic fields measured is from the rotation angle of the polarization plan (Magneto optics, n.d.; GMW – magneto optic systems and sensors, n.d.). One of the major uses is to measure the current in power lines through the magnetic field changes produced by the current flow.

Although several research teams have made devices to map magnetic fields, the results are always restricted to a small active face. One of such devices is from MAGCAM[1] consisting of a miniaturized Hall sensor matrix that produces an image of magnetic fields (Fig. 4.3).

The MAGCAM (MAGCAM, n.d.) is a 2D matrix of Hall sensors that reads the magnetic field intensity in a reduced space. This device creates dynamic images of the magnetic fields placed in front of the active face showing variations of their intensity. The sensor matrix only maps a small part of the space surrounding the sensor, requiring a short distance from the object to the sensors face. The dimension of the magnetic field must fit the active face of the sensor. The device has good sensitivity (0.1 mTesla), and the acquisition rate of 820 ms/image has a spatial resolution of 100 μm.

Fig. 4.3 MAGCAM from MAGCAM

[1] https://www.magcam.com/

There are other prototypes from different entities all with reduced or experimental applications such as (matesy – cmos-magview family, n.d.; Phuong et al., 2014).

The proposed device is part of a research line on magnetic field sensors and applications. It is the third developed prototype following the implementation of a line sensor system with eight Hall sensors coupled to an Arduino. The displacement of magnetic elements in the active sensors reading zone produces an intensity curve of the field sensed at a fixed distance. These values are enough to form a sliding profile of the field, but insufficient for data about the direction. Evaluating the correlation between the Hall sensors distance to the magnetic object, the intensity of the field in space, the distance between sensors was set as 1 cm in the vertical, as horizontal in a regular square lattice.

A 3 × 3 matrix of Hall sensors regularly spaced 1 cm vertically and horizontally, confirmed the possibility of visualization of EMF, although scarce for lack of more sensor for better input data. This lead to an increase in the number of sensors, with the same spacing between in the X/Y directions making this prototype.

4.3 Prototype

The device called *CCMags* (magnetic fields camera) is a two-dimensional sensor Hall array in a square lattice, 1 cm vertically and 1 cm horizontally spaced (c.f. Fig. 4.4). This matrix is connected to an Arduino NANO that reads data, makes a previous treatment and sends the results to the serial port.

The board is shown in Fig. 4.5. It connects to the PC computer through a USB cable, using serial communication.

The sensing hardware is made of Hall sensors array type AH49E. This sensor "*… is small, versatile linear Hall-effect device that is operated by the magnetic field from a permanent magnet or an electromagnet. The output voltage is set by the supply voltage and varies in proportion to the strength of the magnetic field. …*" It as a "*… low noise output, which makes it unnecessary to use external filtering. It also includes precision resistors to provide increased temperature stability and accuracy. The operating temperature range of these linear Hall sensors is -40º C to 85º C…* ".[2]

Figure 4.6 presents the sensor transfer characteristic, according to data retrieved from the datasheet.

The sensors resolution is of 0.326 mT (3.26 Gauss), and the set is assembled on a printed circuit board in a matrix structure, vertically and horizontally spaced 1 cm. This configuration is enough for the magnetic field with usual intensity, and the influence zone is enough to collect data. The board reads continuously in cycles the sensors signals, that is, the tension in each one through an analog port of the Arduino.

In a cycle, data from blocks of five sensors of a line and then by column, thus allowing to have the data from each Hall sensor. Each column is a set of five sensors reads, separated and placed in the array of the intensity of the magnetic field of each

[2] In the AH49E sensor datasheet (BCD Corp.).

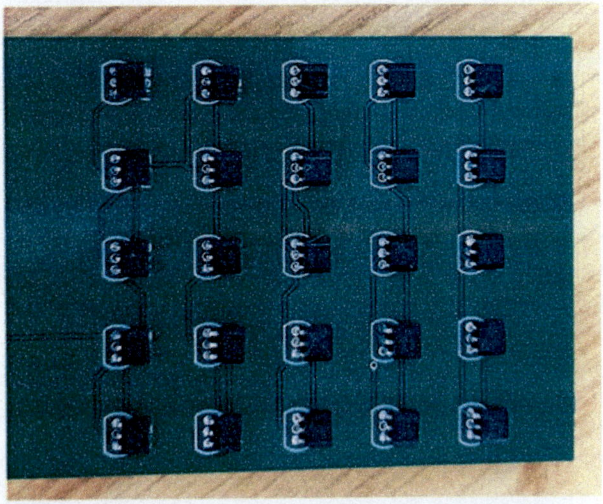

Fig. 4.4 Hall effect sensors matrix

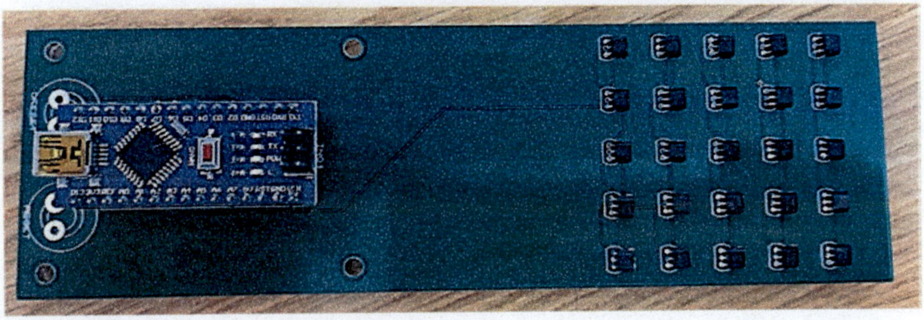

Fig. 4.5 The board plus the sensor matrix

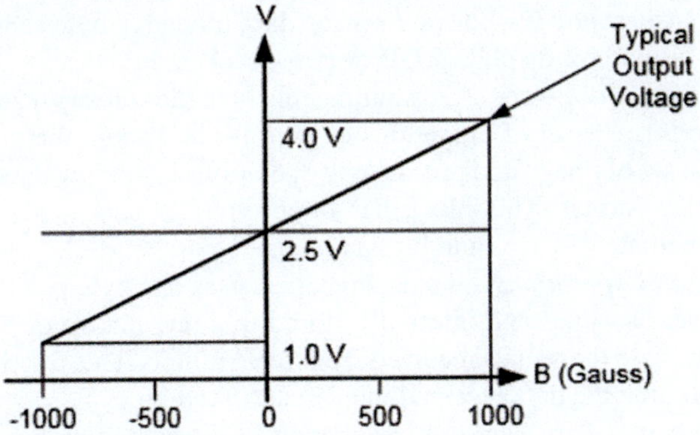

Fig. 4.6 Sensor characteristic. (BCD Corp)

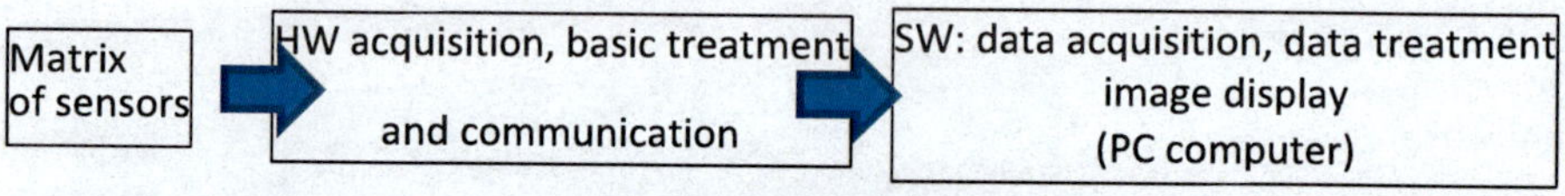

Fig. 4.7 Block diagram of the prototype

one. This data is stored internally in an array that is sent to the computer via serial port. The communication uses a simple handshake protocol that separates each sensor, each line and each block elements of the matrix in the Arduino and sends them to the computer.

The Arduino analog inputs have 10 bits word resolution, coding values in 1024 levels (from 0 to 1023), but a reduction to 256 levels is made, merely for transmission purposes (1 Byte data per sensor). Data from −1000 gauss to 1000 gauss is converted into values that range from 0 to 255 to fit in one byte. These values can be adjusted for higher resolution, but communication protocol must be reviewed. Schematically, Fig. 4.7 presents the basic block diagram.

4.4 Software

Sensors data from the Arduino requires further processing, made in program divided into three main blocks:

- *Data communication* – communication via serial port with the Arduino and transforms the read matrix of strings into a matrix of integer values, which will be the database. These communication blocks repeat at a specific time rate, limited by the hardware specifications and software requirements of processing and display algorithms.
- *Conversion block* – transforms input data into a usable matrix data format for display as a bitmap image. At this level, the program uses two matrices, one that contains the raw data from the 25 sensors and a second the output image. This defines the *sensors matrix* with raw sensor data (coded 1 byte/sensor) and the *image* as the bitmap for display (RGB or grey-level coded).
- *Image processing block* – operates simultaneously in the sensors matrix, as in the image and performs a set of operations to improve and adapt the output image. At this level, a set of image and pre-display operations adapt raw data into a more friendly display format. This block has three types of data matrices – sensor matrix, intermediate matrices and bitmap image.
  - *Sensors matrix operations* – the data received uses one byte per sensor, representing a set of values and internally stored in a raw matrix of 5 × 5 values. These values are the values received from the Arduino. This 5 × 5 data matrix keeps at all time the data received in each data reading cycle.
  - *Intermediate matrices* – are used as support of interpolation, filter and histogram operations, to improve output data. Their dimension is a function of the

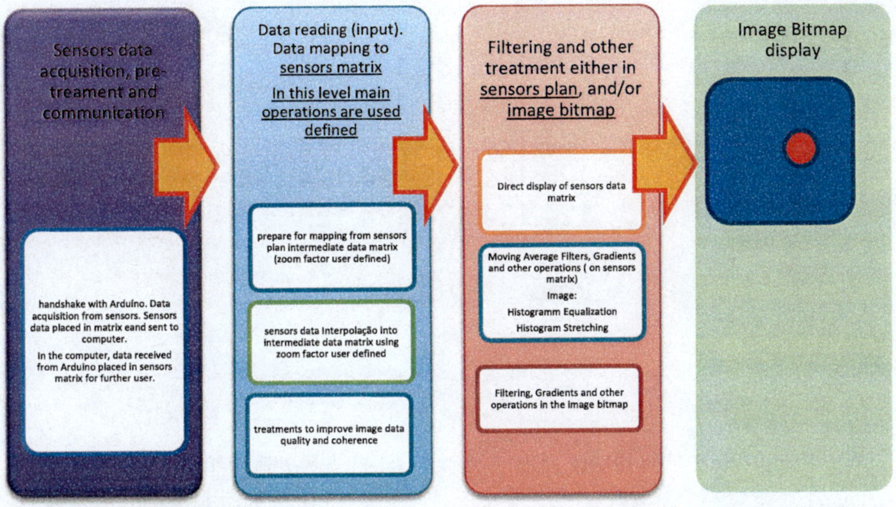

Fig. 4.8 Basic block diagram

user defined zoom factor. Here the main processing techniques (Banerjee & Saha, 2016) applied are interpolation to zoom out data, moving average filters, data histogram operations and gradient operations.

- *Bitmap image* – bitmap array of data used for image display, where the user can apply image processing techniques to improve data quality output. All techniques applied at the intermediate level can be replicated in this level. Yet, due to the dimension of the image bitmap, these operations are time consuming, reducing the refresh rate.
- All these data elements can be overlapped to the bitmap image, presenting information such as values or zoom grids.

- *Image display* – display of bitmap image.

The data communication block replicates data format coming from the serial port, following a communication protocol that assures correct data from each block of sensors at a rate of 115,200 baud. In order to improve communication rate, this block of operations needs change. Figure 4.8 presents the main operating phases of the magnetic field visualization module.

The Arduino hardware sends block of data to the computer. Sensor values received are placed in a 5 × 5 array adjusted to the range 0–255 eliminating eventual noise of the acquisition process. This primary filter consists in eliminating data outside the sensors values band (0–255). Sensors data is then remap into the bitmap image matrix, of 500 × 500 pixels by two interpolation techniques:

- Direct map from sensor data matrix to bitmap image
- By a linear interpolation using bilinear of the nearest neighbour (Russ, 1994) (Fig. 4.9)

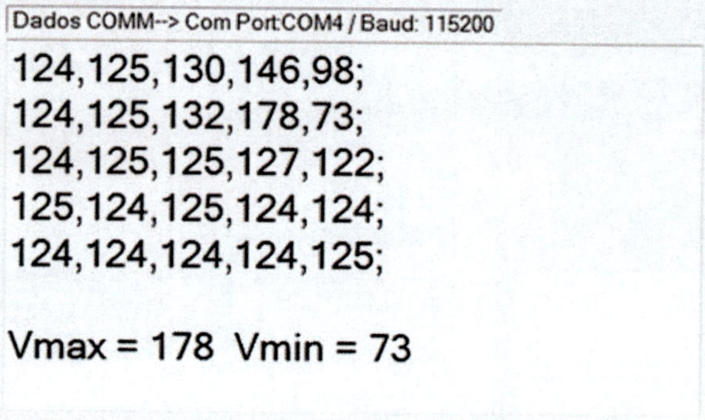

Fig. 4.9 Data sensors received after first treatment

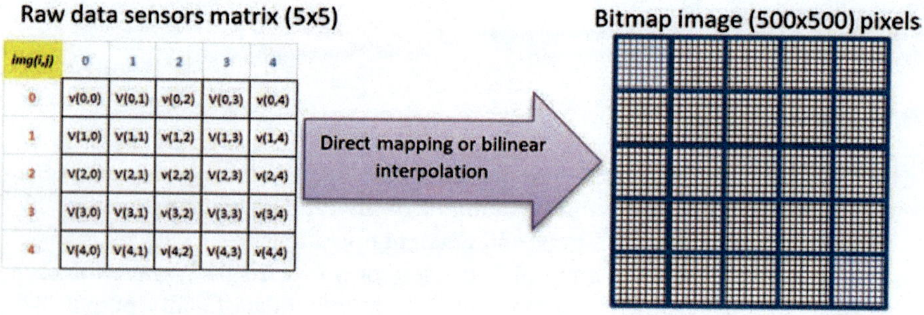

Fig. 4.10 Sensors plan and image canvas

The uses define the zoom factor and the best visual filter for image visualization, where the intermediate matrix of the raw data zoomed stores values and supports filters, histogram manipulations and gradient operators. Figure 4.10 represents the mapping from a small data set to the output image. From the 5 × 5 points, the visualization in the bit image canvas of 500 × 500 pixels made by direct mapping, in this case.

Each group of values as a sample rate of 500 ms, period that takes into account the full set of operations needed to ensure update of raw data, treatment and bitmap image display.

Sensors plan and image canvas superimposed on (a), where a direct mapping is made. On (a), it is possible to observe the received sensor values on the textbox, and on its left on (a) the bitmap image is obtained. In this image, values range from 0 to 255, where 127 means 0 Gauss (grey level/colour), and values between 128 and 255 map intensities in the red colour represent magnetic field values between 0 Gauss and 1000 Gauss. The blue blobs are values between 0 and 126, mapped as (blue) intensities of magnetic field from −1000 Gauss to 0 Gauss (respectively color 0 to 127 values), in a reversed scale. In (b), we have the remap of values to a larger matrix using a zoom factor of 5.

The interpolation algorithm uses the user-defined output zoom level in order to add intermediate values matrix as perceived in Fig. 4.11, to adapt the image to the best possible visualization.

In Figs. 4.11 and 4.12, the zoom factor is 5, mapping each pixel of this image matrix (squares of 5 pixels side) to the bitmap image of 500 × 500 pixels. In Fig. 4.12–(a), input data is presented in colour variation overlapped by the lattice, and the image in (b) presents values from the Hall sensors EMFs in Gauss, not in color code.

In image canvas (a) the same magnetic fields of Fig. 4.11, we have the sensor image zoomed with a factor of 5, overlaid to the values sensor data lattice. In Fig. 4.12–(b), the magnetic field data from the sensors is presented.

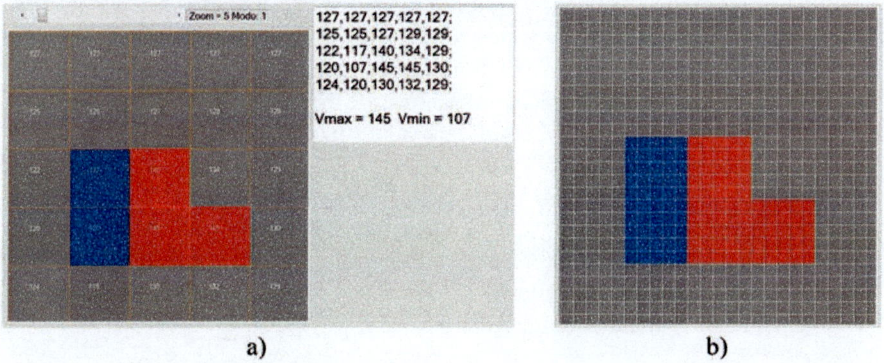

Fig. 4.11 Sensors plan and image canvas. (**a**) Raw data from the 5 × 5 Hall sensors. The grey blobs are neutral (near 0 Gauss EMF intensity). The Blue blobs are EMF from –1000 Gauss to 0 Gauss, mapped from 0 to 117, blue intensities. The Red blobs are EMF from 0 to +1000 Gauss, in mapped to 137 to 255, red intensities. (**b**) the same data diretcly interpolated to a 25 × 25 lattice

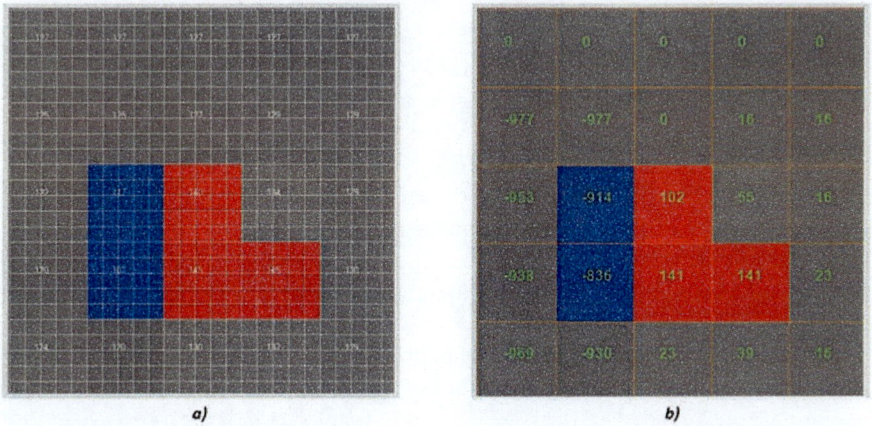

Fig. 4.12 Squares in the image canvas. (**a**) Blobs in a scale from 0 to 255 overlapping the grid of interpolated by a factor 5. (**b**) Scale of values of magnetic fields between −1000 Gauss to 0 (Blue) and from 0 to 1000 Gauss (Red)

Figure 4.12–(a) shows the relation between the large regions that correspond to the values of the Hall sensors and the intermediate resolution matrix on (a) and in (b), it presents the magnetic field values read by the sensor matrix. The user can display data from the magnetic field sensors array in Gauss (values from −1000 Gauss to 0, and from 0 to 1000 Gauss) or as color intensities, reflecting north pole and south pole, and color variations of blue and red, as presented in Fig. 4.12.

There is also the option of displaying as grey-level intensities following the same principles described.

4.4.1 Pseudo-Colour or Grey-Level Imaging

Regarding the colours in the image, the colours map the magnetic poles, which is in line with magnetic field intensity values, or, in option, using grey levels.

In the colour scheme of representation, the *Blue* intensity corresponds to positive/North pole values, and the Red intensities are negative/South Pole magnetic values. This consists in mapping sensor data values (0 @ 255) to a chromatic scale, with the criteria:

- On the scale from 0 to 255, values greater than 127 are converted into colour levels of blue (B), on an RGB scale. The B values range from 0 to 255.
- In the 0 to 255 scale, values bellow 127 are converted into chromatic levels of red (R), on an RGB scale, that range from 0 to 255.

As the values around 127 should be a null magnetic field, and due to variations sensed, there is always a slight drift of values. To compensate this, a *slack* variable redefines a band of values around 127 as null magnetic field. So, there is a gap of *127 ± GAP* considered as "*null magnetic field zone*", where values are mapped in grey levels. So, the colour-mapping algorithm is found to be:

- On a scale 0 to 255, values *above* (127 + *GAP*) become blue intensities.
- On the scale 0 to 255, values *below* (127 – GAP) become red intensities.
- Values inside the band ABS (127 ± gap) are mapped to a *grayscale* scale, but with grey scale variations around the value 127.

Schematically, Fig. 4.13 presents the colour scheme mapping function.

In the image canvas, a colour assignment function assigns pixel the respective value in one of the channels R-G-B, each ranging between 0 and 255. In the case of grey-level image, the same colour is assigned to each of the RGB chromatic components.

Due to the mapping parts of the image in grey levels, blue and red blobs are subject to tonal variations, reflecting intensities of the magnetic field.

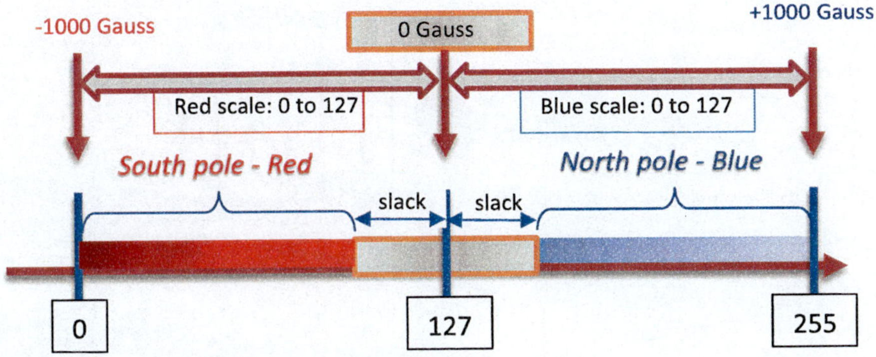

Fig. 4.13 Colour scheme with magnetic field scale

4.4.2 Interpolation and Filtering

The program remaps values one from the raw sensor matrix to the intermediate data matrices and another to the bitmap image. The first is the direct mapping, so variations in values are not very noticeable due to the direct blocks mapping.

To improve display, other interpolation techniques along with moving average filtering, gradient and sensors data histogram equalization are used. These methods improve image quality and visualization.

Interpolation is a technique used to zoom images or regions of them, determining values of pixel colour between base pixels in an image. This can use direct map or blend functions that take into account pixel values of the boundary of the region to be zoomed.

Here zooming must increase the base image resolution of 5 × 5 values (sensor matrix) to a larger image. The initial set of values is scarce and therefore the bitmap image and the magnetic field data must have a variation function similar to the effect that allow the separation of the Hall sensors, even if the basic set of values is always limited to (5 × 5).

So, interpolation aims to find new values between values, such as the example of Fig. 4.14 showing $V(i)$ and $V(i + 1)$ as raw data, and the output must have values from $V'(i)$ to $V'(i + 6)$.

In a dimension function, several methods can be used to find those values.

The problem is the equation to determine new values between the initial point $V(i)$, where $V(-)$ is a vector containing values and final point $V(i + 1)$ introducing new values [from $V'(i)$ to $V'(i + 6)$] that is the equation of the line connecting, or near, the extreme points.

In an image or 2D data, the transformation is done in squares, or blobs, considering four base values, or the corners, of the area to zoom and then find the remaining values, always respecting the border colours (or values) as explained in Fig. 4.15.

The problem is to define a function $f(x,y)$ that allows the distribution of colours from the corners, as shown in the enlarged region. For interpolation, several

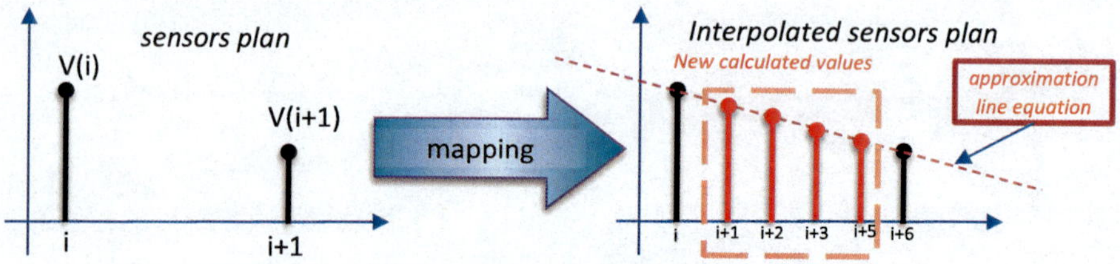

Fig. 4.14 Interpolation basic question

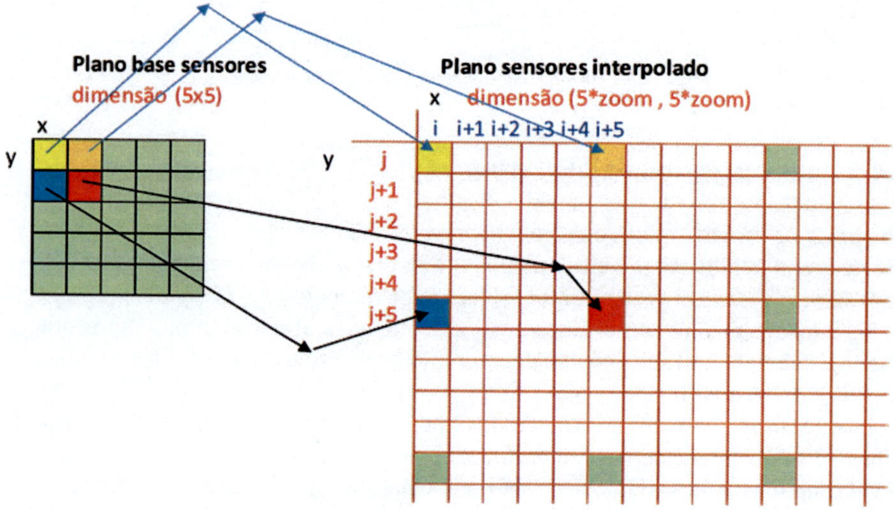

Fig. 4.15 Scheme of 2-dimensional interpolation

algorithms exist using mainly employing three techniques, as stated by Dourado (Dourado 2014):

- Linear
- Nonlinear
- Transformations

All take into account the distances of each new point in the destination plan (interpolated plane) to the pixel colours of the border pixels, assigning a weight to each new element.

Figure 4.16 illustrates the problem with a two-dimensional function of finding pixel values of the new points from a small input set to a larger one.

A surface represents this where the height corresponds to the chromatic levels. Figure 4.17 has the initial set, and Fig. 4.18 has the final set of new interpolated values.

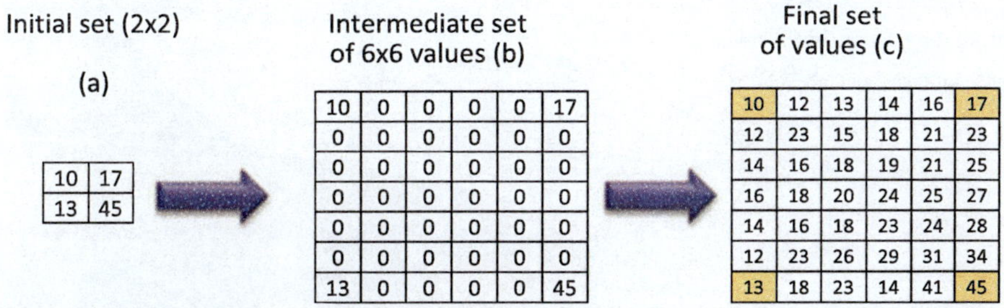

Fig. 4.16 Interpolation principle. Values are merely illustrative

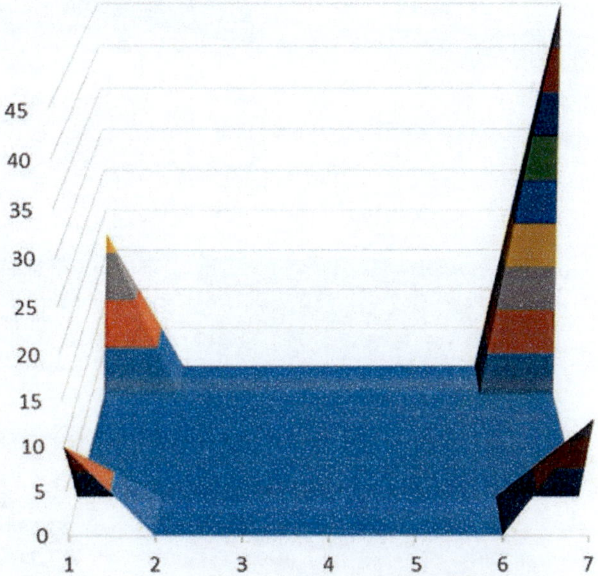

Fig. 4.17 Set of initial values without intermediate values

To attain a redistribution of values for a elaborate visualization of the (limited) data, algorithms must use elementary raw data to find the correct values that should represent the magnetic fields between sensors, considering that EMF are continuous in space (cf. Fig. 4.19a–c).

Figure 4.19-(a) shows the direct blob interpolation that maps basic 5 × 5 sensor data to larger rectangle blobs. In Fig. 4.19-(b), the intermediate matrix is visible, but the shape remains of large rectangles made from small rectangles. In Fig. 4.19-(c), the bilinear interpolation introduces colour variations that reflect the magnetic field between sensors because it is a continuous evolution.

One of the advantages of interpolation is to better visualize images from magnets, even if basic data is limited to the number of sensors.

To further improve image quality, moving average filters, on the sensor matrix and on the bitmap, image can be applied smoothing region contrast of magnetic

Fig. 4.18 Filling by the interpolation algorithm

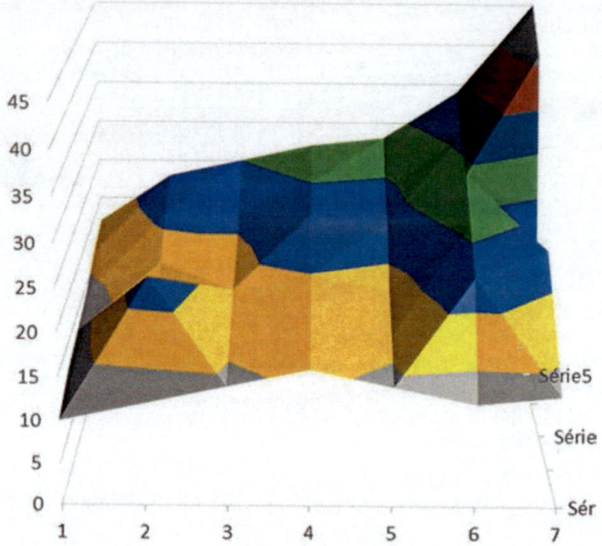

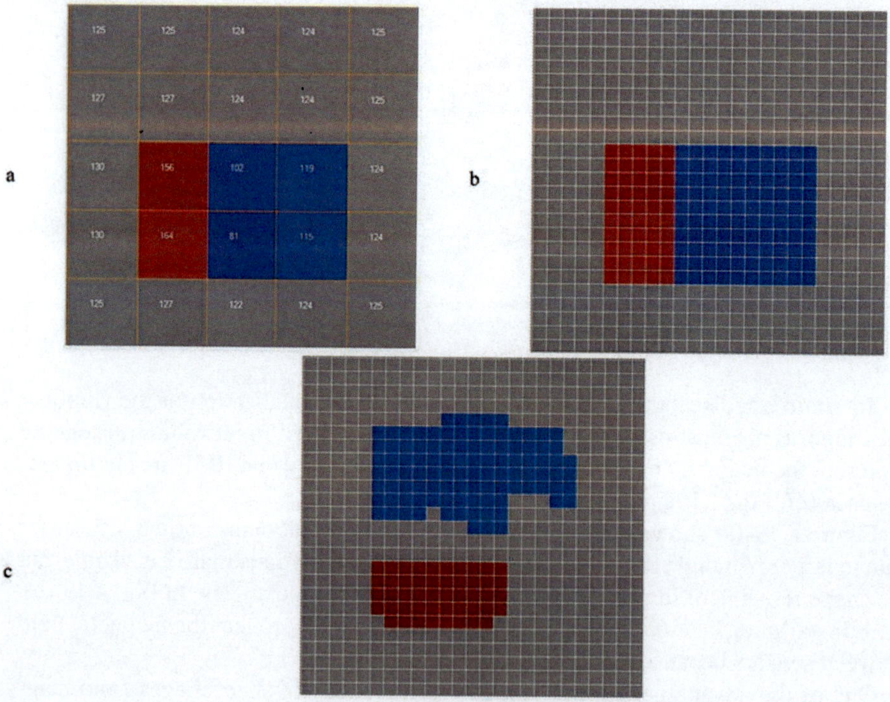

Fig. 4.19 Base plane x interpolated data plane relation. (**a**) is the original image, (**b**) is a simple mapping with 5 × 5 zoom, and **c**) is the same zoom and two-dimensional interpolation

fields. Changing the zoom factor and the size of the moving average window improved the quality of the images considerably.

Another technique used relies on histogram manipulation of the values in the intermediate matrix and/or user defined in the bitmap image. The two main techniques change the colour map and allow a better and more accurate colour set – histogram equalization and histogram stretching that redistribute colours in the intermediate plan allowing a more contrasted bitmap image.

The filters at the sensors and intermediate level are moving average (window size: 3 × 3, 5 × 5, 7 × 7 and 9 × 9) with good results, as well as histogram equalization and stretching of data, to improve contrast between, for a better visual interpretation of the EMF.

The general results in real time are considerably different from a simple interpolation, and the processing time required does not change the refresh rate, as presented in Figs. 4.20 and 4.21.

In addition to the display functionalities, the program determines the gradient of the values, but whose results are inconclusive due to the data scarcity. Another

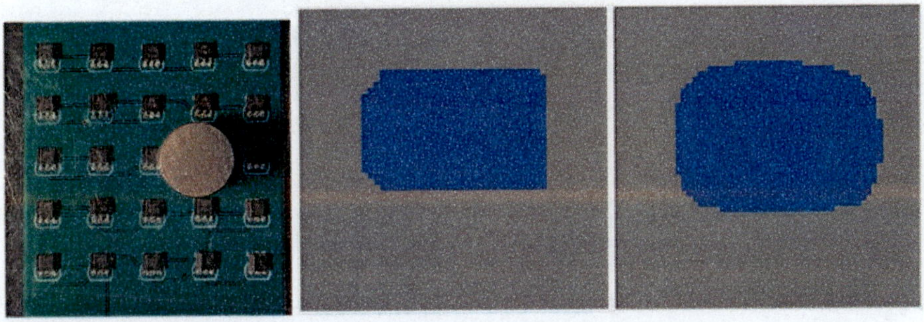

Fig. 4.20 Magnet over the sensors and image with direct mapping (centre), with interpolation, and with 3 × 3 filter (right)

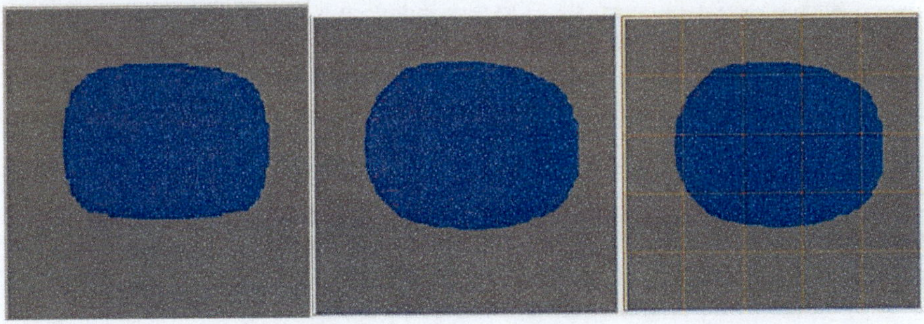

Fig. 4.21 From left to right, image with bilinear interpolation, 12 zoom, 5 × 5 sensor flat filter, and 5 × 5 image flat filter, superimposed on the sensor base mesh

Fig. 4.22 Image, on the left, with a 5 × 5 flat filter in an interpolated image, and in the right the same image in grey level, without any correction or color readjustment

functionality is the grey-level display, where no remap of colour set is made and where the brighter regions are North Pole and the darker South Pole intensities (Fig. 4.22).

4.5 Tests

Several tests with permanent magnets, of different shapes, and electromagnets, powered in either DC or AC at low frequency, were made because of the frequency acquisition limits of the prototype. Figure 4.23 shows a permanent magnet and the respective image.

With permanent magnets, it is possible to discriminate the north and south poles and the displacement over the sensors. With a zoom factor of 5 with bilinear interpolation and a spatial filter of 3 × 3 size window and histogram stretching, the results show variations in the shape position and intensities of some elements (cf. following images Figs. 4.24 and 4.25).

Figure 4.25 is the image of a permanent magnet and the field it causes in the area close to the sensors and in distant regions. The second image is of the same element but turned upside down, which creates a different/reverse colour image.

To carry out additional experiments, a signal power supply was made for high powers (100 W) supporting frequency, amplitude and signal form changes. It works from DC to AC 12 kHz to supply coils with enough current. This device is in the following figure while powering the cylindrical coil (Fig. 4.26).

One of the experiments feed a coil with a sinusoidal voltage of 30 V amplitude at a frequency of 25 Hz. Given the low acquisition frequency of the prototype, it was possible to observe the magnetic field, but not the alternation between poles. The

Fig. 4.23 Image of magnetic field of permanent magnet

Fig. 4.24 Coil and magnetic field obtained, when powered by DC (30 V)

frequency must be lower, and the coil must have a higher inductance value to produce a magnetic field strong enough for the Hall sensors. A higher current is not possible in this initial device for technical issues. A new device is underway for more tests.

An interesting effect, however, was to place the cylindrical coil with the air core and see that the system acquires its image quite well, as long as it is in DC or very low frequency.

a) b) c)

Fig. 4.25 Permanent magnet and images of each of the poles of this element. (**a**) Permanent magnet over sensors matrix. (**b**) Image with zoom factor of 10 and moving average filter of 3 × 3. (**c**) Image after reverse colours and histogram equalization

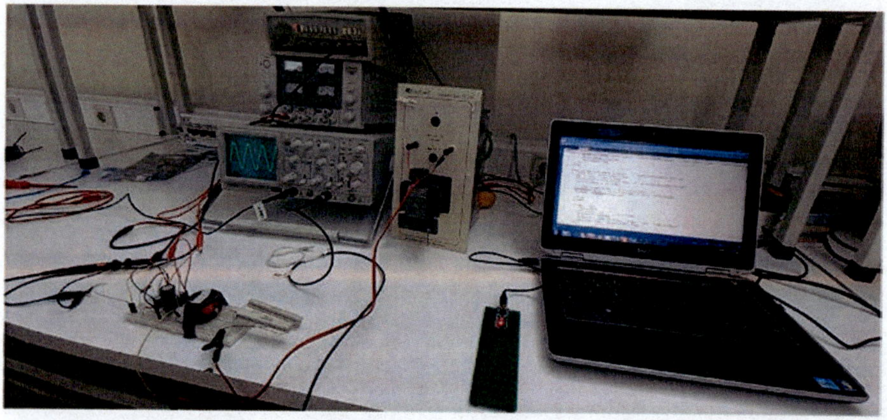

Fig. 4.26 Experimental AC variable signal power system

4.6 Conclusions

This project is still in its initial phase, and issues were detected that will be improved, but the target was to build a tool capable of displaying magnetic fields, either constant or variable, of low frequency. This first step was achieved - being able to produce, visualize and work with images of EMF; however, the most critical element is the sensors number, small in number to allow a consistent image of magnetic fields and slow in frequency response. Therefore, the final image quality needs improvement where software image processing techniques might be a partial answer, but increasing the number of sensors and maintaining the distance between them can produce the same result.

Another aspect is the use of a unique type of sensor or multiple different sensors, with different features, as space dispositions, such as placed in a cubic organization to fetch 3-dimensional data. Different sensors in matrix organization might allow improving data in the base matrix and are used as an auxiliary input for the interpolation algorithms. For this, a new prototype with several magnetic sensors is under development and will be used in addition to the present prototype.

On the other hand, the algorithms are undergoing changes to improve several issues such as communication, processing times, more filtering, and display functionalities.

This prototype served as a proof of concept for the basic magnetic field visualization, essential to improve and test electronic and electromagnetics devices in today's world or just to serve as a display tool for academic teaching.

References

Banerjee, S., & Saha, A. (2016). *Digital image magnification using modified K times image zooming technique*. https://www.researchgate.net/publication/322223477_Digital_Image_Magnification_Using_Modified_K_times_Image_Zooming_Technique

Dourado, W. B. (2014). *Avaliação de técnicas de interpolação de imagens digitais*. MSc Thesis, Universidade Estadual Paulista, Faculdade de Ciências e Tecnologia de Presidente Prudente, Programa de Pós-Graduação em Matemática Aplicada e Computacional, BRASIL.

GMW – magneto optic systems and sensors. (n.d.). https://gmw.com/product/magneto-optic-systems-and-sensors/*impedance*

MAGCAM. (n.d.). https://www.magcam.com/

Magnetic Impedance Sensing. (n.d.). https://dscl.uconn.edu/magnetic-impedance-sensing/#

Magneto Impedance Sensor – JP. (n.d.). https://www.jst.go.jp/chizai/en/cips_details/pdf_2/3-8.pdf

Mangetooptics, M.-S. (n.d.). https://matesy.de/en/products/materials/mo-sensors-magnetooptics

matesy-cmos-magview family. (n.d.). https://www.matesy.de/en/products/magnetic-field-visualization/cmos-magview-magnetic-field-camera-viewer-and-measurement-device

Mohri, K., Uchiyama, T., Panina, L. V., Yamamoto, M., & Bushida, K. (2015). Recent advances of amorphous wire CMOS IC magneto-impedance sensors: Innovative high-performance micromagnetic sensor chip. *Journal of Sensors, 2015*, 1. https://doi.org/10.1155/2015/718069

Phuong, H., Tuan, B. A., & Souza-Daw, d. (2014). *Magnetic camera for visualizing magnetic fields of home appliances. Communications and Electronics* (ICCE). In 2014 IEEE fifth international conference. https://doi.org/10.1109/CCE.2014.6916732

Resistivo, S. D.-E. (n.d.). https://www.sciencedirect.com/topics/chemistry/magnetoresistance

Russ, J. C. (1994). *The image processing handbook*, by John C. Russ, 4th Edition, Boca Raton Imprint, CRC Press. ISBN 9780429118333

Sensor Magneto Resistivo – fabricante. (n.d.). https://www.rechnersensors.com/en/documentations/knowledge/magnetoresistive-sensors

Shi, K., & Uchiyama, T. (2020). A full sample-type magneto-impedance (MI) magnetometer with programmable circuit for high compatibility. *AIP Advances, 10*(1), 015026. https://doi.org/10.1063/1.5129872

Uchiyama, T., Mohri, K., Honkura, Y., & Panina, L. V. (2012). Recent advances of pico-Tesla resolution magneto-impedance sensor based on amorphous wire CMOS IC MI sensor. *IEEE Transactions on Magnetics, 48*(11), 3833–3839. https://doi.org/10.1109/TMAG.2012.2198627

Zhukov, A., Ipatov, M., Corte-León, P., Gonzalez-Legarreta, L., Blanco, J. M., & Zhukova, V. (2020). Soft magnetic microwires for sensor applications. *Journal of Magnetism and Magnetic Materials, 498*, 166180. https://doi.org/10.1016/j.jmmm.2019.166180

5 Sensory Application in Intelligent Monitoring for the Sustainability of Indoor Spaces

Tiago C. Padilha and Nelson Neves

5.1 Introduction

Energy consumption is an increasingly important issue that needs to be raised globally, with the aim of providing methods and technologies that increasingly provide greater energy efficiency. At this time of war-induced energy crisis, it is essential to promote measures to support awareness of excessive energy consumption and to strengthen the market with solutions that help consumers control the consumption and operation of devices without high costs (Avordeh et al., 2022).

The research allows to know more sustainable forms of energy consumption applied in the development of new products (Ahmed et al., 2021). The need to increase comfort, safety, versatility and reduce energy consumption in our daily lives have led to the emergence of new technological solutions (Chu et al., 2017).

In 2018 (European Commission, n.d.), buildings were responsible for 40% of energy consumption and approximately 36% of carbon dioxide (CO_2) emissions in European Union (EU). One of the most relevant reasons in these statistics is because 75% of building are energy inefficient. In the past 30 years, the consumption of electrical energy has increased, especially in China with a representation of 30% of the world's consumption, as can be seen in the graph in Fig. 5.1.

The main goals in reducing energy consumption in buildings and indoor spaces focus on finding the best relationship between comfort and energy efficiency as demonstrated by Yang and Wang (2012) with the demonstration of a decision

T. C. Padilha (✉)
ISPGAYA, Vila Nova de Gaia, Portugal
e-mail: tpadilha@ispgaya.pt

N. Neves
ISPGAYA, Vila Nova de Gaia, Portugal

WIDESKILLS, Inovação, Projetos e Soluções, Lda, Vila Nova de Gaia, Portugal

© The Author(s), under exclusive license to Springer Nature Switzerland AG 2023

F. L. Almeida et al. (eds.), *Multidimensional Sustainability: Transitions and Convergences*, Springer Proceedings in Earth and Environmental Sciences,
https://doi.org/10.1007/978-3-031-24892-4_5

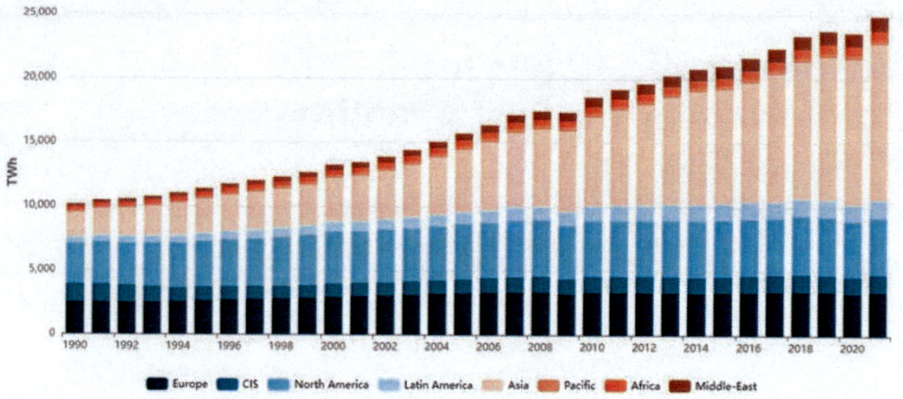

Fig. 5.1 Chart with official ENERDATA statistics on electricity consumption worldwide. (Source: ENERDATA, 2022)

support solution for comfort and energy management in building environments. In the year 2020, the EU implemented a target for zero energy buildings (ZEB) in the building sector (Wang et al., 2021). This measure was implemented due to the large influence that energy consumption has in public and residential buildings, with approximately 20.1% of total energy consumption.

The presented study was supported by the development of an IoT solution with hardware components mounted on PCBs (printed circuit board) and software components for supervision and control of the developed prototypes. The development of this project aims to evaluate the viability of production and use of a device (hardware + firmware). Therefore, it is essential to evaluate a set of parameters, which define the capabilities of the system, which is presented as an IoT product. Features such as security, ease of installation, real-time communication capabilities, size and weight, data processing performance, and maximum communication capacity are factors that this article aims to explore and discuss in comparison with similar products on the market.

This chapter also aims to demonstrate the results obtained through a final graduation project. IoT is part of the industrial and technological revolution, which is driving the way we connect equipment to the internet (Sharma et al., 2019).

5.2 Literature Review

In our daily lives, the electronic equipment that surrounds us is increasingly capable of providing us with greater comfort and flexibility in the various daily tasks, whether professional or personal. The use of electronic equipment is increasing and over the last few years has translated into an exponential increase in energy consumption, especially in buildings. This increase has several direct and indirect consequences, from costs to environmental impact.

Tomat et al. (2020) investigated the various approaches to applying IoT technologies in the past decade. In their studies, the impact of using sensors to increase comfort in buildings through different methods was highlighted. The analysis performed focused on three categories, IoT hardware analysis, dynamic modeling, and crowd sensing. They concluded that the best methodology to adopt is to allow the user to interact and be responsible for controlling the system and to make them aware of energy problems and their attitude toward the environment.

The social interaction of users of spaces with IoT technologies is also mentioned and demonstrated by Franco (2020), who justify the importance of the feedback of a set of parameters that allow for improving comfort and energy saving. They concluded that it is crucial to keep users as a crucial element of IoT systems, just as a sensor and actuator, as well as a system of communication and information technologies.

The control of electrical systems in buildings is essential for sustainability and energy efficiency of common or private spaces, which Casini (2014) highlight through building management, in the perspective of monitoring and control to achieve defined goals and energy savings.

In the scope of IoT solutions implementation, Santos et al. (2018) described the improvements that can be applied in buildings, to allow the implementation of sustainable actions. A data collection for 5 years was carried out at ISCTE-IUL (University Institute of Lisbon), which allowed to understand which classrooms were used more frequently daily, with the support of an internal management system, allowing to apply a predictive control of the HVAC (heating, ventilation, and air conditioning) management system. In the end, they proved that through this predictive management using IoT solutions (sensors to take measurements), they achieved savings of approximately 12% of consumption.

The goal of controlling HVAC systems is a very important factor in several large buildings, which in the case of Elmouatamid et al. (2021), developed a platform adapted for communication and monitoring of buildings with the connection between different IoT devices, called pro MG platform. Silva et al. (2013) proposed to demonstrate the environmental impact of applying smart lighting. They proved that through constant lighting reduction methods in a building, they provided energy savings close to 126 MWh and a reduction of 25T of CO_2 annually emitted.

5.3 Methodology

This work was proposed to investigate and develop sensor-based products for controlling and monitoring spaces using Wi-Fi, allowing their use in buildings with the least possible intervention in their application, even in existing buildings. In the end, it is expected that the products will provide users with increased comfort and flexibility in the use of interior spaces.

The work began to be developed with the perception of the existing needs in the general control of spaces, namely sockets and switches. This first phase involved research and analysis of the possibilities of implementing a small-sized system capable of being incorporated in places with traditional electrical installations. The ideas of the first stages emerged with the greatest needs existing in interior spaces, the remote-control lighting, as well as the monitoring of consumption in each socket and its control.

For the system to be minimally robust in its application, it was essential to develop a solution with sensors, composed of a dedicated processing unit, communication, and power supply. The reason for using each of these features in a dedicated way is to make the application and use process easier, since there is no need to add wires to the electricity pipes. The sustainability of its application has always been a crucial factor in project development, so the minimization of external components is a great advantage that this project aims to prove. In Fig. 5.2, we can observe the general system architecture for the sensor monitoring application.

The techniques for controlling and monitoring the various sensor modules imply the existence of a Wi-Fi network, so that all parts of the system can interact, namely the sensor modules, the server, the web service, the external network, and the ISPGAYA network.

The architecture planning was based on Wi-Fi communications, but to improve the system's ability to be customized, fast, and secure in transmitting data over Wi-Fi, the MQTT (Message Queuing Telemetry Transport) communication protocol was chosen. The MQTT protocol implies the use of a broker, responsible for managing all messages in the communications, so the management of the messages

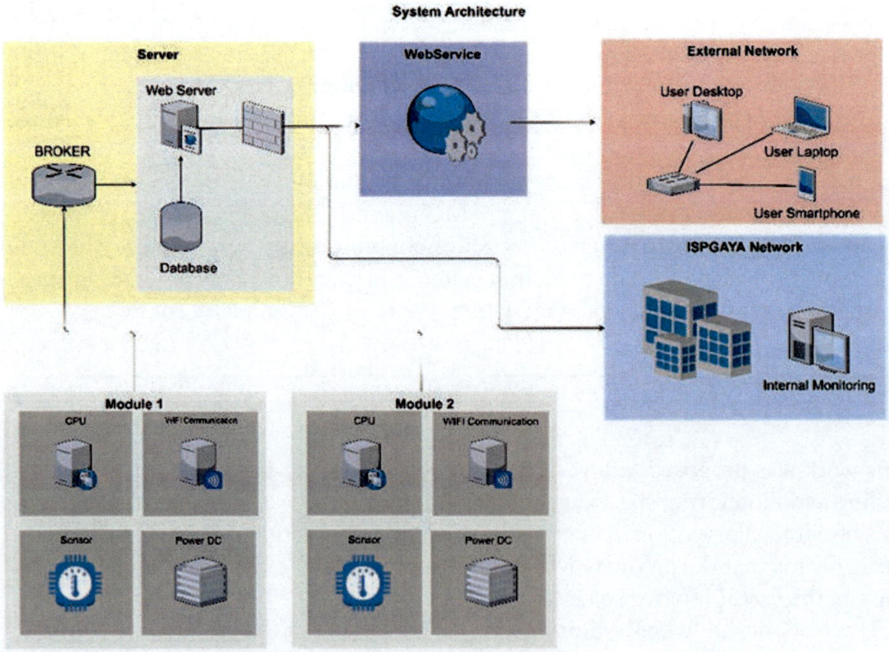

Fig. 5.2 System architecture diagram of the project

is intended to be performed in the server block (see Fig. 5.2). For the server, was used a Raspberry PI 3 with the Debian operating system, enough to be applied as a broker in the communications system. The broker is responsible for the requests to the various sensor modules, so threads are sent that represents a unique string from each module. In Fig. 5.3, we can see the MQTT communication architecture used.

The implementation of sensors was a process that involved predicting the most important data to register in each space, without major complexities in its application and monitoring. The realization of two modules to be incorporated in sockets and switches allows their application without the need for external applications in the surrounding space nor the preparation of an auxiliary power system. The sensors used were the HLW8012 current and power sensor (plug in module) and touch sensor TTP223 (switch module). As for the processing unit, the ESP8266 was used, due to its small dimensions and remarkable capabilities in terms of processing power, Wi-Fi communications, and very low power consumption.

The development of this product involves evaluating many essential factors in today's IoT device market. The developed technology intends to implement the use of low-cost devices for processing and communication, with a versatile and configurable communication protocol. It is intended to analyze a set of essential characteristics in the discussion of the results, as presented in Table 5.1.

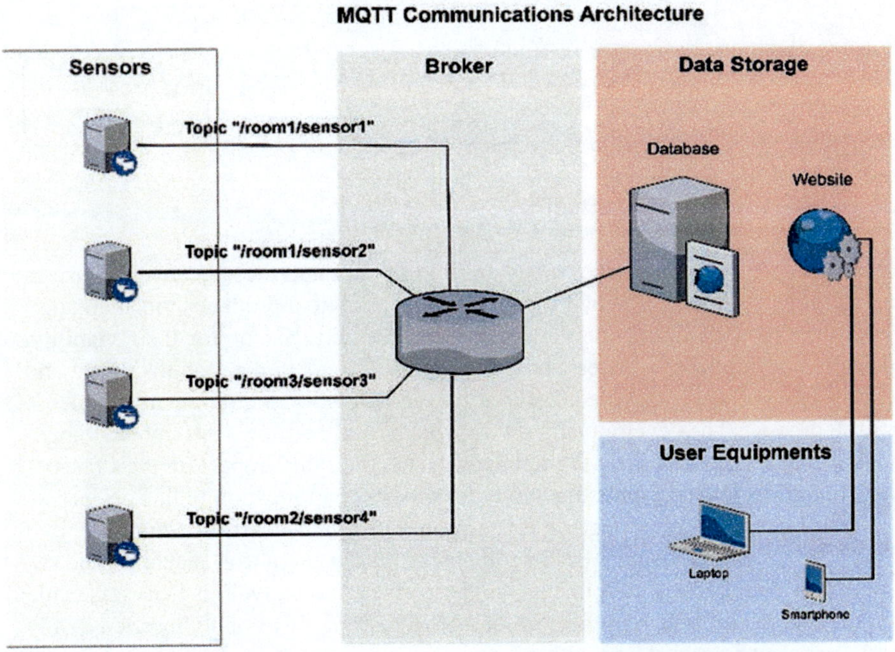

Fig. 5.3 Communication architecture using MQTT protocol, with 4 essential blocks

Table 5.1 Evaluation methods for discussing products features

Hardware features and evaluation methods		
Features	Plug in module	Switch module
Requires internet	Yes	Yes
Max. devices	200 k per broker	200 k per broker
Max. size data messages	Approx. 260 MB	Approx. 260 MB
Dimensions and weight	55 mm × 55 m and 47 gr.	55 mm × 55 m and 41 gr.
Hardware electrical protection	No	No
Real time	Yes	Yes
Battery	No	No
Security communication	Yes, but needs to be encrypted	Yes, but needs to be encrypted
Communication protocol	MQTT through TCP-IP	MQTT through TCP-IP
CPU	Yes	Yes
Required broker	Yes	Yes
Over the air updates	Yes	Yes

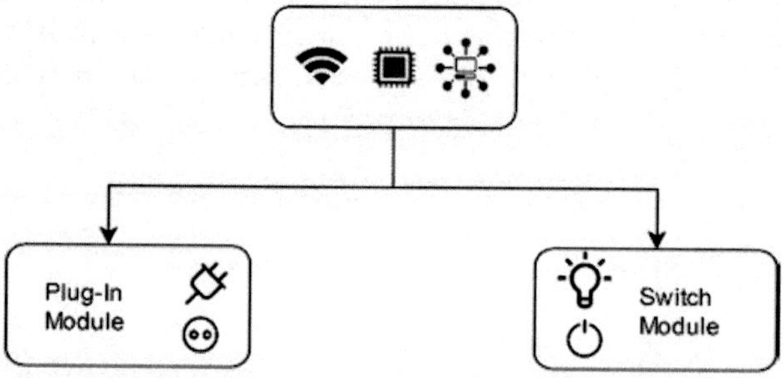

Fig. 5.4 Diagram of the two products developed in the project

5.4 Results and Discussion

A product development implies analyzing the market and the objectives of use in the face of existing needs. When it comes to technological products, the adaptation of sensors for control and monitoring is a determining factor for their viability. During the development process, possible improvements in the manufacturing and welding process were analyzed, facilitating the assembly and automation process for each product. The development of this project resulted in two different and independent products, but in the end intended to have the same impact on their respective use. Figure 5.4 represents the two types of sensor application in this project.

Regarding the development of space monitoring and control technology, two PCBs were developed using a CNC (computer numerical control) machine, and the respective welding of all components was performed manually. This PCB was made with the main goal of having the correct size to be capable of being inserted in any traditional electric socket box, either from a socket or a switch. Both have the same geometries, but different functions, in Fig. 5.5, you can see the result of the PCBs, in the development phase and their finalization.

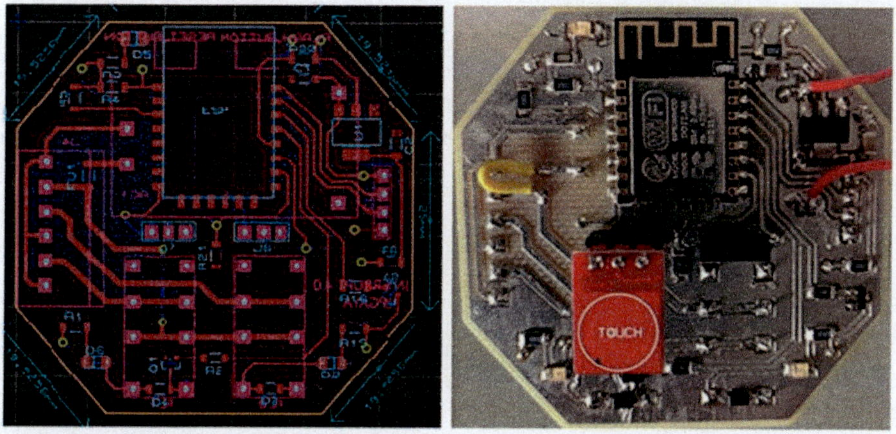

Fig. 5.5 Touch sensor of switch module; view on the left side the PCB project and on the right side the final developed PCB

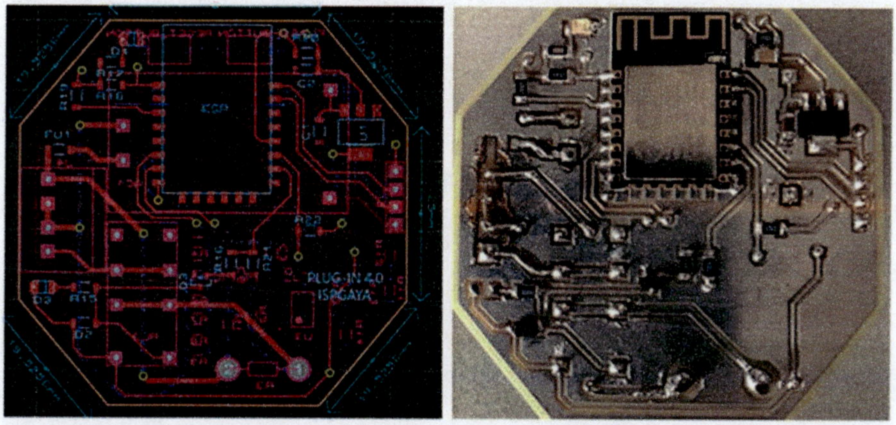

Fig. 5.6 Plug-in sensor module; view on the left side the PCB project and on the right side the final developed PCB

The switch module consists of two local touch connections, which can be applied to the switch itself, as well as controlled remotely. After its placement on site, it is only necessary to connect the phase and neutral cable, to allow the system to be automatically powered. In its development were applied two solid-state relays up to 3 A, a transformer for 5 V, a voltage regulator, indicator LEDs, and finally the ESP8266 processing and communication device. As far as plug-in control and monitoring are concerned, in the following Fig. 5.6, the result of the plug-in module can be seen.

The development of these devices was based on the same power, processing, and communication unit. In this case, a solution was developed that can take measurements of electrical consumption in real time, as well as controlling the status of the outlet, whether it is active or deactivated. For a correct and adequate application of both modules, it implies the application of several improvements in safety and security aspects for the user or technician who intends to install the products. For the whole system to work properly, without failures in its control and monitoring, it is essential that the Wi-Fi network in the surrounding space is relatively stable and even with losses from placing the devices in a hidden area, there are no communication failures.

These first functional prototypes aimed to demonstrate the concept and functionality of the products but are only products for functional testing and proof of development. Simultaneously with the development of the hardware and firmware, a web interface was created for the monitoring and control of the various devices, as well as their configuration for communication and local identification. Figure 5.7 demonstrates an example of the use of the software.

Through the visualization of the software environment, it is possible to observe the ISPGAYA floor plan divided by the various floors, as well as each division has an identification, in which it is possible to select to perform monitoring or control of the various existing sensors. This tool allowed us to demonstrate a little of the front of the project with the visualization of the environment in which the user will be faced with the use of the various products.

There are some limitations in the development of these two products, which make it difficult to produce them. The process of soldering the various components is somewhat complex, as observed in the various images, the components are very close together, due to the need of implementation in a discrete and minimalist format. The general use of the products implies that the application site has Wi-Fi, so this will be a factor that will influence the correct functioning of the entire system. In case of a power failure in the equipment, it is not prepared to store the last commands, since they are not composed of batteries.

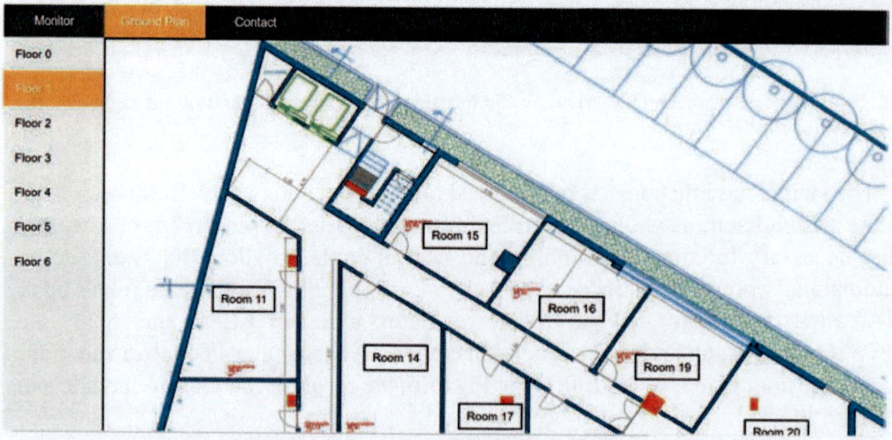

Fig. 5.7 Visualization of one of the software's web interface pages for controlling and monitoring the multiple sensors

In the assembly and handling of the equipment, it has no protection for the user. As well as its communication protocol for a correct operation, it implies the use of a broker. From a sustainability perspective, both products were developed to be low cost, easy to apply, and improve energy savings.

5.5 Conclusions

The development of sensory monitoring solutions and their respective control is the future in the field of sustainability, with increasingly innovative techniques and methods, allowing over the past few years to collect very important data for the understanding of sensory phenomena that were unknown to us until now. Large amounts of data are often complex in their treatment, but after proper processing, they allow the application of measures that provide increased comfort and sustainability in our daily lives.

The implementation of IoT solutions allows applying a set of sustainable actions with data collected during a period or even with data insertion, allowing to predict possible reductions in energy consumption. In the case of an educational institution like ISPGAYA, the promotion of sustainable solutions is essential to demonstrate the potential that the building can provide concerning comfort and energy consumption.

Implementing such a product implies further testing and realizing a sufficiently organized database capable of feeding machines and deep learning algorithms for data analysis and decision support. This product was the first iteration of a more comprehensive solution development. Subsequent iterations will be applied to more divisions of the building to collect data on a larger scale and achieve a more complete analysis. The future of this project and many others related to IoT imply the data storage and implementation of artificial intelligence algorithms that allow meeting user preferences, as well as supporting the sustainability of energy consumption without penalizing comfort.

This project proved that is possible the use of small, low-cost, and easy-to-apply equipment can perform several functions that can be monitored and even controlled, allowing great ease of integration into buildings for the reduction of energy costs and comfort in the various spaces (Mora et al., 2019).

Sustainability is a factor that any project must consider (Khalifeh et al., 2020), mainly because of its impact on its use and the idea it aims to demonstrate. This project promoted the idea of integration in spaces, without an invasive intervention, as well as an easy and feasible application. In future works, some improvements in hardware and firmware optimization should be considered, as well as allowing the collection of large amounts of data for implementation in decision support algorithms. Fundamentally, this project integrates other sensors with easy application and low costs, allowing data collection in larger quantities and the development of user support tools (Guo et al., 2020).

References

Ahmed, Z., Ahmad, M., Rjoub, H., Kalugina, O. A., & Hussain, N. (2021). Economic growth, renewable energy consumption, and ecological footprint: Exploring the role of environmental regulations and democracy in sustainable development. *Sustainable Development, 30*, 595–605. https://doi.org/10.1002/sd.2251

Avordeh, T. K., Gyamfi, S., & Opoku, A. A. (2022). The role of demand response in residential electricity load reduction using appliance shifting techniques. *International Journal of Energy Sector Management, 16*(4), 605–635. https://doi.org/10.1108/IJESM-05-2020-0014

Casini, M. (2014). Internet of things for energy efficiency of buildings. *International Scientific Journal Architecture and Engineering, 2*(1), 24–28. http://architecture.scientific-journal.com/articles/2/4.pdf

Chu, S., Cui, Y., & Liu, N. (2017). The path towards sustainable energy. *Nature Materials, 16*, 16–22. https://doi.org/10.1038/nmat4834

Elmouatamid, A., Alidrissi, Y., Ouladsine, R., Bakhouya, M., Elkamoun, N., Khaidar, M., & Zine-Dine, K. (2021, January). Towards an IoT/Big-data platform for data measurements, collection and processing in micro-grid systems. In *International conference on digital technologies and applications* (pp. 279–289). Springer. https://doi.org/10.1007/978-3-030-73882-2_26

ENERDATA. (2022). *World energy & climate statistics – Yearbook 2022*. https://yearbook.enerdata.net/electricity/electricity-domestic-consumption-data.html

European Union. (n.d.). *The energy performance of buildings directive*. https://ec.europa.eu/energy/sites/ener/files/documents/buildings_performance_factsheet.pdf

Franco, A. (2020). Balancing user comfort and energy efficiency in public buildings through social interaction by ICT systems. *Systems, 8*(3), 29. https://doi.org/10.3390/systems8030029

Guo, Y., Wang, N., Xu, Z. Y., & Wu, K. (2020). The internet of things-based decision support system for information processing in intelligent manufacturing using data mining technology. *Mechanical Systems and Signal Processing, 142*, 106630. https://doi.org/10.1016/j.ymssp.2020.106630

Khalifeh, A., Farrell, P., & Al-edenat, M. (2020). The impact of project sustainability management (PSM) on project success: A systematic literature review. *Journal of Management Development, 39*(4), 453–474. https://doi.org/10.1108/JMD-02-2019-0045

Mora, N., Grossi, F., Russo, D., Barsocchi, P., Hu, R., Brunschwiler, T., Michel, B., Cocchi, F., Montanari, E., Nunziata, S., Matrella, G., & Ciampolini, P. (2019). IoT-based home monitoring: Supporting practitioners' assessment by behavioral analysis. *Sensors, 19*(14), 3238. https://doi.org/10.3390/s19143238

Santos, C., Ferreira, J. C., Rato, V., & Resende, R. (2018, June). Public building energy efficiency-an IoT approach. In *International symposium on ambient intelligence* (pp. 65–72). Springer. https://doi.org/10.1007/978-3-030-01746-0_8

Sharma, N., Shamkuwar, M., & Singh, I. (2019). The history, present and future with IoT. In *Internet of Things and big data analytics for smart generation* (pp. 27–51). Springer. https://doi.org/10.1007/978-3-030-04203-5_3

Silva, F. B., López, G. P., Fontaiña, E. F., & Díaz, A. L. (2013). Energy savings and carbon emission reduction of smart lighting installation in a multipurpose and residential building in Santiago de Compostela, Spain. *Journal of Environmental Science and Management, 16*(2), 56–62. https://doi.org/10.47125/jesam/2013_2/07

Tomat, V., Ramallo-González, A. P., & Skarmeta Gómez, A. F. (2020). A comprehensive survey about thermal comfort under the IoT paradigm: Is crowdsensing the new horizon? *Sensors, 20*(16), 4647. https://doi.org/10.3390/s20164647

Wang, R., Feng, W., Wang, L., & Lu, S. (2021). A comprehensive evaluation of zero energy buildings in cold regions: Actual performance and key technologies of cases from China, the US, and the European Union. *Energy, 215*, 118992. https://doi.org/10.1016/j.energy.2020.118992

Yang, R., & Wang, L. (2012). Multi-objective optimization for decision-making of energy and comfort management in building automation and control. *Sustainable Cities and Society, 2*(1), 1–7. https://doi.org/10.1016/j.scs.2011.09.001

6 The Double Benefit of Environmental Sustainability – The Case of ISPGAYA

Nelson Neves, João Lima, Luis Soveral, and José Carlos Morais

6.1 Introduction

It is common knowledge that the education sector is not considered a polluting sector, however, since in any human activity there is not only consumption of natural resources but also waste production (Matos, 2020). The examples are many and increasingly overwhelming, such as the lack of drinking water quality, pollution of rivers and seas, global warming, ice melting, typhoons and hurricanes, and countless other potentially catastrophic situations (Harvey, 2018).

This is the assumption of all the theoretical production associated with the theme of sustainability (Randers et al., 2018) and the concept of development (Barska et al., 2020; Wyrwa et al., 2020, 2021), and this term is usually associated with that of innovation, in its various domains, which may be that of energy and energy efficiency (Gökgöz & Güvercin, 2018; Kolosok et al., 2020; Sun et al., 2019).

The aim of the study is to show the double benefit (Helepciuc et al. 2018) of the investment of a Portuguese higher education institution, Instituto Superior Politécnico Gaya (ISPGAYA) in environmental issues. As a driving force, it aims to implement environmental certification in the institution according to ISO 14001 (Alnavis et al., 2021; Fonseca & Domingues, 2018; Kuhre, 2018), in line with the understanding of sustainable development (Bravi et al., 2020) and to consolidate the mutual benefit with more environmentally friendly practices.

N. Neves (✉) · J. Lima · L. Soveral
ISPGAYA, Vila Nova de Gaia, Portugal
e-mail: nneves@ispgaya.pt

J. C. Morais
ISPGAYA, Vila Nova de Gaia, Portugal

CEOS.PP, Vila Nova de Gaia, Portugal

© The Author(s), under exclusive license to Springer Nature Switzerland AG 2023

F. L. Almeida et al. (eds.), *Multidimensional Sustainability: Transitions and Convergences*, Springer Proceedings in Earth and Environmental Sciences,
https://doi.org/10.1007/978-3-031-24892-4_6

This study presents a history of data on investment in the institution (building) in technology and innovative solutions rationalizing and optimizing energy consumption related to environmental concerns, such as solar panels and the replacement of traditional light bulbs with low-energy bulbs. This history goes back to 2008 and extends to the current year, 2022 and reveals data that show significant economic benefits resulting from the adoption of environmentally friendly practices, data that serve as a basis for the consolidation of ISPGAYA's Environmental Sustainability Plan (ESP). The research questions posed were are there financial as well as environmental benefits from applying environmentally friendly technology in the building under analysis? What are the specific benefits of using LED technology in the building? What are the specific gains of using solar panel technology?

6.2 Toward an Environment Management System

Sustainability consists of the integration of a set of actions based on the three pillars environmental, social, and economic. Achieving sustainable development means seeking quality of life in a way that disrupts the current pattern of development (Oláh et al., 2018). This concept of development is explicit in the 17 sustainable development goals (SGDs) that drive the 2030 agenda (United Nations, n.d.), seen as the guarantee of the planet itself and the basis of a «smater model of development» (Randers et al., 2018). ISO 14001, published on September 1, 1996, and with a latest edition dated from 2015 (ISO, 1996, 2014), is widely considered the most important environmental certification (Sartor et al., 2019) and goes in and with the contents of SGDs making way for a common vision for humanity in a more executive approach.

This International Standard bases itself on assumption on a better environmental performance that can be achieved by systematically identifying and managing environmental aspects, considering issues such as pollution prevention, improved environmental performance, and compliance with applicable laws, materialized in environmental management systems (EMS) of each institution, which should be the target of certification. The ISO 14001 contains the requirements for EMS: organizational structure and responsibilities; planning of activities; definition of practices, processes, and procedures; allocation of resources to plan, implement, check and improve the environmental policy, with the latest version seeking to expand the organization's controls or influence over environmental management (ISO, 1996, 2014).

Although the norm is not new, there is a concern about the need to evolve toward an international, but more urgently, to a national/internal regulation, forcing specifically Higher Education Institutions (IES) to engage in a most suitable role as drivers for sustainable development. Grindsted (2011) alerts that Sustainability in Higher Education declarations have been signed globally, but are not enough as a measure to ensure executive procedure toward this driver role. Sustainability must be present in education management, in all dimensions, comprising both practices and curricula contents (Blanco-Portela et al., 2017). Brundiers et al. (2021) state the development of key competences regarding sustainability, to be developed within universities. The Portuguese agency for evaluation of higher education – A3ES – intents to make mandatory the demonstration of evidence of the adoption of a strategy and governance in the institutions that places sustainability, in its different accessions, in its educational, scientific, and cultural project. The Institutional Evaluation Manual of the Portuguese evaluation agency on higher education A3ES, still not published and still in public discussion (A3ES, 2022), stabilizes this national internal rule that will ensure executive procedure toward this driver role of higher education toward sustainability and commitment to the sustainable development goals that drive the 2030 agenda. This study presented intents to be a stet more solidifying sustainability as key factor in sustainability attended by the analyzed institution.

Built according to a "double/multiple benefit" approach (Ryan & Campbell, 2013), Fig. 6.1 exposes the possibility of both benefits, financial and environmental. So, in counterpoint to the satisfaction of human needs there is the need for regeneration of the environment. Depending on the will of man and the technological state of the art, it is possible for us to use resources with no or low impact on natured, even with visible economic gains.

Within its limitations, ISPGAYA has adopted for itself this model of environmental sustainability marked by dashes, in which it is possible to obtain resources with economic gains and with the minimization of damage to the environment, that is, environmental sustainability is supported in the profit, with

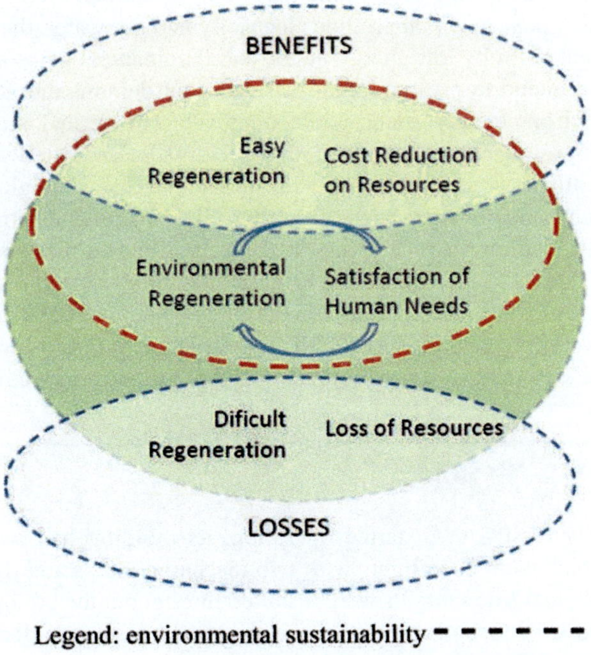

Fig. 6.1 A model for environmental sustainability analysis. (Source: Authors)

cost reduction and in the minimization of adverse impacts to the environment, in short, it is a mutual sum relationship.

6.3 Methodology

What we intend to do with this study is, precisely, to outline the EMS of ISPGAYA. The existing literature (Fonseca & Domingues, 2018) mentions some variables that should be considered in the implementation of these EMS, such as difficulties in measuring their efficiency, bureaucratic issues, higher costs, or lack of available advice or the generic nature of this ISO requirements.

With the presentation of the data, we expect to evolve toward overcoming these difficulties, in addition to working hard on the environmental certification of the institution under analysis, and we have managed to present an EMS, with components of planning, implementation, analysis and continuous improvement, materialized in the investments made in renewable energies and energy efficiency, globally designated as solar use/photovoltaic panels implementation, in a continuous improvement model, of led technology.

In addition to solidifying the institution's EMS, we work to surpass what is known in the literature as the energy-efficiency gap (EEG). The EEG occurs when energy-efficient technologies, which offer considerable promise for reducing the financial costs and environmental damages associated with energy use, may not be adopted by individuals and firms to the degree that might be justified on an environmental and/or financial basis (Gerarden et al., 2017).

By collecting data on the history of investment in environmental issues, such as solar panels, which provide electricity and hot water, the replacement of traditional light bulbs by low energy bulbs, and the financial gains achieved from then to date, we intend to present numerical data that demonstrate the advantages of adopting environmentally friendly practices. This numerical and descriptive presentation allows quantitative comparisons to be made associated with time periods, enabling evaluations and estimates with qualitative and strategic components, referring to items and variables such as: types of lighting units existing in the building and also installed power in lighting on each floor and in the whole building from 2008 to 2022, the actual panorama of LED technology used, distinguishing between LED downlight units and LED tubular units. Data was collected both by the administrative services of the institution and by a student of the course of renewable energies, under a final work/project that addressed the energy optimization of the building.

6.4 Results

When ISPGAYA started operating, its building had two main types of lighting, which were downlights with two discharge lamps and discharge fluorescent tubes. While the downlights were installed throughout the building but more frequently on the 3rd floor and above, the discharge tubes were installed mainly on the 1st and 2nd floors in indirect lighting due to their lower floor height than normal given the building's suitability developed for the ISPGAYA activities.

Table 6.1 presents the characterization of the lighting at the beginning of ISPGAYA's activities in February 2008 and also presents the installed power in the lighting circuits per floor and in the building as a whole.

Since 2012, the replacement of discharge technology lighting by lighting with tubular LED technology units began. The first point of replacement was the elevator cabins with a 24-h operating time. This option was taken as a test of the same technology considered at the time not yet reliable (Meneghini et al., 2010). In 2014, a discharge technology was applied but with electronic ballasts. In this solution, the T8 tubulars were replaced by T5 tubulars in which the adapter included the electronic ballast. This solution was implemented on an experimental basis in the computer lab and in the computer center on the 2nd floor for two reasons, the first because it is cheaper than the solutions of the time for LED tubes, and second because they are intensively occupied and are interiors without natural light, but with an 8-h occupation per day. This solution lasted until the second half of 2020, when it started to have breakdowns, and the replacement was very expensive.

In 2017, the conversion of light points continued, both in the replacement of discharge downlight by LED downlight as well as with the replacement of discharge tubes by LED tubes. This process was concluded at the end of the 2022 first semester.

Thus, alternating between the conversion to LED tubes and downlight from floor 1 to floor 6 (see Fig. 6.2), the sequence was:

- 2017 (first semester) – classrooms of 6th floor not including the library or hallways
- 2017 (second semester) – classrooms of 5th floor with corridors and 6th floor corridor
- 2018 (first semester) – Service, secretariat and 3rd floor in general, except the office corridor
- 2018 (second semester) – LED tubes on floor 2 except computer lab and computer center

Table 6.1 Types of lighting units existence in the building and also power installed in lighting on each floor and all building in 2008

	Discharge downlight units			Discharge tubulars units			
Floor	170 mm 26 W	200 mm 52 W	220 mm 52 W	60 cm 18 W	120 cm 36 W	150 cm 58 W	Total power
1	5	15	28	2′	41	64	8.234 W
2	36	26	0	25	33	59	7.018 W
3	6	109	0	0	0	0	4.014 W
4	11	102	0	0	0	0	3.870 W
5	0	99	0	0	0	0	3.564 W
6	0	59	17	0	0	0	3.220 W
7	1	0	0	0	0	6	330 W
						Total power	*30.140 W*

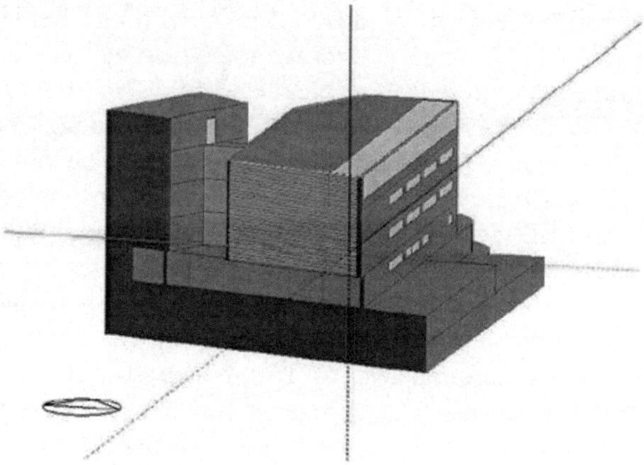

Fig. 6.2 Geometrical model of ISPGAYA building, south view. (Source: Authors)

Table 6.2 Types of lighting units' existence in the building and also power installed in lighting on each floor in 2022

	LED downlight units			LED tubulars units			
Floor	170 mm 20 W	200 mm 40 W	220 mm 40 W	60 cm 9 W	120 cm 16 W	150 cm 24 W	Power by floor
1	5	15	28	2'	41	64	5.542 W
2	36	26	0	25	33	59	4.894 W
3	6	109	0	0	0	0	1.680 W
4	11	102	0	0	0	0	1.629 W
5	0	99	0	0	0	0	1.485 W
6	0	59	17	0	0	0	1.293 W
7	1	0	0	0	0	6	249 W
						Total power	*16.772 W*

- 2019 (first semester) – LED tubes at 1th floor
- 2020 (second semester) – LED tubes in the computer lab and CIISP at 2th floor
- 2021 (second semester) – Library
- 2022 (first semester) – classrooms and corridor of 4th floor and office corridor at 3th floor

Table 6.2 shows the current overview of the characteristics of installed LED lighting and also shows the current panorama of the installed power in LED lighting, where the numbers of light points were not changed but only the discharge technology was converted to LED technology.

This issue forced that the most recently installed downlights, despite having the same electrical characteristics and light color, are of a second generation where the AC/DC drives give more guarantees, having even started to replace the first generation downlights.

Comparing the two solutions, a gain of more than 53% is concluded, as can be seen in Table 6.3, in terms of installed power in the lighting circuit, thus translating into energy savings, making the building more sustainable in terms of lighting.

Table 6.3 Resume of power gain between both solutions

Technology	Installed power
Discharge	30.140 W
LED	16.772 W
Gain	*55,65%*

Table 6.4 E-REDES 'daily cycle' tariff

Tariff period	Sub period	Winter time	Summer time
Floods	(i)	08 h00 to 09 h00	08 h00 to 10 h30
	(ii)	10 h30 to 18 h00	13 h00 to 19 h30
	(iii)	20 h30 to 22 h00	21 h00 to 22 h00
Peak	(i)	09 h00 to 10 h30	10 h30 to 13 h00
	(ii)	18 h00 to 20 h30	19 h30 to 21 h00
Empty	(i)	00 h00 to 02 h00	00 h00 to 02 h00
	(ii)	06 h00 to 08 h00	06 h00 to 08 h00
	(iii)	22 h00 to 00 h00	22 h00 to 00 h00
Super empty		02 h00 to 06 h00	02 h00 to 06 h00

An important aspect is the reliability of the equipment used. There is awareness that technology has evolved ensuring greater reliability. However, this reliability largely depends on the installation points, namely on their ventilation (Richter et al., 2019). This observation is made when one realizes that there are several AC/DC conversion drives that have to exist in LED lighting in which if the lighting point does not have ventilation, the AC/DC drives tend to break down early. Thus, in the installation of ISPGAYA, it was rare to have to replace LED tubes, while the frequency of failure of the AC/DC drives of the LED downlight is much higher because they are placed in a false ceiling where there is no facility for ventilation of the space.

This issue forced that the most recently installed downlights, despite having the same electrical characteristics and light color, are of a second generation where the AC/DC drives give more guarantees, having even started to replace the first generation downlights.

6.4.1 Impact of LED Lighting on Real Consumption

E-REDES is the company in Portugal that distributes electricity[1] independently from the supplier. Through its internet portal, https://balcaodigital.e-redes.pt/login, it is possible to collect the energy consumption history of a point of an electricity consumer in Portugal.

Table 6.4 shows the tariff applied to the CEP/ISPGAYA consumer, which is called the 'daily cycle'.

[1] https://www.e-redes.pt

Applying this tariff to the energy consumption in the lighting circuits, comparing them with the opening hours of the ISPGAYA teaching sessions, it is clear that it is important to analyze the evolution of consumption in the floods (iii), peak (ii), and empty (iii) tariffs corresponding to the timetable from 18:00 to 00:00 of each day of classes from Monday to Friday, where most of the classrooms, laboratories, and other spaces work with lighting on. However, this is not possible because the data collection on the E-REDES portal does not separate the subperiods of (i), (ii), and (iii) in the floods and empty periods, nor the subperiods (i) and (ii) in the peak period. Thus, the analysis will have to incorporate an error of quantification and approximation to the real scenario.

Thus, with the ISPGAYA access credentials, it was possible to obtain data for the period from 2017 (the first year made available) to July of 2022, recording the consumption shown in Table 6.5 and Fig. 6.3.

It is concluded that there is a decrease in energy consumption that accompanies the adoption of LED technology lighting, this fact is supported by the increase in the weight of energy consumed in the period of the super empty tariff while in the other tariffs there is a constant relative decrease.

Table 6.5 Total energy consumed per year and percentage weights of each energy tariff

Year	Power provider (Rates): EDP	Floods (%/year)	Peak (%/year)	Empty (%/year)	Super empty (%/year)
2017	132.55 MWh/year	0.505	0.153	0.236	0.106
2018	120.14 MWh/year	0.457	0.139	0.29	0.114
2019	113.92 MWh/year	0.452	0.128	0.308	0.112
2020	89.72 MWh/year	0.435	0.13	0.31	0.125
2021	94.96 MWh/year	0.421	0.126	0.291	0.162
2022 (July)	42.73 MWh/year	0.416	0.121	0.28	0.183

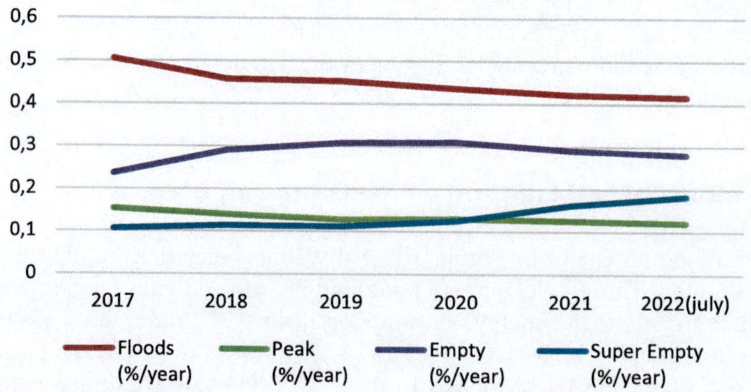

Fig. 6.3 Evolution of the percentage weights of each energy tariff between 2017 and the end of the first half of 2022 at ISPGAYA

6.4.2 LUX Gain

In addition to the improvements felt in terms of energy sustainability, the conversion of lighting from discharge technology to LED technology brought improvements in terms of LUX measured at the level of the work plan, which are classroom tables. Comparing the measurements carried out in 2012 and in 2022, there is a substantial increase in LUX. The variations are not uniform as there are aspects that contribute to potentiate these variations felt with recent measurements. Thus, there are several factors that influence a nonuniform variation:

(i) The color of the walls – The rooms on the 4th, 5th, and 6th floors were repainted and the walls, instead of light cream, changed to a semi-gloss white. The ceilings were also repainted.
(ii) Indirect light – The classrooms on the 1st and 2nd floors do not have direct light, but crown moldings where the room is illuminated by reflection on the classroom ceilings.
(iii) Measurements with natural light component – Measurements made at 18:00 h translate into variations in the orientation and size of the classroom windows. Thus, we have different gains if the room is interior compared to rooms with east, south, or west orientation.
(iv) Furniture change – The furniture of the classrooms on floors 4, 5 and 6 was changed from having tables and chairs in light and dark brown to white tables and chairs in white or light grey.
(v) On the 1st and 2nd floors, the laboratories and interior rooms kept the cream color on the walls, only the ceilings were painted, so there was not such a sharp increase in LUX.

As the analyzed spaces are many, around 35 spaces of classrooms, laboratories, offices, library, study room, snack bar, etc., an aggregation by typology and orientation was chosen. The results of this aggregation are summarized in Table 6.6.

The results show economic gains that demonstrate from a financial point of view the advantages of making investments in environmental issues.

Table 6.6 Lighting improvements by space typology and orientation

Orientation of classrooms in the building	Medium with discharge lighting	Medium with LED lighting	Improvement
Classrooms to the East	291,7	2975,3	491%
Classrooms to the West	400,7	1483,8	370%
Classrooms to the South	557,9	2413,7	432%
Interiors room	267,1	692,6	259%
Laboratories	489,8	1368,2	279%

6.5 Conclusions

The results obtained allow the consolidation of the institutions' ESP (annex 1), in a perspective of short-, medium-, and long-term planning of the set of measures and actions with results in the dimensions of environmental improvement and improvement of the institution.

This study allowed us to realize that ISPGAYA should move the contract to a weekly basis, as we are spending at the weekend as if it were a working day when activities are only from 09:00 to 18:00, and there is almost no use of the facilities after lunch. In future energy audits, a more thorough analysis will be done on the lighting circuits.

As final considerations, we state that humans transform the goods consumed and return to nature the leftovers and all that he does not take advantage of, hence the pertinence of the term "double benefit". It is also increasingly common knowledge that resources are finite and that when the rate of consumption exceeds their regeneration capacity, adverse results arise for all humanity. It is clearly accepted by all that this relationship of mutual survival between man and the environment must be beneficial to both, otherwise both are harmed. The data gathered showed clearly financial benefit on recurring to solar panels and led technology, and the institution's path to prove benefits, is set on solid basis with the ESP. The research question posed in the introduction of the manuscript was answered, demonstrating financial benefits of applying environmentally friendly technology. More difficult to prove is the concept of double benefit, whereby the benefits to the environment are only deducted.

Finally, involving students in projects within curricular units of the courses ministered with processes of environmentally friendly technology implementation states the institution's and scholar community growing compromise with sustainable development goals that drive the 2030 agenda within higher education sustainability purposes, now mandatory according to A3ES standards. The plan for the environmental sustainability of the institution (see Appendix 1) contemplates a strategy for continuous improvement at the technological level, but also the educational aspect and the involvement of the school/educational community in the organization's sustainability strategy. The plan contemplates, concomitantly, dimensions such as: awareness-raising actions for students, employees and teachers; led technology, solar thermic panels, photovoltaic solar panels technology; and Luminosity sensors.

In the future, the study will be further developed by collecting data on consumption and efficiency in terms of space lighting. New solar panel technologies may be considered for implementation, as well as innovations in the led technology used, following the plan for the environmental sustainability of the building.

This manuscript presents, in a descriptive way, financial results of the application of environmentally friendly technology in a building of a higher education institution, according to a time sequence. The analysis could be driven by hypotheses to be tested, based on data regarding other HEIs, allowing for a comparative analysis.

References

A3ES. (2022). *Manual de Avaliação Institucional*. Available on: https://www.a3es.pt/sites/default/files/Manual_de_avaliacao_institucional_A3ES_Consulta_publica.pdf

Alnavis, N. B., Martono, D. N., & Hamzah, U. S. (2021). Recommendations for improving the ISO 14001 certification based on the company's perception analysis of the certification obstacles and benefits. *Journal Kesehatan Lingkungan, 13*(3), 132–141. https://doi.org/10.20473/jkl.v13i3.2021.132-141

Barska, A., Jędrzejczak-Gas, J., Wyrwa, J., & Kononowicz, K. (2020). Multidimensional assessment of the social development of EU countries in the context of implementing the concept of sustainable development. *Sustainability, 12*(18), 7821. https://doi.org/10.3390/su12187821

Blanco-Portela, N., Benayas, J., Pertierra, L. R., & Lozano, R. (2017). Towards the integration of sustainability in Higher Education Institutions: A review of drivers of and barriers to organisational change and their comparison against those found of companies. *Journal of Cleaner Production, 166*, 563–578. https://doi.org/10.1016/j.jclepro.2017.07.252

Bravi, L., Santos, G., Pagano, A., & Murmura, F. (2020). Environmental management system according to ISO 14001: 2015 as a driver to sustainable development. *Corporate Social Responsibility and Environmental Management, 27*(6), 2599–2614. https://doi.org/10.1002/csr.1985

Brundiers, K., Barth, M., Cebrián, G., Cohen, M., Diaz, L., Doucette-Remington, S., et al. (2021). Key competencies in sustainability in higher education—Toward an agreed-upon reference framework. *Sustainability Science, 16*(1), 13–29. https://doi.org/10.1007/s11625-020-00838-2

Fonseca, L. M., & Domingues, J. P. (2018). Exploratory research of ISO 14001: 2015 transition among Portuguese organizations. *Sustainability, 10*(3), 781. https://doi.org/10.3390/su10030781

Gerarden, T. D., Newell, R. G., & Stavins, R. N. (2017). Assessing the energy-efficiency gap. *Journal of Economic Literature, 55*(4), 1486–1525. https://doi.org/10.1257/jel.20161360

Gökgöz, F., & Güvercin, M. T. (2018). Energy security and renewable energy efficiency in EU. *Renewable and Sustainable Energy Reviews, 96*, 226–239. https://doi.org/10.1016/j.rser.2018.07.046

Grindsted, T. (2011). Sustainable universities–from declarations on sustainability in higher education to national law. *Environmental Economics, 2*(2). Available in https://ssrn.com/abstract=2697465

Harvey, L. D. (2018). *Global warming*. Routledge. https://doi.org/10.4324/9781315838779

Helepciuc, C. M., Serbanoiu, A. A., & Serbanoiu, B. V. (2018). Concrete with thermal insulating properties-a double benefit in terms of money and environmental protection. *Advances in Environmental Sciences, 10*(1), 55–61.

ISO 14001. (1996). *Environmental management systems-specification with guidance for use*. International Organization for Standardization.

ISO/DIS 14001. (2014). *Environmental management system: Requirements with guidance for use*. International Organization for Standardization.

Kolosok, S., Pimonenko, T., Yevdokymova, A., Hajiyev, N. O., Palienko, M., & Prasol, L. (2020). Energy efficiency policy: Impact of green innovations. *Marketing and Management of Innovations, 4*, 50–60. https://doi.org/10.21272/mmi.2020.4-04

Kuhre, W. L. (2018). *ISO 14001 certification: Environmental management system*. Prentice Hall.

Matos, A. T. (2020). *Poluição ambiental: impactos no meio físico*. Editora UFV. https://books.google.pt/books?hl=pt-PT&lr=&id=2FEEEAAAQBAJ&oi=fnd&pg=PT4&dq=poluição+ambiental&ots=YB7ErQ-Mqi&sig=t4EmLVmzeWvQQrKjYI1dlfw1Wdk&redir_esc=y#v=onepage&q=poluição%20ambiental&f=false

Meneghini, M., Tazzoli, A., Mura, G., Meneghesso, G., & Zanoni, E. (2010). A review on the physical mechanisms that limit the reliability of GaN-based LEDs. *IEEE Transactions on Electron Devices, 57*(1), 108–118. https://doi.org/10.1109/TED.2009.2033649

Oláh, J., Kitukutha, N., Haddad, H., Pakurár, M., Máté, D., & Popp, J. (2018). Achieving sustainable e-commerce in environmental, social and economic dimensions by taking possible trade-offs. *Sustainability, 11*(1), 89. https://doi.org/10.3390/su11010089

Randers, J., Rockström, J., Stoknes, P., Goluke, U., Collste, D., & Cornell, S. (2018). *Transformation is feasible: How to achieve the sustainable development goals within planetary boundaries. Publisher: A report to the Club of Rome from Stockholm Resilience Centre and BI Norwegian Business School.* Project: SDGinPB. https://www.researchgate.net/publication/343791128_Transformation_is_feasible_How_to_achieve_the_sustainable_development_goals_within_planetary_boundaries

Richter, J. L., Van Buskirk, R., Dalhammar, C., & Bennich, P. (2019). Optimal durability in least life cycle cost methods: The case of LED lamps. *Energy Efficiency, 12*(1), 107–121. https://doi.org/10.1007/s12053-018-9662-4

Ryan, L., & Campbell, N. (2013). *Spreading the net: The multiple benefits of energy efficiency improvements*. OECD/IEA. https://www.researchgate.net/publication/254439556_Spreading_the_Net_The_Multiple_Benefits_of_Energy_Efficiency_Improvements#fullTextFileContent

Sartor, M., Orzes, G., Touboulic, A., Culot, G., & Nassimbeni, G. (2019). ISO 14001 standard: Literature review and theory-based research agenda. *Quality Management Journal, 26*(1), 32–64. https://doi.org/10.1080/10686967.2018.1542288

Sun, H., Edziah, B. K., Sun, C., & Kporsu, A. K. (2019). Institutional quality, green innovation and energy efficiency. *Energy Policy, 135*, 111002. https://doi.org/10.1016/j.enpol.2019.111002

United Nations. (n.d.). *Sustainable development goals*. https://sdgs.un.org/goals

Wyrwa, J., Barska, A., Jędrzejczak-Gas, J., & Siničáková, M. (2020). Industry 4.0 and social development in the aspect of sustainable development: Relations in EC countries. *European Research Studies Journal, 23*, 1068–1097. https://doi.org/10.35808/ersj/173

Wyrwa, J., Barska, A., & Jedrzejczak-Gas, J. (2021). An indicator analysis for the development of EU countries in relation to sustainable development. *European Research Studies Journal, 24*(3), 918–933. https://www.um.edu.mt/library/oar/handle/123456789/85363.

Part III

Energy Efficiency

7 Energy Consumption Analysis from Main Parameters: A Dataset in the Brazilian Scenario

Patrick de Sousa Pessoa, Victor Cesar Costa, Rodrigo Galuzzi Garcia Piva, Arianne Soares do Nascimento Pereira, and Renato Camargo Giacomini

7.1 Introduction

Data analysis and statistical evaluation are based on collecting, exploring and presenting large amounts of data to discover underlying patterns and trends. It is used for reporting, alerting, forecasting, predictive modelling, optimization, prescriptive analysis and actionable knowledge delivery (Cao, 2017). In this respect, data analysis has been used in conjunction with artificial intelligence in order to evaluate relevant data to be used in training and enable the development of more accurate models. In this sense, Jayasutha and Arunachalam (2022) used statistical resources to select an ideal subset of characteristics to develop a model based on deep learning to predict the stress levels of employees of an IT company. The selection of optimal features makes it possible to create more accurate prediction models, as it eliminates unnecessary information that acts as noise.

In "Artificial Intelligence: A Modern Approach" (Russell et al., 2010), intelligence is mainly defined by a rational action, in which a system tries to make the best possible decision within certain conditions. In this sense, AI has advanced very rapidly in recent years due to its ability to find optimal solutions, and the interdisciplinary of the subject has converged on issues of greater relevance, always carrying with it the scientific methodology for decision making and changing approaches. In this scenario, some attempts to improve the demand response system are promising, such as the implementation of artificial intelligence to predict supply and demand in

P. de Sousa Pessoa (✉) · V. C. Costa · R. G. G. Piva · R. C. Giacomini
Electrical Engineering Department, Centro Universitário FEI, São Bernardo do Campo, Brazil

A. S. do Nascimento Pereira
ISPGAYA, Vila Nova de Gaia, Portugal

© The Author(s), under exclusive license to Springer Nature Switzerland AG 2023

F. L. Almeida et al. (eds.), *Multidimensional Sustainability: Transitions and Convergences*, Springer Proceedings in Earth and Environmental Sciences,
https://doi.org/10.1007/978-3-031-24892-4_7

large cities, using parameters such as weather conditions and period of the year (Kurbatsky & Tomin, 2010). Yan et al. (2019) proposed a hybrid deep learning model, combining the structure of a long-term memory neural network (LSTM) with a stationary wavelet transform (SWT) technique. The data was obtained by analysing a dataset on actual household energy consumption in kilowatt-hours, collected every 6 s. The results showed that the proposed method outperformed other state-of-the-art models, such as the support vector regression (SVR) model, long-term memory neural network and convolutional techniques that combine long-term memory techniques. Nespoli et al. (2019) compared different methods for day-ahead production of photovoltaic sources. Both methods are based on artificial neural networks and were trained using the same dataset. The dataset consists of an hourly series of both temperature and measured power production data, and the data was separated into clusters for the system to differentiate sunny and cloudy days for separate training. One of the networks was trained using only temperature and energy production data, while the other uses a hybrid method, with the original dataset added with the daily weather forecast. The results show that the network trained without weather forecast had a more stable response. Porteiro et al. (2022) developed an energy consumption prediction system for industrial and residential facilities. The method applied was the use of machine learning techniques, obtaining an average absolute error of 2.55%. The dataset used for the industrial plant has several parameters, such as year, month, day, hour, working day (checks if it is a working day, if it is a Boolean value representing yes or no) and demand (consumption at that moment). In this regard, it is possible to use different parameters to build the dataset and the predictive models. Yang et al. (2022) work elaborated a model to predict energy demand forecast support-vector regression and also evaluated the most impacting variables in the dataset. However, the study was applied to China's energy system, and there is a lack in actual study to analyse the most impacting variables in a single unit (house or company) perspective.

The objective of this work is to evaluate which variables have the greatest impact in determining a customer's energy demand, avoiding the additional charge for demand contracted with the energy company. This work is the first stage of a study whose aim is to propose a predictive model of demand. For that purpose, two datasets with different parameters and characteristics were analysed in order to understand their behaviours and make model implementations for energy demand forecasting more assertive.

7.2 Tariff Regulation for High-Medium Voltage in Brazil

Demand is the power made available to the consumer, to produce the energy requested by the loads installed in its establishment.

With demand control, customers pay much less for electricity. Most consumers are concerned with reducing electricity costs, consuming less kWh, that is, spending less energy in their facilities. This really brings some benefit, but it is far and far from solving the problem because what weighs on the electricity bill in the Brazilian scenario are both demand (kW) and energy consumption (kWh).

Table 7.1 Tariff categories (ANEEL, normative resolution N° 1.000/2021)

Consumer type	Tariff type
Tension supply equal of superior to 69 kV (any demand)	Blue
Tension supply inferior to 69 kV (demand equal or superior to 500 kW)	Blue or green
Tension supply lower than 69 kV (demand equal of superior to 50 kW and inferior to 500 kW)	Blue, green, or conventional

The tariff structure for consumers connected to high voltage with demand equal to or greater than 50 kW is shown in the following table:

As can be observed in Table 7.1, if demand is exceeded, a different tariff will be applied to the total portion of demand, as it exceeds the contracted amounts.

Power demand tariffs are fixed in reais per kilowatt (R$/kW) and electricity consumption tariffs are fixed in reais per kilowatt-hour (R$/kWh).

There are different forms of pricing, depending on some criteria related to the consumer customer.

In the Brazilian scenario, the lack of awareness impacts both the residential and industrial sectors, in which the customer, unaware about the difference between the contracted energy and the monthly utilized energy, is responsible for paying fines in case of expressive variance.

According to ANEEL (ANEEL, normative resolution N° 1.000/2021), many companies are classified in category "A" which are served in medium and high voltage, above 2300 V. For example, industries and large commercial complexes.

Within this group there are still 6 subgroups:

- Subgroup A1 for voltage level of 230 kV or more
- Subgroup A2 for voltage level from 88 to 138 kV
- Subgroup A3 for the voltage level of 69 kV
- Subgroup A3 for voltage level from 30 to 44 kV
- Subgroup A4 for voltage level from 2.3 to 25 kV
- Subgroup AS for underground system

"Group A" tariffs are based on three supply modes:

- Conventional
- Blue time seasonal
- Green time seasonal

7.2.1 Conventional Tariff Structure

The conventional tariff structure is characterized by the application of energy consumption and/or power demand tariffs regardless of the hours of use of the day and the periods of the year.

Consumers served at high voltage may also opt for the conventional tariff structure, if served at supply voltages below 69 kV, provided that they have contracted a demand of less than 500 kW.

7.2.2 Time-Seasonal Tariff Structure

The time-seasonal tariff structure is characterized by the application of differentiated tariffs for electricity consumption and power demand, according to the hours of use of the day and the periods of the year (ANEEL, normative resolution N° 1.000/2021). The purpose of this tariff structure is to rationalize the consumption of electricity throughout the day and year, motivating the consumer, due to the differentiated value of the tariffs, to consume more electricity at the times of the day and in the periods of the year when it is cheaper.

For the hours of the day, two periods are established, called tariff stations, which were exemplified in item 2.1.1 of this work: the "peak" tariff post and the "off peak" tariff post. Rates during "peak" hours are higher than during "off-peak" hours.

For the year, two periods are established: "dry period," when the incidence of rain is lower, and "wet period" when the volume of rain is greater. Tariffs in the dry season are higher, reflecting the higher cost of producing electricity due to the smaller amount of water in the reservoirs of hydroelectric plants, causing the eventual need to supplement the load by thermal generation, which is more expensive. The hourly seasonal tariff is divided into two different pricing methods: Blue and Green.

7.2.3 Time-Seasonal Tariff (Blue and Green)

Figure 7.1 presents the time-seasonal tariff for blue and green structures. The blue time-seasonal tariff is the structured supply modality for the application of differentiated tariffs for electricity consumption, according to the hours of use of the day

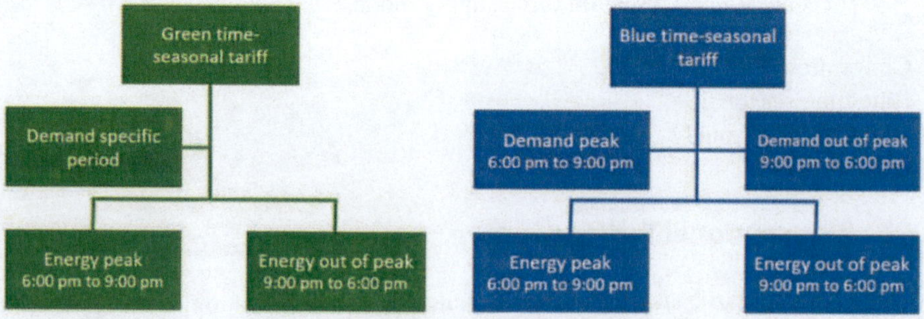

Fig. 7.1 Difference between blue and green tariffs. (ANEEL, normative resolution N° 1.000/2021)

and the periods of the year, as well as differentiated tariffs for power demand according to with the hours of use of the day.

For the blue hourly seasonal tariff is mandatory to apply to consumer units served by the interconnected electrical system, and with a supply voltage equal to or greater than 69 kV, while for the green applies to consumers with less than 69 kV and contracted demand equal to or greater than 500 kW. The blue and green hourly seasonal tariff has the following structure for energy consumption:

- A value for the peak time in wet period (PU)
- A value for the off-peak hours in wet season (FPU)
- A value for the peak time in dry period (PS)
- A value for off-peak dry season (FPS)

7.2.4 Parameters Definition

In the current Brazilian energy supply system, the lack of awareness from the customer side about assertive measurements of energy demand is a common scenario. This is mainly due to the dimensioning of the demand being carried out only in relation to the maximum load of the installation (ANEEL, Normative Resolution n. 414/2010), that is, the amount of electric power required when all equipment is turned on and operating at full power. This analysis is very simplistic because it only considers the installed load, but since these machines and devices do not always work simultaneously, therefore, it is not necessary to contract a demand close to the maximum. Even because the greater the value of contracted demand, the greater the cost to be paid for availability. In addition, this situation can lead to a low efficiency of the electrical system, since the system has low energy use.

In addition, for situations in which demand can be exceeded, one of the ways of acting is to remove loads, either manually, where the plant operator turns off loads in order to reduce power, or automatically. For automatic operation, a study must be carried out on the system loads and to define a group of loads or areas that can be turned off without affecting the company's production or processes.

This situation impacts both the residential and industrial sectors, in which the customer, unaware about the difference between the contracted energy and the monthly utilized energy, is responsible for paying fines in case of expressive variance.

The charging of energy to medium voltage consumers is based on three main components, the TE (energy tariff), the TUSD (tariff for the use of the distribution system) and the contracted demand (ANEEL, Normative Resolution n. 166/2005). Therefore, the following parameters are measured by the energy supplier to compose the price of the customer's energy bill:

- Energy consumption or active energy: Amount of active electrical power consumed in a time interval, expressed in kilowatt-hours (kWh). The consumption can happen during the peak period (PP) or out of the peak period (OPP).

- Contracted demand: Active electrical power to be mandatorily and continuously made available by the concessionaire, at the point of delivery. In general, it is adjusted according to the institution's operating dynamics.
- Measured demand: Highest active power demand, verified by metering, evaluated each 15 min during the billing period.

Tariffs are segmented according to the period in which energy is consumed. There are two differences: daily, according to the variations that take place 24 h a day, and annual, according to the climatic differences between the months of the year.

According to ANEEL, in its Resolution No. 456, the segmentation of the so-called tariff stations is done as follows:

Hours of the day:

- Peak hours (PP): Period defined by the concessionaire and composed of 3 (three) consecutive daily hours, except for Saturdays, Sundays and holidays defined by federal law, considering the characteristics of the electrical system
- Off-peak hours (OPP): Period composed of the set of consecutive daily hours and complementary to those defined during peak hours

Period of the year:

- Dry period (S): 7 (seven) consecutive months, from May to November, inclusive
- Wet period (U): 5 (five) consecutive months, from December of one year to April of the following year

7.3 Materials and Methods

Datasets used in this initial stage of the project will be presented, as well as the respective statistical analyses applied. The first dataset is from a Brazilian scenario, while the second from a Romanian scenario, however, both were analysed under the Brazilian tariff system.

7.3.1 Dataset 1

To build a dataset, a data composed from two different clients (12 months each), in the voltage class of 13.2 kV, were analysed, localized at São Paulo, Brazil.

An important feature from the data is that the demand for energy used throughout the year is mostly lower than the contracted demand. This generates additional expenses for the customer while limiting the utility's ability to supply energy to other sectors. The following features are provided by the data:

- Demand;
- Consumption PP TUSD;
- Consumption PP TE;
- Consumption OPP TUSD;
- Consumption OPP TE.

To enable the analysis of the data provided by the customer, it was initially treated to eliminate outliers, which represented less than 3% of the total amount. The *HeatMap* function (Waskom et al., 2017) was implemented, aiming to evaluate the correlation between the features acquired (positive or negative correlation).

Finally, a linear regressive approach based in the *sklearn* library was applied aiming to stablish an initial model to predict monthly energy demand, implementing a total of 10-fold cross validation. Python native libraries were used, such as *numpy* (Harris et al., 2020), *pandas* (McKinney, 2010), *statesmodels* (Seabold & Perktold, 2010), *sklearn* (Pedregosa et al., 2011) and *seaborn* (Waskom et al., 2017).

7.3.2 Dataset 2

As a second dataset, a public dataset containing Hourly Energy Consumption from a large Romanian hypermarket was analysed (Piedad & Kuo, 2018). The data was acquired during the entire year of 2016 and contains:

- The hour consumption (measured in MWh)
- Day of the week
- Day of the month
- Month
- Temperature (measured in Celsius degrees)

Also, to extract the demand from the dataset, the differential mathematical method was used, since the demand is the derived form of consumption, this made it possible to perform analyses of variance and correlation between the data provided.

This dataset was selected firstly because it allows statistical analysis with a wide range of data, and secondly because it refers to an establishment model that is easily replicated in other countries, including Brazil, in which the hypermarket tariff category would be subgroup A.

The Pearson's correlation coefficient and post hoc test (Pohlert, 2014) are statistical methods which were used to analyse the mentioned data. Those methods were applied using the JASP software (JASP Team, 2022) due to its wide range of possibilities for statistical analysis.

7.4 Results and Discussion

Figure 7.2a presents the heat map result, and Fig. 7.2b presents the score of relevant features for each customer in dataset 1. It is possible to identify a strong positive correlation between the utilized demand and the energy consumption OPP, which is between 6 p.m. and 9 p.m., while compared to the PP counterpart that demonstrated lower correlation with the consumed demand.

The total amount of demand not used during the 12-month period represents an average of 66% unused energy demand during the period for the first customer and 42% to the second, as represented in Fig. 7.3.

A linear regression method was applied by using the random forest method to verify that the initial data provided is sufficient to determine demand. The method was applied by using the number of estimators equal to 500, and the initial results show that the percent absolute mean error is inferior to 8%. However, using a larger amount of samples causes an increase in error, due to the inclusion of new nonstandard sample types. Thus, for predictions based on training with large amounts of

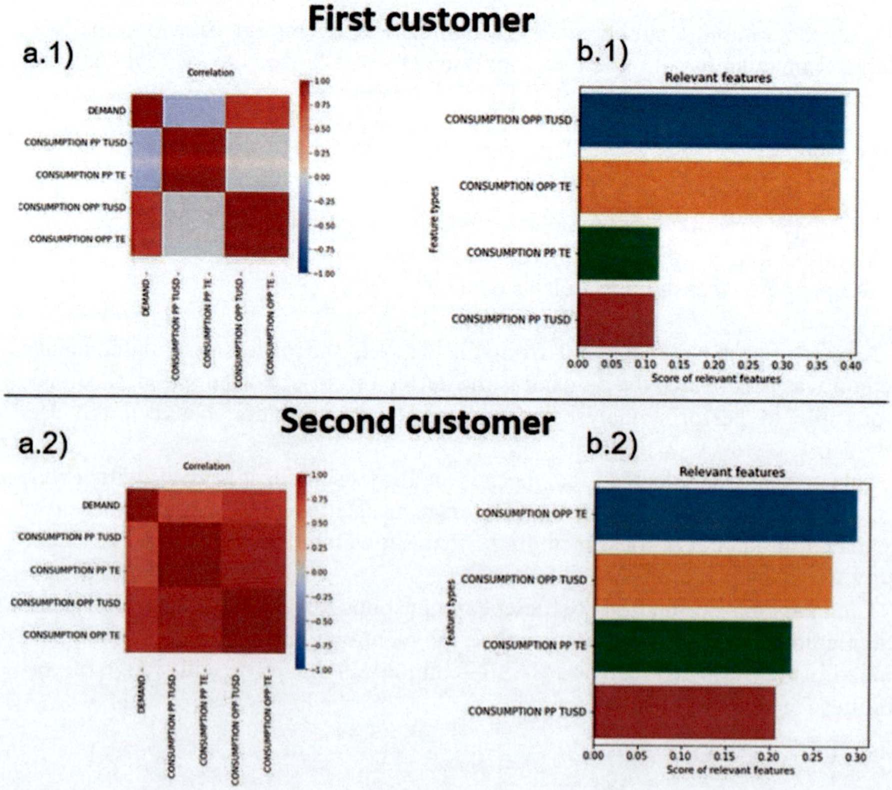

Fig. 7.2 Correlation (**a**) and relevant features (**b**) in Brazilian dataset

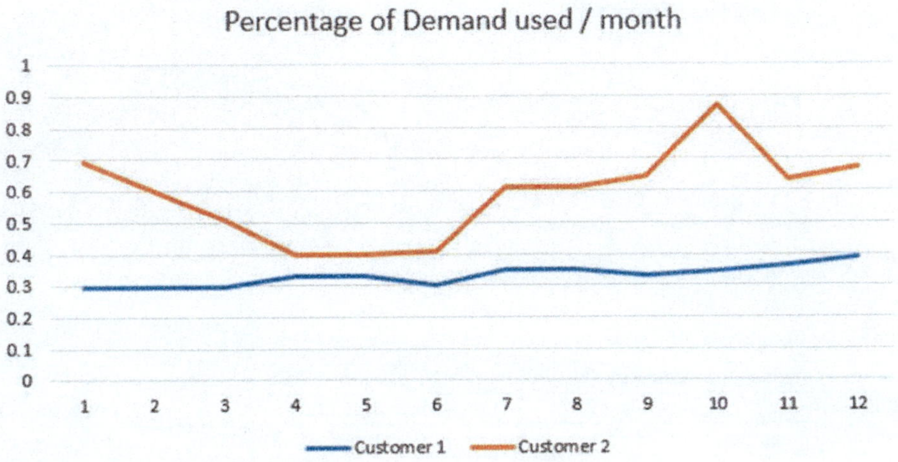

Fig. 7.3 Percentage of the used demand as a function of the month

Fig. 7.4 Correlation between demand and consumption

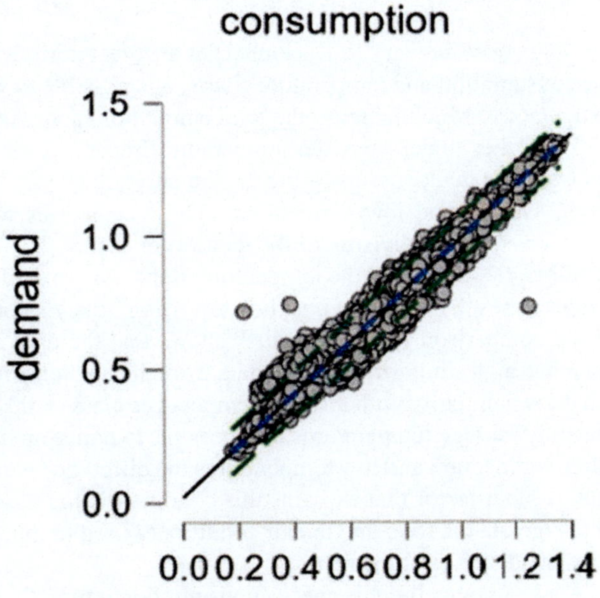

data, it is necessary to use more sophisticated methods that reduce the error, such as artificial intelligence.

For the dataset 2, the Pearson's correlation coefficient was applied, and it was noticed a strong correlation between demand (y axis) and consumption (x axis), representing close to a linear correlation, as shown in Fig. 7.4.

In Fig. 7.5 it is possible to notice an increase in consumption of energy based in the months of the year, with a higher consumption during the summer period (June to August), which demonstrates the consumption (in MWh) from each day through

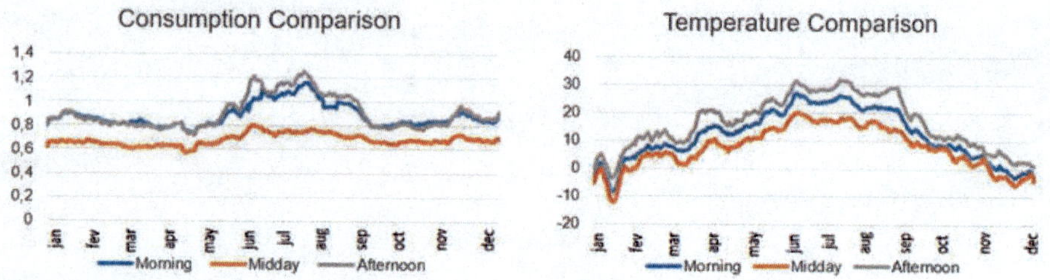

Fig. 7.5 Energy consumption and temperature comparison

Table 7.2 Comparison of different temperatures categories

Post hoc comparisons – category				
Class 1	Class 2	Mean difference	SE	*P*-value
Average temperature	High temperature	−0.234	0.006	<.001
Average temperature	Low temperature	−9.535e-4	0.006	0.986
High temperature	Low temperature	0.233	0.006	<.001

Note. *P*-value adjusted for comparing a family of 3; *SE* Standard Error

the year (from January to December) at different periods of the day. By comparing the consumption and temperature charts, it is possible to verify a direct impact from temperature, which increase the total amount of energy consumption.

To further understand the temperature impact in the energy consumption, an analysis was made by separating the temperature feature into 3 different categories: High, average and low temperature. These categories were separated in order to allow an equitable division of the dataset for each.

Table 7.2 shows that the temperature impact is particularly prevalent during high temperatures, by applying post hoc test using Tukey's correction method. Analysis of the comparison factor "P-value" shows that the data from average temperature class has high similarity with the data from the low temperature class (0.986), while it has low similarity with the high temperature class (<.001). This situation indicates that only for high temperatures, it is possible to notice an increase in energy demand, while for medium and low temperatures the difference generated in demand is minimal. A main factor that explains this condition is the necessity to use more energy to refrigerate the food and the air conditioner used to minimize the higher temperatures inside the establishment.

Also, although there is energy consumption during the night period due to products refrigeration, it is significantly less than during the rest of the day (from 8 am to 9 pm the consumption is higher), and the day of the week has little impact in the consumption variation, as the average consumption for each day of the week has a variation representing less than 2%.

By normalizing the data regarding the temperature, as shown in Fig. 7.6, the energy consumption has little variance, which indicates that the seasonality factor has little impact in the consumption outcome.

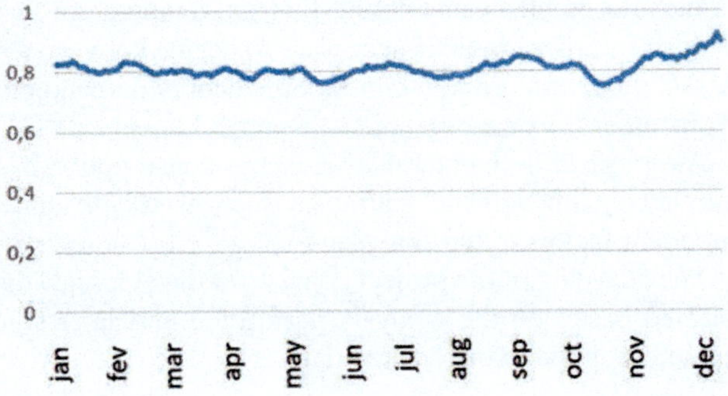

Fig. 7.6 Normalized consumption (in MWh)

7.5 Conclusions

The aim of this work was to discuss the main indicators to be used for prediction of energy consumption in the Brazilian industrial context. Based on the customer's dataset, the utilized energy demand is highly influenced by consumption during the peak period. Also, the implemented model tracks energy consumption, by relating it to the contracted demand, which manages to compare the data and provide an overall situation of the system conditions, making it possible to identify anomalies in the customer load. Thus, the evaluation elaborated using linear regression method is promising, approximately representing the customer's energy demand. However, for large datasets, other methods such as artificial intelligence are more promising to predict customer's demand.

The low level of efficiency of demand usage, shown in Fig. 7.3, demonstrates the need to establish demand prediction mechanisms that are effective and at the same time accessible to different commercial and industrial sectors, since it causes a big difference in the energy bill paid by these companies. In addition, the better match between contracted demand and consume would relieve the load imposed on the energy distributor, enabling a more efficient use of the electricity grid. These are fundamental concepts that are in line with the energy efficiency and sustainability agenda of SGD7, proposed by the United Nations in 2015.

The consumption has a direct correlation with the measured demand and is considered to be an important factor to determine it. The temperature component is also a significant contributor to determine demand, mainly in high temperatures as represented in Fig. 7.5.

In spite of the normalized analysis, Fig. 7.6 shows little impact from seasonal components on the demand for dataset 2, this result is, however, highly dependable on the customer's type and will not determine with precision if this variable is significant to determine demand for other scenarios with companies related to seasonal consumption, like Christmas or Easter.

Considering that temperature is a quantitative variable, and that its variation is directly related to consumption, it is important that the demand and consumption forecast is carried out in such a way that the established demand limits are not exceeded or underutilized since both these conditions may incur additional charges on the energy bill. Considering that the temperature factor is decisive, a rigorous analysis of this variable is necessary to decide between a regulated contract or a free market model, in order to contract the energy demand that is best suited to the type of customer.

Although in both datasets the energy consumption has been shown to be highly correlated with demand, variables such as temperature and seasonality are also important factors in the analysis.

The next step of the project consists of developing a dataset using the parameters selected based on the analysis performed and later studying different predictive methods applied to the created dataset.

References

Brazil. ANEEL. Normative Resolution N° 1000. (2021). Estabelece as Regras de Prestação do Serviço Público de Distribuição de Energia Elétrica. https://www.gov.br/aneel/pt-br/assuntos/tarifas/entenda-a-tarifa/modalidades-tarifarias

Brazil. ANEEL. Normative Resolution N° 166. (2005). Estabelece as disposições consolidadas relativas ao cálculo da tarifa de uso dos sistemas de distribuição (TUSD) e da tarifa de energia elétrica (TE). https://www.legisweb.com.br/legislacao/?id=102484

Brazil. ANEEL. Normative Resolution N° 414. (2010). Estabelece as Condições Gerais de Fornecimento de Energia Elétrica de forma atualizada e consolidada. https://www.legisweb.com.br/legislacao/?id=112868

Cao, L. (2017). Data science: A comprehensive overview. *ACM Computing Surveys (CSUR), 50*(3), 1–42.

Harris, C. R., Millman, K. J., Van Der Walt, S. J., Gommers, R., Virtanen, P., Cournapeau, D., et al. (2020). Array programming with NumPy. *Nature, 585*(7825), 357–362.

JASP Team. (2022). *JASP (Version 0.16.3)* [Computer software]. https://jasp-stats.org/

Jayasutha, V. G., & Arunachalam, T. (2022, April). Deep learning with statistical analysis for stress prediction of remote working IT employees in COVID-19 pandemic. In *2022 International Conference on Sustainable Computing and Data Communication Systems (ICSCDS)* (pp. 179–184). IEEE.

Kurbatsky, V. G., & Tomin, N. V. (2010, August). Application of artificial intelligence technologies for monitoring large power interconnections. In *2010 sixth international conference on natural computation* (Vol. 3, pp. 1360–1365). IEEE.

McKinney, W. (2010, June). Data structures for statistical computing in python. In *Proceedings of the 9th python in science conference* (Vol. 445, No. 1, pp. 51–56). Stefan van der Walt and Jarrod Millman.

Nespoli, A., Ogliari, E., Leva, S., Massi Pavan, A., Mellit, A., Lughi, V., & Dolara, A. (2019). Day-ahead photovoltaic forecasting: A comparison of the most effective techniques. *Energies, 12*(9), 1621.

Pedregosa, F., Varoquaux, G., Gramfort, A., Michel, V., Thirion, B., Grisel, O., et al. (2011). Scikit-learn: Machine learning in python. *The Journal of Machine Learning Research, 12*, 2825–2830.

Piedad, E., & Kuo, C. (2018). *A 12-month data of hourly energy consumption levels from a commercial-type consumer*. Mendeley Data, V1. https://doi.org/10.17632/n85kwcgt7t.1

Pohlert, T. (2014). The pairwise multiple comparison of mean ranks package (PMCMR). *R Package, 27*(2019), 9.

Porteiro, R., Hernández-Callejo, L., & Nesmachnow, S. (2022). Electricity demand forecasting in industrial and residential facilities using ensemble machine learning. *Revista Facultad de Ingeniería Universidad de Antioquia, 102*, 9–25.

Russell, S., Norvig, P., & Davis, E. (2010). *Artificial intelligence: A modern approach* (3th ed.). Prentice Hall.

Seabold, S., & Perktold, J. (2010, June). Econometric and statistical modeling with python skipper seabold 1 1. In *Proceedings of the 9th python in science conference* (Vol. 57, p. 61).

Waskom, M., et al. (2017, September). *Mwaskom/seaborn: v0.8.1*. Zenodo. https://doi.org/10.5281/zenodo.883859

Yan, K., Li, W., Ji, Z., Qi, M., & Du, Y. (2019). A hybrid LSTM neural network for energy consumption forecasting of individual households. *IEEE Access, 7*, 157633–157642.

Yang, Y., Han, L., Wang, Y., & Wang, J. (2022). China's energy demand forecasting based on the hybrid PSO-LSSVR model. *Wireless Communications and Mobile Computing, 2022* (pp. 8–11).

8 New Energies Framework: Hydrogen Ecosystem, Geopolitical and Economic Impacts

Eric Zanghi, Felipe de Brito e Cunha, Maria Mônica de Oliveira, and Carlos Henrique de Oliveira

8.1 Introduction

For decades, human society has been looking for sustainable energy production options, whether in the intelligent use of primary resources, in the logistics of storage and distribution or in terms of the cost of implementation. Several energy sources and technologies have been investigated, improved and consolidated in the past 20 years (e.g. biogas, tidal, solar, and wind power plants), and a few more promising ones are on the way to becoming viable from a technical-economic point of view.

It can be seen that hydrogen is an abundant earth's natural element (Lubitz & Tumas, 2007); however, its naturally available form is strongly connected with other atoms of natural elements. Its capture, storage and deployment present challenges in terms of economic feasibility. Besides, the use of the hydrogen as an energy source is not trivial from the point of view of the holistic sustainability approach advocated by international organizations (Bleischwitz et al., 2018; Falcone

E. Zanghi (✉)
ISPGAYA - Instituto Superior Politécnico Gaya, Vila Nova de Gaia, Portugal

INESC TEC – Instituto de Engenharia de Sistemas e Computadores, Tecnologia e Ciência – Centre for Power and Energy Systems, Lisbon, Portugal
e-mail: ezanghi@ispgaya.pt

F. de Brito e Cunha
ISCTE – Instituto Universitário de Lisboa – Department of Political Science and Public Policy, Lisbon, Portugal

M. M. de Oliveira
Kypséli Data Experience, Lisbon, Portugal

C. H. de Oliveira
Eletrobras Eletronuclear, Rio de Janeiro, Brazil

© The Author(s), under exclusive license to Springer Nature Switzerland AG 2023

F. L. Almeida et al. (eds.), *Multidimensional Sustainability: Transitions and Convergences*, Springer Proceedings in Earth and Environmental Sciences,
https://doi.org/10.1007/978-3-031-24892-4_8

et al., 2021). In order to effectively make hydrogen a significant component of the world energy matrix, international efforts are being made, and increasingly so. These efforts take place from a financial point of view, with extensive funding for research and development like the European initiative (European Commission, 2021), in the social sphere, with the education and awareness of the population about new and clean energy sources (European Commission, 2020; Mazloomi & Gomes, 2012; Schönauer & Glanz, 2022) and in the market sphere (Maggio et al., 2019; Schulte, 2004), encouraging the creation of complete value chains to enable the application of new technological solutions (Lo Faro et al., 2022; Scovell, 2022).

This work aims to consolidate information on important impacts, namely in the technical, geopolitical and economic areas, on the implementation of the use of hydrogen as a primary resource of new technologies of energy production. This article reviews the available literature, and the most recent achievements in the hydrogen economy framework formatted in a logical sequence for a complete contextualization on the topic. This work is structured as follows: Sect. 8.2 presents a review of hydrogen's recent achievements and the consolidating ecosystem. This section still details the technologies, presents geopolitical impacts and goes from business hubs to the general economic aspects. Section 8.3 concludes the work by pointing out the main aspects, challenges and potential benefits that can be achieved.

8.2 Hydrogen Ecosystem

The world energy production based on renewable sources is growing every year according to a report by IRENA (Renewable Energy Agency, 2022), and this growth is not limited to the most socially developed countries. The need to limit carbon emissions and achieve low energy prices in sustainable growth is driving unprecedented changes in the global energy transition. The falling costs of renewable energy and its decentralized nature have increased widespread electrification. Continuous technological advances, well-focused public policy and cost-effective power plant solutions are driving the energy transition accessible to all types of society, even the most socially and economically disadvantaged.

Despite all the advances in electricity production, it is important to note that energy for residential, commercial and industrial use is not limited to just a single secondary energy (i.e. electrical energy). Exclusive dependence on a single secondary energy source is neither a safe nor a reasonable path in technical and economic terms. Other energy sources are needed for the production of heat and movement, which are called end-use energies. At this point, natural gas, methanol, gasoline, diesel and all secondary energy sources play an important role in the energy value chain as a whole. Hydrogen appears as another secondary energy component (Jain, 2009) and can be used basically in the gaseous form (i.e. H_2) or as the main element of hydrogen cells, needing to be combined with other chemical elements (e.g. nickel, platinum, etc.) so that they can produce end-use energy.

All industrial processes for the production of renewabe energy generation solutions, whether for the creation of photovoltaic panels or wind towers, will inevitably have to consume a considerable amount of natural mineral resources and will consume materials based on polymers, that is, based on petroleum. The value chains of these solutions are usually analysed from the moment of their installation until the end of their useful life, discarding the cycles of creation, destruction, recycling or reuse (Jensen, 2019; Tsanakas et al., 2020). The same holistic sustainability analyses and concerns must be made with hydrogen-based solutions, whether they are designed for small-scale or large-scale use.

One of the great advantages of energy production with renewable sources is its ability to use endemic primary energies (e.g. solar, wind, etc.) in a distributed manner on technically possible scales. Regions with high solar intensity may have relatively small areas for the installation of solar panels, which are insufficient to meet their energy demand. However, several endemic sources of primary energies from the same region, being used in a combined and systemic way (e.g. solar, wind and water), could supply this demand. Hydrogen adds great value in this approach to distributed energy production because it is abundant in several natural and industrial substances (e.g. water, methanol and hydrocarbons) that can be endemic or produced in a specific region.

To better understand how complex the hydrogen ecosystem can be, it is important to observe Fig. 8.1 that its use reaches a whole range of applications, such as in industrial and agribusiness resources, through the production of electric energy to the driving energy of large cargo transport equipment.

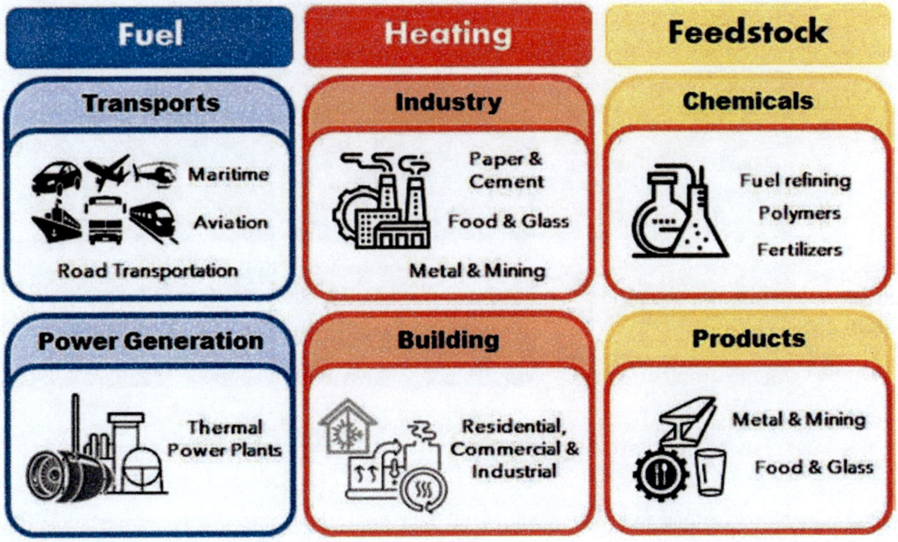

Fig. 8.1 The complex framework of hydrogen usage

8.2.1 Technological Aspects

It is important to emphasize that the use of hydrogen depends primarily on its capture in the molecules of these substances, so that, in its pure and isolated form, it can be used in energy production solutions. Water and fossil substances are the main sources of hydrogen; therefore, a separate analysis of these two types of sources is necessary for a better understanding of the viability of hydrogen in the energy matrix (Fig. 8.2).

8.2.1.1 The Green Hydrogen

Green hydrogen is produced by the electrolysis of water in equipment called electrolyzers. Using electrical energy and water (H_2O) to separate hydrogen (H_2) and oxygen (O_2) molecules, electrolyzers are the key to this process. These equipment are in constant technological evolution in order to optimize their cost-effectiveness (Yue et al., 2021), making them more efficient in terms of energy and monetarily less expensive when using the most abundant catalyst materials on the market. For this hydrogen capture process to be considered fully decarbonized, the electricity used cannot come from fossil sources, such as coal or natural gas, but from renewable sources (e.g. solar, wind or hydro).

It is important to point out that when is it possible the use of renewable energy is directly more efficient than converting it to hydrogen to be used as an energy source.

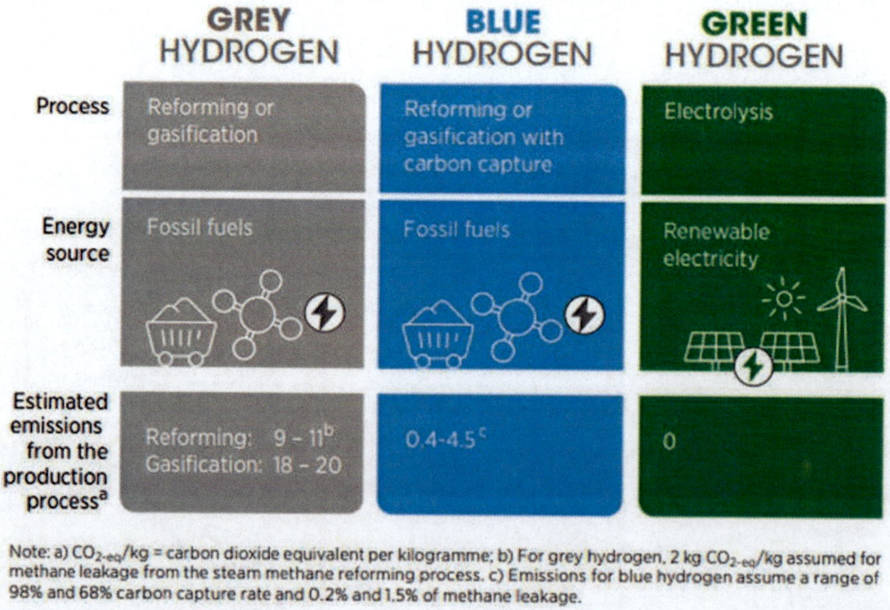

Fig. 8.2 Simplified colour spectrum typology of hydrogen production. (Source: International Energy Agency, 2022)

Using renewable energy to produce hydrogen is about 20–40% less efficient than if renewable energy could be used directly (Energy Transitions Commission, 2021).

8.2.1.2 The Grey Hydrogen

The so-called grey hydrogen is the most used on a world scale for industrial use and is produced from the burning of natural gas, which has increasingly been produced also from coal (so-called brown hydrogen), in a reaction known as steam reforming (SRM). The hydrogen separation process is basically the same for gasoline, naphtha, methanol or even other renewable sources such as ethanol. Regardless of the source, these processes will always emit air pollutants. The process involves the reaction of methane, contained in natural gas and water vapor at high pressure and temperature (around 1000 °C). Steam reform consumes large amount of energy and emits a large amount of carbon dioxide (CO_2) into the atmosphere, which is exactly the main element to be avoided in the new energy transition. An important portion of the world production of grey hydrogen is used to synthesize ammonia and its derivatives (for agriculture) or to carry out oil refining operations.

8.2.1.3 The Blue Hydrogen

Basically, blue hydrogen represents the same capture process as grey hydrogen but with an additional post-processing focused on capturing the CO_2 emitted by the H_2 separation process. This CO_2 capture, called carbon capture and storage (CCS) or CCUS with CO_2 reutilization, can be performed directly in the CO_2 gas outlet or indirectly, capturing CO_2 from the air around the grey hydrogen production plant. The CCS refers to the initiative and its set of technologies (Bui et al., 2018) that aim to prevent the emission of large amounts of CO_2 at the point of origin of its emission pursuing the carbon neutrality for an specific hydrogen plant (Sunny et al., 2020). For this, the CO_2 captured is subjected to a chemical process that produces a relatively pure carbon dioxide stream. This steam is compressed at high pressure to be transported by land and/or sea to its final storage location. This storage is carried out in deep regions, both offshore and onshore (Raza et al., 2019).

The basic technology for the CCS process has been around for many years. However, the high cost of the process and its technical-scientific uncertainties (Koelbl et al., 2014) are the main barriers to the extensive implementation of this technology. In addition, there are studies that consider some risks related to storage techniques in deep geological formations, such as accidental leaks and the occurrence of earthquakes (Zoback & Gorelick, 2012). In a critical analysis, studies show that the methane emission rate from natural gas is reduced at most to about 1.54% (Howarth & Jacobson, 2021), but the greenhouse gas emissions from blue hydrogen are still higher than the combustion of natural gas and between 18% and 25% higher than in the grey hydrogen process.

The composition of the cost of hydrogen production can be understood by taking into account three main factors, namely the price of locally produced electricity, capital expenditure for electrolysis units (CAPEX) and operating costs (OPEX). In cases where hydrogen is not used immediately at your production site, storage,

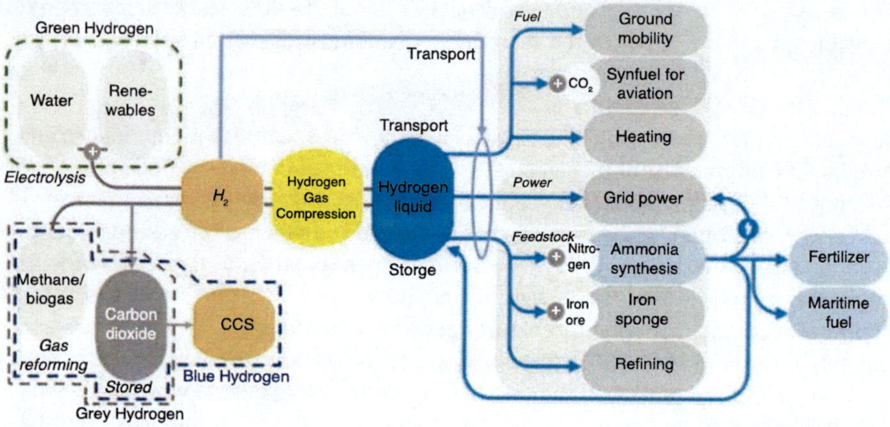

Fig. 8.3 Hydrogen pathways framework

transport and distribution costs would also need to be taken into account, which adds to the complexity of its ecosystem.

Figure 8.3, adapted from (Hydrogen Council and McKinsey & Company, 2021), shows a complete view of the hydrogen pathway since the captation method (e.g. green, grey or blue) of the hydrogen gas (H_2) molecule for use in the final destination, either as a production input or as an energy source. In the process between the capture and the final use of hydrogen, two possible alternatives can be observed. One is the storage and transport of H_2 (gas state) using new gas network infrastructure or using adapted natural gas pipelines already installed, well described in (ACER, 2021; Cerniauskas et al., 2020; Ogden et al., 2018) presenting some technical issues related to the infrastructure adaptation and other challenges. The other alternative is the process of applying low or high compression over the H_2 for its transformation into liquid state. After that, the hydrogen liquid can be transported using long or short distance land or maritime transportation.

8.2.2 The Geopolitical Impacts

The green hydrogen is emerging as an attractive solution for decarbonizing society. This solution allies renewable energy power plants with hydrogen gas production and can help to solve the energy issues that affect the energy-intensive industries that often face challenges in electrification (Balcombe et al., 2018; Espegren et al., 2021). This industry segment is critical to the world economy and comprises long-distance transport (e.g. long-haul aviation and international shipping), heavy industry (e.g. metal mining) and agrochemicals (fertilizers). This analysis is based on the actual technologies and strategies designed or already applied around the world (European Commission, 2019; Marbán & Valdés-Solís, 2007; Neuwirth et al., 2022).

The impact specifically on large maritime freighters and long-distance aviation is a game change in the world demand for fossil fuels, since the great concentration of consumption of these commodities comes from the global logistics chains (Atilhan et al., 2021; Hoelzen et al., 2022). According to the report presented in (Hydrogen Council and McKinsey & Company, 2021), clean hydrogen could reduce cumulative CO_2 emissions by a total of 80 gigatons (GT) by 2050, contributing with 20% of the total abatement forecasted to this period.

All global or local strategies must rank the priorities, especially when the subject, here the hydrogen, is so useful and cappilared into the most important industrial value chains in the world. Figure 8.4 illustrates the maturity level of the hydrogen solutions face to the type of solution, namely centralized or decentralized.

The focus on large impact centralized solution is in line to the technological maturity, which is a consequence of larger investments looking for bigger economical impacts. This strategy is driven by the most important economies in the world (Cuevas et al., 2021) and became an important vector to achieve the challenge of the energy transition (Council, 2017) and the goals for 2030.

Accordingly to the IRENA report (Blanco & Taibi, 2022) shown in Fig. 8.5, the supply and demand of the hydrogen for regions around the world in 2050 will create a new scenario of self-sufficiency. From an optimistic point of view of the technological evolution, the IRENA report shows some emergent countries (e.g. Chile, North Africans countries, etc.) as a potential exporters and some strongly industrialized countries, i.e. Germany, as future importers. Once more in the energy history can be noticed the fragility of the central Europe when natural resources are demanded (e.g. water and petroleum) and the importance of a diverse energy matrix.

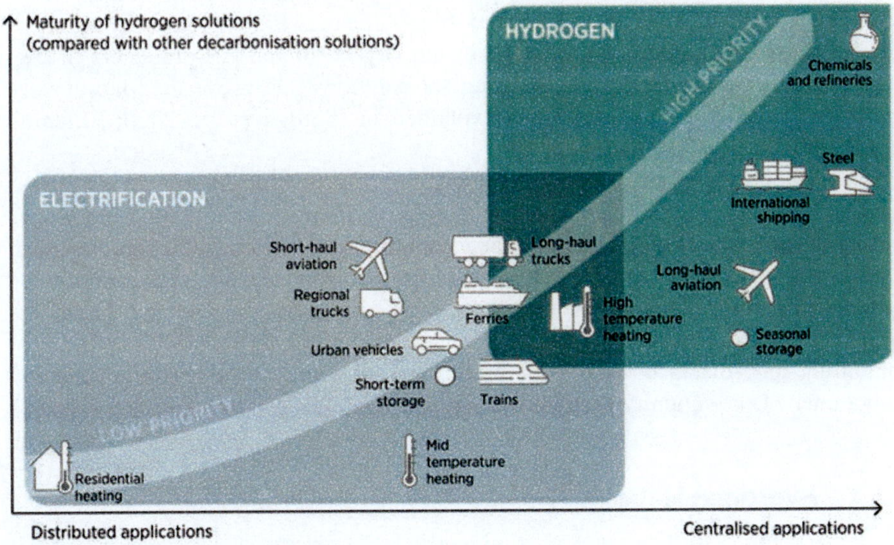

Fig. 8.4 Maturity of hydrogen solutions. (Source: Bianco et al., 2022)

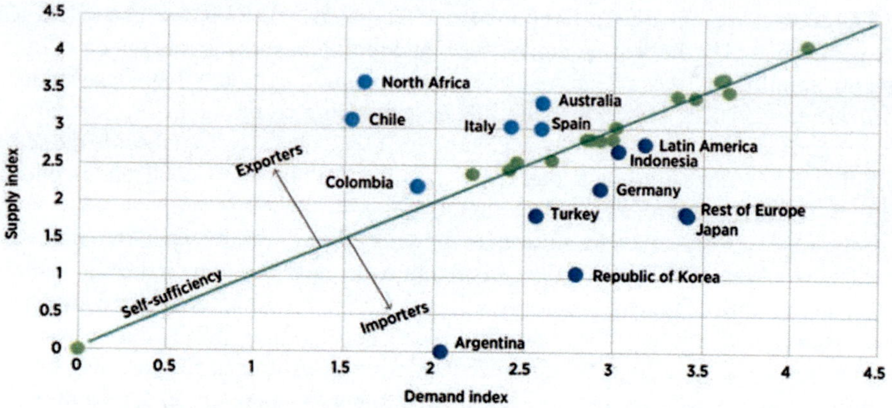

Fig. 8.5 Forecast of hydrogen self-sufficiency. (Source: Blanco & Taibi, 2022)

The Russian invasion of Ukraine in 2022 has pushed the cost of hydrogen from fossil gas to much higher than the cost of hydrogen from renewable sources (i.e. green hydrogen). As fossil fuel prices have risen, so have the costs of grey and blue hydrogen (Fabiola & Pantaleone, 2021). The feasibility of producing green hydrogen from renewable energy sources through electrolysis has become affordable and safe, especially in Europe and countries without natural fossil energy resources.

On 18 May 2022, the European Commission published its "REPowerEU" that is a plan to reinforce the energy independence from Russia as quickly as possible, in principle by 2027 (REPowerEU, 2022). The plan covers three pillars, namely: demand reduction, diversification of fossil fuel suppliers encouraging the construction of new infrastructures and the acceleration of the transition to renewable energy sources. The Russian Federation is one of the largest oil and gas producers in the world, despite this, it has been searching for solutions that prove the viability of hydrogen energy for some years, as explained in (Shulga et al., 2020; Zhiznin et al., 2020).

Focused on the hydrogen, the European Union (EU) is prioritizing energy independency and is developing internal market ecosystems to achieve the optimum level of the green and blue hydrogen. The recent report presented in (Nuñez-Jimenez & De Blasio, 2022) explains the EU strategy for the local and long-distance importation. The future and possible scenarios consider some key strategic variables namely: energy independence, cost (optimization) and energy security, looking for combining long-distance reliable partners (e.g. Australia and the United States) with a inter EU community production.

8.2.3 Hydrogen Hubs

The importance of business hubs in the hydrogen global scenario shows how the use of this natural element are evolving specially as an energy source. The hydrogen energy has emerged from a particular niche of knowledge in the scientific area to a real business ecosystem. The so-called H_2 hubs have been trying to encompass the whole hydrogen usage (i.e. industrial and energetic) in a holistic view of its value chain.

Energy community organizations have emerged and allowed customers to become energy producers and consumers at the same time (i.e. prosumers). These focus on clean energy and uncover opportunities to engage communities, which are milestones in achieving energy goals and carbon footprints. While hydrogen market organizations are gaining momentum and countries are accelerating their adoption through investment policies and cross-border contracts. Likewise, some companies have been consolidating as hydrogen hubs (i.e. H_2 global cluster) and so entire regions (i.e. H_2 sines – Portugal) unlocking the opportunity to accelerate the hydrogen production. Geographic hydrogen hubs are places where hydrogen producers, suppliers and exporters are located. This type of arrangement that combines cross-sector demand and the co-location of the entire hydrogen value chain minimizes the cost of supporting infrastructure, essential for scaling hydrogen production. The H_2 hub companies runs horizontally with actions that engage key resources and shortens the learning curve for stakeholders optimizing the solutions management, decreasing gaps and waste.

The hydrogen hubs must provide the opportunity to develop decarbonization solutions and deploy them at scale. Industrial activity and geological storage capacity in the hub's areas allied to the H_2 hub companies will provide a potential market for producers and consumers of hydrogen and related CCS providers. Figure 8.6, from the American carbon and hydrogen atlas (Abramson et al., 2022), shows a synthetic view of the all roles that can be attributed to the H_2 hubs, not only in United States but also in whole world.

Many aspects of the energy transition are still evolving in the face of challenges and opportunities and some questions often rise such as the position of hydrogen solutions. It could be approaching the peak of inflated expectations (Fig. 8.7) or are still floating over the innovation process? This evaluation is present in many studies such as (Dehghanimadvar et al., 2020) and is important to define the level of trust in the feasibility of this vector in the energy transition agenda.

New global agreements were signed (MECA, 2020), and this trend is growing, as are declarations of interest in intergovernmental cooperation, which have been signed with a focus on practical and objective aspects. The hydrogen hubs initiatives are present in the whole world and are growing towards the industrial and energy supply. Looking at the European H_2 ecosystem, Portugal must accelerate the energy transition and the decarbonization of the economy investing in the production of hydrogen, promoting the substitution of fossil fuels.

The forecasted projects will cover much of the hydrogen value chain, including production, fuel cells, storage, transport and distribution, as well as end-user applications of the mobility sector. Regarding other initiatives, it can be noted the hub "HyDealEspaña" that is the biggest of its kind and focused on green hydrogen. This hub will start in 2025 the H_2 gas supply to the Asturian industrial area to be used as energy source and industrial feedstock. The Netherlands hub called "Heavenn" (H_2 energy applications in valley environments for Northern Netherlands) is a hub

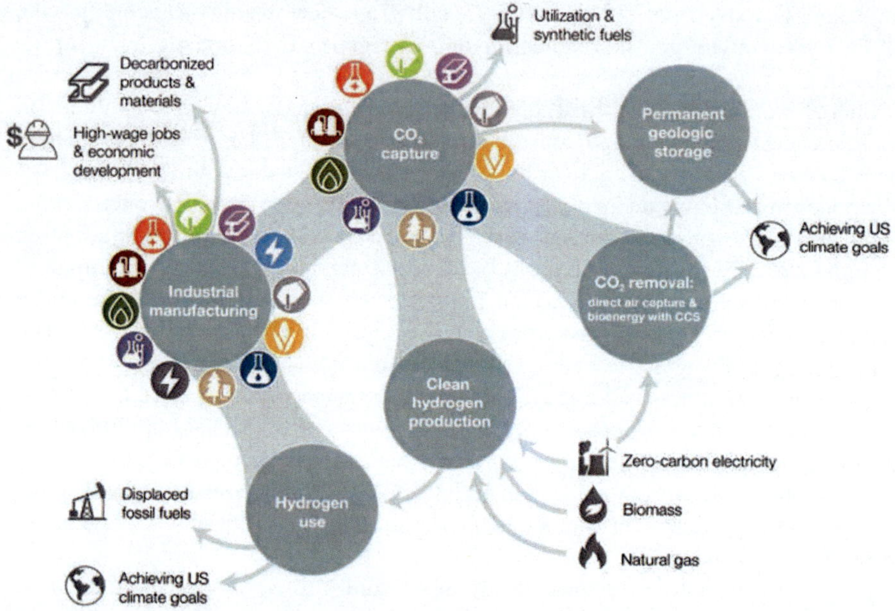

Fig. 8.6 The roles of hydrogen hubs. (Source: Abramson et al., 2022)

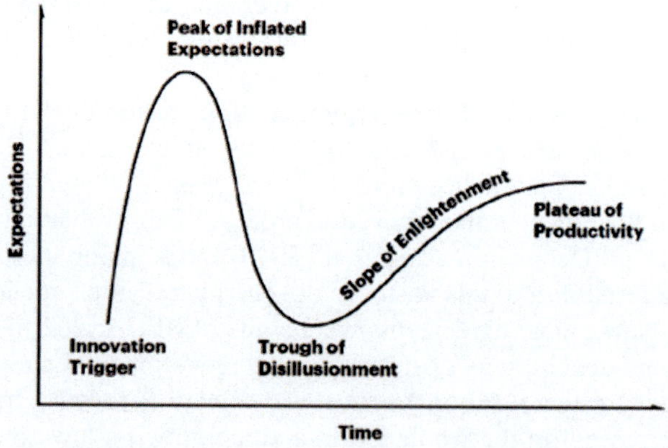

Fig. 8.7 Adapted diagram of a technology and Gartner hype cycle. (Source: Dehghanimadvar et al., 2020)

project that aims the production of green hydrogen through wind energy, decarbonizing the port of Rotterdam and the urban transport, maritime transport and in the countries' infrastructure.

8.2.4 Economical Aspects

Low-carbon hydrogen, amongst its possible varieties, has emerged with the inherent potential to become the driving force towards a more sustainable economy (Fan et al., 2021). For the past decades, it has been considered an essential element by European Union official policies to meet zero emissions target for 2050 and, also, reveals a multiplicity of uses, allowing the implementation of a true circular economy mechanism (Bonciu, 2020) with renewable inputs, reuse of resources and safer outputs.

The current global energy context faces a forced dilemma where the foreseen population growth embracing new cultural standards (technology-based disruption, continually growth-oriented production and digitalized society) is severely increasing energy demand (Saeedmanesh et al., 2018; Sharma et al., 2020) and at the same time the socio-environmental impacts of a linear economy based on hydrocarbons largely compromise the future of the economy itself and the humanity well-being (Lee, 2016). For this reason, the clean energy transition has been reclaimed by most parts of society and widely accepted by policy makers.

In 2016, during the UN Signature Ceremony for the Paris Agreement on Climate Change (COP21), the EU Commissioner for Climate Action and Energy Miguel Arias Cañete, reinforced the mutual commitment between nations as "irreversible and non-negotiable" (European Commission, 2020). Therefore, to achieve this climate challenge, policy makers, scientists, grassroots initiatives and business sectors are creating different pathways to innovate, not just from a technological point of view but also a whole social-economic model, in which energy and feedstock are core functional elements. The European Green Deal, focused on a carbon net-zero economy, has more than 1 trillion Euros budget and highlighted the efficient use of resources by moving to a clean, circular economy (European Commission, 2019). On the other side of the globe is also appearing an inter-state network development towards hydrogen economy, i.e. the Asia-Pacific Hydrogen Valley (Aditiya & Aziz, 2021).

According to the EU – Energy Poverty Observatory Hub, over 34 million people are in an energy poverty situation in the communitarian's countries. In (Bouzarovski et al., 2021) is presented the understanding of the energy poverty as "a condition in which a household is unable to secure a socially and materially needed level of energy services in the home". During the COVID-19 pandemic, for example, more than 50 million people and communities are suffering from energy deprivation in the European Union, that is, 1 in 4 households (Guiteras, 2021). Energy poverty is also in intersection with other forms of asymmetries deepening even more social inequality (Brah & Phoenix, 2004).

Therefore, combining the urge for economic decarbonization (Espegren et al., 2021), democratization of energy access (Burke & Stephens, 2017) and social transition requires systemic change in the way we produce, consume and relate with people and the environment, creating a circular hydrogen-centred and sustainable economy.

8.2.4.1 The Hydrogen Nexus

Clean hydrogen (i.e. green, blue or biohydrogen) can take a central role to a long-term sustainable development shifting from a carbon-based system (Aditiya & Aziz,

2021; Dincer & Rosen, 2011; European Commission, 2020; Saeedmanesh et al., 2018). This capacity is not only related to its intrinsic capacity to achieve zero CO_2 emissions in the energy sector (innovating in power and heat production, storage and transport) but also for its possible interrelationship with a variety of other sectors like fertilizers, chemical compounds, food supply chain and other shared benefits like waste management (Fan et al., 2021; Sartbaeva et al., 2008).

This wider nexus involves also a strong structure on a national level that enables the hydrogen economy to evolve, considering a domestic energy competency, economic security, level of advancement in technology (including R&D capacity) and societal influence as well (Aditiya & Aziz, 2021). According to Aditiya and Aziz (Ibidem), social development is closely related to the cultural aspects – its set of norms, values, beliefs and meanings – which is fundamental for social acceptance and, thus, influencing the social impact of a new energy system.

8.2.4.2 Circular Hydrogen Economy

Contrary to a linear process economics (based on extraction, production, consumption and waste), the circular economy contributes beyond recycling to change the way we design products, produce, consume and circulate materials (Bonciu, 2020) to a more regenerative process, based on renewable energy and material, eliminating waste and pollution (Ellen Macarthur Foundation, 2022).

According to (Bibas et al., 2017), in 1950, a total between 60 and 70 Giga tonnes of waste were generated in the world and probably rising to 100 Giga tonnes in 2030. Around 94% of the waste is discarded precariously only in India, and waste generation is growing at an alarming rate of 1.3% per year (Sharma et al., 2020). Hydrogen losses in industrial waste gas streams, estimated as 10 billion Nm^3 per year only in Europe, constitute a potential source for hydrogen recovery (Yáñez et al., 2019). There are many other examples of renewable feedstock to be introduced in the hydrogen economy through circular processes like the waste-to-energy nexus.

In December 2012, through the document "Manifesto for a Resource Efficient Europe" the EU stated that "the EU has no choice but to go for the transition to a resource-efficient and ultimately regenerative circular economy" (European Commission, 2012). Therefore, hydrogen can definitely be well positioned in this direction, using regular waste and lower impact waste as resource that could be reintroduced in the system.

Biohydrogen, for example, generated from poultry, swine, cattle, brewery, dairy and other industrial, agricultural or urban wastewaters can represent an important contribution to the circular "hydrogen" economy (Ferreira et al., 2018). There are many examples of these biohydrogen production around the globe, like in Italy (Pecorini et al., 2017), India (Venkata Mohan et al., 2018), China, Korea and others. According to Lee (2016), there are also multiple examples of cost-efficient bioeconomy models in which biohydrogen is included and is part of a wider and systemic economic process where renewable resources are interlinked with a circular economy fashion.

8.2.4.3 Sustainability

Beyond the economic bias regarding, for example, the distrust in the long-term viability of the current fossil-based industry and a general better cost efficiency ratio, the motivation towards a hydrogen-based economy is intrinsically linked to sustainability, for instance, the awareness of the ecological pressure on resources, necessity to social equity and welfare in energy access and the development of a low-impact economics.

Although, as explained before, to become a truly sustainable and clean fuel, hydrogen should encompass its blue, green or bio versions, nowadays, 95% of the hydrogen produced and consumed in the world market comes from hydrocarbon sources (i.e. grey and brown hydrogen), which reach about 870 million ton/year of CO_2 emissions. In addition, the process takes place at high temperatures, making it more expensive. This type of production consumes 6% of fossil fuels and 2% of coal produced in the world (Song, 2009).

Combining renewable energy, waste reuse and CCUS can be essential to a successful carbon mitigation and energy transition. In fact, there are not only technological and cost-effective challenges to guarantee this "sustainable hydrogen" but also political, institutional and social acceptance. On this matter, many countries are already evolving with national and regional political strategy as demonstrated previously in this paper.

Hydrogen energy systems can also support a more decentralized production to contribute to local economies, allowing flexibility and facilitating local sustainable development interchanging with regional scale strategies. "Hydrogen energy systems often require relatively small-scale equipment, reducing the time from initial design to operation and providing greater adaptability in responding to changes in energy demand" (Dincer & Rosen, 2011).

8.3 Conclusions

The great challenges of humanity have passed through questions of production and supply of food and energy. This last challenge was thought of and prioritized, in financial and geopolitical terms, as the biggest problem today and demands systemic and clear actions with short-term impacts.

Hydrogen, whether green or any other colour of the spectrum presented in this work, appears as another important vector in the world energy matrix, which tirelessly seeks its diversity and reliability.

Despite bringing great technological and logistical challenges, its use can bring more economic stability in production chains and in global logistics as the main source of fuel and energy.

As it is a source of energy that can be generated in a large part of the planet, its capillarity tends to break the large monopolies in the supply of energy sources, mainly in the form of gas, but not only. This resource could promote the development of large regional and global business hubs, as well as developing countries without fossil energy sources, but with abundant water resources, whether maritime, river or underground. Additionally, all essentially agricultural countries with the potential to turn to the production of alcohol subject to hydrogen capture will benefit from this new market. The latter comprise a large group of countries from poor or emerging economies, which will have an impact on the generation of local wealth and, consequently, on an increase in quality of life indices.

The present world economic moment shows that important public, private and coordinated financial support is being increasingly applied even in countries with free market economic bias. The use of hydrogen as a strategic vector in the decarbonization of the economy has also been strengthened by public initiatives to make the applicable laws more flexible and with tax incentives. These processes facilitate the more intense action of private agents and encourage the natural adhesion of the market to the hydrogen economy, which is important for the development of necessary technologies in present and future value chains. This path was successful in the case of incentives for the use of wind and solar energy and tends to work in the case of hydrogen.

Finally, this work showed that paradigm shifts and the introduction of new solutions on a global scale require a period of technological maturation and financial viability. The hydrogen ecosystem has been moving in this direction at an accelerated rate and is should reach a global scale faster than expected.

References

Abramson, E., Thomley, E., & Mcarlane, D. (2022). *Carbon and hydrogen hubs an Atlas of for United States decarbonization* (Issue February). www.betterenergy.org

ACER. (2021). Transporting pure hydrogen by repurposing existing gas infrastructure: Overview of existing studies and reflections on the conditions for repurposing. In *Report* (Issue July).

Aditiya, H. B., & Aziz, M. (2021). Prospect of hydrogen energy in Asia-Pacific: A perspective review on techno-socio-economy nexus. *International Journal of Hydrogen Energy, 46*(71), 35027–35056. https://doi.org/10.1016/j.ijhydene.2021.08.070

Atilhan, S., Park, S., El-Halwagi, M. M., Atilhan, M., Moore, M., & Nielsen, R. B. (2021). Green hydrogen as an alternative fuel for the shipping industry. *Current Opinion in Chemical Engineering, 31*, 100668. https://doi.org/10.1016/j.coche.2020.100668

Balcombe, P., Speirs, J., Johnson, E., Martin, J., Brandon, N., & Hawkes, A. (2018). The carbon credentials of hydrogen gas networks and supply chains. *Renewable and Sustainable Energy Reviews, 91*(December 2017), 1077–1088. https://doi.org/10.1016/j.rser.2018.04.089

Bianco, E., Diab, S., & Blanco, H. (2022). *Green hydrogen for industry: A guide to policy making*. http://www.irena.org/publications

Bibas, R., Chateau, J., Dellink, R., & Mccarthy, A. (2017). Modelling the circular economy: Designing a global baseline of economic activity and material flows. In *20th annual conference on global economic analysis*. http://www.circle-economy.com/news/knowledge-map-circular-economy/

Blanco, H., & Taibi, E. (2022). *Global hydrogen trade to meet the 1.5 °C climate goal: Part I – Technology review of hydrogen carriers*. https://www.irena.org/-/media/Files/IRENA/Agency/Publication/2022/Apr/IRENA_Global_Trade_Hydrogen_2022.pdf

Bleischwitz, R., Spataru, C., VanDeveer, S. D., Obersteiner, M., van der Voet, E., Johnson, C., Andrews-Speed, P., Boersma, T., Hoff, H., & van Vuuren, D. P. (2018). Resource nexus perspectives towards the United Nations Sustainable Development Goals. *Nature Sustainability, 1*(12), 737–743. https://doi.org/10.1038/s41893-018-0173-2

Bonciu, F. (2020). The European Union Hydrogen Strategy as a significant step towards a circular economy. *Romanian Journal of European Affairs, 20*(2), 36–48.

Bouzarovski, S., Thomson, H., & Cornelis, M. (2021). Confronting energy poverty in Europe: A research and policy agenda. *Energies, 14*(4), 858. https://doi.org/10.3390/en14040858

Brah, A., & Phoenix, A. (2004). Ain't IA woman? Revisiting intersectionality. *Journal of International Women Studies, 5*(3), 75–86.

Bui, M., Adjiman, C. S., Bardow, A., Anthony, E. J., Boston, A., Brown, S., Fennell, P. S., Fuss, S., Galindo, A., Hackett, L. A., Hallett, J. P., Herzog, H. J., Jackson, G., Kemper, J., Krevor, S., Maitland, G. C., Matuszewski, M., Metcalfe, I. S., Petit, C., et al. (2018). Carbon capture and storage (CCS): The way forward. *Energy & Environmental Science, 11*(5), 1062–1176. https://doi.org/10.1039/C7EE02342A

Burke, M. J., & Stephens, J. C. (2017). Energy democracy: Goals and policy instruments for sociotechnical transitions. *Energy Research & Social Science, 33*, 35–48. https://doi.org/10.1016/j.erss.2017.09.024

Cerniauskas, S., Jose Chavez Junco, A., Grube, T., Robinius, M., & Stolten, D. (2020). Options of natural gas pipeline reassignment for hydrogen: Cost assessment for a Germany case study. *International Journal of Hydrogen Energy, 45*(21), 12095–12107. https://doi.org/10.1016/j.ijhydene.2020.02.121

Council, H. (2017). *How hydrogen empowers the energy transition*. Verfügbar Unter: https://Hydrogencouncil.Com/Wp-Content/Uploads/2017/06/Hydrogen-Council-Vision-Document.Pdf. Zugriff Am, 13(January), pp. 1–28.

Cuevas, F., Zhang, J., & Latroche, M. (2021). The vision of France, Germany, and the European Union on future hydrogen energy research and innovation. *Engineering, 7*(6), 715–718. https://doi.org/10.1016/j.eng.2021.04.010

Dehghanimadvar, M., Shirmohammadi, R., Sadeghzadeh, M., Aslani, A., & Ghasempour, R. (2020). Hydrogen production technologies: Attractiveness and future perspective. *International Journal of Energy Research, 44*(11), 8233–8254. https://doi.org/10.1002/er.5508

Dincer, I., & Rosen, M. A. (2011). Sustainability aspects of hydrogen and fuel cell systems. *Energy for Sustainable Development, 15*(2), 137–146. https://doi.org/10.1016/j.esd.2011.03.006

Ellen Macarthur Foundation. (2022). *What is circular economy?* https://ellenmacarthurfoundation.org/topics/circular-economy-introduction/overview

Energy Transitions Commission. (2021, April). Making the hydrogen economy possible: Accelerating clean hydrogen in an electrified economy. https://www.energy-transitions.org/publications/making-clean-hydrogen-possible/

Espegren, K., Damman, S., Pisciella, P., Graabak, I., & Tomasgard, A. (2021). The role of hydrogen in the transition from a petroleum economy to a low-carbon society. *International Journal of Hydrogen Energy, 46*(45), 23125–23138. https://doi.org/10.1016/j.ijhydene.2021.04.143

European Commission. (2012). *Manifesto for a resource-efficient Europe*. https://ec.europa.eu/commission/presscorner/detail/en/MEMO_12_989

European Commission. (2019). *The European green deal*. https://eur-lex.europa.eu/legal-content/EN/TXT/HTML/?uri=CELEX:52019DC0640&from=EN

European Commission. (2020). *European clean hydrogen alliance*. https://ec.europa.eu/growth/industry/policy/european-clean-hydrogen-alliance_en

European Commission. (2021). *Hydrogen public funding compass*. European Commission. https://single-market-economy.ec.europa.eu/industry/strategy/hydrogen/funding-guide_en#:~:text=Whatisit%3F,renewable and low carbon hydrogen

Falcone, P. M., Hiete, M., & Sapio, A. (2021). Hydrogen economy and sustainable development goals: Review and policy insights. *Current Opinion in Green and Sustainable Chemistry, 31*, 100506. https://doi.org/10.1016/j.cogsc.2021.100506

Fan, B. Y. Z., Ochu, E., Braverman, S., Lou, Y., Smith, G., Bhardwaj, A., Mccormick, C., & Friedmann, J. (2021). *Green hydrogen in a circular carbon economy: Opportunities and limits* (Issue August). www.sipa.columbia.edu

Fazioli, R., & Pantaleone, F. (2021). Macroeconomic factors influencing public policy strategies for blue and green hydrogen. *Energies, 14*(23), 7938. https://doi.org/10.3390/en14237938

Ferreira, A., Marques, P., Ribeiro, B., Assemany, P., de Mendonça, H. V., Barata, A., Oliveira, A. C., Reis, A., Pinheiro, H. M., & Gouveia, L. (2018). Combining biotechnology with circular bioeconomy: From poultry, swine, cattle, brewery, dairy and urban wastewaters to biohydrogen. *Environmental Research, 164*, 32–38. https://doi.org/10.1016/j.envres.2018.02.007

Guiteras, N. (2021). *The right to energy in the EU in times of pandemic*. https://righttoenergy.org/

Hoelzen, J., Silberhorn, D., Zill, T., Bensmann, B., & Hanke-Rauschenbach, R. (2022). Hydrogen-powered aviation and its reliance on green hydrogen infrastructure – Review and research gaps. *International Journal of Hydrogen Energy, 47*(5), 3108–3130. https://doi.org/10.1016/j.ijhydene.2021.10.239

Howarth, R. W., & Jacobson, M. Z. (2021). How green is blue hydrogen? *Energy Science & Engineering, 9*(10), 1676–1687. https://doi.org/10.1002/ese3.956

Hydrogen Council and McKinsey & Company. (2021). *Hydrogen for Net-Zero: A critical cost-competitive energy vector*. Hydrogen Council (Issue November). http://www.hydrogencouncil.com/

International Energy Agency. (2022). *Geopolitics of the energy transformation: The hydrogen factor*. IEA Publications. http://www.irena.org/publications

Jain, I. P. (2009). Hydrogen the fuel for 21st century. *International Journal of Hydrogen Energy, 34*(17), 7368–7378. https://doi.org/10.1016/j.ijhydene.2009.05.093

Jensen, J. P. (2019). Evaluating the environmental impacts of recycling wind turbines. *Wind Energy, 22*(2), 316–326. https://doi.org/10.1002/we.2287

Koelbl, B. S., van den Broek, M. A., van Ruijven, B. J., Faaij, A. P. C., & van Vuuren, D. P. (2014). Uncertainty in the deployment of Carbon Capture and Storage (CCS): A sensitivity analysis to techno-economic parameter uncertainty. *International Journal of Greenhouse Gas Control, 27*, 81–102. https://doi.org/10.1016/j.ijggc.2014.04.024

Lee, D.-H. (2016). Bio-based economies in Asia: Economic analysis of development of bio-based industry in China, India, Japan, Korea, Malaysia and Taiwan. *International Journal of Hydrogen Energy, 41*(7), 4333–4346. https://doi.org/10.1016/j.ijhydene.2015.10.048

Lo Faro, M., Cantane, D. A., & Naro, F. (2022). In the path for creating research-to-business new opportunities on green hydrogen between Italy and Brazil. *International Journal of Hydrogen Energy*. https://doi.org/10.1016/j.ijhydene.2022.05.089

Lubitz, W., & Tumas, W. (2007). Hydrogen: An overview. *Chemical Reviews, 107*(10), 3900–3903. https://doi.org/10.1021/cr050200z

Maggio, G., Nicita, A., & Squadrito, G. (2019). How the hydrogen production from RES could change energy and fuel markets: A review of recent literature. *International Journal of Hydrogen Energy, 44*(23), 11371–11384. https://doi.org/10.1016/j.ijhydene.2019.03.121

Marbán, G., & Valdés-Solís, T. (2007). Towards the hydrogen economy? *International Journal of Hydrogen Energy, 32*(12), 1625–1637. https://doi.org/10.1016/j.ijhydene.2006.12.017

Mazloomi, K., & Gomes, C. (2012). Hydrogen as an energy carrier: Prospects and challenges. *Renewable and Sustainable Energy Reviews, 16*(5), 3024–3033. https://doi.org/10.1016/j.rser.2012.02.028

MECA. (2020). Memorandum of understanding in the field of energy – Hydrogen – Between the Minister of Environment and Climate Action of the Portuguese Republic and the Minister of Economic Affairs and Climate Policy of the Netherlands.

Neuwirth, M., Fleiter, T., Manz, P., & Hofmann, R. (2022). The future potential hydrogen demand in energy-intensive industries – A site-specific approach applied to Germany. *Energy Conversion and Management, 252*, 115052. https://doi.org/10.1016/j.enconman.2021.115052

Nuñez-Jimenez, A., & De Blasio, N. (2022). *The future of renewable hydrogen in the European Union Market and geopolitical implications*. Belfer Center for Science and International Affairs (Issue March). www.belfercenter.org/ENRP

Ogden, J., Jaffe, A. M., Scheitrum, D., McDonald, Z., & Miller, M. (2018). Natural gas as a bridge to hydrogen transportation fuel: Insights from the literature. *Energy Policy, 115*, 317–329. https://doi.org/10.1016/j.enpol.2017.12.049

Pecorini, I., Ferrara, G., Galoppi, G., Baldi, F., Bacchi, D., Albini, E., Susini, C., Scappini, A., & Lombardi, L. (2017). Optimization of hydrogen and methane production in two-phase anaerobic digestion. *Procedia Environmental Science, Engineering and Management, 4*, 163–171.

Raza, A., Gholami, R., Rezaee, R., Rasouli, V., & Rabiei, M. (2019). Significant aspects of carbon capture and storage – A review. *Petroleum, 5*(4), 335–340. https://doi.org/10.1016/j.petlm.2018.12.007

Renewable Energy Agency, I. (2022). *Renewable capacity statistics 2022*. www.irena.org

REPowerEU. (2022). *Joint European action for more affordable, secure and sustainable energy*. E.U..

Saeedmanesh, A., Mac Kinnon, M. A., & Brouwer, J. (2018). Hydrogen is essential for sustainability. *Current Opinion in Electrochemistry, 12*, 166–181. https://doi.org/10.1016/j.coelec.2018.11.009

Sartbaeva, A., Kuznetsov, V. L., Wells, S. A., & Edwards, P. P. (2008). Hydrogen nexus in a sustainable energy future. *Energy & Environmental Science, 1*(1), 79. https://doi.org/10.1039/b810104n

Schönauer, A.-L., & Glanz, S. (2022). Hydrogen in future energy systems: Social acceptance of the technology and its large-scale infrastructure. *International Journal of Hydrogen Energy, 47*(24), 12251–12263. https://doi.org/10.1016/j.ijhydene.2021.05.160

Schulte, I. (2004). Issues affecting the acceptance of hydrogen fuel. *International Journal of Hydrogen Energy, 29*(7), 677–685. https://doi.org/10.1016/j.ijhydene.2003.09.006

Scovell, M. D. (2022). Explaining hydrogen energy technology acceptance: A critical review. *International Journal of Hydrogen Energy, 47*(19), 10441–10459. https://doi.org/10.1016/j.ijhydene.2022.01.099

Sharma, S., Basu, S., Shetti, N. P., & Aminabhavi, T. M. (2020). Waste-to-energy nexus for circular economy and environmental protection: Recent trends in hydrogen energy. *Science of the Total Environment, 713*, 136633. https://doi.org/10.1016/j.scitotenv.2020.136633

Shulga, R. N., Petrov, A. Y., & Putilova, I. V. (2020). The Arctic: Ecology and hydrogen energy. *International Journal of Hydrogen Energy, 45*(11), 7185–7198. https://doi.org/10.1016/j.ijhydene.2019.12.122

Song, C. (2009). Introduction to hydrogen and syngas production and purification technologies. In *Hydrogen and syngas production and purification technologies* (pp. 1–13). Wiley. https://doi.org/10.1002/9780470561256.ch1

Sunny, N., Mac Dowell, N., & Shah, N. (2020). What is needed to deliver carbon-neutral heat using hydrogen and CCS? *Energy & Environmental Science, 13*(11), 4204–4224. https://doi.org/10.1039/D0EE02016H

Tsanakas, J. A., Heide, A., Radavičius, T., Denafas, J., Lemaire, E., Wang, K., Poortmans, J., & Voroshazi, E. (2020). Towards a circular supply chain for PV modules: Review of today's challenges in PV recycling, refurbishment and re-certification. *Progress in Photovoltaics: Research and Applications, 28*(6), 454–464. https://doi.org/10.1002/pip.3193

Venkata Mohan, S., Chiranjeevi, P., Dahiya, S., & Naresh Kumar, A. (2018). Waste derived bioeconomy in India: A perspective. *New Biotechnology, 40*, 60–69. https://doi.org/10.1016/j.nbt.2017.06.006

Yáñez, M., Ortiz, A., Brunaud, B., Grossmann, I. E., & Ortiz, I. (2019). The use of optimization tools for the Hydrogen Circular Economy. *Computer Aided Chemical Engineering, 46*, 1777–1782. https://doi.org/10.1016/B978-0-12-818634-3.50297-6

Yue, M., Lambert, H., Pahon, E., Roche, R., Jemei, S., & Hissel, D. (2021). Hydrogen energy systems: A critical review of technologies, applications, trends and challenges. *Renewable and Sustainable Energy Reviews, 146*(June 2020), 111180. https://doi.org/10.1016/j.rser.2021.111180

Zhiznin, S. Z., Timokhov, V. M., & Gusev, A. L. (2020). Economic aspects of nuclear and hydrogen energy in the world and Russia. *International Journal of Hydrogen Energy, 45*(56), 31353–31366. https://doi.org/10.1016/j.ijhydene.2020.08.260

Zoback, M. D., & Gorelick, S. M. (2012). Earthquake triggering and large-scale geologic storage of carbon dioxide. *Proceedings of the National Academy of Sciences, 109*(26), 10164–10168. https://doi.org/10.1073/pnas.1202473109

Part IV

Information Technology & Cybersecurity

9 A Comparative Model for Female and Male Generation Y Individuals Regarding Their Interest and Intention to Use eBooks

Johannes-Hugo van Schalkwyk

9.1 Introduction

There has been an increase in the uptake of eBooks in the academic, consumer, and scientific sectors. There are many reasons for this, as eBooks offer immediacy of access, updating capability, and the ability to store multiple eBooks on one device, which increases portability (Harman, 2018; Romero-Otero et al., 2013). Moreover, consumers can save money when buying eBooks compared to printed books (Wilber, 2022; Martinez-Estrada & Conaway, 2012). These benefits also aid education by offering the above benefits and others, such as enriched media (Harman, 2018; Picciano, 2012). Furthermore, interactive elements are easier to engage with and provide an element of entertainment (Berry et al., 2010, pp. 32–33). However, despite eBook use increasing globally (Carr, 2013), many still prefer printed books to eBooks (Ang, 2021). Therefore, this study aims to ascertain attitudes towards eBooks to assess whether millennials have intention to use eBooks and to assess which factors contribute to this. Thus, to investigate eBooks and the adoption of eBook technology, the technology acceptance model (TAM) was used to analyse intentions. The following factors were deemed essential to examine eBooks, namely, *perceived ease of use* (PEoU) and *perceived usefulness* (PU) (Jin, 2014, p. 472). Moreover, other factors investigated were *subjective norms*, *consumer innovativeness*, *interest and intention to use* eBooks, and *self-reported use*. Several studies have been conducted regarding eBook adoption and these influencing factors (Jin, 2014; Martin & Quan-Haase, 2011; Nwagwu & Okafor, 2014; Van der Velde & Ernst, 2009).

Numerous benefits can be had when using eBooks for both casual use and educational use. Therefore, this study aims to investigate *interest and intention to use*

J.-H. van Schalkwyk (✉)
North West University, Potchefstroom, South Africa

© The Author(s), under exclusive license to Springer Nature Switzerland AG 2023

F. L. Almeida et al. (eds.), *Multidimensional Sustainability: Transitions and Convergences*, Springer Proceedings in Earth and Environmental Sciences,
https://doi.org/10.1007/978-3-031-24892-4_9

eBooks to assess how Generation Y individuals can be persuaded to make use of eBooks. Generation Y are considered those born between 1986 and 2005 (Markert, 2004).

Therefore, this study aims to assess possible structural models to appeal to female and male members of the population separately when promoting eBooks. The study investigates how to achieve a level of interest and intention to use eBooks that would sufficiently convince both females and males to make better use thereof. It is hypothesised that *self-reported use, subjective norms, customer innovativeness, perceived usefulness*, and perceived ease of use regarding eBooks affect the *interest and intention to use* eBooks among individuals. Moreover, it is hypothesised that a different model applies to female individuals than to male individuals. Thus, the following hypotheses apply:

$H_0$1 – Self-reported use does not affect interest and intention to use eBooks
$H_0$2 – Subjective norms does not affect interest and intention to use eBooks
$H_0$3 – Customer innovativeness does not affect interest and intention to use eBooks
$H_0$4 – Perceived usefulness does not affect interest and intention to use eBooks
$H_0$5 – Perceived ease of use does not affect interest and intention to use eBooks
$H_0$6 – There is no difference in the variables that affect female compared to male individuals' interest and intention to use eBooks

9.2 Literature Review

The literature section will broadly investigate what an eBook is and the current landscape of eBooks to clarify terminology and importance. Next, the variables studied are *perceived usefulness, perceived ease of use, subjective norms, self-reported use, consumer innovativeness,* and *interest and intention to use* eBooks.

9.2.1 Defining the eBook

An eBook is the evolution of print to a contemporary digital format. Therefore, eBooks can be accessed through a computer, phone, tablet, or e-reader. Print books are presented as books in electronic form, but enhanced due to their digital nature (Carden, 2008, p. 9). Thus, an eBook is an electronic book which is a digital file that contains text and possibly images and displays them on a screen (Attwell, 2022). However, many different documents are electronic and can be viewed on a screen, thus, to be classified as an eBook, it must have certain characteristics. The required characteristics are that it should not be editable and should resize itself to fit the screen on which it is read (Wahl, 2018).

9.2.2 The Current eBook Landscape

Despite eBooks being an advancement on age-old technology that is printed books, its uptake has been much slower than other digital technologies. However, some countries have progressed quicker in their uptake than others (Kristensen & Lüders, 2021). The availability of both internet connections, devices, and eBooks has increased the consumption of eBooks. However, eBook publishing has been skewed towards only a few countries (Ang, 2021; Wischenbart et al., 2014). In Portugal, 7.5% of people use eBooks, a percentage set to grow slowly each year. Therefore, it is predicted that there will be 0.8 million eBook users in Portugal by 2027 (Statista, 2022a). This is compared to Portugal's predicted population of 10.16 million by 2027 (Statista, 2022b). This shows that eBooks are far from mainstream adoption.

9.2.3 Perceived Usefulness

Perceived usefulness (PU) can be seen as the extent to which an individual believes a particular technology would enhance their actions or be helpful to them Davis (1989). Thus, it is the degree to which a person believes technology or a specific technology, can be useful in their personal or work life. The theory of *perceived usefulness* argues that an individual will use a specific technology if it leads to satisfactory results (Khayati & Zouaou, 2013).

9.2.4 Perceived Ease of Use

Perceived ease of use (PEoU) can efficiently determine whether a person will use a specific technology and whether that technology is compatible with the person's life and ability (Gerlach & Buxmann, 2013). PEoU is thus a subjective assertion of complexity, and more user-friendly technologies will be more likely to be used by a majority as opposed to those that are perceived as too complex (Martin & Quan-Haase, 2013).

9.2.5 Subjective Norms

Subjective norms (SN) gauges an individual's susceptibility to influences from their social environment and how this affects their behaviour (Moan & Rise, 2006). This social network might include friends, family, spouses, co-workers, and neighbours (Fowler & Christakis, 2008).

9.2.6 Self-Reported Use

Self-reported use (SRU) simply measured whether respondents made use of eBooks. The variable asked questions such as "I have used eBooks before"; "I use eBooks for personal reading"; "I have a strong tendency to use eBooks".

9.2.7 Consumer Innovativeness

Consumer innovativeness (CI) explores whether the individual prefers new technology. These individuals will adopt new technologies earlier. Moreover, they will have a greater appreciation for the functional, hedonic, and psychological advantages of eBooks and eBook readers (Antón et al., 2013).

9.2.8 Interest and Intention to Use

Interest and intention to use (IITU) as a variable measure whether a person wants to use a product or service. In terms of eBooks, this can be influenced by age, education, income, need for print, ownership of other digital media, and how innovative they are (Nwagwu & Okafor, 2013). *Interest and intention* were used as dependent variable as it gauges possible future behaviour, which fits into the decision-making process of the individual (Jung et al., 2012).

9.3 Methodology

The study was conducted on Generation Y individuals to assess their intention to use eBooks. A non-probability convenience sampling technique was used, using a self-administered questionnaire. The researcher was nearby to provide aid should there be any confusion regarding the questions. The data was collected as a cross-sectional study.

The questionnaire used was given to individuals to complete. A total of 370 questionnaires were usable after data cleaning. The research instrument focused on demographics such as country of origin, gender, age, and access to electronic equipment. The next section investigated six variables: *interest and intention to use* as the dependent variable, *self-reported use, perceived usefulness, perceived ease of use, subjective norms,* and *consumer innovativeness*. The following section discusses the results of the study.

9.4 Results

The results are divided into five sections, each containing a description, table and/or figure, and an explanation of the results. The first section highlights the demographics of the respondents. Next is a brief overview of factor fit through factor analysis, followed by a multicollinearity test. Proceeding the aforementioned are descriptive statistics for the various variables. The correlation analysis follows to test for nomological validity, and lastly, a structural model was constructed and analysed, followed by a proposed model for females and males.

9.4.1 Demographics

The demographics in the study enquired about age, country of origin, gender, and access to technology such PC, Laptop, eReader, Smartphone, Tablet, and 2-in-1 Hybrid. The following are the results from the analysis (Table 9.1).

The study shows that most of the respondents are from South Africa and are female, though the female/male difference is not significant. Age distribution shows that most individuals are between ages 20 and 23, which consists of 223 of the respondents. However, apart from 18 and those above 24, there is a reasonable spread among the ages. Lastly, most respondents show that they have access to laptops and smartphones. Less than half have access to PCs and tablets, and only a few have access to e-readers and 2-in-1 devices. Thus, this study, although having elements that could be extrapolated to larger populations, would be most accurate in individuals from South Africa, aged 20–23 years, both female and male, who have access to laptops and smartphones.

Table 9.1 Demographic data

Demographic	Item	Frequency	Per cent	Demographic	Item	Frequency	Per cent
Country	South Africa	347	93.8	Age	18	2	1.4
	Other	23	6.2		19	29	7.8
Gender	Female	202	54.6		20	65	17.6
	Male	166	44.9		21	76	20.5
	Other	2	0.6		22	82	22.2
Access	PC	129	34.9		23	59	15.9
	Laptop	253	68.4		24	34	9.2
	eReader	15	4.1		>24	15	4.1
	Smartphone	270	73		Missing	5	1.4
	Tablet	135	36.5				
	2 in 1 Hybrid	11	3				

Where a total is less than 370, there are missing entries; however, these entries do not change the significance of the data. The next section examines the factor loadings from the exploratory factor analysis.

9.4.2 Exploratory Factor Analysis

The exploratory factor analysis was conducted on interest and intention to use as the dependent variable, self-reported use, perceived usefulness, perceived ease of use, subjective norms, and consumer innovativeness. Next, coefficients below 0.4 were suppressed to ensure that the quality of item fit was high enough. The methodology used was Promax rotation, whereby six factors were selected, not according to eigenvalue, as literature had already given credence to five variables. The final variable, self-reported use was accepted due to sufficient values. First, Table 9.2 shows the KMO and Bartlett's test values.

Table 9.2 shows acceptable KMO (KMO = 0.916) and significance lower than 0.5 ($p = 0.000 < 0.05$). Due to the acceptable KMO and Bartlett's levels, the study continued with the exploratory factor analysis. The factor analysis also shows that all items fit into the expected variables. Moreover, factor loadings and communalities show high values, which means that the items fit well into their variables. Lastly, the variance explained by six factors is 69.37%. Next, the study tested for multicollinearity in the data.

9.4.3 Multicollinearity

Multicollinearity assesses whether data is too similar or too different to other variables in the dataset (Hwang & Takane, 2015). To test for multicollinearity, a test was conducted for the Condition Index, VIF, and tolerance of the data.

Table 9.3 shows acceptable levels of VIF and tolerance. The condition index above 10 could be considered high; however, as those variables tested for similar aspects, a higher condition index was expected and acceptable (Gaskin, 2011; Hair et al., 2014; Su, 2016).

Thus, all variables show acceptable CI, VIF, and tolerance levels. The next section briefly investigates the descriptive statistics of the study and focuses more on the variables and items used.

Table 9.2 KMO, chi-square, degrees of freedom, and significance

KMO	0.916
Chi-square	5231.948
Df	325
Sig.	0.000

Table 9.3 Condition index, VIF, and tolerance

Test for multicollinearity	Condition index	VIF	Tolerance
Self-reported use	6.406	1.423	0.703
Perceived usefulness	8.851	1.753	0.571
Perceived ease of use	9.547	2.039	0.491
Subjective norms	12.521	1.368	0.731
Consumer innovativeness	16.056	1.406	0.711

Table 9.4 Mean, standard deviation, and skewness

Variable	*N*	Mean	Standard deviation	Skewness
IITU	370	4.089	1.156	−0.560
SRU	370	2.766	1.358	0.442
PU	370	3.232	1.267	0.134
PEOU	370	4.071	0.996	−0.604
SN	370	3.235	1.128	−0.043
CI	370	4.005	1.116	−0.500

9.4.4 Descriptive Statistics

Descriptive statistics were used to analyse the data's mean, standard deviation, and skewness. This was used to assess and achieve a better understanding of the intricacies of respondent attitudes.

Table 9.4 shows a respondent population size of 370 for each variable. The mean ranges varied from 2.766 to 4.089 on a Likert scale with Likert items from 1 strongly disagree to 6 strongly agree. Analysing each variable indicates high *interest and intention to use* eBooks ($m = 4.089$); however, low actual usage is seen in *self-reported use* ($m = 2.766$). Next, respondents feel somewhat favourable towards the *usefulness* of eBooks ($m = 3.232$) and feel that others at least partly influence them in *subjective norms* ($m = 3.235$). Lastly, respondents indicate that eBooks would be *easy to use* ($m = 4.071$) and feel they are *innovative consumers* ($m = 4.005$). There are signs of deviations, each of which show a standard deviation of at least 1.116, except for PEoU, which shows 0.996. Skewness shows acceptable values in that each value is between 1 and −1. Here, the most interesting part is in *self-reported use*, which has a positive skew. This means that many respondents have even lower *self-reported use* of eBooks than indicated. Despite this fact, the relatively high mean of respondents showing *interest and intention to use* shows promise for this medium.

A few items that stand out amidst the items in the variables are "I want to experience eBooks (m = 4.34)"; "It will be easy to learn how to use eBooks (m = 4.36)"; "I enjoy trying out new communication technologies that come onto the market (m = 4.43)". Next, the study tested for reliability using Cronbach's alpha.

9.4.5 Cronbach's Alpha

Cronbach's alpha was used to test the variables and items' internal reliability. A value of at least 0.7 was sought for each variable (Table 9.5).

The Cronbach's alpha values for each of the variables (testing for the items within) show sufficient values ($a > 0.7$). The lowest Cronbach's alpha value is 0.772. This indicates favourable internal consistency, which means that the study's items and constructs are valid and that the study is reliable. The next section briefly touches on the correlation analysis conducted.

9.4.6 Correlation Analysis

A Pearson's correlation analysis was conducted to ensure that the variables correlated and test nomological validity (Table 9.6).

The Pearson's correlation analysis shows that each variable is correlated and is significant at the 0.01 level, 2-tailed. Thus, each variable influences one another in the expected, positive direction. This indicates that there is nomological validity. Thus, the study's variables are reliable and valid as per Cronbach's alpha and Pearson's correlation analysis. The next section shows the conceptual structural model and individual models for male and female respondents.

Table 9.5 Cronbach's alpha, reliability analysis

Variable	Items in variable	Cronbach's alpha value
IITU	2–6	0.870
SRU	7–10	0.833
PU	11–18	0.917
PEOU	19–23	0.789
SN	24–27	0.772
CI	28–31	0.786

Table 9.6 Example of a table legend

	IITUvar	SRUvar	PUvar	PEOUvar	SNvar	CIvar
IITU	1					
SRU	0.443**	1				
PU	0.562**	0.499**	1			
PEOU	0.559**	0.435**	0.592**	1		
SN	0.398**	0.343**	0.375**	0.474**	1	
CI	0.342**	0.198**	0.355**	0.516**	0.354**	1

Correlation is significant at the 0.01 level (2-tailed)

9.4.7 Structural Model - Overall

The conceptual structural model for the study can be seen in Fig. 9.1 below. This is the overall model that applies to all respondents. Following Fig. 9.1 are model fit and regression weights. After that are comparison models for male and female respondents.

The next section discusses the proposed model for the study and investigates two separate models based on gender.

The conceptual model shows *self-reported use, subjective norms,* and *consumer innovativeness* as independent variables. The independent variables are mediated by *perceived usefulness and perceived ease of use* to the dependent variable *interest and intention to us* (Table 9.7).

Model fit, estimates, and significance are all adequate. Thus, one can ascertain that the model proposed is viable. The model outcomes will be briefly examined and

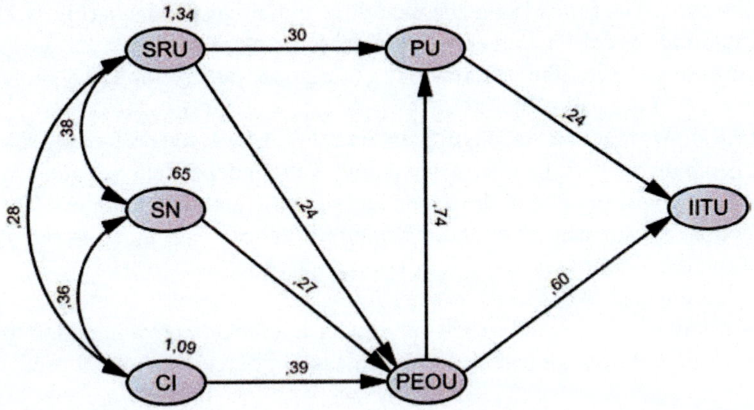

Fig. 9.1 Conceptual model

Table 9.7 Model fit and regression weights

Model fit				
IFI	**TLI**	**CFI**	**RMSEA**	**SRMR**
0.917	0.905	0.916	0.065	0.0586
Regression weights				
Variable	**Variable**	**Estimate**	**C.R**	**Sig.**
PEOU <	**CI**	0.387	7.219	< 0.005
PEOU <	**SRU**	0.235	5.346	< 0.005
PEOU <	**SN**	0.272	3.865	< 0.005
PU <	**SRU**	0.300	4.774	< 0.005
PU <	**PEOU**	0.740	8.309	< 0.005
IITU <	**PU**	0.240	3.677	< 0.005
IITU <	**PEOU**	0.599	6.041	< 0.005

discussed in more detail in the discussion section. The model shows that *self-reported use* does not directly lead *to interest and intention to use* but is mediated by perceived usefulness. *Self-reported* use is also mediated through *perceived ease of use*, which indicates that creating easy to use applications and e-readers should be prioritised. Next, *subjective norms* is mediated through *perceived ease of use*, not *perceived usefulness*. *Consumer innovativeness* is similarly only influenced by *perceived ease of use* and not *perceived usefulness*. Moreover, there appears to be a connection between *perceived ease of use* and *perceived usefulness*. This indicates that higher levels of *perceived ease of use* would lead to eBooks appearing more useful. Lastly, *perceived usefulness* and *perceived ease of use* are directly correlated to *interest and intention to use*.

9.4.8 Structural Model – Females

The model was compared between female and male respondents, which yielded the following results. The following is the model for female respondents (Fig. 9.2).

The structural model for females is similar to the main conceptual model, with the only difference being the difference in correlation values. Table 9.8 shows the values for each of the paths.

Table 9.8 shows that there is a significant correlation between each of the selected paths. This means that with an increase in any of the independent variables, a correlative increase will be seen in the dependent variable and the mediators. The following section investigates the structural mode for males.

Therefore, the hypotheses are as follows:

$H_0 1$ – The alternate is accepted as self-reported use affects interest intention to use, though mediated through perceived usefulness and perceived ease of use.

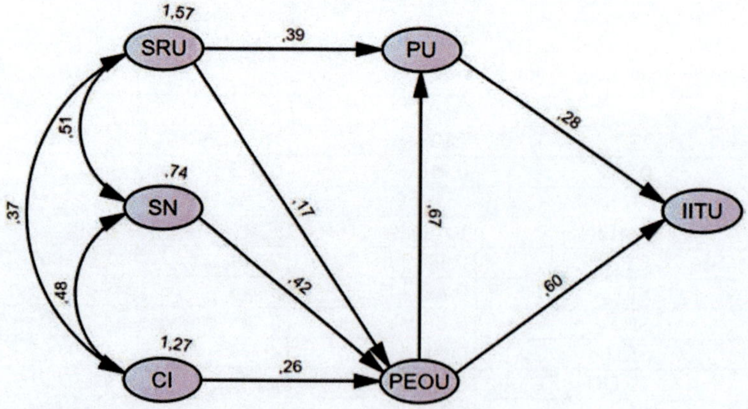

Fig. 9.2 Structural model – females

Table 9.8 Regression weights for female structural model

Regression weights				
Variable	Variable	Estimate	C.R	Sig.
PEOU <	CI	0.261	4.281	< 0.005
PEOU <	SRU	0.165	3.096	**0.002**
PEOU <	SN	0.418	4.166	< 0.005
PU <	SRU	0.385	4.653	< 0.005
PU <	PEOU	0.668	5.523	< 0.005
IITU <	PU	0.281	3.690	< 0.005
IITU <	PEOU	0.599	4.960	< 0.005

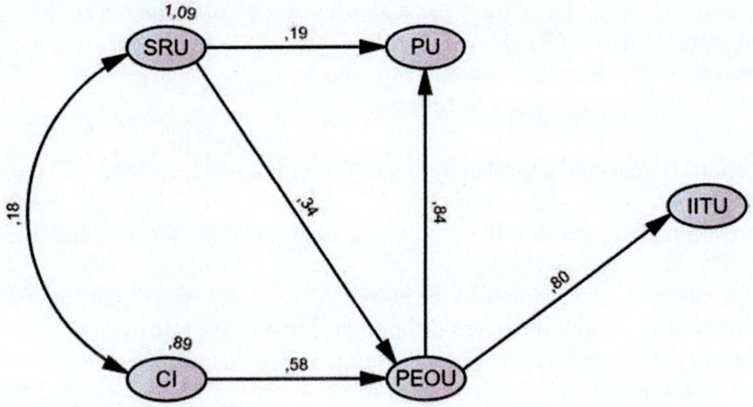

Fig. 9.3 Structural model – males

$H_0$2 – The alternate is accepted as subjective norms affects interest and intention to use, though mediated through perceived ease of use.

$H_0$3 – The alternate is accepted as customer innovativeness affects interest and intention to use, though mediated through perceived ease of use.

$H_0$4 – The alternate is accepted as perceived usefulness does affect interest and intention to use.

$H_0$5 – The alternate is accepted as perceived ease of use does affect interest and intention to use.

9.4.9 Structural Model – Males

The following model shows the path for males for each applicable variable (Fig. 9.3).

The path for males compared to females and the overall conceptual model differ. For example, *subjective norms* do not appear to be significant for male respondents and paths also differ. The following table shows the path results as well as the significance level for each of the paths.

Table 9.9 Regression weights for male structural model

Regression weights				
Variable	Variable	Estimate	C.R	Sig.
PEOU <	CI	0.582	5.810	< 0.005
PEOU <	SRU	0.336	4.510	< 0.005
PU <	SRU	0.187	2.094	**0.036**
IITU <	PEOU	0.797	6.269	< 0.005
PU <	PEOU	0.837	6.359	< 0.005

Table 9.9 shows that there are significant correlations between each of the paths. A robust correlation can be seen between *consumer innovativeness* and *perceived ease of use* (0.58); *perceived ease of use* and *interest and intention to use* (0.80); and *perceived ease of use* and *perceived usefulness* (0.84). A discussion follows in which the results will be explained and expanded.

Therefore, the hypotheses are as follows:

$H_0$1 – The null is accepted as self-reported use does not affect interest and intention to use eBooks.

$H_0$2 – The null is accepted as subjective norms does not affect interest and intention to use eBooks.

$H_0$3 – The alternate is accepted as customer innovativeness affects interest and intention to use, though mediated through perceived ease of use.

$H_0$4 – The null is accepted as perceived usefulness does not affect interest and intention to use eBooks.

$H_0$5 – The alternate is accepted as perceived ease of use does affect interest and intention to use.

$H_0$6 – The alternate is accepted as there is a difference in the structural model comparing female and male individuals' interest and intention to use eBooks.

9.5 Discussion

From the statistics shown, there appear to be optimism towards eBooks; however, organisations such as publishers, retailers, and academic institutions still have much work to do to encourage eBook usage. *Self-reported use*, which assesses actual usage, shows that very few respondents have used eBooks, and the positive skew shows that the data leans towards fewer having used them ($m = 2.766$; skewness = 0.442). This might be explained by the fact that respondents indicated that they partially *perceive usefulness* in using eBooks ($m = 3.232$). Moreover, the data suggests respondents do not have many people around them who make use eBooks or who feel positive about eBooks (*subjective norms* $m = 3.235$). The positive side here is that respondents indicated that they perceive themselves as *innovative consumers* (4.005) and innovative consumers tend to adopt newer technologies faster (Stock et al., 2015). Thus, those in the respondent cohort will adopt eBook technology sooner when it is shown as a useful and compelling tool. Moreover, respondents indicated they are very *interested* in eBooks and *intended to use* them in the future ($m = 4.089$). Therefore, when generalising the population subset, it can be extrapolated that many Generation Y individuals have not made use of eBooks but perceive them as useful and want to use them in the future. The main objective would be to illustrate the *usefulness* of eBooks to convince individuals to adopt them.

The conceptual model, which was validated, shows that *self-reported use* influence both *perceived usefulness* and *perceived ease of use*. Therefore, providing eBooks to university students or showcasing e-readers in stores might be helpful in showing how useful they can be and how easy they are to use. Next, *subjective norms* only influence *perceived ease of use*, which indicates that when those around a person feel eBooks are an excellent tool to use, the person will perceive it as *easy to use*. Therefore, when more people use eBooks, it could create a trend of others using it to a greater extent as well. This only adds to the fact that there should be the encouragement of use to bring this technology to the forefront. Next, *consumer innovativeness* only correlates with *perceived ease of use*. Therefore, the more *innovative* an individual, the easier the technology would seem, which aligns with literature (Fagan et al., 2012). *Perceived ease of use* influences whether the respondents perceive that technology is useful. Therefore, easy to use technologies would be deemed more useful than those that are complex or difficult to learn. This would suggest that eBooks should be simplified for most users to ensure that it is easy to use. There could be options for power users; however, due to the correlation between *perceived ease of use* and *perceived usefulness*, it appears essential that a technology is easy to use for it to be adopted. Lastly, *perceived usefulness* and *perceived ease of use* correlate to *interest and intention to use*. Therefore, there should be a focus on both of those aspects. *Perceived ease of use* has the strongest correlation to *interest and intention to use*, indicating that one of the most important aspects of getting individuals to read eBooks is to make them simple to use. The structure for female students follows the conceptual model closely with minor differences in correlations. The structural model for male Generation Y individuals has several detractions from the main model. Male individuals do not perceive themselves as being influenced by those around them, meaning they do not deem the influence of friends, family, and peers as influential regarding eBooks. Next, although eBooks' *perceived ease of use* influences whether they are deemed useful, the *perceived usefulness* does not affect *interest and intention to use*. The correlation between *perceived ease of use* and *interest and intention to use* is robust. Therefore, when male Generation Y individuals are convinced that eBooks are easy to use, their interest will increase. Thus, the argument for practically showcasing the ease of use of eBooks in retailers and academic institutions might play an important role in increasing eBook use.

9.6 Managerial and Organisational Implications

The usefulness of eBooks can be seen in multiple instances, such as the ease of finding eBooks, ease of storing eBooks, ease of reading eBooks, and the ease of adding extra information to eBooks such as highlights or comments. This can be used both by casual readers as well as those who are in academia, be it teachers or students. It is thus important for those in the casual and academic industries to realise that Generation Y individuals, who have the means to buy eBooks, are innovative and want to use new technologies. It can thus be seen as a shortcoming of current marketing strategies to not showcase how easy it is to use eBooks. Therefore, organisations should take heed that there is a large market of consumers interested in using eBooks; however, they must be convinced. This can perhaps be done by practically showing e-readers in the store where employees can help people find the best e-reader for them and show them how to use them. Moreover, bookstores can accept that the world is orienting towards digital and partner with larger publishers to sell eBooks in-store by simply letting a person walk in, buy it, and upload it to their e-reader. Bookstores can also help people to find e-readers and eBooks and help them use them. Lastly, academic institutions should encourage students to use eBooks to save time and money. This will allow teachers, lecturers, and professors to add information to eBooks, such as hyperlinks, comments, and other contextual information. Students showed better performance in academia when they were given e-readers, locked to the software used for their classes (Ebied & Rahman, 2015). Therefore, organisations should use this data to drive the industry towards its natural evolution, which is eBooks.

9.7 Conclusions

Internet penetration has increased tremendously globally, even in developing nations, which have a history of shortcomings in education. There is still a lot to be done in this regard, but eBooks show the potential to alleviate many problems. However, a concerted effort must be made to let individuals experience eBooks and e-readers for them to make the decision to make use of them.

This study observes the following limitations. The study was conducted in South Africa and can be extended to other countries to compare models and findings. Moreover, the age group was Generation Y individuals between the ages of 18 and 24. This can be extended to assess other age groups. Despite these limitations, the study shows that there is a clear difference in how female and male individuals perceive eBooks. This is important to note as both of these groups stated that eBooks are indeed important, however, convincing them to use this medium varies greatly. It has been shown that the importance of eBooks cannot be understated, which, in world that is advancing towards digital at accelerated speeds, must be accepted. Acceptance thereof can lead to many benefits in various areas such as providing better education, providing education to remote areas, providing up to date information, among a myriad of other advantages presented by eBooks and its digital nature.

However, organisations must take heed of how to appeal to various audiences to convince them to bring about the eBooks revolution that has been postponed for too long, possibly to the detriment of environmental and educational sectors.

References

Ang, C. (2021). *Visual capitalist. Print has prevailed: The staying power of physical books.* https://www.visualcapitalist.com/print-books-versus-e-books

Antón, C., Camarero, C., & Rodríguez, J. (2013). Usefulness, enjoyment, and self-image congruence: The adoption of e-book readers. *Psychology & Marketing, 30*(4), 372–384.

Attwell, A. (2022). *Britannica.* e-book. https://www.britannica.com/technology/e-book

Berry, T., Cook, L., Hill, N., & Stevens, K. (2010). An exploratory analysis of textbook usage and study habits: Misperceptions and barriers to success. *College Teaching, 59*(1), 31–39.

Carden, M. T. (2008). E-books are not books. In *Proceedings of the 2008 ACM workshop on research advances in large digital book repositories* (pp. 9–12). Ingram Digital.

Carr, N. (2013). *Roughtype. The flattening of e-book sales.* http://www.roughtype.com/?p=3590

Davis, F. D. (1989). Perceived usefulness, perceived ease of use, and user acceptance of information technology. *MIS Quarterly, 13*(3), 319–340.

Ebied, M. M. A., & Rahman, S. A. A. (2015). The effect of interactive e-book on students' achievement at Najran University in computer in education course. *Journal of Education and Practice, 6*(19), 71–82.

Fagan, M., Kilmon, C., & Pandey, V. (2012). Exploring the adoption of a virtual reality simulation: The role of perceived ease of use, perceived usefulness and personal innovativeness. *Campus-Wide Information Systems, 29*(2), 117–127.

Fowler, J. H., & Christakis, N. A. (2008). Dynamic spread of happiness in a large social network: Longitudinal analysis over 20 years in the Framingham Heart Study. *British Medical Journal, 1*(337), 1–9.

Gaskin, J. (2011). *YouTube. Detecting multicollinearity in SPSS.* https://www.youtube.com/watch?v=oPXjQCtyoG0

Gerlach, J., & Buxmann, P. (2013). Analyzing electronic book acceptance: A compatibility perspective. In *System sciences (HICSS): Hawaii international conference, Hawaii* (pp. 2813–2822). IEEE.

Hair, J. F., Hult, G. T. M., Ringle, C. M., & Sarstedt, M. (2014). *A primer on partial least squares structural equation modeling (PLS – SEM)* (1st ed.). Sage.

Harman, M. (2018). *Kitaboo. Top 10 advantages of eBooks over printed books.* https://kitaboo.com/top-10-advantages-of-ebooks-over-printed-books

Hwang, H., & Takane, Y. (2015). *Generalized structured component analysis* (1st ed.). CRC Press.

Jin, C. H. (2014). Adoption of e-book among college students: The perspective of an integrated TAM. *Computers in Human Behavior, 41*(1), 471–477.

Jung, J., Chan-Olmsted, S., Park, B., & Kim, Y. (2012). Factors affecting e-book reader awareness, interest, and intention to use. *New Media & Society, 14*(2), 204–224.

Khayati, S., & Zouaou, S. K. (2013). Perceived usefulness and use of information technology: The moderating influences of the dependence of a subcontractor towards his contractor. *Journal of Knowledge Management, Economics and Information Technology, 3*(6), 1–28.

Kristensen, L. B., & Lüders, M. (2021). Convenient and worth the price? Identifying early users and predicting future use of book streaming services. *Convergence, 0*(0), 1–18.

Markert, J. (2004). Demographics of age: Generational and cohort confusion. *Journal of Current Issues and Research in Advertising, 26*(2), 11–25.

Martin, K., & Quan-Haase, A. (2011). Seeking knowledge: An exploratory study of the role of social networks in the adoption of eBooks by historians. *Proceedings of the American Society for Information Science and Technology, 48*(1), 1–10.

Martin, K., & Quan-Haase, A. (2013). Are e-books replacing print books? Tradition, serendipity, and opportunity in the adoption and use of e-books for historical research and teaching. *Journal of the American Society for Information Science and Technology, 64*(5), 1016–1028.

Martinez-Estrada, P. D., & Conaway, R. N. (2012). EBooks: The next step in educational innovation. *Business Communication Quarterly, 75*(2), 125–135.

Moan, I. S. V., & Rise, J. (2006). Predicting smoking reduction among adolescents using an extended version of the theory of planned behaviour. *Psychology and Health, 21*(6), 717–738.

Nwagwu, W. E., & Okafor, J. L. (2014). Diffusion of eBooks among postgraduate students of the University of Ibadan, Nigeria. *Library Review, 63*(1/2), 86–109.

Picciano, A. G. (2012). The evolution of big data and learning analytics in American higher education. *Journal of Asynchronous Learning Networks, 16*(3), 9–20.

Romero-Otero, I., Iglesias-Fernández, E., & Giménez-Toledo, E. (2013). *Use, acceptance and expectations for the eBook in a research library.* http://bid.ub.edu/en/31/romero3.htm

Statista. (2022a). *Statista. eBooks Portugal.* https://www.statista.com/outlook/dmo/digital-media/epublishing/ebooks/portugal

Statista. (2022b). *Statista. Total population of Portugal 2027.* https://www.statista.com/statistics/372162/total-population-of-portugal/

Stock, R. M., Oliveira, P., & von Hippel, E. (2015). Impacts of hedonic and utilitarian user motives on the innovativeness of user-developed solutions. *Journal of Product Innovation Management, 32*(3), 389–403.

Su, X. (2016). *Zapdoc. Model diagnostics.* https://zapdoc.tips/stat-4385-topic-06-model-diagnostics.html

Van der Velde, W., & Ernst, O. (2009). The future of eBooks? Will print disappear? An end-user perspective. *Library Hi Tech, 27*(4), 570–583.

Wahl, J. (2018). *Learng2. What is an eBook? Understanding why they work and how to make your own.* https://learn.g2.com/what-is-an-ebook#what-is-an-ebook

Wilber, J. (2022). *Owlcation. 10 reasons why eBooks are better than paper books.* https://owlcation.com/humanities/10-Reasons-Why-eBooks-are-Better-than-Print-Books

Wischenbart, R., Celaya, J., Carrenho, C., Kovac, M., Mallya, V., & Licher, V. (2014). *Wischenbart. The global eBook report: A report on market trends and developments.* http://www.wischenbart.com/upload/1234000000358_04042014_final.pdf

10 Gamification on Cybersecurity Literacy: Social Sustainability and Educative Projects

Jorge Simões, Justino Lourenço, Sérgio Sargo, and José Carlos Morais

10.1 Introduction

The pandemic stress caused by COVID-19 creates a sudden need for several changes over the daily work routine and specifically in the lecturing playground. The classical classrooms in the university had to quickly move from ordinary presence classes to remote sessions with the support of videoconference software. This sudden shift had to occur in a short period. It happened perhaps without the correct preparation of all those involved, in terms of best practices related to the communications software use and all the panoply of cybersecurity issues that came with those practices.

At the same time, and even after the minimization of the pandemic restrictions, it is clear that the remote communications solutions adopted will stay relevant as enablers of work and lecturing efficiency. This new thinking in the learning environment could be enriched with the lever of gamification. Several relevant

J. Simões (✉)
ISPGAYA, Vila Nova de Gaia, Portugal

INESC-TEC, Porto, Portugal
e-mail: jsimoes@ispgaya.pt

J. Lourenço
ISPGAYA, Vila Nova de Gaia, Portugal

S. Sargo
ISPGAYA, Vila Nova de Gaia, Portugal

Distance Education and E-learning Laboratory – LE@D – Universidade Aberta, Lisbon, Portugal

J. C. Morais
ISPGAYA, Vila Nova de Gaia, Portugal

CEOS.PP, Lisbon, Portugal

© The Author(s), under exclusive license to Springer Nature Switzerland AG 2023

F. L. Almeida et al. (eds.), *Multidimensional Sustainability: Transitions and Convergences*, Springer Proceedings in Earth and Environmental Sciences,
https://doi.org/10.1007/978-3-031-24892-4_10

state-of-the-art articles have shown the clear relevance of the new technologies and processes in the increasing efficiency of the learning ecosystem (King, 2004; Lourenço et al., 2022; Morais et al., 2022; Pearshouse & Sharples, 2000).

If we can link the theme of cybersecurity to that of sustainability through the minimization of associated costs to cyberthreats (Sadik et al., 2020), assuming a financial perspective, both themes are brought together when cybersecurity is linked to the notion of insecurity or uncertainty (Lewallen, 2021) that characterize current societies. The promotion of sustainability in its different meanings should integrate the governance strategy in the educational, scientific, and cultural projects of higher education institutions (HEIs) is about to be mandatory, according to the Institutional Evaluation Manual of the Portuguese evaluation agency on higher education A3ES (2022). We pursue in this study an approach of commitment of the institutions involved in the purposes of sustainability also socially as a path of the institutions' strategy, contributing to a social general access to knowledge in the area of cybersecurity literacy, placing the theme on the horizon of the curricular contents offered to the students.

Within this study, it is intended to answer the questions:

Are students aware of topics such as gamification or cybersecurity?
Are students aware of the potential of recurring to gamification to cybersecurity literacy purposes?

10.2 Gamification: Concepts and Technologies

A strategical introduction of a gaming scenario in a nongaming environment, materialized by solutions such as websites, online communities, learning management systems, or business intranets, aims to increase the participants' motivation, promoting efficiency in the process (González et al., 2021; Lu et al., 2021; Roosta et al., 2016; Trinidad García et al., 2021; Vanduhe et al., 2020).

Gamification, a concept known since 2010 (Dichev & Dicheva, 2017; Krath et al., 2021), is usually defined as the use of game design elements in nongame contexts (Deterding et al., 2011). Gamification uses digital games' features like narrative, feedback, rewards systems, conflict, cooperation, competition, clear objectives, experience points, levels, progressive disclosure of contents, among others. Gamification studies focused on how to use these games' features in other contexts, not directly associated with games (Khaleel et al., 2016; Sailer et al., 2017).

Education has been one of the top nongame contexts of gamification research (Bozkurt & Durak, 2018; Sailer & Homner, 2020; Trinidad et al., 2021). It can be seen as the use of game features in learning environments. Those learning environments often have the support of digital platforms and tools. The purpose of gamified learning environments is to foster students' motivation and engagement by changing their behaviors regarding learning activities, especially those that might not be appealing enough otherwise. Gamification in education should not be confused with game-based learning and the use of serious games. Although these approaches share common elements, gamification does not use fully fledged games but just elements and techniques extracted from games.

Recent literature reviews (Manzano-León et al., 2021; Trinidad et al., 2021) showed that gamification has been applied in different learning settings and at different learning levels. Gamification proved to be a valid learning strategy at different educational levels with a positive impact on students' academic performance, engagement, and motivation. Also, a meta-analysis conducted by Sailer and Homner (2020) found that gamified learning can be an effective approach, but which factors contribute most to successful gamification are still unclear. The academic community is trying to establish gamification as a scientific research discipline, with contributions from other knowledge areas like Psychology, Education Sciences, and Information and Communication Technologies (Trinidad et al., 2021).

The popularity of gamification has increased over the recent years and has encouraged its application in a vast diversity of applications domains such as health, business, society, tourism, and, in particular, education purposes. On the other hand, this gamification push brought a panoply of theatrical, empirical, and technological challenges that urge to be overcome. In terms of technological challenges, urges the need for an adequate set of software tools and new technologic approaches to support and stimulate new gamification tools that bring flexibility, adequate support in the gamification design, proper activities monitoring, and extend the range of game elements besides the conventional score points, badges, and leaderboards. All these topics seek one main purpose: to get faster and more efficient development tools to support rich immersive lecturing environments (Trinidad et al., 2021).

Next are presented new and trendy technological approaches for the technological gamification (expected) evolution.

A gaming solution that could dynamically adapt to the particular learning curve of a student – would promote a more effective Gamification application to a learning scenario. An example is the application as an auxiliary tool for an Artificial Intelligence (AI) course (Draskovic, 2019; Reyes et al., 2020). A tool that could, in an automated form, generate a personalized gamified learning system. A bibliographic source suggests the relevance of the following concepts in gamification: fun, motivation, autonomy, progressiveness, feedback, error tolerance, experimentation, creativity, and adaptation to the specific case (Llorens-Largo et al., 2016). The concept of adaptive gaming cycle is also applied in helping mobile health applications by encouraging user engagement (Llorens-Largo et al., 2016). A different approach is supported by ontological structure to represent gamified collaborative learning (CL) scenarios and demonstrate the utility of this approach in the generation of personalized conceptual models to gamify CL scenario based on students' needs and individual characteristics (Challco et al., 2015).

There is a strong relevance in developing a rich learning environment that truly engages the student into an immersive learning experience. For this purpose, the resources of augmented reality (AR) and virtual reality (VR) can be the success key (Jiang & Zeng, 2019). A much richer experience with a gamified platform could be accomplished by gesture recognition (GR), as proposed in Ekneling et al. (2018) enabling the enhancement of data collection and data annotation threw gamification. The article describes a hand tracking and a GR with the support of an AR and VR application. On the other hand, the Google's AR game called ingress can be used to inform the development of learning from games (Sheng, 2013). To have the support of AR, VR and immersive 3D projections can enrich the students visit to a museum (Pantile et al., 2016).

In some scientific areas, the chance to have an almost physical sense of a form or even an object will enrich the learner experience: astronomic concepts are one possible example (Patricio et al., 2019), an electrochemistry course can also be an application example (Chen & Liao, 2015) or teaching mobile X-ray imaging (Süncksen et al., 2018). The learning experience of new languages will also benefit from the use of AR technologies (Zhenming et al., 2017). At last, all these VR and AR in context of gamifications will leave a major contribution for learners with disabilities (Bozgeyikli et al., 2014; Maidenbaum & Amedi, 2015).

The technology blockchain (BC) is a new trend and with rising popularity. The BC popularity has extended its application beyond a digital currency: from finance, medical, digital marketplace, pharmaceutical, and government agencies. Its relevance is supported by trust, transparency, and integrity, without the need of third-party support entity (Parizi & Dehghantanha, 2018). A presented approach enables the discovering of Da Vinci threw a novel AR that incorporates BC with experiential learning to engage participants in an interactive discovery of Leonardo da Vinci's ouvre (Suvajdzic et al., 2020).

The Internet-of-things (IoT) approach enables in an efficient way build a panoply of ordinary devices that will gain connectivity. Any relevant object in the learning process can be used as a sensor or as an actuator. Besides immersive visual experiences described in the previous sections, with IoT all the objects in the classroom can be used to measure or directly impact the leaner experience.

10.3 Cybersecurity Concerns

A gamified system as learning tool, as described in previous sections, has great potential. At same time all the information flow, processing and storage should keep the integrity and take all the measures to promote the user's privacy. So, a gamification solution should take into consideration the cybersecurity risks (Yonemura et al., 2017; Sharif & Ameen, 2020). On the other hand, gamification is a tool for the lecturing of security awareness (Diakoumakos et al., 2021; Kumar et al., 2017; Martin et al., 2019; Nguyen & Pham, 2020; Raval et al., 2018; Yonemura et al., 2018a, b).

The following theoretical construct details some common issues related to the security of a gamified system. First, we refer the issue of login and password security. The authentication through a login and a password is the most common access procedure to join a gamified tool. If goof security practices are not implemented, this initial procedure is a significant point of failure. A combination of an unsecured password, without an adequate aging mechanism implementation or even the access credentials shared with a co-worker present a serious security risk. Also, of strong relevance the risk of a remote access without the support of a secured connection such as a virtual private network (VPN). A multiple-step authentication is an important step introduced to bypass these vulnerabilities. Several solutions ate proposed to surpass the described vulnerabilities, those include the support of machine learning techniques (Djosic et al., 2020; Misbahuddin et al., 2017), secure login solutions (Waheed et al., 2016), or the support of BC technology (Biswas et al., 2021).

About software vulnerabilities and updates, we can state that the layer of software is also of extreme relevance. All the data input, processing, and output in a usable interface are the result of a secure code execution. Any coding flaw may be the open door for a security issue, and it is an important point of research (Iannone et al., 2022). In particular, over the communication management software module – where all the information flows from and into the communication path: wired or wireless, each of it with its particular vulnerabilities.

There are several proposed solutions: the development of software vulnerability prediction model. This solution enables to forecast whether a software module is vulnerable or not, this way promoting a relevant tool for the security improvement (Shamal et al., 2017). Using automated vulnerabilities detection is also a proposal. With a new approach, a compressive experimental setup is created for accessing the methodology and report the vulnerabilities found (Ghosh et al., 1998; Visalli et al., 2019). The Static Code Analysis (SCA) has, also, a good detection rate and is the central technique for improving the effectiveness of vulnerability detection (Pereira, 2020). As a last bibliographic reference is from the support of machine learning solutions, at last, as relevant tool for the analysis of software vulnerabilities (Peerzada & Kumar, 2021). A good security policy includes the constant software updating is also a simple form to improve software productivity and reduce the menace of software vulnerabilities.

10.4 Methodology

This study implements a predominantly quantitative methodological approach, through the application of questionnaire survey as a research tool, with a set of questions that address two distinct areas, i.e. (i) learning through gamification in online environments and (ii) cybersecurity practices, in a sample that is composed of higher education students from different areas of knowledge academic. The study aims to understand the perceptions of students about active methodologies, in the specific case of gamification in digital learning environments, and the consequent attitude of students regarding cybersecurity practices in these same digital environments.

10.4.1 Statistical Analysis

Statistical analysis involved descriptive statistical measures (absolute and relative frequencies, means, and their standard deviations) and inferential statistics. Exploratory factor analysis (EFS), Cronbach's internal consistency coefficient alpha, Pearson correlation coefficient, and Kruskal–Wallis test were used. The normality of distribution of the variables was analyzed with the Shapiro–Wilk test, and the homogeneity of variances was analyzed with the Levene test. Statistical analysis was performed with the software SPSS (Statistical Package for the Social Sciences) in version 28.

The study included 95 higher education students from two HEIs in Portugal (see Table 10.1). Most students are male (87.4%) and with an academic high school qualification (82.1%). The mean age is 23.3 years (SD = 6.6 years), ranging from a minimum of 18 years to a maximum of 49 years. A considerable proportion of the students have an area of study in the HEI to Science, Mathematics, and Informatics (57.9%) and Engineering, Manufacturing, and Construction Industries (30.5%).

10.4.2 Results

In Table 10.2, we can verify the students' answers to questions related to their experience with gamification (items 1–8) and cybersecurity practices. In gray, we highlight the most frequent responses. However, the answers that motivated higher levels of agreement were "In the virtual learning environments or systems of my Higher Education Institution (HEI), I always use in my student account, a password that has uppercase and lowercase letters, numbers, and symbols." ($M = 3.61$) and "I consider it important to implement different teaching methodologies in digital learning environments, e.g. Gamification." ($M = 3.40$). During the online class

Table 10.1 Sample characterization

	N	%
Age (mean; Std. deviation)	23.6	6.6
Gender		
Male	83	87.4
Female	11	11.5
Other	1	1.1
Academic degree		
High school	78	82.1
Undergraduate	16	16.8
Specialist	1	1.1
Undergraduate degree in progress		
Higher professional course (CTeSP)	33	34.7
Undergraduate	56	58.9
Master's degree	6	6.4
Knowledge area		
Sciences, mathematics, and informatics	55	57.9
Engineering, manufacturing, and construction	29	30.5
Social sciences, business, and law	11	11.6

Table 10.2 Gamification and cybersecurity practices

Subtitle: 1 - Totally disagree (%); 2 - Partly disagree (%); 3 - Partially agree (%); 4 - Totally agree (%) M - Mean SD - Std. Deviation	**1**	**2**	**3**	**4**	**M**	**SD**
1. I have a full understanding, in general terms, of what lessons with Gamification techniques are.	10,5	15,8	49,5	24,2	2,87	0,90
2. I can identify when the teacher implements gamification techniques in class.	10,5	17,9	47,4	24,2	2,85	0,91
3. During the online class period, because of the COVID-19 pandemic, I took classes in which the teacher used Gamification techniques.	35,8	25,3	30,5	8,4	2,11	0,99
4. I have already had classroom lessons in which the teacher used Gamification techniques.	25,3	24,2	34,7	15,8	2,41	1,03
5. Lessons conducted with Gamification techniques are more interesting and motivating than traditional lecture classes.	7,4	3,2	47,4	42,1	3,24	0,83
6. I consider it important to implement different teaching methodologies in digital learning environments, e.g., Gamification.	5,3	3,2	37,9	53,7	3,40	0,79
7. When I use of networked (Internet), gamified, or other digital teaching and learning resources, I check the system's origin and its digital certificates.	13,7	25,3	46,3	14,7	2,62	0,90
8. In the virtual learning environments or systems of my Higher Education Institution (HEI), I always use in my student account, a password that has uppercase and lowercase letters, numbers, and symbols.	1,1	4,2	27,4	67,4	3,61	0,62
9. I am in the habit of periodically changing the passwords for the systems I use.	26,3	31,6	24,2	17,9	2,34	1,06
10. I only remember to change the password when a particular system makes such a request.	15,8	21,1	37,9	25,3	2,73	1,01
11. I try to inform myself about the security procedures used in the systems and applications I use in my HEI.	13,7	30,5	44,2	11,6	2,53	0,87
12. I only use free antivirus systems (among other protection systems) on my personal computer for study and work.	16,8	16,8	26,3	40	2,89	1,11
13. I frequently check that the computer's antivirus is up to date and functional.	7,4	18,9	31,6	42,1	3,08	0,95
14. I periodically check that my computer's firewall is up to date and functional.	9,5	22,1	32,6	35,8	2,94	0,98
15. I try to be informed about the best information security practices to adopt when using the Internet.	3,2	13,7	50,5	32,6	3,13	0,76
16. I have already been reading (even partially) the General Data Protection Regulation (GDPR) and I understand everything I read.	32,6	25,3	30,5	11,6	2,21	1,03

Table 10.3 Cyberattacks suffered by students

		N	%
17. Have you ever been the victim of a cyberattack that caused you any personal, academic, and/or professional harm? If yes, what was the attack, and in what context?			
	Yes	9	9.5
	No	86	90.5

period, because of the COVID-19 pandemic, "I took classes in which the teacher used gamification techniques." ($M = 2.11$) and "I have already been reading (even partially) the general data protection regulation (GDPR) and I understand everything I read." ($M = 2.21$).

The questionnaire presented the students a final question, to verify whether they had already been victims of a cyberattack that had tail damage and if so, describe what the cyberattack suffered. In Table 10.3, we can see that almost all students say that they have never suffered attacks in digital environments. However, the few students who reported having suffered attacks, indicated mostly phishing attacks, and ransomware and some could not specify.

The analysis of the relational structure of the items of the scale about the experience of the students with gamification was performed through exploratory factor

Table 10.4 Variance explained (gamification)

	Initial Eigenvalues			Extraction sums of squared loadings		
Component	Total	% of variance	Cumulative %	Total	% of variance	Cumulative %
1	2.859	40.848	40.848	2.859	40.848	40.848
2	1.279	18.277	59.125	1.279	18.277	59.125
3	1.064	15.205	74.330	1.064	15.205	*74.330*
4	.664	9.486	83.817			
5	.556	7.945	91.762			
6	.368	5.253	97.015			
7	.209	2.985	100.000			

Table 10.5 Variance explained (cybersecurity)

	Initial Eigenvalues			Extraction sums of squared loadings		
Component	Total	% of variance	Cumulative %	Total	% of variance	Cumulative %
1	3.275	36.389	36.389	3.275	36.389	36.389
2	1.209	13.434	49.823	1.209	13.434	49.823
3	1.046	11.618	61.441	1.046	11.618	*61.441*
4	.823	9.145	70.586			
5	.800	8.884	79.470			
6	.640	7.109	86.579			
7	.468	5.200	91.779			
8	.420	4.672	96.451			
9	.319	3.549	100.000			

analysis on the matrix of correlations, with extraction of the factors by the method of the main components followed by Varimax rotation. The common factors retained were those with a higher Eigenvalue 1. The validity of factor analysis was made through the KMO test (0.656 – gamification and 0.751 – cybersecurity) and the Bartlett test (significant in both sets of questions), which indicates acceptable values for the continuation of the analyses. Factor analysis converged to a solution with three main components in the gamification set of questions, which explain 74.3% and three components in the set of questions about cybersecurity, which explains 61.4% of the total variance as detailed in Tables 10.4 and 10.5, and the components presented a good internal consistency in the analysis of internal coefficients through Cronbach Alpha test.

In the rotated matrix, only items that presented saturation levels were considered >0.40. The components generated from the gamification set of questions were designated as (i) "knows gamification," (ii) "uses gamification," and (iii) "importance gamification." The components of the set of cybersecurity questions were designated as (i) "best practices IS personal" (IS means Information Security), (ii) "best practices IS password," and (iii) "best practices IS HEI".

10.5 Conclusions

The pandemic crisis had a dramatic impact over twenty-first-century society. A swift action was needed, in particular, for the lecturing sector. At same time, as described in previous sections, there was a set of new approaches and technologies that could be the necessary lever to bring innovative practices but also to play a relevant role during the COVID-19 pandemic. The bibliography collected in this research clearly suggests that gamification, IoT, AI, or blockchain as innovative and relevant educational practices. The direct implication of the physical distance under the crisis pushes the lecturing players to their connected devices and into remote sections. This article, also, discusses the new security challenges that remote access brought to students and lecturers.

As a methodology, a survey and an in-depth statistical analysis were conducted. Taking into consideration, the most relevant conclusions: most participants partially agree that they understand the topic of gamification in the context of a classroom and that they would identify it. On the other hand, the majority had no feel of the experience of a class occurring with the gamification tool. The majority also do consider to be important to implement different teaching methodologies in digital learning environments, e.g. gamification. The gamification is a tool to be explored that failed its implementation in the pandemic period.

Under the research topic about security exposure and users' literacy, the most positive replies occurred for good practice of using strong passwords, regular concern about virus list update and firewall overall state. Some negative points are the following: there is no practice of a regular password change, the majority uses free antivirus solutions, and there is small knowledge about the principles, implications, and advantages of the GDPR.

Results go to conclusion accordingly with the supposition that gamification can and should be used in cybersecurity literacy objectives, and these objectives must be included in contents offered by HEIs, integrating their strategic planning in the path of best practices, implying directly or indirectly on sustainability related issues.

References

A3ES. (2022). *Manual de Avaliação Institucional*. Available on: https://www.a3es.pt/sites/default/files/Manual_de_avaliacao_institucional_A3ES_Consulta_publica.pdf

Biswas, B. N., Bhitkar, S. D., & Pundkar, S. N. (2021). Secure login: A blockchain based web application for identity access management system. In *2021 2nd International Conference for Emerging Technology (INCET)* (pp. 1–5). IEEE. https://doi.org/10.1109/INCET51464.2021.9456352

Bozgeyikli, E., Bozgeyikli, L., Clevenger, M., Raij, A., Alqasemi, R., & Dubey, R. (2014). Poster: Design and development of a virtual reality system for vocational rehabilitation of individuals with disabilities. In *2014 IEEE symposium on 3D User Interfaces (3DUI)* (pp. 175–176). IEEE. https://doi.org/10.1109/3DUI.2014.6798877

Bozkurt, A., & Durak, G. (2018). A systematic review of gamification research: In pursuit of Homo Ludens. *International Journal of Game-Based Learning (IJGBL), 8*(3), 15–33. https://doi.org/10.4018/IJGBL.2018070102

Challco, G. C., Moreira, D. A., Bittencourt, I. I., Mizoguchi, R., & Isotani, S. (2015). Personalization of gamification in collaborative learning contexts using ontologies. *IEEE Latin America Transactions, 13*(6), 1995–2002. https://doi.org/10.1109/TLA.2015.7164227

Chen, M. P., & Liao, B. C. (2015). Augmented reality laboratory for high school electrochemistry course. In *2015 IEEE 15th international conference on advanced learning technologies* (pp. 132–136). IEEE. https://doi.org/10.1109/ICALT.2015.105

Deterding, S., Dixon, D., Khaled, R., & Nacke, L. (2011). From game design elements to gamefulness: Defining "gamification". In *Proceedings of the 15th international academic MindTrek conference: Envisioning future media environment* (pp. 9–15). https://doi.org/10.1145/2181037.2181040

Diakoumakos, J., Chaskos, E., Kolokotronis, N., & Lepouras, G. (2021). Cyber-range federation and cyber-security games: A gamification scoring model. In *2021 IEEE international conference on Cyber Security and Resilience (CSR)* (pp. 186–191). IEEE. https://doi.org/10.1109/CSR51186.2021.9527972

Dichev, C., & Dicheva, D. (2017). Gamifying education: What is known, what is believed and what remains uncertain: A critical review. *International Journal of Educational Technology in Higher Education, 14*(1), 1–36. https://doi.org/10.1186/s41239-017-0042-5

Djosic, N., Nokovic, B., & Sharieh, S. (2020). Machine learning in action: Securing IAM API by risk authentication decision engine. In *2020 IEEE conference on Communications and Network Security (CNS)* (pp. 1–4). IEEE. https://doi.org/10.1109/CNS48642.2020.9162317

Draskovic, D. (2019). Development of intelligent systems and application of gamification in artificial intelligent learning. In *2019 27th Telecommunications Forum (TELFOR)* (pp. 1–1). IEEE. https://doi.org/10.1109/TELFOR48224.2019.8971360

Ekneling, S., Sonestedt, T., Georgiadis, A., Yousefi, S., & Chana, J. (2018). Magestro: Gamification of the data collection process for development of the hand gesture recognition technology. In *2018 IEEE International Symposium on Mixed and Augmented Reality Adjunct (ISMAR-Adjunct)* (pp. 417–418). IEEE. https://doi.org/10.1109/ISMAR-Adjunct.2018.00124

Ghosh, A. K., O'Connor, T., & McGraw, G. (1998). An automated approach for identifying potential vulnerabilities in software. In *Proceedings. 1998 IEEE symposium on security and privacy (Cat. No. 98CB36186)* (pp. 104–114). IEEE. https://doi.org/10.1109/SECPRI.1998.674827

González, C. M. M., Hernández-Muñoz, G. M., & Leyton, A. E. G. (2021). The effect of gamification on motivation in a virtual classroom. In *2021 XI International Conference on Virtual Campus (JICV)* (pp. 1–4). IEEE. https://doi.org/10.1109/JICV53222.2021.9600324

Iannone, E., Guadagni, R., Ferrucci, F., De Lucia, A., & Palomba, F. (2022). The secret life of software vulnerabilities: A large-scale empirical study. *IEEE Transactions on Software Engineering*. https://doi.org/10.1109/TSE.2022.3140868

Jiang, J., & Zeng, L. (2019). Research on the application of virtual reality technology in the teaching model. In *2019 14th International Conference on Computer Science & Education (ICCSE)* (pp. 145–148). IEEE. https://doi.org/10.1109/ICCSE.2019.8845411

Khaleel, F., Ashaari, N., Wook, T., & Ismail, A. (2016). Gamification elements for learning applications. *International Journal on Advanced Science, Engineering and Information Technology, 6*(6), 868–874. https://media.neliti.com/media/publications/103628-EN-gamification-elements-for-learning-appli.pdf

King, V. (2004). Technology-facilitated online and distance learning student support. In *IEEE international conference on advanced learning technologies, 2004. Proceedings* (pp. 801–803). IEEE. https://doi.org/10.1109/ICALT.2004.1357662

Krath, J., Schürmann, L., & Von Korflesch, H. F. (2021). Revealing the theoretical basis of gamification: A systematic review and analysis of theory in research on gamification, serious games and game-based learning. *Computers in Human Behavior, 125*, 106963. https://doi.org/10.1016/j.chb.2021.106963

Kumar, S. A., Kumar, N. R., Prakash, S., & Sangeetha, K. (2017). Gamification of internet security by next generation CAPTCHAs. In *2017 International Conference on Computer Communication and Informatics (ICCCI)* (pp. 1–5). IEEE. https://doi.org/10.1109/ICCCI.2017.8117754

Lewallen, J. (2021). Emerging technologies and problem definition uncertainty: The case of cybersecurity. *Regulation & Governance, 15*(4), 1035–1052. https://doi.org/10.1111/rego.12341

Llorens-Largo, F., Gallego-Durán, F. J., Villagrá-Arnedo, C. J., Compañ-Rosique, P., Satorre-Cuerda, R., & Molina-Carmona, R. (2016). Gamification of the learning process: Lessons learned. *IEEE Revista Iberoamericana de tecnologías del aprendizaje, 11*(4), 227–234. https://doi.org/10.1109/RITA.2016.2619138

Lourenço, J., Morais, J., Monteiro, J., Santos, J., Neves, N., & Sá, S. (2022). Conceptions on higher education students about use of technologies in the learning process: A comparative study. In *Proceedings of INTED2022 conference* (pp. 3721–3730), 7th–8th March 2022. https://library.iated.org/download/INTED2022TOC

Lu, Y., Mao, X., Zhou, M., Zhang, Y., Li, Z., Wang, T., Yin, G., & Wang, H. (2021). Motivation under gamification: An empirical study of developers' motivations and contributions in stack overflow. In *IEEE transactions on software engineering* (p. 1). IEEE. https://doi.org/10.1109/TSE.2021.3130088

Maidenbaum, S., & Amedi, A. (2015). Blind in a virtual world: Mobility-training virtual reality games for users who are blind. In *2015 IEEE Virtual Reality (VR)* (pp. 341–342). IEEE. https://doi.org/10.1109/VR.2015.7223435

Manzano-León, A., Camacho-Lazarraga, P., Guerrero, M. A., Guerrero-Puerta, L., Aguilar-Parra, J. M., Trigueros, R., & Alias, A. (2021). Between level up and game over: A systematic literature review of gamification in education. *Sustainability, 13*, 2247. https://doi.org/10.3390/su13042247

Martin, S., Diaz, G., Castro, M., & Rodriguez-Artacho, M. (2019). Increasing engagement in a network security management course through gamification. In *2019 IEEE Global Engineering Education Conference (EDUCON)* (pp. 1380–1383). IEEE. https://doi.org/10.1109/EDUCON.2019.8725071

Misbahuddin, M., Bindhumadhava, B. S., & Dheeptha, B. (2017). Design of a risk based authentication system using machine learning techniques. In *2017 IEEE SmartWorld, Ubiquitous Intelligence & Computing, Advanced & Trusted Computed, Scalable Computing & Communications, Cloud & Big Data Computing, Internet of People and Smart City Innovation (SmartWorld/SCALCOM/UIC/ATC/CBDCom/IOP/SCI)* (pp. 1–6). IEEE. https://doi.org/10.1109/UIC-ATC.2017.8397628

Morais, J., Lourenço, J., Santos, J., Monteiro, J., Neves, N., & Sá, S. (2022). Conceptions on higher education professors about use of technologies in the learning process: A comparative study. In *Proceedings of INTED2022 conference* (pp. 3775–3784), 7th–8th March 2022. https://library.iated.org/download/INTED2022TOC

Nguyen, T. A., & Pham, H. (2020). A design theory-based gamification approach for information security training. In *2020 RIVF international conference on computing and communication technologies (RIVF)* (pp. 1–4). IEEE. https://doi.org/10.1109/RIVF48685.2020.9140730

Pantile, D., Frasca, R., Mazzeo, A., Ventrella, M., & Verreschi, G. (2016). New technologies and tools for immersive and engaging visitor experiences in museums: The evolution of the visit-actor in next-generation storytelling, through augmented and virtual reality, and immersive 3d projections. In *2016 12th international conference on Signal-Image Technology & Internet-Based Systems (SITIS)* (pp. 463–467). IEEE. https://doi.org/10.1109/SITIS.2016.78

Parizi, R. M., & Dehghantanha, A. (2018). On the understanding of gamification in blockchain systems. In *2018 6th international conference on Future Internet of Things and Cloud Workshops (FiCloudW)* (pp. 214–219). IEEE. https://doi.org/10.1109/W-FiCloud.2018.00041

Patricio, J. M., Costa, M. C., & Manso, A. (2019). A gamified mobile augmented reality system for the teaching of astronomical concepts. In *2019 14th Iberian Conference on Information Systems and Technologies (CISTI)* (pp. 1–5). IEEE. https://doi.org/10.23919/CISTI.2019.8760658

Pearshouse, I., & Sharples, M. (2000). CETADL: A research and development facility for e-learning. In *Proceedings International Workshop on Advanced Learning Technologies. IWALT 2000. Advanced learning technology: Design and development issues* (pp. 180–183). IEEE. https://doi.org/10.1109/IWALT.2000.890602

Peerzada, B., & Kumar, D. (2021). Analyzing software vulnerabilities using machine learning. In *2021 9th International Conference on Reliability, Infocom Technologies and Optimization (Trends and Future Directions) (ICRITO)* (pp. 1–4). IEEE. https://doi.org/10.1109/IWALT.2000.890602

Pereira, J. D. A. (2020). Techniques and tools for advanced software vulnerability detection. In *2020 IEEE International Symposium on Software Reliability Engineering Workshops (ISSREW)* (pp. 123–126). IEEE. https://doi.org/10.1109/ISSREW51248.2020.00049

Raval, R., Maskus, A., Saltmiras, B., Dunn, M., Hawrylak, P. J., & Hale, J. (2018). Competitive learning environment for cyber-physical system security experimentation. In *2018 1st International Conference on Data Intelligence and Security (ICDIS)* (pp. 211–218). IEEE. https://doi.org/10.1109/ICDIS.2018.00042

Reyes, A. A., Elkin, C., Niyaz, Q., Yang, X., Paheding, S., & Devabhaktuni, V. K. (2020). A preliminary work on visualization-based education tool for high school machine learning education. In *2020 IEEE Integrated STEM Education Conference (ISEC)* (pp. 1–5). IEEE. https://doi.org/10.1109/ISEC49744.2020.9280629

Roosta, F., Taghiyareh, F., & Mosharraf, M. (2016). Personalization of gamification-elements in an e-learning environment based on learners' motivation. In *2016 8th International Symposium on Telecommunications (IST)* (pp. 637–642). IEEE. https://doi.org/10.1109/ISTEL.2016.7881899

Sadik, S., Ahmed, M., Sikos, L. F., & Islam, A. K. M. (2020). Toward a sustainable cybersecurity ecosystem. *Computers, 9*(3), 74. https://doi.org/10.3390/computers9030074

Sailer, M., & Homner, L. (2020). The gamification of learning: A meta-analysis. *Educational Psychology Review, 32*, 77–112. https://doi.org/10.1007/s10648-019-09498-w

Sailer, M., Hense, J., Mayr, S., & Mandl, H. (2017). How gamification motivates: An experimental study of the effects of specific game design elements on psychological need satisfaction. *Computers in Human Behavior, 69*, 371–380. https://doi.org/10.1016/j.chb.2016.12.033

Shamal, P. K., Rahamathulla, K., & Akbar, A. (2017). A study on software vulnerability prediction model. In *2017 International Conference on Wireless Communications, Signal Processing and Networking (WiSPNET)* (pp. 703–706). IEEE. https://doi.org/10.1109/WiSPNET.2017.8299852

Sharif, K. H., & Ameen, S. Y. (2020). A review of security awareness approaches with special emphasis on gamification. In *2020 International Conference on Advanced Science and Engineering (ICOASE)* (pp. 151–156). IEEE. https://doi.org/10.1109/ICOASE51841.2020.9436595

Sheng, L. Y. (2013). Modelling learning from Ingress (Google's augmented reality social game). In *2013 IEEE 63rd annual conference International Council for Education Media (ICEM)* (pp. 1–8). IEEE. https://doi.org/10.1109/CICEM.2013.6820152

Süncksen, M., Bendig, H., Teistler, M., Wagner, M., Bott, O. J., & Dresing, K. (2018). Gamification and virtual reality for teaching mobile x-ray imaging. In *2018 IEEE 6th international conference on Serious Games and Applications for Health (SeGAH)* (pp. 1–7). IEEE. https://doi.org/10.1109/SeGAH.2018.8401364

Suvajdzic, M., Oliverio, J., Barmpoutis, A., Wood, L., & Burgermeister, P. (2020). Discover DaVinci–A gamified blockchain learning app. In *2020 IEEE International Conference on Blockchain and Cryptocurrency (ICBC)* (pp. 1–2). IEEE. https://doi.org/10.1109/SeGAH.2018.8401364

Trinidad García, M., Calderón Sánchez, A., & Ruiz Carreira, M. (2021). GoRace: A multi-context and narrative-based gamification suite to overcome gamification technological challenges. https://doi.org/10.1109/ACCESS.2021.3076291

Trinidad, M., Ruiz, M., & Calderón, A. (2021). A bibliometric analysis of gamification research. *IEEE Access, 9*, 46505–46544. https://doi.org/10.1109/ACCESS.2021.3063986

Vanduhe, V. Z., Nat, M., & Hasan, H. F. (2020). Continuance intentions to use gamification for training in higher education: Integrating the technology acceptance model (TAM), social motivation, and task technology fit (TTF). *IEEE Access, 8*, 21473–21484. https://doi.org/10.1109/ACCESS.2020.2966179

Visalli, N., Deng, L., Al-Suwaida, A., Brown, Z., Joshi, M., & Wei, B. (2019). Towards automated security vulnerability and software defect localization. In *2019 IEEE 17th International Conference on Software Engineering Research, Management and Applications (SERA)* (pp. 90–93). IEEE. https://doi.org/10.1109/SERA.2019.8886795

Waheed, A., Shah, M. A., & Khan, A. (2016). Secure login protocols: An analysis on modern attacks and solutions. In *2016 22nd International Conference on Automation and Computing (ICAC)* (pp. 535–541). IEEE. https://doi.org/10.1109/IConAC.2016.7604975

Yonemura, K., Yajima, K., Komura, R., Sato, J., & Takeichi, Y. (2017). Practical security education on operational technology using gamification method. In *2017 7th IEEE International Conference on Control System, Computing and Engineering (ICCSCE)* (pp. 284–288). IEEE. https://doi.org/10.1109/ICCSCE.2017.8284420

Yonemura, K., Komura, R., Sato, J., Hoga, T., Takeichi, Y., Chida, E., & Matsuoka, M. (2018a). Effect of security education using KIPS and gamification theory at KOSEN. In *2018 IEEE Symposium on Computer Applications & Industrial Electronics (ISCAIE)* (pp. 255–258). IEEE. https://doi.org/10.1109/ISCAIE.2018.8405480

Yonemura, K., Sato, J., Takeichi, Y., Komura, R., & Yajima, K. (2018b). Security education using gamification theory. In *2018 International Conference on Engineering, Applied Sciences, and Technology (ICEAST)* (pp. 1–4). IEEE. https://doi.org/10.1109/ICEAST.2018.8434432

Zhenming, B., Mayu, U., Mamoru, E., & Tatami, Y. (2017). Development of an English words learning system utilizes 3D markers with augmented reality technology. In *2017 IEEE 6th Global Conference on Consumer Electronics (GCCE)* (pp. 1–2). IEEE. https://doi.org/10.1109/GCCE.2017.8229353

Relevance and Characteristics of Responsible Innovation Assessment Tools

11

Carla Caldeira, Diana Pereira, José Duarte Santos, Cristina Guimarães, and Fernando Almeida

11.1 Introduction

The growing importance of innovation for competitiveness is widely recognized both in academia and science, as well as in business. There are several authors such as Lichtenthaler (2016), Roach et al. (2016), and Tavassoli and Karlsson (2016), who have explored the need for organizations to innovate to achieve sustainable success in the markets in which they operate or even to reinvent such markets. Formulating goals and plans and implementing actions and monitoring their execution are activities that companies need to undertake to remain competitive. Therefore, it is up to the organizations to ensure that this process is correctly conducted so that it is possible to achieve the intended goals. Furthermore, it is important for an organization to evolve over time, otherwise it runs the risk of disappearing. Thus, the current strategic guidelines are increasingly based on innovation. By developing new technologies, products, services, processes, and systems in a sustained way, the organization creates much more value and differentiates itself in the market (Meissner & Kotsemir, 2016; Pozo et al., 2019).

C. Caldeira (✉) · D. Pereira
ISPGAYA, Vila Nova de Gaia, Portugal
e-mail: ispg2019102354@ispgaya.pt

J. D. Santos
CEOS.PP and ISCAP/P.PORTO, Porto, Portugal

C. Guimarães
INESC TEC, Porto, Portugal

F. Almeida
ISPGAYA, Vila Nova de Gaia, Portugal

INESC TEC, Porto, Portugal

© The Author(s), under exclusive license to Springer Nature Switzerland AG 2023

F. L. Almeida et al. (eds.), *Multidimensional Sustainability: Transitions and Convergences*, Springer Proceedings in Earth and Environmental Sciences,
https://doi.org/10.1007/978-3-031-24892-4_11

In a world faced with socio-environmental problems, the combination of innovation with sustainability becomes increasingly necessary. Besides being a way to preserve the environment, it is also a path to achieve new market opportunities (Andrade, 2017). With the increasing concern of the market with socio-environmental aspects, it is necessary to re-evaluate the performance aspects of companies. From this perspective, innovation is seen differently, which leads to the idealization of sustainable innovation (Kneipp et al., 2019). Innovation and sustainability are increasingly interconnected, since for the development of a sustainable product, process, or service, it is essential to use innovation as a driving force, and these two terminologies are the foundations of the construction of a sustainable innovation context for the advances and evolution of companies.

It is in this context that the concept of responsible innovation emerges in the market intending to promote a more responsible development of innovation in terms of inclusion of a wider society and so that the results of innovation are ethically acceptable, sustainable, and socially desirable (EC, 2011). Responsibility in research and innovation is driven by global concern for the planet, its natural resources, as well as a just and inclusive society. This concern concerns governments, which can act through policies and regulations, even if they are still in the developmental stage (Scherer et al., 2006). However, responsible innovation is also a concern of the private sector with the aim to align R&D activities with society's values, needs, and expectations (Burget et al., 2017).

Knowing the concept of responsible innovation and how to apply it is relevant in the business context, but insufficient if there are no ways to measure it that can be standardized and comparable. This study addresses this challenge by identifying, categorizing, and exploring a set of responsible innovation assessment tools (RIATs) that can be used to measure responsible innovation. The manuscript is organized into five main sections. In the first phase, an introduction to the theme is given, followed by a theoretical framework of the concepts related to strategy, innovation, and responsible innovation. After that, the methodological process in the identification and evaluation of RIATs is described. Next, the results obtained are presented and discussed considering their technical and scientific contributions. Finally, the main conclusions of this study are listed.

11.2 Background

11.2.1 The Relevance of Strategy

As markets are increasingly competitive, managers must accept the challenge of applying strategic thinking and the skills of an effective leader, i.e., it is necessary to go beyond the efficient planning, organization, execution, and control of business activities. The definitions of the concept of strategy are quite numerous, the literature on strategic management defines strategy as a multifaceted concept, with different dimensions, such as: strategic thinking, strategic process, and strategic change (Meyer et al., 2012).

According to Pasquale (2012), to achieve the objectives it is necessary to determine which strategy to adopt. Therefore, strategies consist of the actions that the organization must take to achieve its goals, these are established based on the objectives to be achieved and influenced by the mission, vision, beliefs and values, micro-environment, and situation of the organization. In environments of constant change, it becomes common to carry out restructuring processes for survival and competitiveness. As discussed in Balogun (2007) and Srivastava and Mushtaq (2011), strategic restructuring can be organizational, financial, but also of products and services offered in the market.

The strategic management process is based on the shared importance of strategy development and strategy in action. In this sense, the phases of implementation and strategic control will be fundamental to achieve the proposed objectives (Chung et al., 2016). Strategic management requires the manager's ability to monitor and interpret the reality of organizational environments, both inside and outside, to use them in the development of the company's strategic posture, translated through the statement of values, vision, mission and strategic objectives. From this point on, it will be possible to challenge the overall business strategy and subsequently propose specific objectives and actions. Consequently, strategic alignment throughout the process is understood (Bora et al., 2017).

11.2.2 The Relevance of Innovation

Innovation has been widely regarded as the core process that drives economic growth and sustainable competitive advantages of both companies and nations, besides driving global sustainable growth (Chen et al., 2018). This has undoubtedly become an important issue in economic and social development, all developed economies are aware that only innovation can continuously stimulate new economic growth, while developing countries are also pushing to continuously upgrade their industrial structure through innovation to improve their national competitiveness (Acs et al., 2017).

Innovation is crucial for companies' profitability and long-term survival because it allows a company to adapt to the dynamic needs of the market (Hauser et al., 2006). Innovation is not only based on the development of new products, but it also helps to promote new business models, offers new services, and improves processes to make people's lives easier. But above all, it makes sure that these processes reach those who need them. Another important factor of innovation is encouraging the emergence of collaborative knowledge networks by exploring the concept of open innovation, which becomes particularly relevant for companies with less installed capacity (Almeida, 2021; Oduro, 2019).

It is important that innovation is valued and properly recognized and, instead of being seen as an expense, be understood as an investment. To this end, society must be open to new ideas and developments and take advantage of the opportunities that are offered to improve the quality of life (Aguirre et al., 2021). It is also important to encourage new scientific skills and the acceptance of new technologies, something that requires the commitment of all governments, industries, schools, and universities (Morawska-Jancelewicz, 2021).

Innovation, always present in society, is one of the priorities of industrialized countries and can dictate the difference between organizations and economies. It is through it that organizations can build the foundations of their development and growth in a sustainable way. Innovation is an extremely relevant factor for a successful modern economy, providing high levels of return on investment and leading to economic growth, the creation of high-quality jobs and the acquisition of a higher standard of living (Gupta, 2008).

11.2.3 The Need of Responsible Innovation Practices

Long et al. (2020) state that, innovation leads to new products, business models and even changes in socio-economic systems. However, innovation must have the 'right impacts', so responsible innovation can help achieve this. An innovation is therefore defined as responsible when it generates relevant and sustainable value for all stakeholders: society, users, and shareholders. Responsible Innovation (RI) is about delivering value to all stakeholders, the issue of managing these stakeholders with integrity must be treated with care.

RI can be defined according to Von Schomberg (2013, p. 19) as "a transparent, interactive process by which societal actors and innovators become mutually responsive to each other with a view to the (ethical) acceptability, sustainability and societal desirability of the innovation process and its marketable products (in order to allow a proper embedding of scientific and technological advances in our society)".

RI offers a way to support innovation processes in a way that builds stakeholder trust (Sutcliffe, 2011), and a method to make innovations more effective. This is because RI should enable more and better connections to be made with people who can enrich the innovation process by increasing demand articulation and improving forecasting and anticipation (Long et al., 2020). However, the adoption of RI also has its drawbacks and one of the main ones is the lack of unity, recognized approaches, and professional standards for implementing and evaluating RI.

RI is also synonymous with sustainable innovation and both concepts cohere. Innovation is looked at as a crucial approach to address present and future sustainability challenges. In Tello and Yoon (2008) sustainable innovation is seen as the development of new products, processes, services, and technologies that contribute to the development and well-being of human needs and institutions while respecting natural resources and regenerative capacities. This view is further explored by Bos-Brouwers (2010) who explains sustainable innovation by defining it as innovations in which the renewal or improvement of products, services, technological or organizational processes not only provides improved economic performance, but also social performance, both in the short and long term, have the capacity to generate positive social and environmental impacts. A sustainable approach to innovation should guide all business choices regarding products and services as well as the new business and organizational models that need to be adopted (Boons & Lüdeke-Freund, 2013).

An innovation can be seen as responsible when the value created benefits not only shareholders and customers, but also society, in a lasting and sustainable way. An innovation that destroys value for society, for consumers, or does not generate economic profit for shareholders is unlikely to be sustainable (Dreyer et al., 2020). RI encompasses multiple dimensions that seeks on the one hand to avoid causing harm, also to do good, without forgetting that it must be motivated by responsible governance (Bacq & Aguilera, 2022). Through RI, companies can effectively use resources more efficiently, reducing costs, or develop a product that gives them a competitive advantage (Almeida & Wasim, 2022).

11.3 Data and Methods

Responsible innovation is an emerging field of research that has been framed in multiple scientific research domains. It is a pertinent topic for all research domains and its philosophical discussion on its importance is unequivocal as recognized in Chen et al. (2022), Stahl et al. (2019), and Tijdink et al. (2021). Several tools for measuring responsible innovation have emerged in recent years. These tools do not measure responsible innovation in the same way, and consequently it is important to recognize and synthesize the various dimensions of responsible innovation considered in this study. In this sense, this study follows a methodology design based on an exploratory qualitative approach. Exploratory research seeks to explore a problem to provide information for a more precise investigation. This approach aims to get closer to the topic, which can be built on hypotheses or intuitions. It is a qualitative, unstructured process that uses literature searches and case studies as data collection (Saunders et al., 2009). This approach offers several benefits such as an early and timely analysis of the relevance of the subject being discussed, allows for a sense of the need for new research and new joint areas to be explored, and helps maintain focus on the topic (Silverman, 2010). This study is composed of a literature review and a comparative analysis of the various tools for assessing RI. The choice of this methodological model has to do with the very nature of the study, insofar as it aims to understand an issue and not exactly measure the object of study. In this sense, the accounting of the data has only comparative purposes.

Figure 11.1 visually presents the three fundamental phases of this process. Initially, the goal is to identify responsible research innovation tools (RRITs). Generic platforms with potential for application and measurement of RI were considered, even if this was not their initial design focus. Tools were also considered regardless of their application in specific RI sectors. Finally, frameworks and tools were also considered, even if some of these approaches are excessively conceptual and without a technical implementation framed in a software solution (e.g., web application, excel, etc.). However, proprietary frameworks of companies with commercial objectives were not considered, such as workshops launched by these organizations. After this, a conceptual map of the dimensions for analysing RI was built,

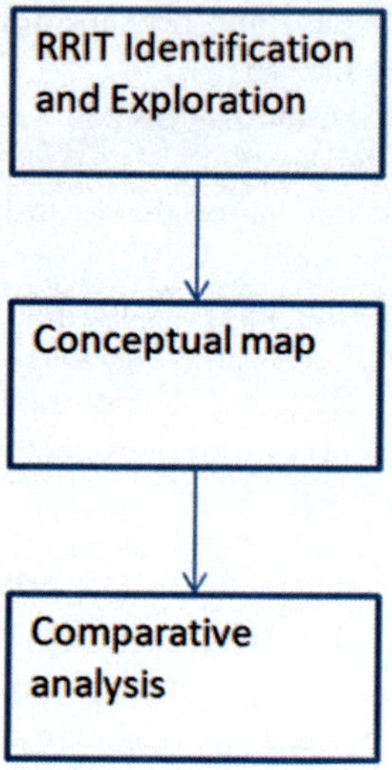

Fig. 11.1 Phases of the methodological process

which allows for generic information about the various dimensions that RI covers. Finally, the last step explicitly seeks to assess comparatively the similarities and differences between the various tools. Each tool is analysed according to the previously identified dimensions, which allows the identification of the most common and least common dimensions.

11.4 Results and Discussion

The presentation of the results begins with the identification of RRITs. Table 11.1 presents this list and a brief description of each tool's objective. The data are organized alphabetically, in which each tool is identified by its acronym. The description of each tool highlights its unique characteristics, the main objectives, and the author that supports it. In total, 18 tools were identified. Most of these tools were launched in the last 3 years, which shows a high concern of the scientific and business community in finding ways to measure RI. We also found tools that emerged in specific business sectors such as health, but whose potential applicability may be broader. The role that European projects have played in the emergence of these initiatives is also noted. These European projects have also been essential to promote a closer relationship between the scientific community and the business fabric, through the launch of technology transfer projects (Deloitte, 2020). It is worth highlighting the emergence of proposals for measuring RI that result from the reuse of existing networks of European projects funded under FP-6, FP-7 and Horizon 2020 and that have allowed expanding the area of action of these consortia and maintaining transnational collaboration networks. It is also evident that addressing the challenges of measuring RI is not specific to a given economy, and across the board all countries are faced with the challenge of effective and sustained growth in the future. As

Table 11.1 List of RRITs

Tool	Description
Exploring Natural Capital Opportunities, Risks and Exposure (ENCORE)	ENCORE is a tool that allows users to visualize and understand the impact of environmental change on the economy. It explores how business activities impact nature and looks at how dependencies and impacts can also pose a risk to business (ENCORE, 2022).
Environment and Social Impact Assessment (ESIA)	The goal is to assess and predict potential adverse social and environmental impacts and develop appropriate mitigation measures, which are documented in an Environmental and Social Management Plan (Dreyer et al., 2020).
Four Dimensions of Responsible Innovation Framework (F4-RIF)	The four dimensions of RI (anticipation, reflexivity, inclusiveness, and responsiveness) provide a framework for raising, discussing, and responding to questions such as the purposes, motivations, social and political constitutions, trajectories, and directions of innovation (Stilgoe et al., 2013).
Ethical Matrix (EM)	This tool is a structured approach to analyze the impacts of technologies according to stakeholder groups and the ethical principles of justice, autonomy, and well-being (Kaiser & Forsberg, 2001).
Health Technology Assessment (HTA) Core	It offers many opportunities for different types of comparison between health technologies. It considers ethical, cultural, social, legal, and regulatory issues in the form of a list, and has a high degree of transparency (Thorstensen, 2019).
KARIM	It serves to help companies reconsider their business model, develop new products and services, new technologies, or even improve their production processes. It combines a self-diagnostic tool with a summative analytical grid composed of 24 criteria (Hin et al., 2014).
Life Cycle Assessment Anticipatory (LCAA)	Tool to support environmentally RI. It seeks to provide environmental criteria for R&D decision-makers to broaden the range of values used in formulating hypotheses and experimental research agenda and thus support RI of emerging technologies (Wender et al., 2014).
Monitoring the Evaluation and Benefits of responsible Research and Innovation (MoRRI)	Develops a list of RRI indicators for proper measurement of research and innovation responsibility, which could serve as KPIs (Gurzawska, 2021).
ORBIT Self-Assessment Tool	Its goal is to spark ideas for actions that will help put RRI into practice. It provides innovators with a series of questions organized in a 4 × 4 matrix: Process, Product, Purpose, People × Anticipate, Reflect, Engage, Act (Lehoux et al., 2020).
PRISMA	It aims to help companies implement RRI in their innovation and social responsibility strategies. It is a toolbox composed by a self-assessment survey, a 5-criteria impact analysis, 10 KPIs, and a roadmap template (Lehoux et al., 2020).
Product Impact Tool (PIT)	It was developed with the aim of raising awareness of the possibility of influencing behavior through design. Its function is analysis and design support, both applicable to RI, in terms of identifying key issues and impacts and in terms of improving impacts, for example through design changes (Long et al., 2020).

(continued)

Table 11.1 (continued)

Tool	Description
Responsibility Navigator (RN)	It serves to support strategic-level reflections on ways to promote different accountability-related goals throughout the organization. It is important for addressing how innovators working in large organizations can be adequately supported by higher level managers to implement RRI at the operational level (Lehoux et al., 2020).
Responsible Innovation COMPASS self-check tool	Developed with the intention of helping SMEs determine to what extent their practices align with IR principles, how to improve their innovation processes and outcomes, and how they compare to other companies (Gurzawska, 2021).
Responsible Innovation in Health (RIH)	This tool directs the attention of policymakers "upstream," where they can promote innovations that can address significant system-level challenges and support more equitable and sustainable health services. This generates an overall accountability score that can be used to compare the respective value of different innovations (Silva et al., 2021).
Responsible Management of Innovation (RMoI)	It aims to provide innovators with a systematic way to identify and consider socioethical risks and opportunities. It is administered through a workshop and incorporates three distinct stages (Long et al., 2020).
RRI Maturity Model (RRI-MM)	The model proposes a combination of the maturity levels with the activities, processes, or artifacts that are associated with each level. Three categories are listed as purpose, process, and product (Stahl et al., 2017).
ROSIE Responsible Innovation Toolbox	Tool designed to support SMEs and other organizations to understand, assess and implement RI (Interreg Central Europe – ROSIE, 2020).
SDG Compass (SDG-C)	The objective of the SDG compass is to guide companies in the process of aligning their corporate strategies with the sustainable developing goals of the United Nations. Furthermore, it gives indications on how companies can measure and manage their contribution to the SDGs (SDG Compass, 2015).

argued in Balland et al. (2019) and Schütz and Strohmaier (2022), European countries need to enhance their active presence in international innovation and competitiveness networks, with all the consequences in terms of impact on their economic and social matrix. Moreover, European projects have also made significant contributions to the development of the Global Sustainable Development Goals (SDGs) through their collaborative and social transformation potential (Moczek et al., 2021).

For each RRIT tool, a set of dimensions were identified that they use to categorize RI. An attempt was made to group synonyms within the same dimension, such as transparency with visibility. This approach allowed the identification of 16 dimensions as shown in the conceptual map presented in Fig. 11.2. Studies indicate that RI cannot be segmented, and the various dimensions should be interconnected, even as Voegtlin and Scherer (2017) state that economic, environmental, and social perspectives are associated with the sustainable development paradigm. This view is also adopted in Landeweerd et al. (2015) when advocating the interconnection

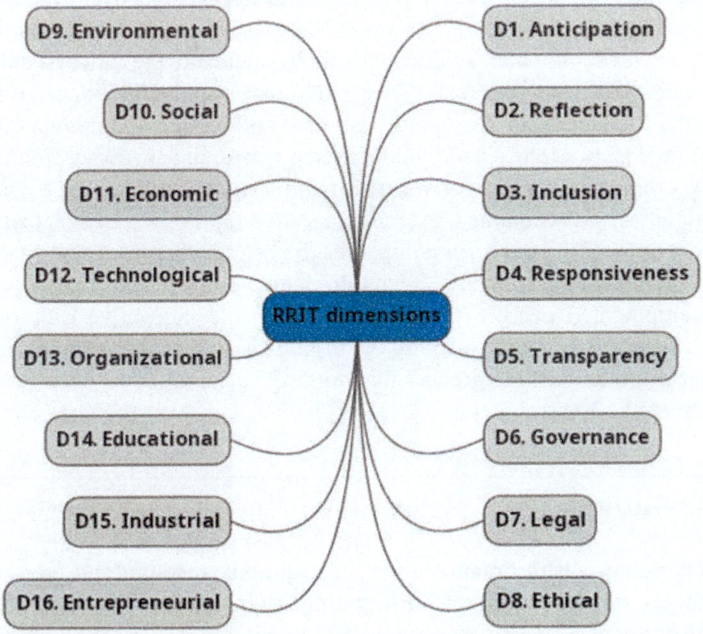

Fig. 11.2 RRIT dimensions

and incorporation in the governance of science and innovation. The dimensions identified in the conceptual map can be mutually reinforcing (e.g., greater reflexivity tends to promote greater inclusion, and vice versa). However, they can also be in tension and generate conflicts (e.g., anticipation can generate greater participation, but can also be resisted by scientists seeking to defend their autonomy, or their prior commitments to particular innovation trajectories). Making such tensions visible and bringing them into negotiation are important steps in making RI responsive. In short, it is a matter of integrating the dimensions and strategies for RI into a coherent and legitimate governance approach.

Finally, the Table 11.2 in Appendix performs a comparative analysis of the RRIT tools. The most common dimensions identified in the RI frameworks are the social and educational dimensions. It is argued that the proposal of new products and services must be aligned with user needs but must also have the potential to alleviate social and environmental challenges. Consequently, RI necessarily has to look at social innovation with the aim of developing solutions to the sustainability challenge (Dacin et al., 2010; Phillips et al., 2015). While social innovation is pursued by organizations or individuals with pro-social motivations, it is recognized that companies must have a corporate social responsibility (CSR). The concept of CSR is related to companies' commitment to society. A socially responsible company should trace its actions in an ethical and transparent management. To do so, it needs to involve issues such as the quality of life and well-being of the company's internal public, the relationship with stakeholders, and the reduction of negative impacts on the community and the environment (Sheehy & Farneti, 2021).

The educational dimension is also relevant from two perspectives: science education and learning. In Bhaduri and Talat (2020), it is recognized that a public dialogue between academia and science can help increase the understanding of innovations and their positive effects. This would increase the likelihood of market acceptance of a product or service. Companies can collaborate with universities and other educational institutions, attend public events, participate in discussion forums, etc. There is a relevant set of initiatives, short and long term, that can be promoted to bring companies and academia closer together. In another perspective, learning is a core element of RI. The goal is for companies to improve their processes and make them more efficient and sustainable. Technological evolution has helped build sustainable development. However, the exploitation of one innovative technology is insufficient. RI practices should also look at organizational processes and make the company learn throughout this process by adopting sustainable products and practices (Bianchi et al., 2022).

11.5 Conclusions

In an ever-changing world, organizations must quickly respond and adapt to the constant changes perpetuated in the surrounding environment. Organizations must have the ability to create, acquire and transfer knowledge, so that they can modify their behaviour to reflect new knowledge and ideas. Innovation assumes the challenge of presenting something new or improved by an organization, through the different possible types of innovation and in all areas of the business, and which has as its main objective to transform this novelty into something with recognized value, for all its possible users and preferably for society in general. Innovation should be consequently responsible and generate relevant and sustainable value for all audiences (e.g. society, users and shareholders). In this way, innovation can have the right impacts with the help of RI, as this is a framework that enables the governance and evaluation of innovations concerning their potentially harmful consequences and positive contributions to societal challenges.

There are several tools for measuring RI. However, given the emergence of the topic, there is no single tool that allows for a complete and comprehensive measurement of all their dimensions. This study identified 18 RIATs and found a total of 16 dimensions. The social and educational dimensions stand out as the most relevant considering the number of different frameworks that incorporate these dimensions. It was also possible to conclude that RI dimensions cannot be seen in isolation since there is a strong interconnection between them.

This work presents some limitations that it is important to address. First, the identification of the frameworks resulted from an exploratory qualitative research process. It is relevant to systematize this process by conducting a systematic literature review to understand in greater depth the characteristics of these tools and their scientific framework. It is also relevant to explore the applicability of these tools considering several activity factors. The health field was highlighted as an important activity sector for the application of these tools, but there are potentially other activity sectors in which RI is also a priority.

Appendix

Table 11.2 Comparative analysis of RRIT tools

Tool	D1	D2	D3	D4	D5	D6	D7	D8	D9	D10	D11	D12	D13	D14	D15	D16
ENCORE									X							
ESIA							X			X				X	X	
F4-RIF	X	X	X	X												
EM		X			X			X		X						
HTA							X	X		X	X		X			
KARIM						X			X	X	X					
LCAA		X					X					X		X		X
MoRRI			X		X	X				X				X		
ORBIT	X	X	X	X	X	X		X		X				X		
PRISMA					X	X		X		X				X		
PIT						X						X				
RN		X	X	X	X		X			X				X		X
COMPASS			X		X	X		X		X				X	X	
RIH									X		X		X			
RMoI	X	X	X	X	X	X										
RRI-MM	X	X	X	X	X			X						X		
ROSIE			X			X	X	X	X	X	X	X		X	X	
SDG-C			X				X	X	X	X	X	X		X	X	
Count	4	7	9	5	8	8	6	8	5	11	5	4	2	10	4	2

References

Acs, Z. J., Audretsch, D. B., Lehmann, E. E., & Licht, G. (2017). National systems of innovation. *The Journal of Technology Transfer, 42*(5), 997–1008. https://doi.org/10.1007/s10961-016-9481-8

Aguirre, J. C., Peralta Zuñiga, M. L., Mora, P., & Blanco, F. (2021). Innovative entrepreneurship and quality of life. The case of Ecuador. *Academia Revista Latinoamericana de Administración, 34*(1), 143–164. https://doi.org/10.1108/ARLA-04-2020-0076

Almeida, F. (2021). Open-innovation practices: Diversity in Portuguese SMEs. *Journal of Open Innovation: Technology, Market, and Complexity, 7*(3), 1–19. https://doi.org/10.3390/joitmc7030169

Almeida, F., & Wasim, J. (2022). Eco-innovation and sustainable business performance: Perspectives of SMEs in Portugal and the UK. *Society and Business Review, 18*, 28. https://doi.org/10.1108/SBR-12-2021-0233

Andrade, V. (2017). *Como promover inovação com foco na sustentabilidade?* https://blog.waycarbon.com/2017/06/inovacao-com-foco-na-sustentabilidade/

Bacq, S., & Aguilera, R. V. (2022). Stakeholder governance for responsible innovation: A theory of value creation, appropriation, and distribution. *Journal of Management Studies, 59*(1), 29–60. https://doi.org/10.1111/joms.12746

Balland, P. A., Boschma, R., & Ravet, J. (2019). Network dynamics in collaborative research in the EU, 2003–2017. *European Planning Studies, 27*(9), 1811–1837. https://doi.org/10.1080/09654313.2019.1641187

Balogun, J. (2007). The practice of organizational restructuring: From design to reality. *European Management Journal, 25*(2), 81–91. https://doi.org/10.1016/j.emj.2007.02.001

Bhaduri, S., & Talat, N. (2020). RRI beyond its comfort zone: Initiating a dialogue with frugal innovation by 'the vulnerable'. *Science, Technology and Society, 25*(2), 273–290. https://doi.org/10.1177/0971721820902967

Bianchi, G., Testa, F., Boiral, O., & Iraldo, F. (2022). Organizational learning for environmental sustainability: Internalizing lifecycle management. *Organization & Environment, 35*(1), 103–129. https://doi.org/10.1177/1086026621998744

Boons, F., & Lüdeke-Freund, F. (2013). Business models for sustainable innovation: State-of-the-art and steps towards a research agenda. *Journal of Cleaner Production, 45*, 9–19. https://doi.org/10.1016/j.jclepro.2012.07.007

Bora, B., Borah, S., & Chungyalpa, W. (2017). Elaborando objetivos estratégicos: examinando o papel da visão de negócios e declarações de missão. *Órgão Gerenciar, 6*(1), 1e6.

Bos-Brouwers, H. E. J. (2010). Corporate sustainability and innovation in SMEs: Evidence of themes and activities in practice. *Business Strategy and the Environment, 19*(7), 417–435. https://doi.org/10.1002/bse.652

Burget, M., Bardone, E., & Pedaste, M. (2017). Definitions and conceptual dimensions of responsible research and innovation: A literature review. *Science and Engineering Ethics, 23*(1), 1–19. https://doi.org/10.1007/s11948-016-9782-1

Chen, J., Yin, X., & Mei, L. (2018). Holistic innovation: An emerging innovation paradigm. *International Journal of Innovation Studies, 2*(1), 1–13. https://doi.org/10.1016/j.ijis.2018.02.001

Chen, J., Nichele, E., Ellerby, Z., & Wagner, C. (2022). Responsible research and innovation in practice: Driving both the 'how' and the 'what' to research. *Journal of Responsible Technology, 11*, 1–5. https://doi.org/10.1016/j.jrt.2022.100042

Chung, C.-C., Chao, L.-C., Chen, C.-H., & Lou, S.-J. (2016). A balanced scorecard of sustainable management in the Taiwanese bicycle industry: Development of performance indicators and importance analysis. *Sustainability, 8*(6), 1–21. https://doi.org/10.3390/su8060518

Dacin, P. A., Dacin, M. T., & Matear, M. (2010). Social entrepreneurship: Why we don't need a new theory and how we move forward from here. *Academy of Management Perspectives, 24*, 37–57. https://doi.org/10.5465/amp.24.3.37

Deloitte. (2020). *A quick guide to EU Funding 2014–2020*. https://www2.deloitte.com/content/dam/Deloitte/cy/Documents/finance/CY_Finance_EUFunding2014-2020Guide_Noexp.pdf

Dreyer, M., von Heimburg, J., Goldberg, A., & Schofield, M. (2020). Designing responsible innovation ecosystems for the mobilisation of resources from business and finance to accelerate the implementation of sustainability. A view from industry. *Journal of Sustainability Research, 4*(3), e200033. https://doi.org/10.20900/jsr20200033

ENCORE. (2022). *Exploring natural capital opportunities, risks and exposure*. https://encore.naturalcapital.finance/en

European Commission (EC). (2011). *Horizon 2020 – The framework programme for research and innovation*. EC.

Gupta, P. (2008). *Inovação empresarial no século XXI*. Vida Económica.

Gurzawska, A. (2021). Responsible innovation in business: Perceptions, evaluation practices and lessons learnt. *Sustainability, 13*(4), 1826. https://doi.org/10.3390/su13041826

Hauser, J., Tellis, G. J., & Griffin, A. (2006). Research on innovation: A review and agenda for marketing science. *Marketing Science, 25*(6), 687–717. https://doi.org/10.1287/mksc.1050.0144

Hin, G., Daigney, M., Haudebault, D., Raskin, K., Bouché, Y., Pavie, X., & Carthy, D. (2014). *Introduction to responsible innovation criteria*. https://www.nweurope.eu/media/1118/guide_online.pdf

Interreg Central Europe – ROSIE. (2020). *About Rosie Ri toolbox*. https://www.ciseonweb.it/progetti-europei/rosie/

Kaiser, M., & Forsberg, E. M. (2001). Assessing fisheries – Using an ethical matrix in a participatory process. *Journal of Agricultural and Environmental Ethics, 14*(2), 191–200. https://doi.org/10.1023/A:1011300811590

Kneipp, J. M., Gomes, C. M., Bichueti, R. S., Frizzo, K., & Perlin, A. P. (2019). Sustainable innovation practices and their relationship with the performance of industrial companies. *Revista de Gestão, 26*(2), 94–111. https://doi.org/10.1108/REGE-01-2018-0005

Landeweerd, L., Townend, D., Mesman, J., & Van Hoyweghen, I. (2015). Reflections on different governance styles in regulating science: A contribution to 'responsible research and innovation'. *Life Sciences, Society and Policy, 11*(8), 1–22. https://doi.org/10.1186/s40504-015-0026-y

Lehoux, P., Silva, H. P., Oliveira, R. R., & Rivard, L. (2020). The responsible innovation in health tool and the need to reconcile formative and summative ends in RRI tools for business. *Journal of Responsible Innovation, 7*(3), 646–671. https://doi.org/10.1080/23299460.2020.1844974

Lichtenthaler, U. (2016). Toward an innovation-based perspective on company performance. *Management Decision, 54*, 66–87. https://doi.org/10.1108/MD-05-2015-0161

Long, T. B., Blok, V., Dorrestijn, S., & Macnaghten, P. (2020). The design and testing of a tool for developing responsible innovation in start-up enterprises. *Journal of Responsible Innovation, 7*(1), 45–75. https://doi.org/10.1080/23299460.2019.1608785

Meissner, D., & Kotsemir, M. (2016). Conceptualizing the innovation process towards the 'active innovation paradigm' – Trends and outlook. *Journal of Innovation and Entrepreneurship, 5*(14), 1–18. https://doi.org/10.1186/s13731-016-0042-z

Meyer, J. V., Pascucci, L., & Mangolin, L. (2012). Gestão estratégica: Um exame de práticas em universidades privadas. *Revista de Administração Pública, 46*(1), 49–70. https://doi.org/10.1590/S0034-76122012000100004

Moczek, N., Voigt-Heucke, S. L., Mortega, K. G., Cartas, C. F., & Knobloch, J. (2021). A self-assessment of European citizen science projects on their contribution to the UN Sustainable Development Goals (SDGs). *Sustainability, 13*(4), 1–18. https://doi.org/10.3390/su13041774

Morawska-Jancelewicz, J. (2021). The role of universities in social innovation within quadruple/quintuple helix model: Practical implications from Polish experience. *Journal of the Knowledge Economy, 13*, 1–42. https://doi.org/10.1007/s13132-021-00804-y

Oduro, S. (2019). Examining open innovation practices in low-tech SMEs: Insights from an emerging market. *Journal of Science and Technology Policy Management, 10*(3), 509–532. https://doi.org/10.1108/JSTPM-03-2019-0036

Pasquale, P. P. (2012). *Comunicação Integrada de Marketing – a Teoria na Prática*. Elsevier.

Phillips, W., Lee, H., Ghobadian, A., O'Regan, N., & James, P. (2015). Social innovation and social entrepreneurship: A systematic review. *Group & Organization Management, 40*, 428–461. https://doi.org/10.1177/1059601114560063

Pozo, H., Akabane, G. K., & Tachizava, T. (2019). Innovation and technology processes in micro and small business. *Cogent Business & Management, 6*(1), 1–21. https://doi.org/10.1080/23311975.2019.1588088

Roach, D. C., Ryman, J. A., & Makani, J. (2016). Effectuation, innovation and performance in SMEs: An empirical study. *European Journal of Innovation Management, 19*, 214–238. https://doi.org/10.1108/EJIM-12-2014-0119

Saunders, M., Lewis, P., & Thornhill, A. (2009). *Research methods for business students*. Pearson Education.

Scherer, A. G., Palazzo, G., & Baumann, D. (2006). Global rules and private actors: Toward a new role of the transnational corporation in global governance. *Business Ethics Quarterly, 16*(4), 505–532. https://doi.org/10.5840/beq200616446

Schütz, M., & Strohmaier, R. (2022). Power relations in European RDI-collaboration networks. Disparities in policy-driven opportunities for knowledge generation in ICT. *Economics of Innovation and New Technology, 31*(3), 209–230. https://doi.org/10.1080/10438599.2020.1799139

SDG Compass. (2015). *The SDG Compass provides guidance for companies on how they can align their strategies as well as measure and manage their contribution to the realization of the SDGs*. https://sdgcompass.org

Sheehy, B., & Farneti, F. (2021). Corporate social responsibility, sustainability, sustainable development and corporate sustainability: What is the difference, and does it matter? *Sustainability, 13*(11), 1–17. https://doi.org/10.3390/su13115965

Silva, H. P., Lefebvre, A. A., Oliveira, R. R., & Lehoux, P. (2021). Fostering responsible innovation in health: An evidence informed assessment tool for innovation stakeholders. *International Journal of Health Policy and Management, 10*(4), 181–191. https://doi.org/10.34172/ijhpm.2020.34

Silverman, D. (2010). *Doing qualitative research: A practical handbook*. Sage.

Srivastava, V., & Mushtaq, G. (2011). Corporate restructuring – A financial strategy. *Asian Journal of Technology & Management Research, 1*(1), 1–9.

Stahl, B. C., Obach, M., Yaghmaei, E., Ikonen, V., Chatfield, K., & Brem, A. (2017). The responsible research and innovation (RRI) maturity model: Linking theory and practice. *Sutainability, 9*, 1–19. https://doi.org/10.3390/su9061036

Stahl, B. C., Chatfield, K., Holter, C. T., & Brem, A. (2019). Ethics in corporate research and development: Can responsible research and innovation approaches aid sustainability? *Journal of Cleaner Production, 239*, 1–14. https://doi.org/10.1016/j.jclepro.2019.118044

Stilgoe, J., Owen, R., & Macnaghten, P. (2013). Developing a framework for responsible innovation. *Research Policy, 42*(9), 1568–1580. https://doi.org/10.1016/j.respol.2013.05.008

Sutcliffe, H. (2011). *A report on responsible research and innovation*. MATTER and the European Commission.

Tavassoli, S., & Karlsson, C. (2016). Innovation strategies and firm performance: Simple or complex strategies? *Economics of Innovation and New Technology, 25*, 631–650. https://doi.org/10.1080/10438599.2015.1108109

Tello, S. F., & Yoon, E. (2008). Examining drivers of sustainable innovation. *International Journal of Business Strategy, 8*(3), 164–169.

Thorstensen, E. (2019). Responsibility for assistive technologies: Product assessment frameworks and responsible research and innovation. *Nordic Journal of Applied Ethics, 13*(1), 55–80. https://doi.org/10.5324/eip.v13i1.2525

Tijdink, J. K., Horbach, S., Nuijten, M. B., & O'Neill, G. (2021). Towards a research agenda for promoting responsible research practices. *Journal of Empirical Research on Human Research Ethics, 16*(4), 450–460. https://doi.org/10.1177/15562646211018916

Voegtlin, C., & Scherer, A. G. (2017). Responsible innovation and the innovation of responsibility: Governing sustainable development in a globalized world. *Journal of Business Ethics, 143*(2), 227–243. https://doi.org/10.1007/s10551-015-2769-z

Von Schomberg, R. (2013). A vision of responsible research and innovation. In R. Owen, J. Bessant, & M. Heintz (Eds.), *Responsible innovation: Managing the responsible emergence of science and innovation in society* (pp. 51–74). https://doi.org/10.1002/9781118551424.ch3

Wender, B. A., Foley, R. W., Hottle, T. A., Sadowski, J., Prado-Lopez, V., Eisenberg, D. A., Laurin, L., & Seager, T. P. (2014). Anticipatory life-cycle assessment for responsible research and innovation. *Journal of Responsible Innovation, 1*(2), 200–207. https://doi.org/10.1080/23299460.2014.920121

12 Sustainability in Erasmus+ Projects: StartIT

Hiram Bollaert, Johanna Hautamäki, João C. Monteiro, Krzysztof Podlaski, Dereck O'Reilly, and Katharina Simbeck

12.1 Introduction

Sustainability can be defined as "meeting the needs of the present without compromising the ability of future generations to meet their own needs" (United Nations Brundtland Commission, 1987). This was the first time the term sustainability was mentioned and its importance was highlighted. The concept focuses on the satisfaction of needs in a responsible way. The sustainability concept is largely related to the definition of economics. In fact, economics science is also essentially focused on the satisfaction of needs. As a matter of fact, we can define economics as being "the study of how people and society choose, with or without the use of money, to employ scarce productive resources which could have alternative uses, to produce various commodities over time and distribute them for consumption now and in the future among various persons and groups of society" (Samuelson & Nordhaus, 2022). From this definition, we can infer that economics is largely related to the

H. Bollaert
AP Hogeschool Antwerpen, Antwerp, Belgium

J. Hautamäki
Centria University of Applied Sciences, Kokkola, Finland

J. C. Monteiro (✉)
ISPGAYA, Vila Nova de Gaia, Portugal
e-mail: jmonteiro@ispgaya.pt

K. Podlaski
Uniwersytet Lodzki, Lodz, Poland

D. O'Reilly
DkIT, Dundalk, Ireland

K. Simbeck
HTW Berlin, Berlin, Germany

© The Author(s), under exclusive license to Springer Nature Switzerland AG 2023

F. L. Almeida et al. (eds.), *Multidimensional Sustainability: Transitions and Convergences*, Springer Proceedings in Earth and Environmental Sciences,
https://doi.org/10.1007/978-3-031-24892-4_12

choice we make about the use of scarce resources to satisfy unlimited human needs. Immediately associated with this concept is the concept of economic growth. Economic growth is an increase in the economic production. A continuous increase in production is needed to satisfy a higher number of needs and a higher percentage of the population. A higher production, although sometimes possible without it, should imply a higher, more intensive, use of resources. A correlated concept to economic growth is the concept of economic development. Economic development is primarily concerned with the well-being and quality of life. Hence, it is not only focused on higher production but also on the impact of that increased production in a wide variety of aspects. Consequently, improvements in health and education, for example, are relevant for the economic development of societies. As such, increase in life expectancy and literacy rates are essential elements of the economic development concept. But other aspects related to the global quality of life are also relevant. Among those, the ones connected to the environment are amid the most impactful. The quality of the air we breathe and the quality of the water we use are just a few examples of the importance of the environmental aspects to the concept of economic development. Economic growth is a purely quantitative concept. Economic development includes qualitative aspects. Sustainability adds to this qualitative insight, a significant long-time perspective. It does not focus solely on the present, but it looks ahead. Precisely because resources are scarce, the present use of resources must be appropriate, so that future use of those same resources should still be possible. Awareness of its relevance caused a shift in the way economic concepts should be addressed. This need for a different approach to economic growth and economic development is a key element to address sustainability and is the ground basis for sustainable development. This new approach to these concepts started to be dealt with only recently, but it has gained significant relevance.

In Europe, the European Union (EU) sustainable development strategy, launched by the European Council in Gothenburg in 2001, aims for the continuous improvement of quality of life for current and future generations. The different programs that the EU launches should also address this issue. Erasmus+ is the EU program to support education, training, youth, and sport in Europe. "The general objective of the programme is to support, through lifelong learning, the educational, professional and personal development of people in education, training, youth and sport, in Europe and beyond, thereby contributing to sustainable growth, quality jobs and social cohesion, to driving innovation, and to strengthening European identity and active citizenship" (Erasmus+ program guide, 2022). As it can be noted, the reference to sustainable development/sustainability, by the term sustainable growth, is one key element of the program. The priorities that are set to achieve such general objectives are inclusion and diversity, digital transformation, environment and fight against climate change, and participation in democratic life. All of which, either straightforward or indirectly, are aimed at a better quality of life for European citizens but also to support sustainable development. To achieve such goals, the Erasmus+ program aims to implement three different key actions. The first supports the general, and more known, mobilities of individuals. The second one supports cooperation among organizations and institutions under partnerships for cooperation, partnerships for excellence, and partnerships for innovation. The third one supports policy development and cooperation. Each one of these key actions should comply with the general rules established for the program. Naturally, it should also address the general objective established by the EU for the program. Therefore, the EU is setting up a procedure so that higher education institutions (HEIs) can properly address sustainability and might benefit from support to promote projects and actions that pursue sustainable development.

This chapter describes one project to be implemented under key action two. The StartIT got funding under Erasmus+ program. It started in March 2022, with the first activities associated to the project to be implemented shortly after. In this chapter, the intention is to describe the general objective of the project, its main activities, and how, with its proper implementation, it contributes to sustainable development in the EU. The next chapter makes such a description. It starts with a description of the project, and it continues with its focus on sustainability. Final remarks about the expected outcomes are presented.

12.2 StartIT

StartIT is a project developed by a consortium of six European Union HEI, led by HTW Berlin (Germany), together with AP Hogeschool Antwerpen (Belgium), Centria University of Applied Sciences (Finland), DKIT (Ireland), ISPGAYA (Portugal) and University of Lodz (Poland). The project is financed by Erasmus+ under contract no. ° 2021-1-DE01-KA220-HED-000023215. As the budget code suggests, it falls under key action two, being a strategic partnership project for cooperation.

12.2.1 Project Background

In today's global economy, soft skills, future skills, entrepreneurial skills, and IT skills are increasingly important. HEI need to invest in providing their students with the skills that are highly sought-after in the labor market. COVID-19 has had a devastating effect on the lives of many EU citizens. One group that has been particularly badly hit is young people. By moving online, third-level education across the EU has mainly remained open. New teaching methodologies emerged. However, despite the development of online methods, HEI focused on the essential technical/scientific contents. The above-mentioned skills were, most of them, neglected in the process. In fact, as they have moved online, many HEIs have focused their efforts on ensuring the successful online delivery of their core disciplinary subjects. As a result of this, there is an even greater deficit in soft and future learning among students. Online learning does not lend itself to the teaching of soft-skill competencies, such as communication, leadership, decisiveness, teamwork, responsibility, time management, problem solving, ability to work under pressure, flexibility, negotiation, and conflict resolution. Nonverbal communication can support a speaker's ability to communicate effectively in a group, using hand gestures to indicate the importance of an idea to your audience, nodding your head to express solidarity with the speaker, conveying feelings such as disappointment, relief, or happiness, offering support with a simple smile or pat on the back or deescalating tension by using a calm tone of voice. Nonverbal cues can even highlight that a person is being dishonest, unengaged, excited, or aggressive. According to the OECD/EU (2020), about 40% of youth express an interest in entrepreneurship but very few youths are self-employed. This gap signals untapped entrepreneurial potential. Youth face many barriers, including a lack of entrepreneurship skills. Other key barriers include a lack of entrepreneurship role models, little entrepreneurship and work experience, few financial resources, limited business networks, and market barriers such as low credibility with potential customers. Common public policy responses to these barriers should include entrepreneurship training, grants, and loans for business start-ups, coaching and mentoring, and support in network building. It is also important for public policy to go beyond helping youth start businesses by helping them develop and grow their businesses. Youth have a high level of interest in self-employment but only 6.5% of working youth (20–29 years old) in the European Union (EU) were self-employed in 2018. Furthermore, the number of self-employed youths has declined over the past decade, falling from 2.7 million in 2009 to 2.5 million in 2018. The gender gap in self-employment starts early. In 2018, young women in the EU were only about 60% as likely as young men to be self-employed (4.8% vs. 8.0%). According to the EU (2021), the COVID-19 pandemic highlighted the importance of digitization for European society. The digital compass sets out objectives to achieve the EU's vision for the digital future, called the four points of the compass to identify the main goals to reach over the next decade: digitally skilled population and highly skilled digital professionals; – secure and substantial digital infrastructures; digital transformation of businesses; and digitization of public sectors. StartIT is an innovative approach to teaching, through the application of an active learning methodology, these skills.

12.2.2 Project Description

The main goal of the project is the promotion of digital transformation in higher education students while addressing climate change. The project promotes innovative learning and teaching methodologies, bringing together students from the six member countries to work in diverse, international, interdisciplinary teams, thus acquiring skills that will enable them a higher social well-being. They will be challenged to develop a mobile app intended to mitigate climate change. At the same time, they will also have to develop an associated business plan. With this joint development, they will contribute to the promotion of the digital economy and environmental protection. Three different student mobilities will be organized within the next 3 years to achieve these objectives.

Being financed by Erasmus+, the project must contribute to some of the program's priorities. This project contributes to the following program priorities:

Higher education–specific priorities
Stimulating innovative learning and teaching practices

All partnering higher education institutions (HEI) include project activities and team tasks in their regular curricula. However, those curricular activities focus on exercising specific capabilities that have been taught before. Start IT will develop this teaching concept further by putting students specifically in situations where they need to identify missing skills and acquire them in the process of working on the outcome. The new teaching experience will also need to be linked to different gradable activities (i.e., lab books and learning diaries). We are developing and using an active learning procedure to teach students about soft and digital skills. The participating HEI will benefit from increased capabilities in the application of active learning.

Horizontal priorities:

Addressing digital transformation through development of digital readiness, resilience, and capacity.
Environment and fight against climate change.

The skills that we teach our students today will be outdated in 20 years from now. Therefore, students not only need to learn to master the foundations of their discipline but also to acquire new skills and knowledge. We are developing a concept to teach this important skill by tasking the students to work in international, interdisciplinary teams to create a digital entrepreneurial platform to tackle major environmental challenges. Building the digital platform (i.e., mobile app or web app) will give students the opportunity to acquire new digital skills and understand current technological trends. This proposal is about enhancing soft skills in higher education students. We shall organize events during which students work in teams. The task, to implement a business plan in an Information and Communication Technologies (ICT) related business concept or idea, is merely a flag to cover the crucial skills, which need to be stimulated and nurtured. The task will foster inherently digital literacy, whereas the format of the event activates this crucial set of skills. Being challenged to develop a specific task, students are mixed in international and multidisciplinary teams. They will need to actively communicate within their team and outside to build a picture of the needs of potential clients and users. By doing so they encounter team roles to organize themselves, and they practice ways of giving and receiving feedback to stimulate each other toward a proper outcome of their collaboration. As a result, they become an interdisciplinary team. Skills, knowledge, and expertise flow over from one student to another, as they discuss and share ideas, promoting creativity and critical thinking, and judging ideas for their value and feasibility. From scenarios, outlining how certain decisions have a positive or negative impact on reaching their potential clients, the student teams develop a story filled with persona (characterization of users) and actions (feedback cycles between application and user). The outcome is logical and coherent. At the same time, and because the focus of all the mobilities and of all the apps to be developed is the environment and the fight against climate change, the project itself is also related to sustainable development. The digital platforms to be developed are with a local focus but must allow for global implementation.

It is planned to involve local partners in the teaching and learning activities in Porto, Berlin, and Antwerp. Those will contribute with guest lectures, mentoring, and coaching. For Porto, the local partners will be INOVAGAIA and the Vila Nova de Gaia city hall. The first partner is a business center for digital start-ups that regularly promotes entrepreneurship seminars and cooperates with HEI in seminars about this topic. The city hall environmental and tourism department will cooperate in the mobility by setting up the goals/standards/aims for the application in what relates to the environment. For Berlin, the partners will be two multinational companies that have offices in Berlin and strongly support both education and environmental activities. For Antwerp, it will be Ecohuis Antwerpen. The local partner will bring climate change domain expertise to mobility.

12.3 Conclusions

The objective was to demonstrate that sustainability can be a part of projects designed to tackle other priorities. StartIT is essentially promoted to teach nontechnical skills to higher education students, with a new active methodology. However, despite this focus, sustainability is an essential element of the process. The subject of the learning process will be the environment and climate change thus promoting sustainable practices. In fact, and because sustainability practices will be developed and put into practice, the outcomes of the project are not only for the ones directly involved in it, but for a wider audience. The impact should be felt at a local, regional, national, European, and international level. With the promotion of sustainable practices, sustainable development is closer to being a reality. The project started to be developed in 2022 and will extend itself until 2025. A review of its implementation and a proper evaluation of its results are the steps that follow.

References

European Commission. (2021). *Digital strategy of EU – The digital compass*. https://digital-strategy.ec.europa.eu/en/policies/digital-compass

European Commission. (2022). *Erasmus+ program guide*. https://erasmus-plus.ec.europa.eu/programme-guide/erasmus-programme-guideintroduction

OECD/European Commission. (2020). *Policy brief on recent developments in youth entrepreneurship* (OECD SME and Entrepreneurship Papers (No. 19)). OECD Publishing. https://doi.org/10.1787/5f5c9b4e-en

Samuelson, P., & Nordhaus, W. (2022). *Economics* (19th ed.). McGraw Hill.

United Nations Brundtland Commission. (1987). *Report of the world commission on environment and development: Our common future*. https://www.un.org/en/academic-impact/sustainability

The Challenges of Teaching in a Context of Concern for Sustainability and Strong Technological Innovation: A Literature Review

13

Justino Lourenço, Susana Sá, and José Carlos Morais

13.1 Introduction

A new era in the twenty-first century is changing the way we interact in our daily routine. In parallel, research work is being conducted to understand the best innovative educational practices to meet the educational needs of a new generation (Li & Zhang, 2021).

C3 (Creativity, Capability, and Challenge) is a new concept, which in particular is being tested and evaluated in engineering courses. At the Innovation Centre for Engineering Education (ICEE), a project is underway: Cultivating challenging twenty-first-century students with creative thinking, practical skills, and a future-oriented spirit of challenge. This can be simplified as C3: Creativity Fostering Education, Practical Skills Enhancement Education, Future-Oriented Challenge Spirit, and Engineering Education Support System Innovation are categories of the ongoing project (Lim et al., 2013).

Innovation in education is happening with the impetus of technology introduction. The general concern for environmental sustainability is a current issue in discussion and research. In addition to the scientific community, citizen groups around the

J. Lourenço (✉)
ISPGAYA, Vila Nova de Gaia, Portugal
e-mail: jml@ispgaya.pt

S. Sá
Research Center, Instituto de Estudos Superiores, Medelo, Portugal

J. C. Morais
ISPGAYA, Vila Nova de Gaia, Portugal

CEOS.PP, Porto, Portugal

© The Author(s), under exclusive license to Springer Nature Switzerland AG 2023

F. L. Almeida et al. (eds.), *Multidimensional Sustainability: Transitions and Convergences*, Springer Proceedings in Earth and Environmental Sciences,
https://doi.org/10.1007/978-3-031-24892-4_13

world are also demanding much more practical and feasible attitudes toward a more sustainable world.

The new path that is emerging will bring a natural fusion of new educational practices, with a panoply of technological support allied to the concern for the planet's sustainability.

The research question is: How does the University contribute to a community formed by citizens that are aware of the sustainability of the planet?

13.2 Methodology

The main objective of this study is to carry out an integrative review to find out in a global way how the University contributes to a community formed by citizens that are aware of the sustainability of the planet.

The integrative review method, proposed by Whittemore and Knafl (2005), was used to guide this study, as it is considered the most appropriate to achieve the central objective, from a variety of scientific sources. This method allows determining the current state of knowledge on a given topic, through the investigation of the literature and interconnection between the different contributions of the state of the art. "The integrative review method is the only approach that allows for the combination of several methodologies (e.g. experimental and non-experimental research) and has the potential to offer a major contribution in evidence-based practice" (Whittemore & Knafl, 2005, p. 547).

To ensure the methodological rigor of this study, the authors propose four steps: problem identification (previously performed), literature search, data evaluation, and data analysis, presented below.

The literature search was conducted using the Scielo, Web of Science, and Google Scholar databases. This search took place during the month of July 2022. Research published in English, giving preference to the last 20 years of publication, was included, using as keywords: "Sustainable education", "Sustainability", "Learning environments," and "University education".

The Search in the databases was carried out by all the authors of the study, and the information was cross-checked at the end of the stage.

To ensure the scientific rigor of this study, we elected the following inclusion criteria: (a) it is a scientific study; (b) it uses primary data sources; (c) articles in Portuguese and English language, and (d) peer review. Scientific research, using quantitative, qualitative, and mixed methods, was also included. Thus, research that did not directly focus on the topic under study, grey literature (unpublished studies), as well as articles of a reflexive nature were excluded.

We began by analyzing the articles that met the established inclusion criteria. A total of 38 documents were selected, and 361 were rejected because they did not focus on the defined characteristics or because they corresponded to publication dates prior to 2002. The articles were subsequently obtained in full and analyzed. Of the 117 remaining articles, 111 were rejected for not complying with the objective of the study and for not being of an investigative nature, but rather reflective. From this evaluative process, 38 studies resulted.

13.3 The Analysis of Data

13.3.1 Technology Sustainability

Since the 1970s, advances in technology have affected everything including education, training, and business environment, and jobs have always been challenged between educational, economic, social, and environmental disruptions in the best possible way to ensure the continuity of educational business. Therefore, it is necessary to recognize and manage today's challenges and according to the 3D model of Socio-Environmental SMEs to 1-) predict, 2-) prevent, and 3-) cope with tomorrow's crises (Mohammadian, 2022).

On the other hand, the same Internet of Things (IoT) technology is a powerful tool to fuel the sustainability theme (Siddula et al., 2018; Chen et al., 2020; Schmitt et al., 2018; Ramakala et al., 2017). The University's practice and knowledge can be applied in hindsight in real sustainable applications: real-time air quality monitoring system based on multiple LoRa sensor nodes to implement an Internet of Things (IoT) network (Od et al., 2021) in a water treatment system (Tretjakova et al., 2022).

13.3.2 Energy Sustainability

A major concern is the energy savings in educational buildings (Mylonas et al., 2019). Existing studies neither satisfy the demand for fast response in time-constrained IoT applications nor fundamentally solve the problem of energy sustainability. Therefore, discussing the problem of energy sustainability and timeliness in IoT systems is a relevant research topic (Deng et al., 2019). The automatic lightning discharge control system in a University can also bring a relevant contribution to sustainability (Cruz et al., 2021). Finally, as a relevant theme, the new paradigm of smart cities may have a huge contribution to the sustainability issue (Doster & Chavis, 2021).

13.3.3 Indicators about Artificial Intelligence Application and Sustainability

The theme artificial intelligence (AI) is also a strong new course in universities. A productive and effective learning environment should expose students to information, but also to experimentation. Instead of a common model of repetitive exercise, the incorporation of neural networks would bring a very rich learning experience, thus leading to better teaching practices. Thus, AI will bring a positive revolution for learners. Learning will be a personalized experience with a pace tailored to the learner's profile. The literature points to the following areas that will be most impacted: personalized learning, experiential learning, performance-linked learning, lifelong learning, and virtual assistance. This topic is detailed in Sinha and Huraimel, Transforming Education with AI (2021).

AI also has several other fields of application in university curricula: enhancer of foreign language learning (Zhu, 2020), multidisciplinary study models (Laato et al., 2020), informational literacy training and information security education (Jun & Yu, 2012), computer science learning (Kandlhofer et al., 2016), and as an enhancer of health education (Jiang et al., 2021).

In addition to providing a tool for academic purposes and real-life enhancement tools, AI can and should be integrated into the issue of sustainability. AI and sustainability is a nascent area, and elaborating its brief history with indicators of recent and notable activities is the main goal of this paper, as well as contextualizing AI and sustainability within the larger computing and sustainability movement (Fisher, Computing and AI for a Sustainable Future, 2011) and all its latest advances (Dauvergne, 2020).

In the concept of smart cities, particularly in terms of energy efficiency and Intelligent Transportation Systems (ITS), AI plays an important role (NICODEME, 2021; Sinha & Huraimel, 2021).

The alarming rate of environmental destruction in the near future will undermine any attempt to reconcile human well-being and development with respect for the basic rules of sustainability. It is clear that International Institutions are not oriented at an adequate speed toward a global convergence of environmental sustainability. Some AI models and simulations can help humans understand the "point of no return" in the environmental crisis (Lahsen, 2020).

A totally new design of our industrial system of production and consumption around the circular patterns of resources versus energy consumption observed in mature ecosystems is a real need. Thinking and designing a regenerative economy model brings a whole new and complex model that requires a huge amount of data to be analyzed. The integration of data science – together with AI as tools to accelerate the rapid transition needed for regenerative solutions – is a matter of urgency (Kadar & Kadar, 2020).

The agriculture sector is under extreme pressure to converge toward a balanced environment of production and consumption. There is an incredible shortage of resources such as water and soil and, on the other hand, serious climate change problems. At the same time, there is a known problem in attracting young people to this sector. Developing a solution that provides a communication path between producers and buyers that enables a better balance in production versus consumption of agricultural goods with the support of AI would be a big step toward an environmentally friendly approach to agriculture in the twenty-first century, this could be implemented with a mobile app supported by a shared platform of buyers and consumers (Ogubuike et al., 2021).

Water wastage and declining water availability in some countries are of extreme concern. Water desalination is a global strategy to suppress the problem. Strategic planning and various technical decisions have a huge impact. AI can provide a tool for decision makers and thus achieve greater optimization in terms of environmental, social, and economic impacts (AlZu'bi et al., 2019).

On the other hand, research has shown the increasing carbon footprint of machine learning (ML)-oriented applications due to the extraordinary growth in the size of deep learning models, which are estimated to have grown by a factor of 300,000 in the last 6 years. It is relevant to understand the preliminary exploration of the energy usage profile of ML training in the cloud and demonstrate how transfer learning can be used to reduce this energy consumption (Walsh et al., 2021).

The AI and the digital learning approach has a broad range of application from enhancement of the learning process to sustainabilty, and also some crossed positive relations (Kastornova et al., 2022), several cutting-edge researches point in the same direction (Alsyouf, 2020; Bian, 2009; Zhiming & Hai-ying, 2009; Gao, 2021; Zhang & Xia, 2021; Gomes, 2021; Guerra & Costa, 2018; Kadnár & Hrmo, 2010; Peters et al., 2020; Prins et al., 2008; Tang, 2021; Yunos et al., 2016).

13.4 Conclusions

The University promotes the sustainability of its citizens in a way that can be discussed as a form of cross-cutting computational issues and AI challenges with the help of optimization, machine learning, multi-agent reasoning, citizen science, and crowdsourcing.

The analysis of international experience in the use of digital technologies in the activities of educational institutions in foreign countries with high-quality education and analysis of international experience in the use of digital technologies in the management of general education educational institutions were tested in the following countries: Australia, Singapore, Sweden, France, Russia, and the United States. The results show a promising path.

Admittedly, the education stage should take these concerns and bring teaching practices and course themes that deepen the discussion about a more sustainable world to the entire young generation.

Students and teachers, both at the level of formal education or formal construction, which promotes the use of useful and usable knowledge in the context and utilities of life, allow each of them to improve reality with natural reality, providing greater motivation to education for sociability, collaborative work and the promotion of sustainable education, even in a context of strong technological innovation, for example, with the use of active methodologies, such as artificial intelligence. The approach for studying transformative learning for sustainable development presented in this paper calls for a multidisciplinary approach.

This paper has argued for moving beyond simplistic, win–win understandings of learning for sustainable development. To enable sustainable transformations of modern societies, we need a more elaborate understanding of learning; one which accounts for the institutional, social, and conflictual dimensions of learning. We need insights on how transformative learning could be managed. The approach outlined here constitutes an important step in such direction. It calls for transformative learning rooted in disciplines across the natural and social sciences taking a holistic approach that acknowledges how power permeates differing notions of sustainability across the world's globalized societies. Learning for sustainable development is too important for anything less.

References

Alsyouf, I. (2020). Sustainability circles the way to sustainability excellence in institutions of higher education. In *2020 advances in science and engineering technology international conferences (ASET)*.

AlZu'bi, S., Alsmirat, M., Al-Ayyoub, M., & Jararweh, Y. (2019). Artificial intelligence enabling water desalination sustainability optimization. In *2019 7th international renewable and sustainable energy conference (IRSEC)*.

Bian, L. (2009). Modern education technologies in higher special education. In *2009 first international workshop on education technology and computer science*.

Chen, W., Deligiannis, N., Andreopoulos, Y., & Wassell, I. J. (2020). *Wei Chen; Nikos Deligiannis; Yiannis Andreopoulos; Ian J. Wassell*. China Communications.

Cruz, I. Z., Bernabe, J. G., Benítez, R. A., & Toxqui, J. M. (2021). Design of a prototype for automatic lighting control in classrooms at Faculty of Accounting and Administration Xalapa based on the IoT paradigm. In *2021 IEEE international conference on engineering Veracruz (ICEV)*.

Dauvergne, P. (2020). II the prospects and limits of AI. In *AI in the wild: Sustainability in the age of artificial intelligence*. MIT Press.

Deng, Y., Chen, Z., Yao, X., Hassan, S., & Ibrahim, A. M. (2019). Parallel offloading in green and sustainable mobile edge computing for delay-constrained IoT system. *IEEE Transactions on Vehicular Technology, 68*, 12202.

Doster, M. K., & Chavis, J. S. (2021). IoT & smart cities: "Smartainability". In *2021 IEEE integrated STEM education conference (ISEC)*.

Fisher, D. H. (2011). Computing and AI for a sustainable future. *IEEE Intelligent Systems, 26*, 14.

Gao, X. (2021). Discussion on the integration of information technology into the practical application of college education and teaching under the background of the new era. In *2021 international conference on education, information management and service science (EIMSS)*.

Gomes, C. P. (2021). Keynote 2 – Computational sustainability: Computing for a better world and a sustainable future. In *2021 IEEE international conference on smart computing (SMARTCOMP)*.

Guerra, C., & Costa, N. (2018). Educational innovations in engineering education: Sustainability of funded projects developed in Portuguese higher education institutions. In *2018 3rd international conference of the Portuguese Society for Engineering Education (CISPEE)*.

Jiang, T., Li, W., Wang, J., & Wang, X. (2021). Using artificial intelligence-based online translation website to improve the health education in international students. In *2021 2nd international conference on artificial intelligence and education (ICAIE)*.

Jun, Z. A., & Yu, W. C. (2012). Research on information literacy training and information security education. In *2012 7th international conference on computer science & education (ICCSE)*.

Kadar, T., & Kadar, M. (2020). Sustainability is not enough: Towards AI supported regenerative design. In *2020 IEEE international conference on engineering, technology and innovation (ICE/ITMC)*.

Kadnár, J., & Hrmo, R. (2010). The implementation of information and communication technologies in life-long education at the Faculty of Materials Science and Technology in Trnava (Slovakia). In *2010 2nd international conference on education technology and computer*.

Kandlhofer, M., Steinbauer, G., Hirschmugl-Gaisch, S., & Huber, P. (2016). Artificial intelligence and computer science in education: From kindergarten to university. In *2016 IEEE Frontiers in education conference (FIE)*.

Kastornova, V. A., Kozlov, O. A., Mukhametzianov, I. S., Poliakov, V. P., Robert, I. V., & Shikhnabieva, T. S. (2022). International experience of digital technologies applying in secondary education organizations activities. In *2022 2nd international conference on technology enhanced learning in higher education (TELE)*.

Laato, S., Vilppu, H., Heimonen, J., Hakkala, A., Björne, J., Farooq, A., et al. (2020). Propagating AI knowledge across university disciplines – The design of a multidisciplinary AI study module. In *2020 IEEE Frontiers in education conference (FIE)*.

Lahsen, M. (2020). Should AI be designed to save us from ourselves?: Artificial intelligence for sustainability. *IEEE Technology and Society Magazine, 39*, 60.

Li, H., & Zhang, Y. (2021). Practice exploration of innovation education mode based on science and technology innovation team under new engineering background. In *2021 2nd international conference on education, knowledge and information management (ICEKIM)*.

Lim, J.-M., Kim, S.-C., & Lee, C.-H. (2013). C3 (creativity, capability, and challenge) activities of innovation center for engineering education. In *Proceedings of 2013 IEEE international conference on teaching, assessment and learning for engineering (TALE)*.

Mohammadian, H. D. (2022). Mapping the future sustainable, through the theory of comprehensive everything or the 5th wave/tomorrow age theory, with a focus on hybrid SMEs/SME 5.0 educationally. In *2022 IEEE global engineering education conference (EDUCON)*.

Mylonas, G., Amaxilatis, D., Tsampas, S., Pocero, L., & Gunneriusson, J. (2019). A methodology for saving energy in educational buildings using an IoT infrastructure. In *2019 10th international conference on information, intelligence, systems and applications (IISA)*.

Nicodeme, C. (2021). AI legitimacy for sustainability. In *2021 IEEE conference on technologies for sustainability (SusTech)*.

Od, S., Huang, H.-H., & Wei, J.-B. (2021). Apply LoRa technology to construct an air quality monitoring IoT system. In *2021 IEEE 3rd Eurasia conference on biomedical engineering, healthcare and sustainability (ECBIOS)*.

Ogubuike, R., Adib, A., & Orji, R. (2021). Masa: AI-adaptive mobile app for sustainable agriculture. In *2021 IEEE 12th annual information technology, electronics and mobile communication conference (IEMCON)*.

Peters, A.-K., Bengtsson, S., Cajander, Å., Daniels, M., Grande, V., Lönngren, J., & Salminen-Karlsson, M. (2020). Care ethics to develop computing and engineering education for sustainability. In *2020 IEEE Frontiers in education conference (FIE)*.

Prins, R., Kander, R., Moore, T., Pappas, E., & Pierrakos, O. (2008). Special session – Engineering for a sustainable world: How do we incorporate sustainability in undergraduate engineering education? In *2008 38th annual Frontiers in education conference*.

Ramakala, R., Thayammal, S., Ramprakash, A., & Muneeswaran, V. (2017). Impact of ICT and IOT strategies for water sustainability: A case study in Rajapalayam-India. In *2017 IEEE international conference on computational intelligence and computing research (ICCIC)*.

Schmitt, C., Steiner, Y., Herzog, R., Terrenghi, N., & Stiller, B. (2018). A 4-step tool-chain for IoT business model and sustainability recommendations. In *2018 IEEE Globecom workshops (GC Wkshps)*.

Siddula, S. S., Babu, P., & Jain, P. C. (2018). Water level monitoring and management of dams using IoT. In *2018 3rd international conference on internet of things: Smart innovation and usages (IoT-SIU)*.

Sinha, S., & Huraimel, K. A. (2021). Transforming education with AI. In *Reimagining businesses with AI*. Wiley.

Tang, W. (2021). Research on the application and practice of continuing education in college of continuing education based on WeChat official account: Guangdong University of Science and Technology. In *2021 international conference on internet, education and information technology (IEIT)*.

Tretjakova, R., Halvorsen, H.-P., Viumdal, H., Mylvaganam, S., Timmerberg, J., & Thiriet, J. M. (2022). Water care system (WCS) for numeric water quality criteria (NWQC) with IoT/IoUT – Educational implications of sustainability goal SDG 14. In *2022 31st annual con-*

ference of the European association for education in electrical and information engineering (EAEEIE).

Walsh, P., Bera, J., Sharma, V. S., Kaulgud, V., Rao, R. M., & Ross, O. (2021). Sustainable AI in the cloud: Exploring machine learning energy use in the cloud. In *2021 36th IEEE/ACM international conference on automated software engineering workshops (ASEW).*

Whittemore, R., & Knafl, K. (2005). The integrative review: Updated methodology. *Journal of Advanced Nursing, 52*(5), 546–553.

Yunos, J. M., Sern, L. C., & Hamdan, N. H. (2016). Changes and challenges in sustainability of technical and vocational education and training – Teacher education programme: A case study. In *2016 IEEE 8th international conference on engineering education (ICEED).*

Zhang, M., & Xia, Y. (2021). The effective research of computer multimedia technology in auxiliary higher vocational physical education teaching. In *2021 international conference on internet, education and information technology (IEIT).*

Zhiming, Q., & Hai-ying, H. (2009). Application of performance technology in vocational technology education based on comprehensive fuzzy evaluation. In *2009 second international conference on education technology and training.*

Zhu, A. (2020). Application of AI identification technology in foreign language education. In *2020 international conference on artificial intelligence and education (ICAIE).*

Part V

Management & Administration

Exploring the Determinants of Social Entrepreneurship Intention

14

Fernando Almeida and José Milton de Sousa-Filho

14.1 Introduction

Entrepreneurship is a topic that has been the subject of intense study and research in recent decades. It is present both in the academic perspective and in the main political, economic, and social discussions. According to Kumar and Raj (2019), entrepreneurship is a factor of economic balance, contributing to innovation, generation of new companies, and increased dynamism in the market and in society. Entrepreneurship is also associated with risk, rupture, and paradigm shift (Marquez & Ortiz, 2021; Zhao, 2005). Young people are attracted to this type of business philosophy in which they can propose innovative solutions and create value in a short period of time.

One of the areas of entrepreneurship that has gained great visibility due to its contribution to a more balanced and sustainable social development is social entrepreneurship. Ruskin et al. (2016) point out that an individual involved in social entrepreneurship is a person concerned with developing relationships of trust and respect with the community. In this sense, this individual is an actor who contributes to the sustainable development of that community. Therefore, social entrepreneurship has as its main objective the involvement of local communities in a set of activities aimed at promoting their well-being and reducing the risk of harmful behaviors particularly in adverse situations such as unemployment, economic recession, or natural disasters (Barki et al., 2015; Chell et al., 2010; Dwivedi & Weerawardena, 2018).

F. Almeida (✉)
ISPGAYA & INESC TEC, Porto, Portugal
e-mail: falmeida@ispgaya.pt

J. M. de Sousa-Filho
University of Fortaleza, Fortaleza, Brazil

© The Author(s), under exclusive license to Springer Nature Switzerland AG 2023

F. L. Almeida et al. (eds.), *Multidimensional Sustainability: Transitions and Convergences*, Springer Proceedings in Earth and Environmental Sciences,
https://doi.org/10.1007/978-3-031-24892-4_14

Studies conducted by Cohen et al. (2019) and Satar and Natasha (2019) seek to profile a social entrepreneur. The results allow us to understand some key factors such as the ability to take initiative, recognize opportunities, and take risks. Equally relevant is that the proposed solutions are low cost and high cost. The concern for the sustainability of the projects presented is also highlighted by Kamaludin et al. (2021). Individuals characterized as social entrepreneurs can also emerge in the context of a classical business framework (Hadad & Cantaragiu, 2017). These change agents can stimulate organizations to step out of their comfort zone and identify new lines of action outside their traditional area of action. This new paradigm opens up new opportunities for both individuals and companies and currently represents one of the emerging trends in terms of sustainability.

This study looks at the role that higher education can play in the emergence of entrepreneurial individuals in the social area as recognized in Haugh and Talwar (2016) in which it is argued that education can play a relevant role in enhancing skills in individuals to understand the specificities of their communities and territories. Furthermore, Tran and Von Korflesch (2016) recognize the need to academically explore the entrepreneurial intentions of students in higher education and characterize the entrepreneurial potential of these individuals. In this regard, empirical studies have emerged, like Hockerts (2018) and Naveed et al. (2021), that explore how individual social entrepreneurship orientations and entrepreneurial education have contributed to the increased entrepreneurial intention of these students. Despite the relevance of these studies, their replication potential for diverse student profiles is limited as they have difficulties in characterizing students' entrepreneurial intention in a sufficiently comprehensive way because their profile is very homogeneous and only focus on the creation of new social entrepreneurship projects at the individual level. In addition, it is also relevant to investigate the effect of other constructs in the model such as organizational and contextual factors that are equally relevant to characterize entrepreneurial intention (Ahadi & Kasraie, 2020; Looi, 2020; Sansone et al., 2021). This study addresses this challenge and research gap and seeks to gauge entrepreneurial intention considering these three constructs. This research was conducted in the context of a higher education institution in Portugal, in which students attend an entrepreneurship course in the final year of their undergraduate degree in management and computer science. The heterogeneous and multidisciplinary profile of the students gathers unique characteristics that allow exploring the entrepreneurial intention considering diversified characteristics of each student. The rest of this chapter is organized as follows: Initially, a theoretical contextualization is made about entrepreneurship teaching, also considering the specific case of social entrepreneurship. Next, the methodology adopted is presented. After that, the results of the study are analyzed and discussed considering the three main constructs of the model (i.e., individual, organizational, and contextual). Finally, the main conclusions of this work are highlighted and some suggestions for future work are given.

14.2 Background

14.2.1 Entrepreneurial Education

The European Commission establishes in the New Skills Agenda for Europe (EC, 2016) the need to promote entrepreneurship. This is a phenomenon still very much linked to the business view of the phenomenon in which the aim is to encourage the emergence of new startups. From this perspective, the focus is on turning ideas into viable and commercially successful businesses (Murdock & Varnes, 2018). However, this overly focused view on the emergence of new businesses does not always prioritize entrepreneurial education.

Entrepreneurship education should aspire to be more ambitious and develop fundamental marketable skills. Unequivocally the concept is related to launching an innovation in the market, but it is also equally fundamental to stimulate the development of skills that are common to the entrepreneur. This view of entrepreneurship education is based on the idea that it is a constant learning process. There are several studies that build on this conceptualization of entrepreneurship teaching to explore its effects on students' entrepreneurial intention (Amofah & Saladrigues, 2022; Kara et al., 2022; Passoni & Glavam, 2018; Sohu et al., 2022).

The entrepreneur must possess a diverse and harmonious set of knowledge, skills, and entrepreneurial attitude. From the knowledge perspective, the identification of business opportunities that allow them to grow personally and professionally is fundamental (Clausen, 2020). Added to this element is the concern with business ethics (Harris et al., 2009). In terms of skills, proactive project management stands out in which it is essential to know how to organize, lead, delegate, and communicate (Felix et al., 2019; Odewale et al., 2019). It is important to be able to handle the moments when the entrepreneur will be isolated but also to communicate and work in a team (Almeida & Devedzic, 2022). Finally, entrepreneurial attitude is characterized as a set of characteristics assumed in personal, social, and work life (Farrukh et al., 2018).

In the literature it is further argued that entrepreneurship education aims to strengthen individuals' abilities to see and exploit opportunities in an economic, cultural, and social context (Rusok et al., 2017). Therefore, entrepreneurship education implies more than introducing new activities. It also implies the development of higher education institutions according to a maturity model proposed by Almeida and Amaral (2019). One of the key points is argued by Curtis et al. (2021) in which the adoption of active learning methodologies is advocated. In this model, one no longer looks exclusively at the direct transmission process from teaching to learning, but students are given a leading role in the learning process. Crisol-Moya et al. (2020) add that this teaching model encourages the academic community to develop the capacity to absorb content autonomously and participatively.

14.2.2 The Specific Case of Social Entrepreneurial Education

Social entrepreneurship addresses the goal of building a better, fairer, and more balanced society. As Mandrysz (2020) points out, the basic ideology of social entrepreneurship seeks to involve society in the search for transformative solutions. In practice, the social entrepreneur is responsible for creating their own businesses and seeking new resources. The social entrepreneur seeks to create organizations to fill social gaps that cannot be fully addressed by governments. At this level, Prim et al. (2019) highlight the role that collaborative networks can play in bringing the social entrepreneur's initiatives to fruition. Here too, education emerges as a key element, as skills related to soft skills are relevant elements in establishing these networks (Klyver & Arenius, 2020).

Innovation is an indispensable element in social entrepreneurship since the projects developed in this area project significant changes in the society where it is implemented (Bataglin & Kruglianskas, 2022; Boughzala, 2021). Innovation, both from the perspective of products and services, stems from the search for methodologies that seek to find solutions to social problems, and the measure of success of these initiatives can be given by the impact caused in society. Entrepreneurship education can also play a relevant role here by providing methodologies for an individual to identify different ways to change the living conditions of the less privileged in society.

Qualifying an individual and securing decent employment are conditions of great relevance in combating poverty and social exclusion (Halleröd & Larsson, 2008; Pohlan, 2019). This is a challenge that cannot be addressed exclusively by governments but must call for the participation of other entities. In this context, higher education institutions can play a relevant role in entrepreneurship education by promoting the development of skills and methodologies that allow students to become more entrepreneurial, and that these projects have a strong impact on the development and transformation of society.

Social entrepreneurship education is also an element that can play a relevant role in promoting education with sustainability concerns as argued in the studies by Ashari et al. (2022) and Del Vecchio et al. (2021). Social entrepreneurship is strongly related to the idea of inclusive and sustainable development by simultaneously looking at the development needs of societies while considering the limits of the planet. The conjunction of both factors is fundamental for societies to develop in harmony and respect local communities (Esteves et al., 2021).

14.3 Data and Methods

The data for this study were obtained from students taking an entrepreneurship course in higher education. This course is part of the BSc in Computer Science and the BSc. in Management which allows us to include in this sample a relatively broad and rich student profile. The objective of this course was to formulate new business projects in the field of social entrepreneurship. The course includes the phases from the ideation of a new business to the formulation of a business plan, which includes a financial plan that aims to ensure the viability and sustainability of each business. The students were divided into working groups of four to six students from both courses. We sought to include students from both courses in each group to strengthen interdisciplinarity among the students. The quantitative data regarding entrepreneurial intention was obtained from a survey that received a total of 177 valid responses. A response is valid when at least 50% of the survey is filled out.

The structure of the survey can be seen in Table 14.1. A total of four constructs were considered: (i) control variables with sociodemographic information on the respondents' profile; (ii) individual with dimensions related to the characteristics of each respondent; (iii) organizational with information on the characteristics of entrepreneurship support programs and means of support for this activity; and (iv) contextual with information on the context in which the entrepreneurial activity takes place. The first construct works only as an element to collect socio-demographic information about the students. The other three constructs were defined based on studies by Mair and Noboa (2006), Hockerts (2017), Prieto et al. (2012), and Dobele (2016), in which an individual's entrepreneurial intentions arise from the interaction between individual, organizational, and contextual factors.

Data analysis was performed using descriptive statistical analysis and correlational analysis using Excel 365 and SPSS v.21. The descriptive analysis aims to synthesize the analysis of a large number of values of similar nature, which allows for an overview of the behavior and variation of these values. The correlational analysis complements the previous analysis and aims to understand the analysis of sample data to find out how two or more variables are relational (Reid, 2013). Parametric correlation and nonparametric correlation were used in the correlational analysis.

14.4 Results and Discussion

14.4.1 Individual Construct

Table 14.2 presents the statistical descriptive analysis and correlational analysis of the dimensions of this construct. The descriptive analysis is given by calculating the mean, median (med), and standard deviation (SD); while the correlational analysis determines the value of the Pearson coefficient (Pr), Kendall tau_b coefficient (Kr), and Spearmam rho coefficient (Sr). The representation of the relative importance of each dimension is presented in Fig. 14.1. The findings show that empathy, moral obligation, and self-efficacy are dimensions those students consider very relevant to social entrepreneurship. The pentagonal pyramid in Fig. 14.1 highlights this occurrence. The correlation between the dimensions of the construct and entrepreneurial intention is weak with all values below 0.30. Thus, despite the relative importance given to these three dimensions, they do not stand alone as good variables for estimating students' entrepreneurial intention.

Table 14.1 Survey's structure

Construct	Dimension	Proposition
Control	Age Gender Working experience Experience in volunteer activities	More than 18 (intervals of 5). Masculine or feminine. Without professional experience of more than 3 years. Not at all, once or twice, 4–5 times, and 6 or more times. Reference period of last 6 months.
Individual	Empathy Moral obligation Self-efficacy Perceived social support Prior experience	When thinking about socially disadvantaged people. I try to put myself in their shoes. Seeing socially disadvantaged people triggers an emotional response in me. I feel compassion for socially marginalized people. It is an ethical responsibility to help people less fortunate than ourselves. We are morally obliged to help socially disadvantaged people. Social justice requires that we help those who are less fortunate than ourselves. It is one of the principles of our society that we should help socially disadvantaged people. I am convinced that I personally can contribute to address societal challenges if I put my mind to it. I could figure out a way to help solve the problems that society faces. Solving societal problems is something each of us can contribute to. People would support me if I wanted to start an organization to help socially marginalized people. If I planned to address a significant societal problem people would back me up. It is possible to attract investors for an organization that wants to solve social problems. I have some experience working with social problems. I have volunteered or otherwise worked with social organizations. I know a lot about social organizations.

(continued)

Table 14.1 (continued)

Construct	Dimension	Proposition
Organizational	Curriculum Critical pedagogy Mentor Social network	The curriculum offers specific courses for the development of entrepreneurial skills. The course addresses the topic of creating a new business plan. The course encourages the generation of business ideas with strong social impact. Throughout the course, issues of social exclusion were addressed. Minorities are included in the pedagogical process. Creativity aligned with responsible attitudes is encouraged in teaching entrepreneurship. The mentors have experience in teaching entrepreneurship The mentors have already developed entrepreneurial projects The mentors are involved in projects to fight social exclusion External elements are involved (e.g., successful entrepreneurs, social entrepreneurs, and venture capitalists) in conducting workshops and seminars. Social networks are established with entities in the social sector. The development of social networks involving minorities is promoted.
Contextual	Political Economic Social Technological	There is legal recognition of social entrepreneurship. Government authorities support social entrepreneurship activities. Political stability promotes continuous support for social entrepreneurship activities. External funds are available to support social entrepreneurship activities. There are tax incentives for social entrepreneurial activity. There are economic benefits for companies that engage in social support activities. Access to credit is facilitated for social entrepreneurial activity. Activities promoting social entrepreneurship are easily accessible. Cultural norms and values are aligned with social entrepreneurship practices. There are entities and associations formed to support social entrepreneurship. There is collaboration among the various stakeholders. R&D activity aims to promote social inclusion. Innovative technology-based practices are included in social entrepreneurship. Intellectual property issues do not restrict the emergence and development of social innovation.

Table 14.2 Statistical analysis for the "individual" construct

Dimension	Mean	Med	SD	Pr	Kr	Sr
Empathy	4.82	5	0.341	0.261	0.208	0.237
Moral obligation	4.89	5	0.274	0.255	0.197	0.221
Self-efficacy	4.49	4.67	0.594	0.173	0.139	0.170
Perceived social support	3.74	3.67	0.528	0.145	0.100	0.128
Prior experience	3.38	3.33	0.585	0.215	0.100	0.128

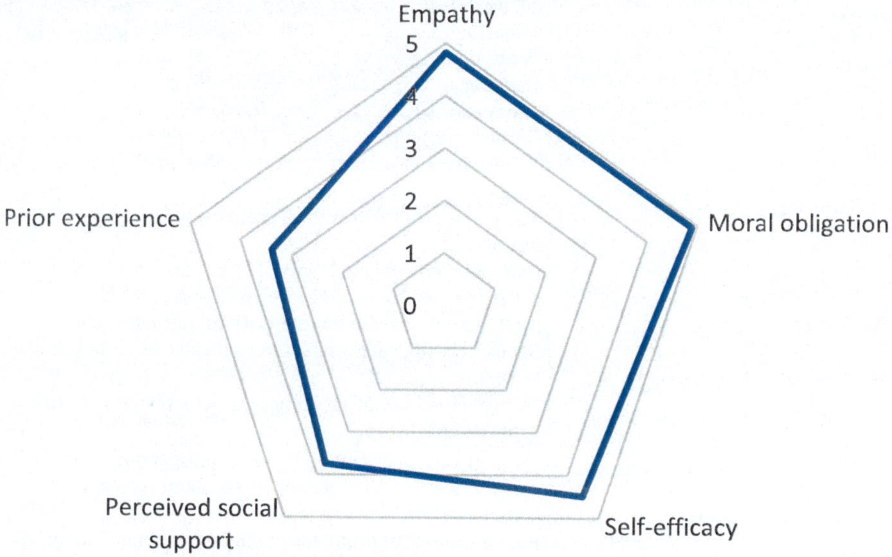

Fig. 14.1 Relative importance of the dimensions of the "individual" construct

The motivations of the individual to become a social entrepreneur represent a relevant bridge to understand the path between intention and action. As Santoso and Oetomo (2018) recognize, motivation plays an important role in the creation of new organizations since it influences decision making. In the analysis of individual factors that explain entrepreneurial intention, the push/pull theory proposed by Gilad and Levine (1986) stands out. This conceptual model argues that individuals are pushed toward entrepreneurship by external push factors (e.g., performance, divorce) and by pull factors that result from the characteristics of each person (e.g., self-realization, search for independence). Intrinsic motivation refers to personal interest in the entrepreneurial task, while extrinsic motivation is related to an external reward that follows a certain behavior. Accordingly, it was concluded that the intrinsic motivations of the individual in starting a new social entrepreneurship business are determining elements. This conclusion is also in line with the evidence provided by the study developed by Segal et al. (2005) in which it is evidenced that pull factors are more prevalent than push factors and, in addition, entrepreneurs motivated by pull factors are more likely to succeed in their businesses.

14.4.2 Organizational Construct

Table 14.3 and Fig. 14.2 present a similar statistical analysis but considering the organizational construct. The findings reveal that the curriculum is built to offer a specialized training course in entrepreneurship. This is an element that is widely recognized by the students as also shown by the lower SD value compared to the other dimensions. However, the three correlation coefficients are practically null and even slightly negative, which indicates that the curriculum is not a determining element for increasing entrepreneurial intention. The other dimensions also point in the same direction. However, the social network is the one that shows a slightly higher correlation, which indicates that it is an area that can enhance the increase of entrepreneurial intention among some students, although it is not a finding that can be generalized to the entire community.

The organizational environment in which entrepreneurial activity takes place is recognized by Turro et al. (2014) as an influential element of entrepreneurial

Table 14.3 Statistical analysis for the "organizational" construct

Dimension	Mean	Med	SD	Pr	Kr	Sr
Curriculum	4.61	4.33	0.345	−0.011	−0.022	−0.026
Critical pedagogy	4.14	4.33	0.543	0.107	0.082	0.101
Mentor	3.69	3.33	0.508	0.172	0.136	0.158
Social network	3.94	4.33	0.615	0.364	0.238	0.281

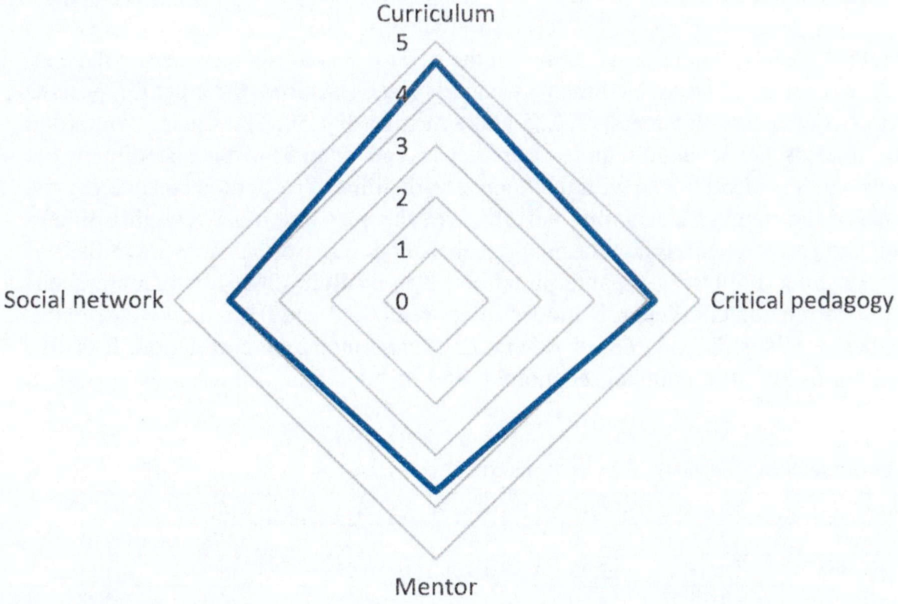

Fig. 14.2 Relative importance of the dimensions of the "organizational" construct

intention. In a higher education institution, the organizational context is related to the existing policies to support entrepreneurship initiatives launched by students and faculty. As Wright et al. (2004) point out, an entrepreneurial university facilitates technology dissemination and transfer. This view is also confirmed by Etzkowitz et al. (2021) when they indicate that one of the main characteristics of an entrepreneurial university is its strong connection with the market and industry, which simplifies the practical application of research conducted and knowledge acquired in the classroom. This knowledge regarding entrepreneurial practice can be embedded in the course curriculum. The students participating in this study highlighted the two components regarding curriculum and critical pedagogy. The social network is an element that also emerges as important for social entrepreneurship. It is noted that research in the field of entrepreneurship has shifted its focus from the individual behaviors of entrepreneurs to a more general understanding of the social dynamics of entrepreneurship (de Bruin et al., 2017). The social network emerges as an important element for entrepreneurs to expand their field and gain access to other people's resources. The literature in this area, highlighting the studies carried out by Chen et al. (2018) and Moghaddam et al. (2018), points out that entrepreneurial social networks are determining elements in the characterization of the intention to start a new business and its organizational performance. Therefore, the growth and sustainability of these businesses depend on them. This study confirms the importance given to it by students but does not allow us to conclude that in isolation it is a factor that increases entrepreneurial intention.

14.4.3 Contextual Construct

Finally, Table 14.4 and Fig. 14.3 present the statistical analysis considering the contextual construct. The social dimension clearly stands out from the other dimensions of this construct with a mean of 4.35 and a median of 4.50. The square pyramid in Fig. 14.3 is consequently quite asymmetric and extends over this dimension. However, it is also the one with the highest variability of responses considering the value of the standard deviation. Nevertheless, the correlation between this dimension and entrepreneurial intention is higher than all other dimensions regardless of the construct. Following the interpretation given by Reid (2013), which stipulates that a correlation coefficient is moderate between 0.40 and 0.60, then we can conclude that this is the only construct that presents a moderate correlation. It is also noteworthy that the political, economic, and technological dimensions appear to

Table 14.4 Statistical analysis for the "contextual" construct

Dimension	Mean	Med	SD	Pr	Kr	Sr
Political	2.93	3.00	0.452	0.234	0.178	0.214
Economic	2.24	2.25	0.577	0.218	0.085	0.106
Social	4.35	4.50	0.702	0.454	0.330	0.392
Technological	2.74	3.00	0.636	0.166	0.082	0.100

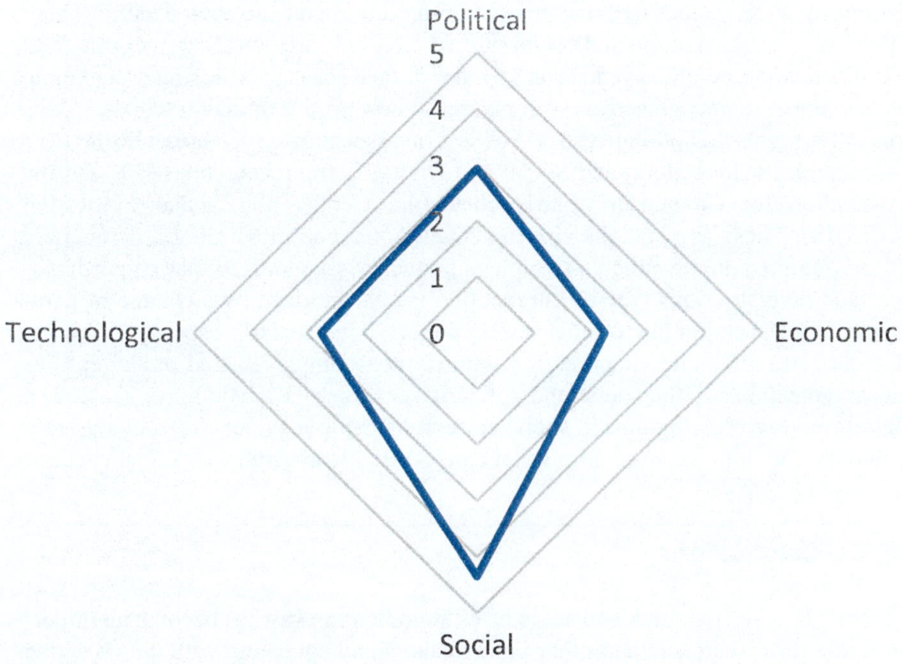

Fig. 14.3 Relative importance of the dimensions of the "contextual" construct

have much less relevance for the launch of new social entrepreneurship projects when compared to the social dimension. It can also be concluded considering the three constructs that the parametric correlation is more significant than the nonparametric one, which indicates that the students' perception of each dimension is significant in the interpretation of the results. On the other hand, the economic dimension, which included answers such as financial support and access to credit, is the least relevant.

Despite the importance of individual and organizational characteristics in the entrepreneurial process, it is assumed that entrepreneurship is not only the result of these two constructs. There are also external contextual factors that play an equally relevant role in the entrepreneurship process, such as the political, economic, social, and technological dimensions. Nevertheless, the findings indicate that the social dimension is clearly more relevant for social entrepreneurial activity than the other dimensions. This dimension includes elements related to the promotion of social entrepreneurship activities, the alignment of cultural norms with social entrepreneurship, the existence of entities and associations that support this type of activity, and collaboration among the various stakeholders. In the literature, there are studies that mainly highlight the role of cultural norms in social entrepreneurship activity (Emami & Khajeheian, 2019) and the role that associations

(Defourny & Nyssens, 2010) can play in promoting social entrepreneurship. These new forms of organization and social intervention have assumed an emerging field in political and scientific agendas on an international scale, in which their economic and social role is recognized and offers a set of new proposals and practices. In this area in Portugal, we highlight the role of several associations such as the Portuguese Association for Innovation and Social and Digital Entrepreneurship (AI9) and the Association for Citizenship, Entrepreneurship, Gender and Social Innovation (ACEGIS). These associations aim to actively intervene in building a democratic society founded on the ethical principles of equality, freedom, solidarity, participation, and diversity. This type of intervention is characterized by a change of paradigm and intervention models that foster economic and social prosperity, in a logic of global and articulated intervention with people, companies, and public entities and organizations of the social and solidarity economy (Kamaludin et al., 2021). Therefore, training programs in social entrepreneurship may benefit from a greater articulation with these associations that can locally identify relevant social causes.

14.5 Conclusions

Knowing how people think and act in an entrepreneurial way has become an important issue for researchers and other policymakers and educators with the objective of knowing and supporting entrepreneurial activities. Entrepreneurship has been investigated on how to create new businesses with financial purposes, but also the creation of new projects to mitigate social imbalances and support the underprivileged classes. These social businesses adopt a model that proposes transformation and social and environmental improvement in a sustainable way.

This study allows understanding of the relative importance of the constructs related to the individual, organizational, and contextual components for entrepreneurial activity. The individual and organizational factors appear to be more integrated in the social entrepreneurship process than the contextual factors, with the specific exception of the social dimension, which has a remarkable importance. Factors like curriculum specifically designed for social project development are highlighted positively by students but then have no impact on entrepreneurial intention. It is also highlighted that the social component is the only factor that shows a moderate correlation with entrepreneurial intention. The other dimensions of the model in isolation have a low and not very significant correlation.

This work offers relevant contributions to understanding the phenomenon of social entrepreneurship in higher education. From a conceptual perspective, it was possible to assess the relative importance of each dimension of the model and understand its isolated impact on entrepreneurial intention. However, the various dimensions of the model were not grouped together, which does not allow us to conclude on the impact of each construct. In this sense, and as future work, it is necessary to investigate using the structural equations method to understand this phenomenon. This approach would allow us to explore the correlation between the 1st and 2nd order constructs. From a practical perspective, the outcomes of this study are relevant for educational institutions to define support structures to foster entrepreneurial intention. The findings of this study highlight the role of the social dimension, which means that the involvement of external social stakeholders can be a determining element for increasing entrepreneurial intention. In addition, this would allow students to know the needs of their community and also improve their network of contacts. In future work, it is also suggested to build social entrepreneurship training programs that can integrate these stakeholders in training programs that should not only be focused on curriculum design.

References

Ahadi, S., & Kasraie, S. (2020). Contextual factors of entrepreneurship intention in manufacturing SMEs: The case study of Iran. *Journal of Small Business and Enterprise Development, 27*(4), 633–657.

Almeida, F., & Amaral, M. (2019). Conceptual framework for assessment entrepreneurial education. *International Journal of Knowledge, Innovation and Entrepreneurship, 7*(3), 5–25.

Almeida, F., & Devedzic, V. (2022). The relevance of soft skills for entrepreneurs. *Journal of East European Management Studies, 27*(1), 157–172.

Amofah, K., & Saladrigues, R. (2022). Impact of attitude towards entrepreneurship education and role models on entrepreneurial intention. *Journal of Innovation and Entrepreneurship, 11*(36), 1–30.

Ashari, H., Abbas, I., Abdul-Talib, A. N., & Zamani, S. (2022). Entrepreneurship and sustainable development goals: A multigroup analysis of the moderating effects of entrepreneurship education on entrepreneurial intention. *Sustainability, 14*(1), 1–13.

Barki, E., Comini, G., Cunliffe, A., Hart, S., & Rai, S. (2015). Social entrepreneurship and social business: Retrospective and prospective research. *RAE – Revista de Administração de Empresas, 55*(4), 380–384.

Bataglin, J. C., & Kruglianskas, I. (2022). Social innovation: Field analysis and gaps for future research. *Sustainability, 14*(3), 1–17.

Boughzala, Y. (2021). Toward a collective approach to social innovation: The case of social entrepreneurship in Tunisia. *Journal of Innovation Economics & Management, 3*(36), 139–166.

Chell, E., Nicolopoulou, K., & Karataş-Özkan, M. (2010). Social entrepreneurship and enterprise: International and innovation perspectives. *Entrepreneurship and Regional Development, 22*(6), 485–493.

Chen, F. W., Fu, L. W., Wang, K., Tsai, S. B., & Su, C. H. (2018). The influence of entrepreneurship and social networks on economic growth – From a sustainable innovation perspective. *Sustainability, 10*(7), 1–19.

Clausen, T. H. (2020). Entrepreneurial thinking and action in opportunity development: A conceptual process model. *International Small Business Journal: Researching Entrepreneurship, 38*(1), 21–40.

Cohen, H., Baruch, O. K., & Katz, H. (2019). The social entrepreneur puzzle: The background, personality and motivation of Israeli social entrepreneurs. *Journal of Social Entrepreneurship, 10*(1), 1–21.

Crisol-Moya, E., Romero-López, M. A., & Caurcel-Cara, M. J. (2020). Active methodologies in higher education: Perception and opinion as evaluated by professors and their students in the teaching-learning process. *Frontiers in Psychology, 11*, 1703–1713.

Curtis, V., Moon, R., & Penaluna, A. (2021). Active entrepreneurship education and the impact on approaches to learning: Mixed methods evidence from a six-year study into one entrepreneurship educator's classroom. *Industry and Higher Education, 35*(4), 443–453.

de Bruin, A., Shaw, E., & Lewis, K. V. (2017). The collaborative dynamic in social entrepreneurship. *Entrepreneurship and Regional Development, 29*(7–8), 575–585.

Defourny, J., & Nyssens, M. (2010). Social enterprise in Europe: At the crossroads of market, public policies and third sector. *Policy and Society, 29*(3), 231–242.

Del Vecchio, P., Secundo, G., Mele, G., & Passiante, G. (2021). Sustainable entrepreneurship education for circular economy: emerging perspectives in Europe. *International Journal of Entrepreneurial Behavior & Research, 27*(8), 2096–2124.

Dobele, L. (2016). A new approach in higher education: Social entrepreneurship education. In *Proceedings of the 11th international conference on management, enterprise and benchmarking (MEB 2016), Budapest, Hungary* (pp. 227–238).

Dwivedi, A., & Weerawardena, J. (2018). Conceptualizing and operationalizing the social entrepreneurship construct. *Journal of Business Research, 86*, 32–40.

EC. (2016). *Entrepreneurship education.* https://ec.europa.eu/growth/smes/supporting-entrepreneurship/entrepreneurship-education_en

Emami, A., & Khajeheian, D. (2019). Social norms and entrepreneurial action: The mediating role of opportunity confidence. *Sustainability, 11*(1), 1–18.

Esteves, A. M., Genus, A., Henfrey, T., Penha-Lopes, G., & East, M. (2021). Sustainable entrepreneurship and the Sustainable Development Goals: Community-led initiatives, the social solidarity economy and commons ecologies. *Business Strategy and the Environment, 30*(3), 1423–1435.

Etzkowitz, H., Dzisah, J., & Clouser, M. (2021). Shaping the entrepreneurial university: Two experiments and a proposal for innovation in higher education. *Industry and Higher Education, 36*(1), 3–12.

Farrukh, M., Alzubi, Y., Shahzad, I. A., Waheed, A., & Kanwal, N. (2018). Entrepreneurial intentions: The role of personality traits in perspective of theory of planned behaviour. *Asia Pacific Journal of Innovation and Entrepreneurship, 12*(3), 399–414.

Felix, C., Aparicio, S., & Urbano, D. (2019). Leadership as a driver of entrepreneurship: An international exploratory study. *Journal of Small Business and Enterprise Development, 26*(3), 397–420.

Gilad, B., & Levine, P. (1986). A behavior model of entrepreneurial supply. *Journal of Small Business Management, 24*(4), 45–54.

Hadad, S., & Cantaragiu, R. (2017). Corporate social entrepreneurship versus social intrapreneurship: Same idea, different trajectories? *Management & Marketing: Challenges for the Knowledge Society, 12*(2), 252–276.

Halleröd, B., & Larsson, D. (2008). Poverty, welfare problems and social exclusion. *International Journal of Social Welfare, 17*(1), 15–25.

Harris, J. D., Sapienza, H. J., & Bowie, N. E. (2009). Ethics and entrepreneurship. *Journal of Business Venturing, 24*(5), 407–418.

Haugh, H. M., & Talwar, A. (2016). Linking social entrepreneurship and social change: The mediating role of empowerment. *Journal of Business Ethics, 133*, 643–658.

Hockerts, K. (2017). Determinants of social entrepreneurial intentions. *Entrepreneurship Theory and Practice, 41*(1), 105–130.

Hockerts, K. (2018). The effect of experiential social entrepreneurship education on intention formation in students. *Journal of Social Entrepreneurship, 9*(3), 234–256.

Kamaludin, M. F., Xavier, J. A., & Amin, M. (2021). Social entrepreneurship and sustainability: A conceptual framework. *Journal of Social Entrepreneurship*. https://doi.org/10.1080/19420676.2021.1900339

Kara, A., Spillan, J. E., Mintu-Wimsatt, A., & Zhang, L. (2022). The role of higher education in developing entrepreneurship: A two-country study. *Latin American Business Review, 24*(1), 1–22.

Klyver, K., & Arenius, P. (2020). Networking, social skills and launching a new business: A 3-year study of nascent entrepreneurs. *Entrepreneurship Theory and Practice, 46*(5). https://doi.org/10.1177/1042258720967063

Kumar, R., & Raj, T. (2019). Role of entrepreneurship in boosting economic growth and employment in India. *SEDME (Small Enterprises Development, Management & Extension Journal): A Worldwide Window on MSME Studies, 46*(4), 273–281.

Looi, K. H. (2020). Contextual motivations for undergraduates' entrepreneurial intentions in emerging Asian economies. *The Journal of Entrepreneurship, 29*(1), 53–87.

Mair, J., & Noboa, E. (2006). Social entrepreneurship: How intentions to create a social venture get formed. In J. Mair, J. Robinson, & K. Hockerts (Eds.), *Social entrepreneurship* (pp. 121–136). Palgrave Macmillan.

Mandrysz, W. (2020). Community-based social economy – Social capital and civic participation in social entrepreneurship and community development. *Management Dynamics in the Knowledge Economy, 8*(1), 81–93.

Márquez, F. O. S., & Ortiz, G. E. R. (2021). The disruptive triad and entrepreneurship: A theoretical model. *Journal of Innovation and Entrepreneurship, 10*(45), 1–21.

Moghaddam, K., Rustambekov, E., Weber, T., & Azarpanah, S. (2018). Transnational entrepreneurship, social networks, and institutional distance: Toward a theoretical framework. *New England Journal of Entrepreneurship, 21*(1), 45–64.

Murdock, K. A., & Varnes, C. J. (2018). Beyond effectuation: Analysing the transformation of business ideas into ventures using actor-network theory. *International Journal of Entrepreneurial Behavior & Research, 24*(1), 256–272.

Naveed, M., Zia, M. Q., Younis, S., & Shah, Z. A. (2021). Relationship of individual social entrepreneurial orientations and intentions: Role of social entrepreneurship education. *Asia Pacific Journal of Innovation and Entrepreneurship, 15*(1), 39–50.

Odewale, G. T., Rani, S. H., & Migiro, S. O. (2019). Does communication skills matter to a nascent entrepreneur? *Journal of Contemporary Management, 16*, 209–225.

Passoni, D., & Glavam, R. B. (2018). Entrepreneurial intention and the effects of entrepreneurial education: Differences among management, engineering, and accounting students. *International Journal of Innovation Science, 10*(1), 92–107.

Pohlan, L. (2019). Unemployment and social exclusion. *Journal of Economic Behavior & Organization, 164*, 273–299.

Prieto, L. C., Phipps, S. T., & Friedrich, T. L. (2012). Social entrepreneur development: An integration of critical pedagogy, the theory of planned behavior and the ACS model. *Academy of Entrepreneurship Journal, 18*(2), 1–16.

Prim, M. A., Dandoline, G. A., da Silva, S. M., & de Souza, J. A. (2019). Collaboration networks for social innovation: A case study in a technological incubator of popular cooperatives. *International Journal for Innovation Education and Research, 7*(12), 147–164.

Reid, H. M. (2013). *Introduction to statistics: Fundamental concepts and procedures of data analysis*. SAGE.

Ruskin, J., Seymour, R. G., & Webster, C. M. (2016). Why create value for others? An exploration of social entrepreneurial motives. *Journal of Small Business Management, 54*(4), 1015–1037.

Rusok, N. H. M., Kumar, N., & Manaf, K. A. (2017). The economic and social value of fostering entrepreneurship education. *International Journal of Asian Social Science, 7*(9), 783–799.

Sansone, G., Ughetto, E., & Landoni, P. (2021). Entrepreneurial intention: An analysis of the role of student-led entrepreneurial organizations. *Journal of International Entrepreneurship, 19*, 399–433.

Santoso, S., & Oetomo, B. S. (2018). Influence of motivation and self-efficacy on entrepreneurial intention to run a business. *Expert Journal of Marketing, 6*(1), 14–21.

Satar, M. S., & Natasha, S. (2019). Individual social entrepreneurship orientation: Towards development of a measurement scale. *Asia Pacific Journal of Innovation and Entrepreneurship, 13*(1), 49–72.

Segal, G., Borgia, D., & Schoenfeld, J. (2005). The motivation to become an entrepreneur. *International Journal of Entrepreneurial Behaviour & Research, 11*(1), 42–57.

Sohu, J. M., Junejo, I., Khuwaja, F. M., Qureshi, N. A., & Dakhan, S. A. (2022). The impact of entrepreneurial education on entrepreneurial intention during the COVID-19 pandemic: An empirical study from Pakistan. *The Journal of Asian Finance, Economics and Business, 9*(3), 95–103.

Tran, A., & Von Korflesch, H. (2016). A conceptual model of social entrepreneurial intention based on the social cognitive career theory. *Asia Pacific Journal of Innovation and Entrepreneurship, 10*(1), 17–38.

Turro, A., Urbano, D., & Peris-Ortiz, M. (2014). Culture and innovation: The moderating effect of cultural values on corporate entrepreneurship. *Technological Forecasting and Social Change, 88*, 360–369.

Wright, M., Birley, S., & Mosey, S. (2004). Entrepreneurship and university technology transfer. *The Journal of Technology Transfer, 29*(3/4), 235–246.

Zhao, F. (2005). Exploring the synergy between entrepreneurship and innovation. *International Journal of Entrepreneurial Behavior & Research, 11*(1), 25–41.

Check for updates

15 Private Agricultural Insurance and Its Current Adhesion by Rural Producers: A Partial Analysis in Paraná State

Thaís Martha Welter de Moura
and Fabricio Pelloso Piurcosky

15.1 Introduction

It is well known that Brazilian agribusiness is extremely important, both internally, such as the contribution of 20.4%, 26.4%, and 27.4% of the country's Gross Domestic Product (GDP) in the years 2019–2021, respectively, according to data from the Centre for Advanced Studies in Applied Economics (CEPEA) and the National Confederation of Agriculture and Livestock (CNA) (n.d.), and externally, by promoting the supply of food product, raw materials, and energy to several countries.

Analyzing agribusiness only from an internal perspective, sometimes the agricultural sector can be the main regional economic activity, from which it can be concluded that any damage suffered by it indirectly affects other sectors of the economy, such as commerce and industry, and may affect the economy as a whole if the effect of an eventual accident persists for a few years.

And, specifically regarding agriculture, it is a vulnerable activity, dependent on a series of climatic events that are out of the producer's control and difficult to anticipate, among which are frost, excessive rain, hail, drought, and biological agents, which can lead to a reduction and even loss of agricultural productivity, which directly impacts the availability of food for the general population.

Contextualizing, in the 2021/2022 harvest, there was a severe drought in the states of Parana, Santa Catarina, Rio Grande do Sul, Mato Grosso do Sul, and Sao Paulo, resulting in losses of up to 70% (seventy percent) of production. Such

T. M. W. de Moura (✉)
Business, Universidade Cesumar, Maringá, Brazil
e-mail: fabricio.pelloso@grupointegrado.br

F. P. Piurcosky
Grupo Educacional Integrado: Campo Mourão, Campo Mourão, Brazil

© The Author(s), under exclusive license to Springer Nature Switzerland AG 2023

F. L. Almeida et al. (eds.), *Multidimensional Sustainability: Transitions and Convergences*, Springer Proceedings in Earth and Environmental Sciences,
https://doi.org/10.1007/978-3-031-24892-4_15

damage could have been minimized with the previous contracting of agricultural insurance by the rural producer, the main mechanism for mitigating the risk of bad weather.

The present study has as its object the optional private agricultural insurance, which guarantees the investment made in the rural property; the adhesion rates to the agricultural insurance contract, whether they are in decline, stability, or rise; and the reasons why they are so, giving special attention to Parana State.

15.2 Development

Agricultural insurance consists of a contract whereby, through the payment of the premium by the insured – in this case, the farmer, the insurance company undertakes to indemnify him for any damage that may be verified, mainly resulting from meteorological phenomena. According to the Superintendence of Private Insurance (SUSEP) (n.d.), aforementioned insurance covers the life of the plant, from its emergence to harvest, protecting the rural producer against risks such as fire, lightning, waterspout, strong wind, frost, hail, excessive rain, drought, and extreme temperature variation.

Through agricultural insurance, the growth and development of agriculture can be encouraged, with an increase in the technological standard and productivity, expanding the cultivated area, since the risks of production are reduced. In addition, it maintains investment stability and the sector's competitiveness, even if there is a loss of capital or crop failure, reducing the uncertainty existing in rural activity and the need to create cash reserves. It also helps to generate jobs.

Although the importance of agricultural insurance is recognized by the rural producer, some factors prevent its adhesion to be broader, highlighting the following, pointed out by Santos (2017): (a) high price – the smaller number of policyholders prevents the decrease in value of the premiums charged, because only producers in high-risk regions or who cultivate more sensitive crops, such as off-season corn and wheat, are willing to take out insurance; (b) absence of an insurance culture, once it is understood as a part of the cost, and not as an investment against future losses; (c) restricted territorial coverage; and (d) lack of innovative products that attract the farmer's attention. Added to these is the possibility of debt restructuring, sometimes assisted by the Federal Government after the occurrence of adverse phenomena, as well indicated by Ozaki (2007), with probable forgiveness of part of the amount due.

Regarding the amount needed to subscribe to agricultural insurance, the Rural Insurance Premium Subsidy Program (PSR), created by Law no. 10.823/2003, whose operation began at the end of 2005, helped to reduce the amounts charged until then. The mentioned program consists of financial support from the Federal Government to rural producers who contract this type of insurance, bearing part of the cost of its acquisition, which, from 2022, consists of a fixed level of 40% (forty percent) for all crops and/or activities, except for soybean, whose percentage remains at 20% (twenty percent), observing the individual financial limit of R$ 60,000 (sixty thousand reais) per group of activities, limited to the total of R$

120,000 (one hundred and twenty thousand reais) each calendar year, according to information from the Ministry of Agriculture, Livestock and Supply (2022).

In order to obtain the subsidy in question, it is essential that the rural producer in compliance with the Federal Government and that the Agricultural Zoning of Climate Risk (ZARC) is observed, which consists of guidelines from the Ministry of Agriculture, Livestock, and Supply on the sowing periods by municipality, crop and soil type, identifying the best time for planting crops, with a reduction in accident rates.

It should be noted that, although the Agricultural Zoning of Climate Risk (ZARC) guidelines do not consider climatic effects such as El Niño, according to Buainain et al. (2008), their results have been so positive, with up to 80% (eighty percent) of a guarantee of success, that some farmers dispense with taking out agricultural insurance, considering the credibility of the system (Assad et al., 2005).

With the subsidy, which can be federal, state, and municipal at the same time, there is the sharing of the cost of agricultural insurance with the state entity and the consequent reduction of expenses related to it, which leads to an increase in its offer and, given the higher prices affordable, of its demand.

In practice, there has already been an increase in the agricultural insurance market, which has been growing in Brazil since the beginning of the Rural Insurance Premium Subsidy Program (PSR) at the end of 2005 (Miquelluti et al., 2022).

However, this growth has been verified specifically in the South, Southeast, and Central-West regions, due to the greater exposure to the risk of weather conditions, especially in the South region, which is most affected by these climatic events, which leads to constant falls in crops, allied to the culture of insurance that rural producers have and the greater presence of insurance companies, among other reasons.

This information is corroborated by the fact that Parana is the state that contracts the most rural insurance in Brazil, accounting for, on average, 38% (thirty-eight percent) of total contracts (Gestão..., 2022).

In the North and Northeast regions, as weather conditions are, as a rule, more regular and policy prices are high, there is a low adherence to agricultural insurance.

In general, from January to August 2021, there was a 43% (forty-three percent) increase in rural insurance, according to information from the National Confederation of General Insurance Companies (Bueno, 2021).

Furthermore, "According to Mapa, more insurers were qualified to operate in the PSR, reaching 18 companies in 2022. At the same pace, the number of experts practically doubled in three years, totaling 1,178 professionals" (Gestão..., 2022).

Advancing the issue of the absence of an insurance culture, it should be noted that 15.45% (fifteen point forty-five percent) of rural producers submitted to the 2017 agricultural census never attended school, which allows us to assume that, even if the most may have been informally literate, there might be some difficulty in understanding agricultural insurance, its terms, and conditions (Brazilian Institute of Geography and Statistics (IBGE) (n.d.)).

Agricultural insurance should be understood by the farmer, according to Silva (2008, p. 50), "as a basic input for the business, as well as seeds, fertilizers, pesticides, limestone, animal feed, among many others," with large-scale contracting, before the start of the harvest, behold, it is in that soil, vulnerable to weather conditions, that it is depositing its investments (Aos leitores…, 2022).

It is equally important for rural producers to be aware that "the risks of unsuccessful harvests can even make certain crops unfeasible, or else, as a way of reducing costs, lead to the perpetuation of rudimentary and backward technologies" (Ferreira & Ferreira, 2009).

Considering that private agricultural insurance in Brazil is incipient, with greater prominence after the implementation of the Rural Insurance Premium Subsidy Program (PSR) in 2005, as stated elsewhere, it is worth analyzing the conduct adopted in the world, especially by the United States and Spain, considered as the countries whose rural insurance market is among the most developed.

15.2.1 Rural Insurance Worldwide

Rural insurance has been existing for over 80 (eighty) years in the United States, and its government subsidy program, started in 1981, allows the reduction of costs for rural producers by up to 100% (one hundred percent) in cases of catastrophe coverage (Ferreira & Ferreira, 2009). Even the operational and administrative cost of insurance companies is subsidized by the government, not limited to the value of the insurance premium (Medeiros, 2013).

The government subsidy extends to reinsurance, carried out by the public entity and to whom the risk is transferred, thus allowing insurers to operate in risk areas in which they would not normally offer insurance (Ferreira & Ferreira, 2009).

According to Buainain et al. (2008), "In the USA, despite the reduction in the perception of risk that public participation can generate, the monitoring mechanisms are exacerbated, and the rural producer responds with his patrimony in case of opportunism."

It can be seen that in that country up to 70% (seventy percent) of its production is insured, with emphasis on large producers, with a high level of education and well capitalized, with the great exclusion of low-income, small-scale producers with specialty products in adhering to agricultural insurance (Fornazier et al., 2014).

On the one hand, in Spain, where agricultural insurance in its current form has approximately 38 (thirty-eight) years of existence, according to Ozaki (2006 as cited in Ozaki, 2007), we can see the same picture of a strong public sector subsidy on the awards, differing from the United States because it also uses the assistance of autonomous institutions (Buainain et al., 2008). Thus, there is a public–private model, in which insurers come together in a "pool" to reduce administrative and operational expenses, as well as to share risks.

Spanish agricultural insurance covers all types of crops, and all rural producers have access to it, in all regions of the country, respecting the norms and conditions of interest of agricultural policy (Buainain et al., 2008).

In Canada, agricultural insurance has the great differential of aiming at stabilizing the income of the farmer as a whole. There is a government subsidy, with costs being shared between the federal government, the province, and the producer. However, state participation has been gradually decreasing, with a greater insertion of the rural producer (Ferreira & Ferreira, 2009).

According to Fusco et al. (2018), in Italy the adherence to agricultural insurance is still low despite significant subsidies in agricultural insurance premiums, pointing to the need to increase rural producers' information about the usefulness of this instrument.

For comparative purposes, it is necessary to present data from Ghana, a developing country (Freire, 2020). According to Adarkwah et al. (2021), in general terms, access to and acceptance of agricultural insurance by small producers is low, although the vast majority understand its importance as an effective instrument to deal with agricultural risks. The obstacles to adhering to agricultural insurance in that country lie in the lack of knowledge about existing products, the lack of agricultural insurance in the areas where rural properties are located, the high costs and the lack of knowledge about how insurance policy works. It was also found that the contracting of agricultural insurance is greater and faster among rural producers with a higher educational level.

Finally, in Europe, there is a movement to create an agricultural insurance program for the entire European Union, with indications that a public–private partnership will be adopted. To this end, it is important to note that member countries differ, either because of the different types of climate risks they face, or because of their cultural, political, and historical diversity, or because of the time they joined the European Union, or because of the number of risks covered for crops, which, as a rule, is limited, except for Spain, Greece, and Italy, which have developed more comprehensive insurance policies (Van Asseldonk et al., 2018).

15.3 Materials and Methods

To collect data from the study under examination, bibliographic research and documentary research were carried out, with emphasis on data obtained from websites of official bodies, as well as field research. The field research was divided into two moments: first, an interview was carried out, on March 30, 2022, with the agronomist Fernando Welter de Moura, and, later, from April 6 to April 16, 2022, a questionnaire was presented, with open answers, to 11 (eleven) rural producers.

15.4 Results and Discussion

15.4.1 Interview

In the interview conducted, information was obtained that, because they are crops more sensitive to climatic events and subject to greater risk, with an accident at the level of 80% (eighty percent), wheat and off-season corn crop have about 95% (ninety-five percent) of the crop insured, being known to the interviewee that some rural producers do not even venture to plant such crops if they are not guaranteed by insurance.

According to him, the soybean crop has a normally low risk of loss of 20% (twenty percent), hence the demand for insurance ends up being minimal. However, due to the atypical scenario of the drought in the 2021/2022 crop, which significantly affected the soybean harvest, rural producers are more accepting of the idea of contracting private agricultural insurance, with a change in position regarding the adhesion of this type of insurance, even by the elderly. In other words, a shift in insurance culture is taking place.

In addition, he informs that the company he works for, which sells agricultural inputs, has a partnership with an insurance broker; it is up to agronomists to fill in the farmer client's data and send them to the insurance company, so that the entire crop is insured. It is, according to him, the so-called invoicing insurance, a way of expanding the reach of this instrument that guarantees the investments made.

Through this approach, at least two of the problems that prevent the expansion of agricultural insurance in Brazil are minimized, namely, the absence of an insurance culture and the restricted territorial coverage. In this case, the agronomist, in conversation with the rural producer, explains to him about the existence of agricultural insurance in the region of the rural property and the benefits arising from contracting the insurance.

15.4.2 Questionnaire

On the other hand, a survey was carried out by a questionnaire, answered by 11 (eleven) rural producers, with properties between 10 (ten) and 1000 (one thousand) "alqueires,"[1] all in Parana State, more specifically in the Central-Eastern, West and Central North regions of the state (Secretary of Education and Sports).

Most respondents – 27.3% (twenty-seven point three percent) – have been engaged in rural activity for about 50 (fifty) years, followed by those who have been engaged in rural activity for 30 (thirty) years – 18.2% (eighteen point two percent) (Fig. 15.1).

Among those surveyed, the vast majority (81.8%) see an advantage in using agricultural insurance, while 9.1% (nine point one percent) understand that the advantage is partial and 9.1% (nine point one percent) do not see the advantage in its use (Fig. 15.2).

Regarding the 11 (eleven) surveyed, only 02 (two) do not have insurance, one because understands it is unnecessary, since it has its resources, and another because of the price and coverage, with the record of 01 (one) rural producer who has only part of the current insured crop (Fig. 15.3).

The type of insurance used, currently or in the past, is or was private, with the register of the use of Proagro, as illustrated in Fig. 15.4.

[1] 1 "alqueire" = 2.42 hectares.

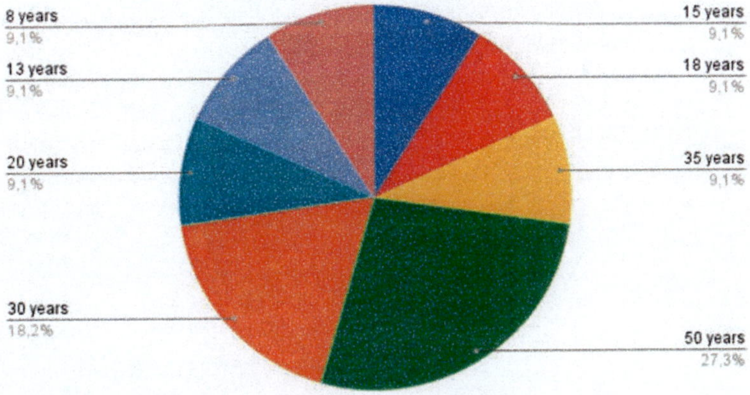

Fig. 15.1 Time spent in rural activity. (Source: Author)

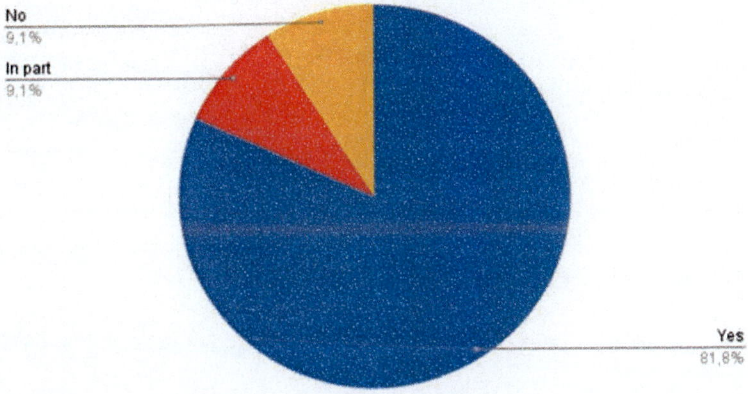

Fig. 15.2 Advantage in agricultural insurance. (Source: Author)

Furthermore, the insurance coverage, past or present, in 54.5% (fifty-four point five percent) of the cases covers only the financing obtained from the banking institution, considering that 45.5% (forty-five point five percent) remaining are intended to guarantee the activity of the rural producer (Fig. 15.5).

Regarding insured crops, it's been noticed the constantly taken out for wheat and corn, with emphasis on the latter crop, which, as it has been adapted to produce between summer and winter crops, is more subject to claims, as informed by the interviewee (Fig. 15.6).

Finally, the biggest motivator for the purchase of agricultural insurance by the surveyed rural producers was the climate, which can also be included in the topic "activity risks" (Fig. 15.7).

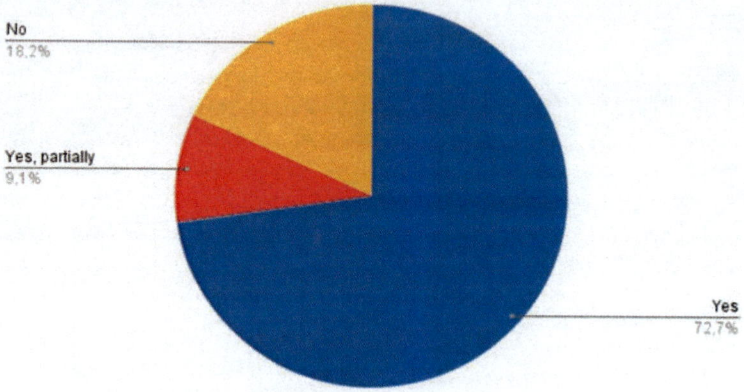

Fig. 15.3 Current insured crop. (Source: Author)

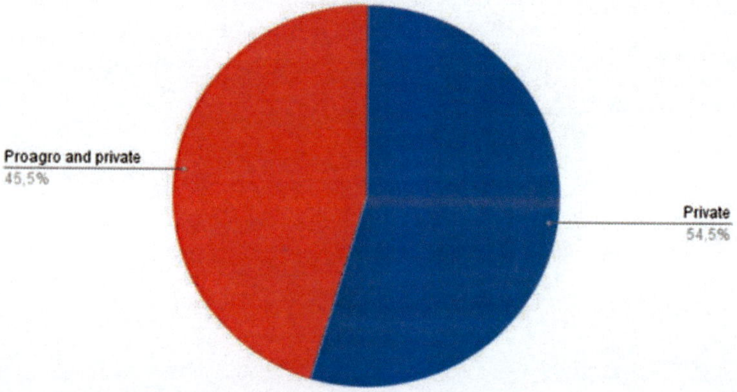

Fig. 15.4 Type of insurance used. (Source: The author 2022)

Both the interview and the questionnaire corroborated, considering only Parana State, the following information brought by the theoretical reference research: (a) the climate is the most important factor that leads rural producers to adhere to agricultural insurance; (b) there is a change in the thinking of farmers, who are beginning to understand the usefulness of the aforementioned insurance in their activity; and (c) the agricultural insurance market has been increasing year by year.

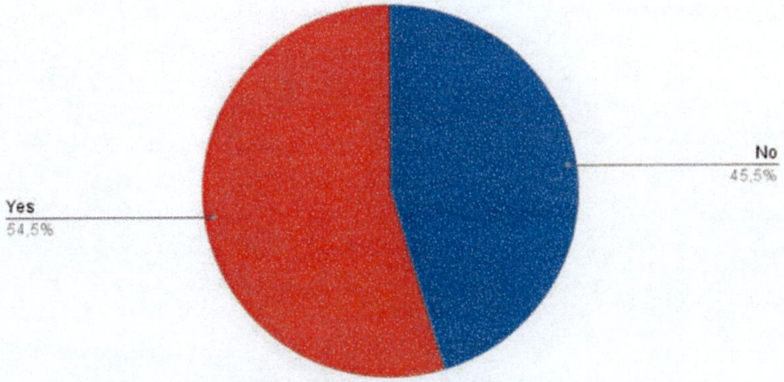

Fig. 15.5 Insurance coverage. (Source: Author)

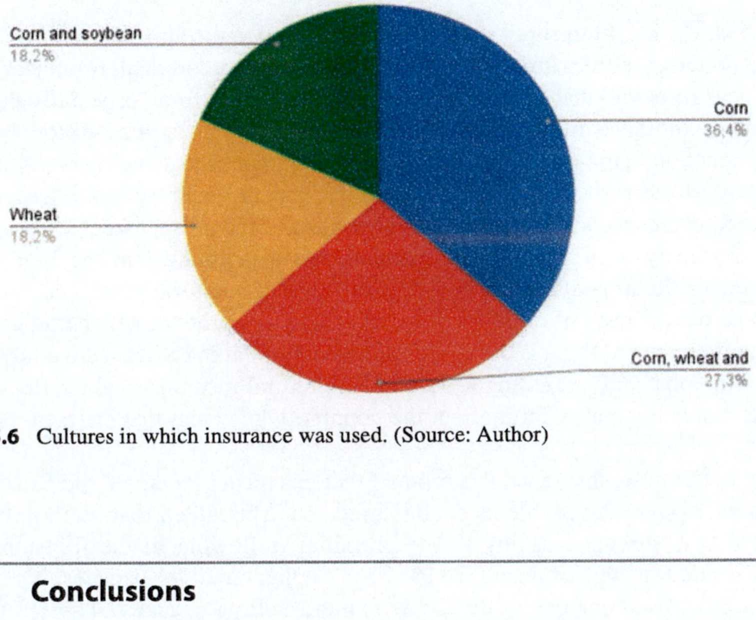

Fig. 15.6 Cultures in which insurance was used. (Source: Author)

15.5 Conclusions

Considering the importance presented by agribusiness in Brazil, as it accounts for 27.4% (twenty-seven point four percent) of the country's Gross Domestic Product (GDP) in 2021, plus its role as a food provider, raw material, and energy, sometimes exported, and considering its vulnerability to the climate, which depends on it to obtain high productivity and, as a corollary, profitability, the use of agricultural insurance as an instrument to guarantee the high investments made by rural producers is increasingly shown most essential.

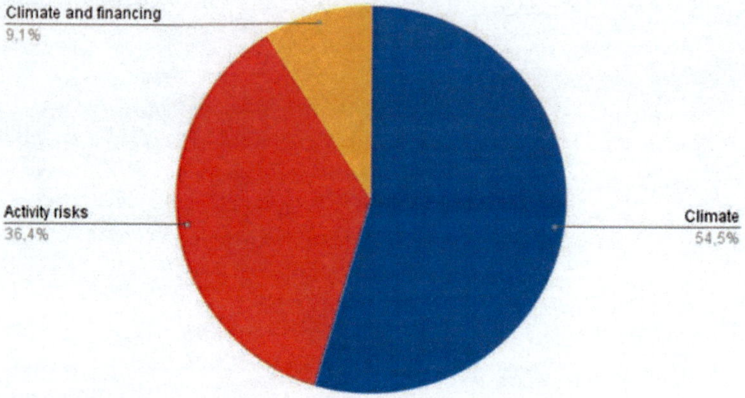

Fig. 15.7 Reasons that led to the acquisition of agricultural insurance. (Source: Author)

With this, we sought to study the rates of adherence to the agricultural insurance contract, noticing, either through the analysis of the theoretical framework or through field research, that they have been increasing over time, especially due to the existence of the Rural Insurance Premium Subsidy Program (PSR), which allows a reduction in the cost of the insurance policy, thanks to state intervention, as well as due to the series of climatic variations that the country has been going through recently, especially Parana State, the object of the present study.

From the analysis of the field research, it was also found that the high cost remains one of the impeding factors for adhering to agricultural insurance, having also pointed out the need of offer, by the market, and of a product with better coverage. It should be noted that, as only rural producers in Parana State were analyzed, there were no obstacles regarding access to agricultural insurance, which does not mean that this is the reality throughout the country, whose regional differences are evident.

Private agricultural insurance is a product that has recently entered the Brazilian market. And, despite the problems encountered, it is concluded that, with government assistance, through subsidy to the premium, following in the footsteps of countries where this type of insurance has been implemented for decades, it's been noticing a behavioral change by the farmers, more willing to take out agricultural insurance.

We understand that, through agricultural insurance, an agricultural policy instrument provided for in Article 187, item V, of the Federal Constitution, Brazilian agribusiness can be strengthened even more, mitigating eventual investment losses arising mainly from extreme weather events that occur particularly in the South of the country.

We believe that this chapter is of value to insurance companies that have agricultural insurance in their portfolio or that intend to implement it, to companies that resell agricultural inputs, since it is possible for them to intermediate the contracting of agricultural insurance with a view to guarantee payment for the products it sells, as well as for rural producers, by providing them an idea of the current situation of the referred insurance in part of Parana State.

Finally, it must be considered that the study on agricultural insurance deserves to continue, given the recent and not yet consolidated change in the insurance culture by the rural producer, and it is interesting to examine their future behavior and their consequent satisfaction or not with agricultural insurance products available on the market. Still, research that analyzes past experiences and what their impact is on future contracts, whether for those who are users of this means of mitigating losses, or for those who are not yet, considering other states of the federation, including, turns out to be equally interesting.

References

Adarkwah, F., Ankrah, D., Boateng-Gyambiby, D., Eghan, D., & Kwapong, N. A. (2021). Agricultural insurance access and acceptability: Examining the case of smallholder farmers in Ghana. *Agriculture & Food Security, 10*. https://doi.org/10.1186/s40066-021-00292-y

Almeida, W. (2007). Massificação das operações do seguro rural: o grande desafio brasileiro. *Revista de Política Agrícola, 16*(4), 21–26. https://seer.sede.embrapa.br/index.php/RPA/article/view/484

Aos Leitores. (2022, February). *Boletim Informativo – A Revista do Sistema*, (1555), p. 2, Year XXXVII.

Assad, E., Biudes, F., & Castillo, R. (2005). O Seguro agrícola a partir do zoneamento de riscos climáticos. In *Congresso Brasileiro de Agrometeorologia*, Campinas, Brazil. https://www.alice.cnptia.embrapa.br/alice/bitstream/doc/9062/1/PL-Seguro-Biudes-CBAgro-2005.pdf

Buainain, A., Madi, M, Ojima, A., Silveira, J., Souza, R., Vieira, A., & Vieira Junior, P. (2008, July). Dimensões e Perspectivas do Seguro Rural: O caso brasileiro e algumas experiências internacionais. In *Congresso da Sociedade Brasileira de Economia, Administração e Sociologia Rural – SOBER*, Rio Branco, Brazil (pp. 20–23). https://ageconsearch.umn.edu/record/109437

Bueno, D. (2021). *Valor: Insurtechs prometem ofertas personalisadas para pequeno produtor*. Sonho Seguro. https://www.sonhoseguro.com.br/2021/11/valor-insurtechs-prometem-ofertas-personalizadas-para-pequeno-produtor/

Caffagni, L. (1998). *Seguro rural no Brasil: evolução, alternativas e sugestões*. Masters dissertation, Escola Superior de Agricultura Luiz de Queiroz, Piracicaba, Brazil. https://www.teses.usp.br/teses/disponiveis/11/11132/tde-20181127-160134/en.php

Caridade, A., Guimarães, T. C., & Igari, A. T. (2019). Mudanças climáticas, seleção adversa e o seguro agrícola no Brasil. In *Encontro Internacional sobre Gestão Empresarial e Meio Ambiente*. http://engemausp.submissao.com.br/17/anais/arquivos/253.pdf

Centro de Estudos Avançados em Economia Aplicada (CEPEA) e Confederação Nacional da Agricultura e Pecuária (CNA). (n.d.). *PIB do agronegócio brasileiro de 1996 a 2021*. https://www.cepea.esalq.usp.br/br/pib-do-agronegocio-brasileiro.aspx

FAEP pede ações ao Mapa contra perdas pela seca. (2022, January). *Boletim Informativo – A Revista do Sistema* (1554), p. 12, Year XXXVII.

Ferreira, A., & Ferreira, L. (2009). Experiências internacionais de seguro rural: as novas perspectivas de política agrícola para o Brasil. *Economica, 11*(1), 131–156. https://periodicos.uff.br/revistaeconomica/article/view/34875

Fornazier, A., Souza, P., & Ponciano, N. (2014). A Importância do Seguro Rural na Redução dos Riscos da Agropecuária. *Revista de Estudos Sociais, 14*(28), 39–52. https://search.ebscohost.com/login.aspx?direct=true&db=edsdoj&AN=edsdoj.70af76177c4e413bab0ca4f020bd9b6e&lang=pt-br&site=eds-live

Freire, D. (2020). *Veja o ranking completo dos 189 países por IDH*. São Paulo, Brazil, CNN Brasil. https://www.cnnbrasil.com.br/internacional/veja-o-ranking-completo-de-todos-os-paises-por-idh

Fusco, G., Miglietta, P., & Porrini, D. (2018). How drought affects agricultural insurance policies: The case of Italy. *Journal of Sustainable Development, 11*(2). https://pdfs.semanticscholar.org/e3b3/99978088cedc293406183860b880f14b5306.pdf

Garrido, A., & Zilberman, D. (2007, July). Revisiting the demand of agricultural insurance: The case of Spain. In *European Association of Agricultural Economists (EAAE)*, Berlin, Germany. https://ageconsearch.umn.edu/record/9266/

Gestão de risco em tempos de crise. (2022, February). *Boletim Informativo – A Revista do Sistema* (1555), pp. 14–19, Year XXXVII.

Hazell, P., Hess, U., & Kuhn, S. (2015, September). Innovations and emerging trends in agricultural insurance. In *GPFI SME Finance Subgroup*. https://www.researchgate.net/profile/Peter-Hazell/publication/283089244_Innovations_and_emerging_Trends_in_Agricultural_Insurance/links/567ebcdd08ae051f9ae657d2/Innovations-and-emerging-Trends-in-Agricultural-Insurance.pdf

Instituto Brasileiro de Geografia e Estatística (IBGE). (n.d.). *Censo Agro 2017: resultados definitivos*. https://censoagro2017.ibge.gov.br/templates/censo_agro/resultadosagro/produtores.html

Lobato, T. (2021). *Análise espacial para o seguro agrícola: um estudo de caso no Paraná*. Doctoral thesis, Escola Superior de Agricultura Luiz de Queiroz, Piracicaba, Brazil. https://www.teses.usp.br/teses/disponiveis/11/11132/tde-05102021-085301/en.php

Matos, M., & Ninaut, E. (2008). O seguro rural no Brasil: situação atual e perspectivas. *Informações Econômicas, 38*(7), 43–51. http://www.iea.sp.gov.br/ftpiea/publicacoes/tec5-0708.pdf

Medeiros, E. (2013). Avaliação da implementação do programa de subvenção do prêmio do seguro rural. *Revista de Economia e Sociologia Rural, 51*(2), 295–308. https://doi.org/10.1590/S0103-20032013000200005

Medeiros, J., & Mitidieri, F. (2015). Zoneamento agrícola de risco climático: ferramenta de auxílio ao seguro rural. *Revista de Política Agrícola, 17*(4), 33–46. https://seer.sede.embrapa.br/index.php/RPA/article/view/435/386

Ministério da Agricultura, Pecuária e Abastecimento. (2022). *Limites/percentuais de subvenção: confira os limites e percentuais de subvenção de seguro rural*. https://www.gov.br/agricultura/pt-br/assuntos/riscos-seguro/seguro-rural/limites-percentuais-de-subvencao

Miquelluti, D., Mota, A., & Ozaki, V. (2022). Métodos de previsão de prêmios para o seguro agrícola e destinação de recursos públicos ao Programa de Subvenção ao Prêmio do Seguro Rural. *Revista de Economia e Sociologia Rural, 60*. https://doi.org/10.1590/1806-9479.2021.249013

Mota, A. (2019). *Previsão de prêmio e a ocorrência de sinistros no mercado de seguro agrícola brasileiro*. Masters dissertation, Escola Superior de Agricultura Luiz de Queiroz, Piracicaba, Brazil. https://doi.org/10.11606/D.11.2019.tde-19112019-181327

Ozaki, V. (2007). O papel do seguro na gestão do risco agrícola e os empecilhos para o seu desenvolvimento. *Revista Brasileira de Risco e Seguro, 2*(4), 75–92. https://www.rbrs.com.br/arquivos/RBRS4-5%20Vitor%20Ozaki%20On%20Line.pdf

Santos, L. (2017). *O seguro agrícola como instrumento de incentivo e proteção da atividade agrícola brasileira: aspectos limitadores e custos* (Final paper). Universidade Federal do Paraná, Curitiba, Brazil. https://www.acervodigital.ufpr.br/handle/1884/54619

Secretaria da Educação e do Esporte. *Regiões Geográficas do Paraná*. http://www.geografia.seed.pr.gov.br/modules/galeria/detalhe.php?foto=1589&evento=8

Silva, C. (2008). O dilema do seguro rural. *Agroanalysis, 28*(2), 49–50. http://bibliotecadigital.fgv.br/ojs/index.php/agroanalysis/article/view/36430/35202

Superintendência de Seguros Privados (SUSEP). *Seguro rural: o que cobre cada uma das modalidades do seguro rural?* http://www.susep.gov.br/menu/informacoes-ao-publico/planos-e-produtos/seguros/seguro-rural

Van Asseldonk, M., Meuwissen, M., & De Mey, Y. (2018). Prospects for agricultural insurance in Europe. *Agricultural Finance Review, 78*(2), 174–182. https://doi.org/10.1108/AFR-04-2018-093

Integrating Sustainability into Decision-Making: An Analysis of Different Types of Management Control Systems

16

Anna Katharina Meyer, Barbara E. Weißenberger, and Andrew Isaak

16.1 Introduction

In order to meet the grand challenges of climate change and resource scarcity, the private sector must not only play a less harmful role but be proactively supportive and contribute to solutions (Schaltegger, 2018). Here, an orientation toward absolute sustainability, which refers to taking the global and local environmental capacities into account, is not yet widely applied but is considered highly relevant by scholars who have empirically examined its implementation (Bjørn et al., 2020; Dahm, 2019).

Simultaneously, reporting on environmental, social, and governmental aspects of sustainability has rapidly increased in importance, and environmental boundaries are frequently referenced (Bjørn et al., 2016). Enhancements in reporting, though, are not tantamount with changes in accounting and decision-making practices. While researchers ask for further research on how companies can incorporate absolute sustainability, it remains unclear how sustainability makes its' way into their management accounting and control practices. The aim of this study is to answer the research question on how sustainability is incorporated into managerial decision-making and control today and which adjustments would facilitate further integration of sustainability.

By qualitatively exploring and analyzing the levers of control for sustainability as well as key factors involved in transformational management in the context of

A. K. Meyer (✉)
Manchot Graduate School "Competitiveness of Young Enterprises", Düsseldorf, Germany
e-mail: anna.meyer@hhu.de

B. E. Weißenberger · A. Isaak
Heinrich-Heine University of Düsseldorf, Düsseldorf, Germany

© The Author(s), under exclusive license to Springer Nature Switzerland AG 2023

F. L. Almeida et al. (eds.), *Multidimensional Sustainability: Transitions and Convergences*, Springer Proceedings in Earth and Environmental Sciences,
https://doi.org/10.1007/978-3-031-24892-4_16

firm's investment decisions, this study contributes to the literature on management control systems, leadership, and sustainability. In particular, the results highlight the interplay between the levers of control framework and transformational leadership.

16.2 Theoretical Background

16.2.1 Management Control Theory

The widely adopted Levers of Control (LOC) framework, developed by Simons (1994, 1995), includes four essential components: belief systems, boundary systems, diagnostic control systems, and interactive control systems. Belief systems are those that establish, communicate, and reinforce the basic values and direction of an organization, while boundary systems establish limits for decision-making. Diagnostic control systems are used to compare current performance against targets. Finally, interactive control systems are those which focus on dialogue between the top management team and lower-level staff members. We examine evidence on which role the different types of control are playing to foster the integration of sustainability into management practices.

16.2.2 Transformational Management

The concept of transformational leadership (or management) was originally coined by Burns (1978) and further extended by the works of Bass (1999). It is different from transactional leadership in the sense that transformational leaders watch out for initiatives that bear the potential to add value to the organization. They are often visionary and imaginative, looking toward the future of the company within the larger ecosystem in which it operates and taking a formative approach for strategic development. Even though both types of leaders profit from the use of management controls, they are likely to use them in a different way. While transactional leaders typically use controls to address complexity in a given strategic setting, thus aiming to manage toward efficiency more effectively, transformational leaders can be thought of as using management controls to instigate change, i.e., guiding the organization toward a new (common) vision with the target of adding to firm value beyond its' status quo (Kotter, 1990).

16.3 Literature

In 2012, Gond et al. set the scene for research on sustainability in management accounting when writing that traditional management control systems are "limited (…) in addressing environmental and social issues as well as their interrelationships with financial issues" (p. 208). Soon after, Hartmann et al. (2013) conducted a literature review on the integration of strategy and sustainability in control systems, while Management Accounting Research published a special issue on "Sustainable development, management and accounting: Boundary crossing" (Bebbington & Thomson, 2013).

Rodrigue et al. (2013) mapped sustainability-related controls onto the four levers of control dimensions of Simons (1995), which also informed our view on control systems, and which helped structuring the information deriving from our interviews. Lueg and Radlach (2016) as well as Johnstone (2019) have written about so-called sustainability control systems which are "providing information and direction, as well as monitoring and motivating employees to continually develop sustainable practices and procedures for future improved sustainability performance" (p. 34). Nevertheless, having sustainability control systems in place does not ensure that sustainability is successfully integrated into corporate strategy and business activities (Beusch et al., 2022).

Therefore, we highlight three key research streams here. The first is about the assessment of impacts of business activities and how these can be measured and incorporated in decision-making. Here, Cuckston (2013) conducted a literature review on the crossroads of management accounting for carbon and management control, while Pondeville et al. (2013) focused on decision-making when examining environmental control systems and the role of contextual and strategic factors. More recently, Taïbi et al. (2020) make use of intervention research in order to describe accounting mechanisms for strong sustainability.

The second research stream examines whether sustainability objectives conflict with financial objectives. Here, Epstein et al. (2010) wrote about the challenge of integrating sustainability into operational and capital investment decision-making. Schaltegger et al. (2012) distinguish between managers who have a collaborative business case frame integrating financial, social, and environmental aspects and rather operational managers who strive to improve profitability by making incremental sustainability-related improvements to products and processes. In 2020, Siegrist et al. developed a conceptual framework embedding environment and sustainability into corporate financial decision-making.

The third related literature stream is based on Crutzen and Herzig (2013), who describe that responding to sustainability requires strategic renewal and extensive organizational learning and change. Rodrigue et al. (2013) and Johnstone (2019) underline that companies oftentimes fail to enable an effective dialogue between strategic and operating levels to ensure the integration of sustainability in decision-making on all levels. Here, Beusch et al. (2022) provide evidence that dialogue between managers with different business case frames may hamper subordinating sustainability to financial concerns. The ground for Beusch was paved by Sharma and Jaiswal (2018), who examined the circumstances in which dialogue mitigates the marginalization of sustainability initiatives.

16.4 Data and Methodology

In our study, we follow the research paradigm of analytic induction, hereafter AI (Robinson, 1951; Znaniecki, 1934), starting with the tentative definition of our research subject based on pre-existing knowledge of theory and empirical findings and then go on to test this based on findings from explorative interviews. We conducted our interviews with a group of mostly C-level executives, all of them being responsible for high-level investment decisions in their organizations. Company size ranged from less than 10 employees to more than 100.000, coming from divergent types of industries, e.g., Manufacturing, Services, and Energy.

The research logic of AI is used to guide data collection and analysis as well as to organize the presentation of findings (Katz, 2001): "Its objective is causal explanation, a specification of the individually necessary and jointly sufficient conditions for the emergence of some part of social life. AI calls for the progressive redefinition of the phenomena to be explained" (Katz, 2001, p. 484). Authors have invoked AI to conduct more rigorous qualitative analyses and to bridge the qualitative-quantitative divide (Hammersley & Cooper, 2012). Examples of such work in management and entrepreneurship literature include Bansal and Roth's (2000) model of ecological responsiveness, Busch's (2011) work on organizational adaptation to disruptions of the natural environment (e.g., climate change) and Hoffmann et al.'s (2009) analysis of regulatory uncertainty, investments, and resources.

Following this approach and guided by extant knowledge of management control theory as well as transformational management, both comprising the theoretical scaffolding of our study (Wener & Woodgate, 2013). Interviews were conducted and analyzed in an iterative manner, beginning with an initial set of questions, and following the principle of the hermeneutic circle, with the goal of refuting, supporting, or further developing the initial theoretical approach.

Following Spence et al. (2010, p. 86), we concentrate on observing key activities/managerial decisions rather than on the reporting level. To understand the context of investment decisions, we explore the process of information use in (sustainability) decision-making, a step that is strongly dependent on the accounting approaches in place. Following the research approach of AI, we categorized the thematic codes along the lines of the initial theoretical framework. That is, we first pulled together all quotes relevant to the four key components of management control theory, as well as for transactional and transformational management, each as they relate to sustainability accounting and decision-making in the organizations under observation.

16.5 Findings

16.5.1 Belief Systems

As a first result, we found in our interviews that belief systems can be an important driver of sustainability accounting (or inhibit it). First, I1 recounted that "sustainability has always been very important to me. But this is often not lived in large corporation (I1:3)." Not only can inertia be challenging, but having a vision is quite different than living it. A partial explanation for this can be found in the further remarks of I1. For instance, in context of the harsh daily business reality, I1 remarked that, strikingly, "the people, the managers, they are measured by their royalties (share of the sales) and not by environmental goals (I1:25)." Therefore, mindset in the first place, but also the setup of incentive systems is crucial for managers' buying-in to moving from vision to actual execution.

I11 gave some insights into hurdles for integrating sustainability: "The business case must pay off. But the topic of profit and how do I get that together with the topic of sustainability, that's a nut that has to be cracked (11:15)."

The last statement by I11 also underlines how belief systems can collide – those based on profit vs. social welfare maximization, or rather that having room for sustainability goals requires first meeting the company's basic needs in terms of profitability or even finding a business case for sustainability.

16.5.2 Boundary Systems

Several interviewees mentioned regulations as forcing them to change their internal processes toward sustainability. I1 described: "If there was a need (strong regulatory pressure) to invest now in order to comply with the regulations in one- or two-years' time, then that's what we did (otherwise not)" (I1:17). Furthermore, they described the need to account for the environmental impact of their production (e.g., on CO_2 emissions, pollution, etc.). I2 remarked: "This means that when we talk about sustainability, it is also important not only to assess everything financially, but also to consider what impact, so to say, we are making (I2:10)."

16.5.3 Interactive Control Systems

According to the work of Simons (1995), interactive control systems are used to stimulate organizational learning and the emergence of new ideas and strategies.

A few companies established specific processes to make sure interaction takes place and different opinions are heard: "(...) we have adapted this in the past by saying that we will make this assessment broader. And that we should also bring together the most diverse views internally as possible (I3:21)." Asked if he believes that this is the right way to handle potential conflicts, he says: "If you don't allow these rooms, the conflicts are still there, right? (I3:42)."

To stimulate organizational learning and the emergence of new ideas, employees of a rather small company established a task force for sustainability (I6:21). Ideas which are generated in the group are communicated to the top management and interaction is called upon by the employees. Similarly, in a company of a bigger size, "a working group of employees was summoned to develop indicators for sustainability (I8:25)." One interviewee describes: "We are dependent on dialogue.

(…) But you could tell that they weren't immediately and voluntarily getting involved in such issues. (…) Here a great deal of sensitivity is required. It doesn't happen overnight (I6:9)."

16.5.4 Diagnostic Control Systems

Diagnostic control systems, as one element of Simons' levers of control framework, are used to motivate, monitor, and reward the achievement of specific goals.

The monitoring of certain indicators, of course, increases their importance which can be perfectly related to monitoring sustainability. "Because I believe that the impact will be stronger if we can underpin this purpose with actual measurable effects (I2:34)." Another interviewee pointed out accordingly that facts and figures are quite important in his business context to move the topic of sustainability further: "I work in a company that is very strongly driven by numbers. Where controlling values also have a certain value in a double sense (I3:13)." Here, "it would help to establish true costs, I mean what that really means for the future, if you had that business models or certain processes (I3:47)." In another interview, a similar critique was formulated: "Before, it was often like this, yes, self-descriptions, but which were more like a report, without many numbers. Or where you set your own goals. […] Well, the risk of window dressing was already considerable (I7:25)."

16.5.5 Transformational Management

Parallel to focusing on control systems, we identified the following three subthemes describing what is needed for companies to step out and leave the sphere of simply operational management toward a higher focus on sustainability aspects in decision-making.

16.5.5.1 Value Orientation

A value orientation apart from financial aspects was tentatively named by one of our interviewees with saying: "I don't want to call it, maybe idealistic, but something like this is considered apart from strictly financial terms (I2:26)." Apart from a strategic orientation, personal beliefs seem to be important for decision makers to act apart from fixed and solely financial criteria: "One managing director in the past said: I want it this way. This is what I want to leave behind when I am gone (I2:48)." Another interviewee adds that "if the understanding, the self-perception (of being a company that cares for the environment) is there, then ways are identified, and decisions are made. Then people find additional finances for investments in the environment (I1:25)."

16.5.5.2 Willingness to Pay

While solely one of the companies we interviewed has an extra budget in place to enable investment decisions where taking environmental aspects into account would otherwise kill the business case, most talked about the fact that there is no willingness to pay for environmental aspects. Other than the lack of an extra budget, the following was criticized: "No one clearly says that it is worth it to put one percentage point of return into the environment. (I1:25)." Similarly, another interviewee pointed out that he believes such a willingness to pay "must come from the very top, (…) saying that x percent of the companies' investments or division must be ESG measures, where there is no business case or meeting a KPI needed (I2: 52)."

16.5.5.3 Risk Taking and Innovation/R&D

Whenever investments are made to enter new businesses, sustainability may play a larger role compared to short-term investments or so-called maintenance investments. "The economic viability must also be given, otherwise the money is not tackled, but then other criteria come into play. You ask yourself how sustainable the product is, because it will only cross the profitability threshold for this investment in several years. (…) Sustainability is then a big point, because a product that does not comply with sustainability cannot be economical in the future. Yes, these are such criteria (I5:9)."

16.6 Conclusion and Future Research Opportunities

The objective of our research was to examine how sustainability issues are addressed by different types of controls using Simon's (1994, 1995) seminal levers of the control framework and what might strengthen the successful integration of sustainability into business activities using Burn's (1978) pivotal concept of transformational leadership.

Our interviews document that for integrating sustainability into decision-making, all four types of control are considered relevant and examples of them playing a significant role were identified. There was not one of the four considered more important than the other ones. In order for transformational change to occur, we identified the need for a value orientation toward sustainability aspects, some degree of willingness to accept additional costs, and an openness for risk-taking, e.g., in the form of research and development.

Furthermore, we describe pathways to incorporate transformational management for changing business activities toward "deep" or "strong" sustainability, both being terminus technici for an approach stating that "critical parts of the natural capital must be preserved as to guarantee present and future human wellbeing" (Buriti, 2018, p. 8).

This study is subject to several limitations. First, as this is work-in-progress, we have not yet reached the full sample size envisioned; nonetheless, we are already close to theoretical saturation. Second, due to the short paper format, we cannot include many of the informative quotes that we otherwise would have. Finally, although our interviewee sample is comprised of experts from different sectors and firm sizes, they are all from Germany; therefore, our study's results may not be fully generalizable to other countries with starkly different national cultures or economies.

In line with Beusch et al. (2022), our research calls for future studies that more strongly incorporate stakeholder influences for integrating sustainability into investment decisions. For example, our work could be extended by studying how to measure the distance of current decision-making practices from a scientifically based measurement of companies' environmental impact, i.e., the need to identify patterns and business practices that facilitate or impede a resulting net positive worth for the environment.

References

Bass, B. M. (1999). Two decades of research and development in transformational leadership. *European Journal of Work and Organizational Psychology, 8*(1), 9–32.

Bansal, P., & Roth, K. (2000). Why companies go green: A model of ecological responsiveness. *Academy of Management Journal, 43*(4), 717–736.

Bebbington, J., & Thomson, I. (2013). Sustainable development, management and accounting: Boundary crossing. *Management Accounting Research, 24*(4), 277–283.

Beusch, P., Frisk, J. E., Rosén, M., & Dilla, W. (2022). Management control for sustainability: Towards integrated systems. *Management Accounting Research, 54*, 100777.

Bjørn, A., Margni, M., Roy, P.-O., Bulle, C., & Hauschild, M. Z. (2016). A proposal to measure absolute environmental sustainability in life cycle assessment. *Ecological Indicators, 63*, 1–13.

Bjørn, A., Chandrakumar, C., Boulay, A.-M., Doka, G., & Fang, K. (2020). Review of life-cycle based methods for absolute environmental sustainability assessment and their applications. *Environmental Research Letters, 15*(8), 083001.

Buriti, R. (2018). "Deep" or "strong" sustainability. In W. L. Filho (Ed.), *Encyclopedia of sustainability in higher education*. Springer.

Burns, J. M. (1978). *Leadership*. Harper & Row.

Burritt, R. L., & Schaltegger, S. (2010). Sustainability accounting and reporting: Fad or trend? Accounting. *Auditing & Accountability Journal, 23*(7), 829–846.

Busch, T. (2011). Organizational adaptation to disruptions in the natural environment: The case of climate change. *Scandinavian Journal of Management, 27*(4), 389–404.

Crutzen, N., & Herzig, C. (2013). A review of the empirical research in management control, strategy and sustainability. In L. Songini (Ed.), *Studies in managerial and financial accounting: Vol. 26. Accounting and control for sustainability* (pp. 165–195). Emerald.

Cuckston, T. (2013). Bringing tropical forest biodiversity conservation into financial accounting calculation. *Accounting, Auditing & Accountability Journal, 26*(5), 688–714.

Dahm, D. J. (2019). *Benchmark Nachhaltigkeit. Sustainability Zeroline: Das Maß für eine zukunftsfähige Ökonomie*. transcript Verlag.

Epstein, M. J., Buhovac, A. R., & Yuthas, K. (2010). Implementing sustainability: The role of leadership and organizational culture. *Strategic Finance, 4*, 41–47.

Hammersley, M., & Cooper, B. (2012). Analytic induction versus qualitative comparative analysis. In B. Cooper, J. Gaessner, R. Gomm, & M. Hammersley (Eds.), *Challenging the qualitative-quantitative divide: Explorations in case-focused causal analysis*. Continuum Publishing Corporation.

Hartmann, F., Perego, P., & Young, A. (2013). Carbon accounting: Challenges for research in management control and performance measurement. *Abacus, 49*(4), 539–563.

Hoffmann, V. H., Trautmann, T., & Hamprecht, J. (2009). Regulatory uncertainty: A reason to postpone investments? Not necessarily. *Journal of Management Studies, 46*(7), 1227–1253.

Johnstone, L. (2019). Theorising and conceptualising the sustainability control system for effective sustainability management. *Journal of Management Control, 30*, 25–64.

Katz, J. (2001). Analytical induction. In N. S. Smelser & P. B. Baltes (Eds.), *International encyclopedia of the social and behavioral sciences* (pp. 480–484). Elsevier BV.

Kotter, J. P. (1990). What leaders really do. *Harvard Business Review, 68*(3), 103–111.

Lueg, R., & Radlach, R. (2016). Managing sustainable development with management control systems: A literature review. *European Management Journal, 34*(2), 158–171.

Pondeville, S., Swaen, V., & de Rongé, Y. (2013). Environmental management control systems: The role of contextual and strategic factors. *Management Accounting Research, 24*(4), 317–332.

Robinson, W. S. (1951). The logical structure of analytic induction. *American Sociological Review, 16*(6), 812–818.

Rodrigue, M., Magnan, M., & Boulianne, E. (2013). Stakeholders' influence on environmental strategy and performance indicators: A managerial perspective. *Management Accounting Research, 24*(4), 301–316.

Schaltegger, S. (2018). Linking environmental management accounting: A reflection on (missing) links to sustainability and planetary boundaries. *Social and Environmental Accountability Journal, 38*(1), 19–29.

Schaltegger, S., Freund, F. L., & Hansen, E. G. (2012). Business cases for sustainability: The role of business model innovation for corporate sustainability. *International Journal of Innovation and Sustainable Development, 6*(2), 46944–46995.

Sharma, G., & Jaiswal, A. K. (2018). Unsustainability of sustainability: Cognitive frames and tensions in bottom of the pyramid projects. *Journal of Business Ethics, 148*(2), 291–307.

Siegrist, M., Bowman, G., Mervine, E., & Southam, C. (2020). Embedding environment and sustainability into corporate financial decision-making. *Accounting and Finance, 60*(1), 129–147.

Simons, R. (1994). How new top managers use control systems as levers of strategic renewal. *Strategic Management Journal, 15*(3), 169–189.

Simons, R. (1995). *Levers of control: How managers use innovative control systems to drive strategic renewal*. Harvard Business School Press.

Spence, C., Husillos, J., & Correa-Ruiz, C. (2010). Cargo cult science and the death of politics: A critical review of social and environmental accounting research. *Critical Perspectives on Accounting, 21*(1), 76–89.

Taïbi, S., Antheaume, N., & Gibassier, D. (2020). Accounting for strong sustainability: An intervention-research based approach. *Sustainability Accounting, Management and Policy Journal, 11*(7), 1213–1243.

Wener, P., & Woodgate, R. L. (2013). Use of qualitative methodological scaffolding process to design robust interprofessional studies. *Journal of Interprofessional Care, 27*(4), 305–312.

Znaniecki, F. (1934). *The method of sociology*. Rinehart & Company.

Sustainable Cities? Insights from Brazil and Portugal

17

Silvio Roberto Stefani, João Paulo Peixoto, Paula Santos, Reinaldo Ferreira, Claudio Luiz Chiusoli, and Cláudia Susana Ribeiro Rodrigues

17.1 Introduction

Cities are the main form chosen by the human being to live in society, where there are more than half of the human population and by 2050 there will be 70% (UN, 2019). In cities are built the most miscellaneous economic and social relationships, demonstrating the importance cities hold for the modern world. Cities end up adopting a predatory position in the environment once they consume resources from the environment and they end up presenting a deficit with the environment because of the pollution and degradation.

The Sustainable Development Goals (SDG) from UN 2030 Agenda bring in its 11th goal the sustainable cities and communities with 11 goals and various indicators. Another perspective is the evaluation of sustainable cities through ISO 37120, which brings 100 indicators, divided into 16 themes, covering topics such as Economy, Energy, Health, Education, Governance, Public Security, and Solid Waste, among others. The city of Porto (PT) has already got the certification and other important cities such as London (ENG), Barcelona (SPA), Paris (FRA), Rotterdam (NED), and Milan (ITA) (Anthopoulos, 2017; Intedya, 2021).

Added to cities importance and sustainability importance for society there is a whole set of interests toward cities transformation, being these interests recognized as stakeholders and their citizens. The evolution from a regular city to a sustainable

S. R. Stefani (✉) · C. L. Chiusoli
Atlântico Business School, Vila Nova de Gaia, Portugal

UNICENTRO, Guarapuava, Brazil
e-mail: silvio.stefani@abs.pt

J. P. Peixoto · P. Santos · R. Ferreira · C. S. R. Rodrigues
Atlântico Business School, Vila Nova de Gaia, Portugal

© The Author(s), under exclusive license to Springer Nature Switzerland AG 2023

F. L. Almeida et al. (eds.), *Multidimensional Sustainability: Transitions and Convergences*, Springer Proceedings in Earth and Environmental Sciences,
https://doi.org/10.1007/978-3-031-24892-4_17

city is a necessity, notice the large group of agents and interests that these changes may bring.

The UN SDGs are the result of more than 3 years of research and negotiation including three items from sustainable development, and the ISO 37120 is related to sustainability pillars: environmental, social, and economic. SDG 11th brings goals and indicators aiming the sustainability and life quality. Whereas the ISO introduces indicators (not goals) related to cities' sustainability (indicators for urban services and life quality) (Alfaro et al., 2017; Ribeiro, 2019).

Increasingly growing cities can generate high energy consumption, water and air pollution, environmental degradation, poor housing, and working conditions, as well as social inequality (Bibri, 2018a; Raspotnik et al., 2020), with this, sustainable and smart cities are increasingly investing in innovation systems. Although innovation systems are complex, it is possible to verify the contribution in minimizing social, environmental, and economic impacts, focusing on the creation, diffusion, and use of knowledge and technology that can be visualized through architectural and urban projects, ecosystem services, seeking to improve people's quality of life, environmental protection, and long-term social well-being (Anthopoulos, 2017; Bibri, 2018a, b).

It is possible to notice a gap about ISO 37120 in cities because the systematic review made by Stefani and Viante (2020) in periodicals from Capes, Spell, Scielo, and Google Scholar from 2015 to 2019 there were identified 417 full articles in Portuguese and English about sustainable cities, but only 23 about ISO 37120 in their various indicators and aspects of sustainable cities, presenting as an opportunity of empirical and analyze research.

Thus, the research question: What are the citizens from Brazil and Portugal assessments in relation to ISO 37120? To answer such a question, the main goal was chosen: analyzing ISO 37120 indicators in a city located in Brazil and Portugal, sustainable development strategies are present in public policies. It is important to highlight the importance of the constant presence of governmental and organizational actions, and community participation, as these can be considered protagonists for sustainable development (Galleli & Hourneaux, 2019). The research justifications focus on the application of indicators in any city, the transformation of urban environments, and the importance of sustainability for the future of cities.

17.2 Theoretical Framework

The concern with sustainable development, as well as sustainable cities, has been the object of studies since the elaboration of the Brundtland Report in 1987, a document entitled Our Common Future, which deals with matters pertaining to social, economic, and social aspects and environmental issues such as poverty, environmental resources, pollution, and energy consumption. In January 1992, the United Nations World Conference on Environment and Development, held in Rio de Janeiro, the Agenda 21 document was approved, so studies and governmental interests on the subject began to have more prominence and visibility in institutions of education, public and private organizations, governments (Bibri, 2018a, b).

Kobayashi et al. (2017, p. 17) understand that: "Smart and Sustainable Cities are territories that use Information and Communication Technologies and sustainable urban development practices in order to provide a better quality of life for their citizens," so that they meet the needs of their citizens. Included in the Brundtland report (Jarrar & Al-Zoabi, 2008).

Thus, new technologies are also linked to development, as Kniess et al. (2019) claim that the development of new technologies combined with the challenges of the urban environment can create new models for the reality of cities. Thus, such measures are in accordance with what was exposed by Junior and Duenhas (2020), who point to a greater concern with negative environmental externalities, such as air, soil, river, and other pollution. In this sense, Kobayashi et al. (2017) emphasize that cities that seek to reinvent themselves must pay greater attention to social inclusion and the use of green technologies, factors that, together with an intelligent management of the urban territory, are capable of providing sustainable urban development.

When defining sustainable development, urban planners and specialists use the concept proposed in the document Our Common Future or the Brundtland Report (RWCED 1988, p. 46): "it is one that meets the needs of the present without compromising the ability of future generations to meet the needs of the future," "their own needs." However, studies indicate that the most vulnerable populations are still not viewing this concept in a broad way, due to the social, economic, and environmental differences that still exist (Bento et al., 2018).

It should be considered that the concept of sustainable cities goes beyond the economic aspects, as cities when planned need to preserve green areas, without changes in natural ecosystems, aiming to provide the urban population with a better quality of life, preservation of air quality, air conditioning, and recovery of water systems, without compromising the ability of people or future generations to meet their needs (Bibri, 2018b; Ferreira et al., 2018; Kobayashi et al., 2017).

Cities, also called centers of population concentration, end up attracting various economic activities, thus presenting a great productive activity and high levels of energy and raw materials consumption, generating waste and pollution (Sarubbi & Moraes, 2019). For Junior and Duenhas (2020), cities concentrate a large part of the population, wealth, and productive activities worldwide, but they are also responsible for serious social and environmental problems.

In the search for sustainability, whether through social, environmental, and economic policies, Fabris et al. (2020, p. 2016) state that the "challenge of a sustainable city lies in the diversification of its dynamic, creative and sustainable local economy, based on the creation of development alternatives to traditional chains" that may reverse local actions in improvements and quality of life, with the active participation of citizens in the improvement processes (Aina, 2017).

Sustainable cities need urban planning, which involves elaborating, planning, evaluating, and predicting the development of a city, community, or region, through organized and standardized systems, making the city sustainable, and capable of generating benefits for the existing population (Bibri, 2018a, b; Stefani & Viante, 2020).

In this way, sustainable cities require changes in cultural, social, environmental, economic, and governmental contexts, changes that can be based on ethics, citizenship, solidarity, and reciprocity (Weisenfeld & Hauerwaas, 2018), since sustainability requires the participation of all.

Cities around the world, as an important incremental innovation, can use the Brazilian technical standard ISO 37120 as a management parameter, which deals with the Sustainable Development of Communities, with indicators for urban services, quality of life, as well as those to help municipalities measure and compare sustainable development.

The ISO 37120 standard is structured into three types: essential indicators, supporting indicators, and profile indicators, which are described below: (1) Economy: The indicators are linked to fundamental issues of economic growth in cities, such as unemployment rate, percentage of the population below the poverty line and number of companies in the region. (2) Education: The indicators encourage the participation of students at all stages, from primary education to the training of professionals with higher education. (3) Energy: The indicators are linked to the consumption of energy by the inhabitants and also other forms of energy, especially those from renewable sources. (4) Environment: Directly linked to the emission of harmful gases to the environment, especially those that increase the greenhouse effect. The Norm also treats noise pollution as an aspect of the environment, which is very present in the urban environment. (5) Finance: Point out the expenses and financial health of public management, even more with factors such as the collection of taxes and own source of revenue. (6) Responses to Fires and Emergencies: Point out the activities of the city's fire department, with emphasis on its contingent and the number of volunteer firefighters.

The indicator also collects information such as the number of deaths related to natural disasters and fires, an aspect directly related to the impact of the environment on urban settlements. (7) Governance: They are linked to popular participation in elections, the number of women elected, and also the number of civil servants convicted of corruption. (8) Health: The importance of the number of health professionals, such as doctors, nurses, and psychologists within the city. (9) Recreation: They are mainly linked to the presence of recreational spaces open to the population, a factor related to leisure and the quality of life of citizens. (10) Security: Reinforces the importance of public security within a sustainable city, with a view to reducing crime and homicides. (11) Housing: The housing indicator presents aspects related to citizens' housing, such as the percentage of the population in slums, the number of homeless people, and illegally occupied housing within the territory. (12) Solid Waste: The standard indicator establishes the study of the percentages of hazardous waste, waste incinerated and dumped in the open. (13) Telecommunications and Innovation: The sustainable city needs connected, agile, and globally accessible communication systems, whether over the internet or by telephone; connectivity is essential for the application of communication tools. Thus, the standard seeks the ratio of the number of landline and cell phone lines, as well as the number of internet connections per inhabitant. (14) Transport: Urban infrastructure is essential within a city that seeks better connectivity and cleaner means of transport; in this way, the standard presents aspects of the public transport system, number of cars and the use of alternative means of transport, the measurement of the number of kilometers of cycle lanes. (15) Urban Planning: Provision of more natural environments to citizens, mainly due to the number of trees planted and green areas available. A greater contact of citizens with green areas can provide a better quality of life and moments of leisure within the territory of the municipality. (16) Sewage: Indicators related to sewage deal with the number of people that are served by the collection system, as well as the treatment that such sewage receives. (17) Water and Sanitation: The latest indicators of the standard deal with the quality of water available to citizens, as well as the percentage of the population with access to drinking water and other services relevant to its distribution (ABNT, 2017).

ISO 37120 points out that the indicators can be applied to any city, municipality, or government that seeks to measure its performance in a comparable way, regardless of its size and location (ABNT, 2017).

17.3 Methodological Procedures

This research fits in a quantitative methodology, making use of online survey, by having a descriptive character to the fulfillment of the main purpose and to answer the research question. The units of analysis of this research are the municipalities of Guarapuava, located in the central-south region of the State of Paraná – Brazil, with more than 167 thousand inhabitants, and Matosinhos in the region of Porto, North of Portugal with more than 175 thousand people, that is, with approximate populations.

Due to the uniqueness of the cases under study, we define our research as instrumental (Stake, 2003), since it is a particular case that was examined to provide insights.

The cases of Guarapuava and Matosinhos Cities that excels when regional hub of cities in the region, so it is possible to understand the resultant complexities of municipal management practices. Stake (2003) emphasizes that cases allow us to understand their deepest complexities. The author also explains that even when performing quantitative studies there is relevance and contribution that can be extracted from the deviant or illustrative. A research protocol was elaborated where the concepts of sustainable cities and sustainable city indicators were aligned with data collection and analysis techniques.

As a quantitative technique for data collected, a standard indicator questionnaire was adapted (Norm ISO 37120) to the citizens in this study, with 65 items taken from ISO on a Likert scale of five points, from totally disagree to totally agree and seven more questions about economic and social profile. This instrument was validated by four specialists with PhD.

For Creswell (2010), the survey process can use non-probability random sampling. The survey application happened in 2021 and 2022. It collected 345 questionnaire answers and validated from Brazil and 201 questionnaire answers and validated from Portugal, treated as a random sample, not probabilistic by convenience. The questionnaire was available through Google Forms® and shared in social media, e-mails, and groups of instant messaging. The quantitative data was made highest and lowest averages (mean) with the software SPSS® (v. 21).

17.4 Results

Through quantitative research, the citizens have presented a general perception of Education, Health, and Security indicators, sometimes in a positive or negative way. Respondents were asked to assign a score from 1 to 5 (strongly disagree to strongly agree) that represented their opinion regarding the ISO 37120 (SC) issues. By analyzing the descriptive statistics of the indicators, it was possible to identify the variables that had the lowest averages among the questions surveyed, as shown in Table 17.1.

Of the 66 variables surveyed, variable Q55 was the one that had the highest average in both cities studied and is linked to Transport indicators (own cars). It would be very interesting for citizens to use public transport as a means of transportation: buses, subways, and trains, as they are less polluting and have lower per capita costs, contributing to air quality, reduced gas emissions, reduced traffic, lack of parking, and among other things.

The second-highest average in both cities was Q36 (The city offers outdoor leisure facilities). It demonstrates that the cities surveyed have outdoor leisure spaces in their spaces, as well as accessible parks and green areas, woods, and parks. We can still consider that the averages are related to the fact that in recent years several public leisure spaces have been restored and created in various points and neighborhoods of both cities, contributing to the socialization and quality of life of citizens.

In variable Q9, on quality education in higher education in the city, it was the fourth highest average in Guarapuava and the eighth highest average in Matosinhos, highlighting the expressive number of Universities in Guarapuava with five institutions and Matosinhos with three, in addition to the proximity to several others in the metropolitan region.

Also noteworthy are the high averages on quality air, quality electrical energy, quality treated water, and the satisfactory number of teachers in Guarapuava. In Matosinhos, in turn, quality landline and mobile telephones, water, and sanitation stand out.

When analyzing the descriptive statistics of the indicators, it was also possible to identify the ten variables that had the lowest average among the questions surveyed, which are presented in Table 17.2.

Among the questions with the lowest averages presented, it was possible to verify that despite the averages being different between the city of Guarapuava and Matosinhos. The questions that presented the lowest average cities in Guarapuava

Table 17.1 Questions with the highest averages (mean): Guarapuava X Matosinhos

Guarapuava			Matosinhos		
Questions	Mean	Deviation	questions	Mean	Deviation
Q55 – The city of Guarapuava has a large number of cars	4319	0,798	Q55 – The city of Matosinhos has a high number of cars	4060	0,807
Q36 – The city of Guarapuava offers places for outdoor leisure	4313	0,866	Q36 – The city of Matosinhos offers places for outdoor leisure	4000	0,857
Q34 – The city of Guarapuava has parks and accessible green areas	4206	0,938	Q63 – The city of Matosinhos offers quality treated water	3930	0,83
Q9 – The city of Guarapuava offers quality education in Higher Education	4197	0,919	Q52 – The city of Matosinhos offers a quality landline service	3890	0,817
Q56 – The city of Guarapuava has a large number of motorcycles	4110	0,952	Q64 – The city of Matosinhos has no interruptions in water and sanitation services	3880	0,848
Q14 – The city of Guarapuava has quality air	3893	1,020	Q62 – The city of Matosinhos has quality basic sanitation	3860	0,855
Q11 – The city of Guarapuava offers a quality electricity service	3814	1,040	Q51 – The city of Matosinhos offers a quality mobile phone service	3850	0,841
Q64 – The city of Guarapuava offers a quality treated water service	3571	1,080	Q9 – The city of Matosinhos offers quality education in the 3rd cycle	3840	0,783
Q10 – The city of Guarapuava has a satisfactory number of teachers	3522	1,120	Q34 – The city of Matosinhos has accessible green parks	3840	0,817

Source: Survey data

were Q42 (The city does not have people living in slums) and Q33 (The city does not have suicides). In Matosinhos, the lowest averages obtained were related to not having politicians arrested for corruption, not having crimes against property and not having homeless people (urban settlements).

It is noteworthy that in Agenda 2030 (2016), goal 11 aims to make cities and human settlements inclusive, safe, resilient, and sustainable, being one of the concerns in the pursuit of sustainable goals, in addition to SDG 1, which aims to end poverty in all its forms, in all places, demonstrating that this variable is one of the great concerns of the citizens.

The variable Q6, which deals with innovation, was one of the variables that obtained intermediate averages in both cities, a factor that may be related to the lack of innovation on the part of organizations and the lack of industries. Thus, it is

Table 17.2 Questions with the lowest averages (mean): Guarapuava X Matosinhos

Guarapuava			Matosinhos		
Questions	Mean	Deviation	Questions	Mean	Deviation
Q42 – The city of Guarapuava does not have people living in slums	1520	0,839	Q28 – The city of Matosinhos has politicians arrested for corruption	2160	1080
Q33 – The city of Guarapuava has no suicides	1540	0,828	Q39 – The city of Matosinhos has no property crimes	2500	1140
Q43 – The city of Guarapuava has no homeless people (urban settlements)	1600	0,904	Q43 – The city of Matosinhos has no homeless people (urban settlements)	2520	0,960
Q38 – The city of Guarapuava has no crimes against life	1650	0,899	Q33 – The city of Matosinhos has no suicides	2540	1040
Q41 – The city of Guarapuava does not have violent crimes	1700	0,919	Q17 – The city of Matosinhos has no noise pollution	2620	0,893
Q39 – The city of Guarapuava has no property crimes	1810	1,010	Q21 – The city of Matosinhos has no noise pollution	2760	0,961
Q62 – The city of Guarapuava does not have irregular occupations with lack of public services	1990	1,060	Q38 – The city of Matosinhos has no crimes against life	2820	0,949
Q25 – The city of Guarapuava has a satisfactory number of elected women	2080	1,100	Q16 – The city of Matosinhos does not emit a large amount of greenhouse gases	2870	0,835
Q28 – The city of Guarapuava offers a satisfactory number of hospital beds	2120	1,100	Q41 – The city of Matosinhos has no violent crimes	2880	1,000

Source: Survey data

highlighted that innovation has been one of the priorities in developing countries (Huatuco & Ball, 2019); generating value, whether economic or strategic, factors that are essential for the development of a city or region (Zen et al., 2017).

The results earned show the ISO 37120 indicators can assist in the integrated monitoring of city governance and advise some specific goals for city improvement in many aspects. The Norm can bring benefits for the governance and effective services provisions in many areas, decision taking support, urban planning, cities master plans, and others.

The application contribution of ISO 37120 indicators presented rates and values about some economic, social, and environmental aspects inside the city. In this survey, it is possible to verify indicators that were analyzed and in future research, other indicators related to SDG11 or other metrics may be studied. This data collection is essential to public management to take this information as a basis for public policy and in the future chase the ISO certification. Because ISO 37120 presented 100 indicators, other contributions are linked with the necessity of complementary research about sectors and indicators inside the city, this other research can be developed in interdisciplinary manner to other cities.

The ISO indicators study allows the recognition about stronger and weaker indicators that can be improved in the city, function like a tool to help the public management. This analyzed information helps to identify sustainability indicators in the city. This fact can conduct a set of actions from interested agents and bring long-term benefits for the citizens.

In general, studies advocate only the relationship between sustainability and municipal management. This research brings a practical and useful tool, ISO 37120, which can be used in several countries as indicators benchmarking of evaluation of municipal management and strategic planning of cities.

17.5 Conclusions

It is understood that the objective of the research was achieved, as it was analyzed how the citizens evaluate the indicators of ISO 37120 in the municipalities studied. It is noteworthy that citizens are the main Stakeholders of cities and should be investigated for the formulation of public policies, the municipal master plan, and the participatory budget, among others. The results show that ISO 37120 indicators are perceived by the respondents, some with more, others with less intensity. These indicators, when positive, can significantly contribute to achieving the sustainable development goals of the UN 2030 Agenda (2016).

Answering the research question: Do Guarapuava and Matosinhos have positive indicators in ISO 37120 to be considered sustainable cities? It was possible to state that yes, the municipalities have several positive indicators, but they cannot be considered fully sustainable cities in the view of the surveyed citizens, as several indicators had a negative perception.

These indicators that have a negative perception could be used to influence public and private projects aimed at improving quality of life, reducing poverty, improving education, and generating jobs and income for the population, among other sustainable development goals (SDGs) of the UN 2030 Agenda (2016) in the respective cities.

It can be concluded that the research brings contributions to the academic community, as it sought to deepen the theme based on national and international studies, in addition to relating the study to the SDGs of the UN 2030 Agenda (2016) and presenting gaps for research and development of future public policy studies.

For society, the study contributes toward demonstrating the presence of the indicators of ISO 37120 in the view of those surveyed, since sustainable development encompasses social, environmental aspects and the reduction of inequality in a population, whether local, regional, or national in scope (Agenda 2030, 2016). The survey can also be used as an incremental innovation in public management, serving as a basis for management to act to improve the areas of greatest dissatisfaction pointed out by respondents, such as public policies on health, safety, education, inclusion of women, and innovation.

In this context, the results can serve as a basis for further research, as the present study points to important indicators for the development of cities, based on the perception of the citizens themselves; it can also contribute to the formulation of public policies and the municipal master plan.

References

Abnt Nbr ISO 37120. (2017). *Sustainable development of communities – Indicators for urban services and quality of life, 87*. ISO. https://www.normas.com.br/visualizar/abnt-nbr-nm/11880/abnt-nbriso37120-cidades-e-comunidades-sustentaveis-indicadores-para-servicos-urbanos-e-qualidade-de-vida

Agenda 2030. (2016). *Transforming our world: The 2030 Agenda for sustainable development*. https://sdgs.un.org/2030agenda

Aina, Y. A. (2017). Achieving smart sustainable cities with GeoICT support: The Saudi evolving smart cities. *Cities, 71*, 49–58. https://doi.org/10.1016/j.cities.2017.07.007

Alfaro, J. L., López, V. R., & Nevado, D. (2017). A new sustainability city index based on intellectual capital approach. *Sustainability, 9*, 860. https://doi.org/10.3390/su9050860

Anthopoulos, I. (2017). Smart utopia vs smart reality: Learning by experience from 10 smart city cases. *Cities, 63*, 128–148. https://doi.org/10.1016/j.cities.2016.10.005

Bento, S. C., Conti, D. M., Baptista, R. M., & Ghobril, C. N. (2018). The new guidelines and the importance of urban planning for the development of sustainable cities. *Rev Management Ambient Sustainability, 7*(3), 469–488. https://rgsa.emnuvens.com.br/rgsa

Bibri, S. E. (2018a). A foundational framework for smart sustainable city development: Theoretical, disciplinary, and discursive dimensions and their synergies. *Sustainable Cities and Society, 38*, 758–794. https://doi.org/10.1016/j.scs.2017.12.032

Bibri, S. E. (2018b). Backcasting in futures studies: A synthesized scholarly and planning approach to strategic smart sustainable city development. *European Journal of Futures Research, 6*(1), 1–27. https://doi.org/10.1186/s40309-018-0142-z

Creswell, J. W. (2010). *Research project: qualitative, quantitative and mi methods*. Porto Alegre: Artmed.

Fabris, J., Bernardy, R. J., Sehnem, S., & Piekas, A. A. S. (2020). Sustainable cities: Paths and possibilities. *International Journal of Professional Business Review, 5*(2), 214–233. https://doi.org/10.26668/businessreview/2020.v5i2.218

Galleli, B., & Hourneaux, F. J. (2019). Human competences for sustainable strategic management: Evidence from Brazil. *Benchmarking: An International Journal*. https://doi.org/10.1108/BIJ-07-2017-0209

Huatuco, L. H., & Ball, P. D. (2019). The quest for achieving United Nations sustainability development goals (SDGs) infrastructure and innovation for responsible production and consumption. *RAUSP Management Journal, 54*(3), 357–362. https://doi.org/10.1108/RAUSP-04-2019-0068

Intedya. (2021). *Consulting, auditing and training*. City of Porto certified in ISO 37120. http://www.intedya.pt/internacional/intedya-noticias.php?id=572

Jarrar, O. M., & Al-Zoabi, A. Y. (2008). The applicability of sustainable city paradigm to the city of Jerusalem: Criteria and indicators of efficiency. *Building and Environment, 43*(4), 550–557. https://doi.org/10.1016/j.buildenv.2007.01.025

Junior, M. P., & Duenhas, R. A. (2020). Smart cities and sustainable cities: Convergence of actions or mere advertising? *Brazilian Journal of Planning and Development, 9*(2), 317–328. https://doi.org/10.34117/bjdv6n8-333

Kobayashi, A. R. K., Kniess, C. T., Serra, F. A. R., Ferraz, R. R. N., & Ruiz, M. S. (2017). Smart Sustainable Cities: bibliometric study and patent information. *International Journal of Innovation (IJI Journal), São Paulo, 5*(1), 77–196. Jan/April.

Kniess, C. T., Aguiar, A. O., de Conti, D. M., & Philipi, A., Jr. (2019). Urban Innovation and human resources for sustainable cities management. *Advanced Studies, 33*(97), 4.

Lamano Ferreira, M., De Souza, L. C., Conti, D. D. M., Capellani Quaresma, C., Reis Tavares, A., Gonçalves da Silva, K., et al. (2018). Soil biodiversity in urban forests as a consequence of litterfall management: Implications for São Paulo's ecosystem services. *Sustainability, 10*(3), 684. https://doi.org/10.3390/su10030684

Raspotnik, A., Gronning, R., & Hermann, V. (2020). A tale of three cities: The concept of smart sustainable cities for the Arctic. *Polar Geography, 43*(1), 64–87. https://doi.org/10.1080/1088937X.2020.1713546

Ribeiro, T. S. V., Cortese, T. T. P., Kniess, C. T., & Conti, D. de Melo (2019). What is the role of indicators as a governance instrument tool to help cities become more sustainable? *Brazilian Journal of Management - ReA UFSM, 12*(1), pp. 580.

Sarubbi, M., & de Moraes, C. S. B. (2019). Comparative evaluation of indicator methodologies for urban sustainability. *Notebooks Zygmunt Bauman, 8*(18), 211–231.

RWCED (1988) Report of the World Commission on Environment and Development: Our Common Future. https://sustainabledevelopment.un.org/content/documents/5987our-common-future.pdf

Stake, R (2003) Case studies. In N. K. Denzin & Y. S. Lincoln (Eds.), *Strategies of qualitative inquiry* (2nd Ed.) (pp. 134–164). Thousand Oaks, CA: Sage.

Stefani, S. R., & Viante, M. (2020). Sustainable cities: A systematic review of the literature. *Atlântico Business Journal, 1*(4), 63–68.

UN - United Nations. (2019). The 2030 Agenda for Sustainable Development. A/RES/70/1. https://sustainabledevelopment.un.org/content/documents/21252030%20Agenda%20for%20Sustainable%20Development%20web.pdf

Weisenfeld, U., & Hauerwaas, A. (2018). Adopters build bridges: Changing the institutional logic for more sustainable cities. From action to workset to practice. *Research Policy, 47*, 911–923. https://doi.org/10.1016/j.respol.2018.02.015

Zen, A. C., Machado, B. D., López, A. I. J., Borges, M. C., & Menezes, D. C. (2017). Innovation route: A proposal for innovation management methodology. *RAC, 21*(6), 875–892.

Part VI

Marketing

The Relationship Between Corporate Social Responsibility and Customer Satisfaction: Literature Review

18

Inês Veiga Pereira, Joana Araújo, and José Duarte Santos

18.1 Introduction

Corporate social responsibility (CSR) has been an important focus of companies and for the ones concerned about their sustainability. Despite being focused on improving society and their performance, organizations want their customers to be aware of their behaviour and be more satisfied with the brand.

Literature has researched about CSR, its concepts and use. Also, research focused on testing some consequents of CSR, but it could not be found a paper which summarized the main consequents of CSR. Furthermore, the focus of CSR has been on how it affects the brand, but further understanding of its relationship with customer satisfaction is required. Therefore, this chapter aims to analyze if literature shows if both constructs can be connected.

Through a systematic literature review of the main databases (Scopus and WOS), this research addresses CSR concept and consequents, leading the research to three major themes. The first theme is related to CSR and includes research on the concept, evolution, and dimensions, as well as CSR consequents – brand image, brand value, and consumer satisfaction.

In a second moment, two concepts will be analyzed, among them Brand Image and Brand Value, as well as the existing relations with CSR. To this end, it will be necessary to briefly present some studies by authors who have addressed these constructs. Lastly, research relating to Consumer Satisfaction will be exposed.

I. V. Pereira (✉) · J. D. Santos
CEOS.PP and ISCAP/P.PORTO, Porto, Portugal

J. Araújo
ISCAP/P.Porto, Porto, Portugal

© The Author(s), under exclusive license to Springer Nature Switzerland AG 2023

F. L. Almeida et al. (eds.), *Multidimensional Sustainability: Transitions and Convergences*, Springer Proceedings in Earth and Environmental Sciences,
https://doi.org/10.1007/978-3-031-24892-4_18

18.2 Methodology

18.2.1 Research Agenda

The research agenda is about the consequents of CSR. The search key for finding articles, books, and documents related to the research agenda was, at first: CSR concept; importance of CSR; and CSR consequents. The main aim of the research is to understand the main consequents of CSR and its relationship with customer satisfaction. So, after a previous analysis of the papers, other keywords were added: Brand Image; Brand image and CSR: Brand value; Brand value and CSr; Consumer satisfaction; Consumer satisfaction and CSR; Antecedents of Consumer Satisfaction; Consumer satisfaction and Brand Image; and Consumer Satisfaction and Brand Value. These key works help to identify the articles which are most likely to explain the previously identified consequents of CSR.

18.2.2 Literature Search Criteria

In search of relevant articles, the search will consist of journal articles and conference proceedings and other relevant work with peer review. The search of literature will be conducted by using major multi-purpose databases such as Web of Science (Thomson Reuters), ProQuest, Emerald, Science Direct, and EBSCO. A search for more articles using the same search keywords will be conducted on the Internet using Google Scholar to increase the coverage of the literature search. The search criterion for the publication period is up to May 2022.

18.2.3 Literature Search Procedure

The initial searches revealed that a total of 475 articles were found from various sources. Then these articles' content would be analyzed for the relevance about CSR. When the articles were found to be relevant to the study agenda, they would be assessed in more detail of their purposes, methodologies, and findings. The citation criteria were applied to get the articles that are most valuable to the research topic. Research was conducted on the relationship between brand value and customer satisfaction and 104 papers were found; and brand image and customer satisfaction, where 380 papers were found. The same procedure explained before was implemented. When all four concepts were added, 17 papers were found and analyzed.

18.3 Corporate Social Responsibility

Currently, there are many definitions attributed to CSR. According to Dahlsrud (2008), they are mostly congruent and therefore the absence of a single universally accepted definition is not relevant. However, this absence caused by the broad nature of the concept and the different views of scholars (Brunk, 2010) causes the lack of a normative basis to explain, in practice, the concept of CSR. It is then necessary to create a common ground that relates the various concepts of CSR (Okoye, 2009).

Votaw (1972) mentions that CSR does not have the same meaning for everyone, since for some it means legal responsibility, for others it can be understood as socially responsible behaviour at the ethical level, and for others, it can mean responsibilities in general. The author also adds that CSR is often seen as a charitable contribution or as a duty that imposes higher standards of behaviour on entrepreneurs than on citizens in general.

The concept has, in fact, been studied by several authors. Carroll, in 1999, studied the evolution of the concept from the 1950s to the 1990s. Some of the definitions included in the literature are:

CSR can be understood as the obligations of businesspeople towards desirable decisions and attitudes, based on the goals and values of society (Bowen, 2013).

CSR is understood as the way in which companies act in the social system regarding certain issues other than economic, technical, and legal ones. The company should evaluate the effect of its decisions in order to obtain social benefits beyond the usual economic gains of the company (Davis, 1973).

CSR relates to a company's concerns for the needs and goals of society, beyond the economic, as it can only survive when embedded in a free and effectively functioning society (Eells & Walton, 1974).

In short, there are several aspects mentioned by different authors. In general, the authors associate CSR with economic, environmental, social, ethical, philanthropic, and legal factors, among others. Carroll (1999) concludes, in the study in question, that the concept is constantly under construction and that more attention needs to be paid to measuring the concept.

Subsequently, Lantos (2001), following on from the previously presented model, suggested three types of CSR: ethical, altruistic, and strategic. Briefly, ethical CSR is morally obligatory and goes beyond fulfilling economic and legal obligations. Altruistic CSR is equivalent to the philanthropic responsibilities that are described in Carrol's (2000) model and presupposes contributing to the welfare of stakeholders that exist in society, even if it means a loss in business profitability. On the other hand, strategic CSR involves business community service activities that fulfil certain strategic business objectives that will benefit the company through positive publicity and goodwill.

Still regarding the definition of CSR, Dahlsrud (2008) analyzed 37 definitions proposed by the literature from 27 different authors. The author grouped the definitions found into five different dimensions, including environmental, social, economic, voluntary, and stakeholder. The study concludes that all dimensions are important in defining CSR and that at least three dimensions are almost always included in a random definition of CSR, making it impossible to separate the definitions in different schools of thought.

Thus, based on the aforementioned, the present research essentially relied on the research of Martínez et al. (2014), who studied the influence of CSR on brand image and loyalty in the hotel industry. For this, the author measured three dimensions of CSR, among them the environmental, social, and economic dimensions.

> The economic dimension is considered as the basic dimension because, without economic results, it would not be possible to implement practices in the other dimensions (Carroll, 1991). This dimension is associated with the fulfilment of companies' duties and obligations towards their economic activity, to promote growth and prevent possible problems with stakeholders (Torugsa et al., 2013). In addition to profit maximization, the firm is also expected to be profitable in the long term, ensure stable employment, and offer high-quality goods and services (Carroll, 1999; Currás-Pérez et al., 2018).

The social dimension refers to the company's relationship with the sociocultural environment, its support for social causes and also its involvement in community issues (Choi & Ng, 2011). Companies are increasingly aware of the needs of society and therefore these practices go beyond the need to comply with the mere law. Companies play an important role in solving social problems and implementing solutions aimed at increasing social well-being (Potočan et al., 2021). CSR practices at the social level can also bring benefits not only to the company but also to the stakeholders through the creation of a social dialogue that involves ethical and social issues of interest to both parties and is important in decision-making (Bansal, 2005). Thus, CSR activities can contribute to increased value and competitive advantage (Husted et al., 2015).

The environmental dimension of CSR can be seen as the environmentally 'friendly' behaviours, such as optimizing the use of natural resources, improving waste management, promoting product ecology, and others (Choi & Ng, 2011). Mijatovic et al. (2019) note that environmental practices aim to minimize environmental damage and promote environmental sustainability efforts beyond corporate boundaries.

According to Tate et al. (2011), companies implement voluntary environmental practices for two reasons: first, because they can gain a competitive advantage through better access to and management of natural resources, and second, because they want a positive reputation among stakeholders in addition to the competitive advantage they can achieve (Mijatovic et al., 2019).

The following table serves as a summary of the aforementioned and presents 23 variables proposed by the authors Chwiłkowska-Kubala et al. (2021) and organized into three major dimensions. Thus, for each dimension, the authors propose CSR practices respectively.

In 2015, Carroll studied the concept of CSR again through a paper entitled "Corporate social responsibility: The centrepiece of competing and complementary frameworks." The author noted that as long as the world economy is growing, so is CSR. Thus, it can be concluded that CSR has grown over the years and has

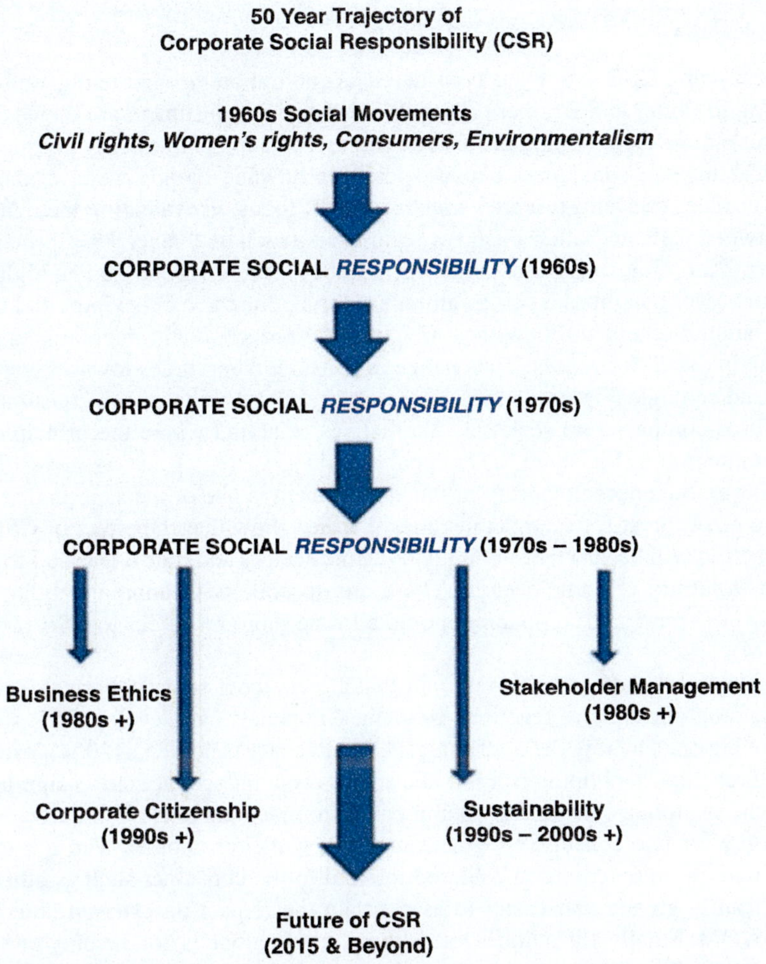

Fig. 18.1 Evolution of the CSR concept. (Source: Carroll, 2015, p. 91)

accompanied the surrounding social context. Through Fig. 18.1, it is possible to analyze this same relationship in a schematic way.

This section aimed at exploring the concept and the evolution of the CSR concept with a brief consideration of what Carroll (2021) believes to be the future of CSR. In fact, Carroll (2021) highlights some of the most important principles and themes regarding CSR, including purpose, innovation, engagement, and collaboration with all partners, shared and integrated value, and the Sustainable Development Goals (SDGs).

18.3.1 CSR Relevance

In recent years, CSR has been considered a good strategic marketing tool and, according to Porter and Kramer (2006), this is because consumers are increasingly demanding more from companies and expect more than a high-quality product at a low price. In fact, consumers bet on socially renowned brands when evaluating similar products. Several research studies on CSR focus on evaluating the relationship between CSR and other business parameters (such as financial performance). However, according to Aguinis and Glavas (2019), more recent studies highlight individual-level outcomes, such as attitudes, perceptions and behaviour, and satisfaction, among others and how they are currently proactive and intentional agents engaging in the CSR process. The change in consumers' attitudes towards environmental and ecological issues, social responsibility, and the changes in consumption habits force companies to seek new alternatives of action where the principles of CSR are integrated (Virvilaite & Daubaraite, 2011).

Moreover, the unpredictability of the environment is one of the aspects that represents a great threat for many companies. Studies show that the effect of CSR on risk reduction will be stronger in highly dynamic sectors and thus it is easier to deal with the volatility of some sectors. Therefore, to cope with unpredictability and changing environments, companies should allocate more resources to CSR (Sun & Cui, 2014).

Consumers are then an important part that affects these same initiatives, and by understanding consumers' reactions to them, companies can develop CSR strategies. Moving on to a more detailed analysis, Bhattacharya and Sen's (2004) research reveals three important findings. First, the authors concluded that there is significant heterogeneity among consumers when it comes to reactions to CSR initiatives, i.e., what works for one consumer segment may not work for another. The impact of CSR initiatives on outcomes considered internal to the consumer such as attitudes is significantly greater and easier to assess than the impact on external outcomes such as WOM. Finally, the authors found that the firm alone is not the only one that benefits from CSR initiatives, as both consumers and society in general, take advantage of these activities. Thus, the various CSR initiatives were grouped into six major domains:

1. Community support includes, for example, support for health programs or educational initiatives for economically disadvantaged individuals.
2. Support for diversity with, for example, gender and sexual orientation.
3. Support for workers with, for example, concern for safety and building union relations.
4. Support for the environment with, for example, environmentally friendly products and hazardous waste management.
5. Operations outside the United States, and this includes foreign labour practices and operations in countries with human rights violations.
6. Product support can be, for example, product safety and R&D practices.

Thus, the company is expected to develop CSR activities that reach one or more of these six domains, and to do so in a consumer-centric way, companies need to identify which CSR activities are most valued.

18.3.2 CSR Consequents

CSR has an influence on several aspects, including corporate reputation, consumer trust, and consumer loyalty (Stanaland et al., 2011). Furthermore, positive associations based on CSR contribute to a more favourable attitude of stakeholders towards the company (Sen et al., 2006). Several studies have been conducted in relation to CSR. As an example, Chauhary et al. (2016) studied consumers' perceptions of CSR activities and concluded that the antecedents of CSR are repurchase intention, consumer attachment and loyalty, and perceived corporate performance.

In general, a commitment to socially responsible actions leads to positive outcomes for a company. In fact, companies may be able to improve their reputation and corporate image through CSR activities (Waddock & Graves, 1997). In addition to brand image, based on the literature, there are reasons to believe that CSR can also positively affect brand value (Zhao et al., 2021).

According to Prayag et al. (2019), consumer satisfaction may also result from CSR activities and thus when they are successfully achieved, they positively affect consumer satisfaction (Mohammed & Rashid, 2018).

By analysing literature about CSR, several consequents have arisen. Therefore, the next chapters will discuss these and what previous research has found about the way they may be connected to customer satisfaction.

18.4 Brand Image

Brand image refers to the personality traits of a company or one of its brands operating in the market (Dong, 2016) and to consumer perceptions of that same brand and/or a product (Malmelin & Moisander, 2014) reflected through the associations stored in consumers' memory (Leone et al., 2006). Brand image is indivisible from the brand itself and reflects the strength and essence of the brand (Dong, 2016) relating, then, to a series of associations (Aaker, 1991).

Cho and Fiore (2015) studied brand image based on three dimensions: (1) cognitive associations that relate to consumers' personal beliefs, thoughts, and evaluations regarding a brand's attributes (Keller, 2001); (2) emotional associations that involve subjective feelings such as excitement, happiness, and joy (Keller, 2001); and, finally, (3) sensory associations that reflect the involvement of consumers' physical senses, i.e., sight, smell, touch, among others (Schmitt, 1999). The results confirmed the importance of including cognitive, emotional, and sensory items to measure brand image.

Meanwhile, Martínez et al. (2014), regarding brand image, studied two dimensions: the affective and the functional. The functional dimension is related to tangible characteristics that can be easily measured, while the emotional dimension is associated with a psychological dimension manifested through feelings and attitudes towards the company (Kennedy, 1977) cited by Martínez et al. (2014). It should be noted that the present research was based on the study of these two dimensions to measure brand image.

Regarding CSR and its relationship with brand image, there are several studies that associate these two concepts (Mohammed & Rashid, 2018; Porter & Kramer, 2006). The level of the brand image of products or services is affected by CSR actions, that is, when CSR works for the benefits of society, environment and living conditions of its employees and society in general consumers favour the products and services created by these same companies and, consequently, there is a significant improvement in the brand image (Maldonado-Guzman et al., 2017) and in the retention of customers who express intentions to relate to it again (Othman & Hemdi, 2015). Thus, if the company's goal is to improve its brand image through CSR initiatives, it should take more into account the emotional and social aspects, such as ethics, community support, responsible environmental behaviour, and fair treatment of employees, among others, than the functional aspects, such as price and quality (Martínez et al., 2014). In addition to CSR improving brand image, it also provides financial advantages. From Sun and Cui (2014) perspective, CSR creates a positive image not only for customers but also for other stakeholders such as shareholders and debt holders.

In short, the existence of a brand whose image is attractive to consumers is a guarantee for the success of companies (Martínez et al., 2014).

18.5 Brand Value

The definition of brand value can be very broad. The concept first gained more importance in the 1980s when there was a need to define the relationship between brands and consumers and when it became apparent that the purchase price paid by many companies largely reflected the value of their brands (Leone et al., 2006).

The study of brand value is important for two reasons. First, one of the motivations relates to the financial area, as brand value provides information for accounting purposes or for merger, acquisition, or divestment purposes. According to Keller (1993), the study of brand value can be fundamental to improve marketing productivity, since costs are higher, there is greater competition and a decrease in demand in many markets.

For Aaker (1996), brand value supposes a set of assets and, therefore, brand management must invest in the creation and improvement of these assets. The brand creates value not only for consumers but also for the company itself through, for example, the brand name and symbol. The author conceptualizes brand value based on four dimensions: brand recognition, brand loyalty, brand associations, and perceived quality. Burmann et al. (2009) define brand value as the present and future valuation of the brand. This value is explained by internal and external performance and includes three fundamental categories, among them brand value at the psychological, behavioural, and financial levels.

Brand value has a different meaning for different stakeholders. As an example, what is important to a brand owner may not be relevant to society (Naidoo & Abratt, 2018). Brand value is defined in various ways, depending on the specific purpose. Keller (1993) was another of the great scholars in this area. The author conceptualized brand value by basing it on the consumer so that managers have more specific data regarding the marketing program and how it increases the value of brands. Keller (1993, p. 8) regarding brand value states that: 'although the ultimate goal of any marketing program is to increase sales, it is first necessary to establish knowledge structures for the brand so that consumers respond favourably to the brand's marketing activity'.

Regarding brand value, Keller (1993) presents a model known as the customer-based brand equity (CBBE) model, that is, it is based on the individual consumer's perspective. Thus, a brand has a positive or negative value (based on the customer) when consumers react favourably, respectively, to an element of the brand's marketing mix than to that same element when it is associated with a fictitious or unnamed version of the product or service. Thus, when there is a favourable consumer response, the brand can increase its revenues, have lower costs, and earn higher profits. There are two approaches to measuring customer-based brand value. The indirect approach assesses potential sources of customer-based brand value by measuring brand awareness and is useful in identifying them. As for the direct approach, it measures customer-based brand value by relying on the impact of brand awareness on consumer response to different elements of the company's marketing program and is useful in determining the nature of the differential response. These approaches are complementary and should be used together.

There are several models that explain the foundation of brand value. Burmann et al. (2009) propose a brand management approach based on brand identity and incorporating external and internal perspectives of value creation through the analysis of behavioural and financial variables. Leone et al. (2006) state that brand value can be assumed as the 'added value' of a product in the thoughts, words, and actions of consumers. This value can increase companies' profitability and consumers' loyalty (Beig & Nika, 2019).

Regarding CSR, it can be stated from the research conducted that, in general terms, CSR positively impacts brand value (Zhao et al., 2021). Esa et al. (2020) also analyzed the positive relationship between the three CSR activities (environmental, community, and labour activities) and brand value among the top 100 brands in Malaysia. The objective was to prove that brand value increases if the company communicates its CSR activities and operations. In addition to the aforementioned, the company starts to benefit from competitive advantages that lead to its success.

Lin and Chung (2019) studied the impact of CSR on brand value in the restaurant industry. The aspects of brand value studied were perceived quality, brand recognition, brand image, and brand loyalty and, according to the authors, companies that practice CSR activities gain advantages over companies that do not. It is also important to mention the study of Bhattacharya et al. (2020). The authors demonstrated that CSR plays an important role during economic downturns and has a positive effect on brand value.

18.6 Consumer Satisfaction

Most researchers do not provide a specific definition on the conceptualization of satisfaction (Prayag et al., 2019). However, satisfaction can be characterized as an emotional state that results from an evaluation that the consumer makes concerning a service or product and their response to it (Westbrook, 1987), and its totality can only be achieved when customers' desires and preferences are prioritized (Khudhair et al., 2019).

Giese and Cote (2002) note that all definitions share some common ideas, these being:

1. Consumer satisfaction is an emotional or cognitive response. However, more recent definitions attribute greater significance to the emotional response, since, according to the authors' study, 64% of respondents relate the concept of satisfaction with more affective terms. It should also be noted that this affective response varies in intensity depending on the situation, that is, it can range from strong to weak.
2. The response is based on a particular focus. The focus identifies the object of consumer satisfaction and usually involves comparing performance to a standard (which can be more specific or more general) and relates, for example, to a product, consumption experience, purchase decision, among others. Determining an appropriate focus for satisfaction varies from context to context, and without a focus, any definition of satisfaction has little meaning.
3. Response occurs at a particular point in time. Typically, consumer satisfaction is assessed after purchase, however, consumer satisfaction can occur before choice or even in the absence of purchase or choice. Thus, it can be concluded that none of the time frames mentioned are entirely appropriate, as satisfaction can vary dramatically over time. Thus, satisfaction is only determined at the time the evaluation occurs.

Although there are several definitions for the term, the model that Oliver (1980) developed, known as the expectation-disconfirmation model, remains a reference. According to the theory, consumer satisfaction with a product or service results from the subjective comparison between expectation and perception. In other words, consumer satisfaction depends on the comparison between initial expectations and actual results. Thus, satisfaction arises when expectations are confirmed, and dissatisfaction arises when these expectations are not met. In addition to the above, previous literature reviews have shown that, regarding satisfaction, there are also studies about other theories, including attribution theory, dissonance theory, and contrast theory.

The study of the main antecedents of satisfaction has become a strategic issue. In fact, consumers are increasingly autonomous, reflexive, and critical. Thus, the antecedents of satisfaction should be analyzed in a deeper way to predict some consumer behaviours and subsequently obtain a series of beneficial results for organizations, such as word of mouth (WOM) communication, loyalty, and financial profitability (Palací et al., 2019).

Szymanski and Henard (2001) based on Oliver's (1997) model present, in their article, the antecedents they consider to be important for consumer satisfaction. These include (1) expectations, which can be addressed at two different times: through anticipation or, later, through comparison. Regarding expectations now of anticipation, it can be mentioned that they have a direct influence on satisfaction levels, since there is no evaluation or comparison of the actual results. Another antecedent referred to in the study is (2) the disconfirmation of expectations which, through comparison, is evaluated relative to actual performance results, i.e., consumers are satisfied when results exceed expectations and are dissatisfied when expectations exceed results and, finally, they are only satisfied when results meet expectations. In addition to the above, the authors mention that (3) performance can affect satisfaction. This relationship is explained through the value–perception relationship, that is, consumers are satisfied when there is a good performance of the company that can be demonstrated through the ability to offer consumers what they need, want, or desire. Another antecedent referred to in the authors' study is (4) affect which, through the emotions aroused during consumption, is prone to leave affective traces in consumers' memories that are subsequently included in satisfaction evaluations. Finally, (5) fairness is also considered an antecedent of satisfaction, in that consumers consider themselves satisfied when their fairness ratio is proportionally higher than the ratio achieved by the reference person or the group.

The antecedents of consumer satisfaction were also studied by authors Prayag et al. (2019). Indeed, they addressed in their article four important antecedents of consumer satisfaction, among them (1) perceived justice and fairness, (2) emotions, (3) the influence of CSR, (4) sustainable practices, and, finally, (5) employee–customer interactions. Among these five antecedents, it is important to highlight the influence of CSR on consumer satisfaction. Thus, both CSR and organizational reputation have a positive impact on consumer satisfaction (Su et al., 2015) thus there is a relationship between CSR associations and consumer satisfaction (Martínez & Rodríguez del Bosque, 2013). Mohammed and Rashid (2018), based on Carrol's (1991) four dimensions of CSR, concluded that CSR positively affects consumer satisfaction.

Indeed, one of the main objectives of CSR is the possible advantages that companies can gain by being socially responsible towards stakeholders. However, consumers seem to need special attention, as CSR activities have a significant effect on consumer-related outcomes (Bhattacharya & Sen, 2004). Currently, consumers are more satisfied with products developed by socially responsible companies. First, it is important to note that a company's actions appeal to the multidimensionality of the consumer as not only an economic agent but also a representative member of various stakeholder groups. Second, CSR activities create a favourable context that positively drives consumer evaluations and thus satisfaction. Finally, the antecedents of consumer satisfaction are also relevant, as consumers get a better perception of value and consequently a higher satisfaction with the product that is made by a socially responsible company (Luo & Bhattacharya, 2006).

Hsieh et al. (2018), in their study concerning the catering field, studied the relationship between various concepts. In their research, they concluded that quality service improves not only brand image but also consumer satisfaction. Still within the catering area, the author Cuong (2020) proved the influence of brand image of fast-food restaurants in Vietnam on consumer satisfaction. In fact, considering the results obtained, it can be stated that brand image is an antecedent of consumer satisfaction and positively impacts it (Cuong & Long, 2020).

In fact, consumers who recognize a positive brand image tend to believe that the brand succeeds in offering high satisfaction (Mohammed & Rashid, 2018).

Regarding value, Aaker (1992) states that it is important for consumer satisfaction in three aspects. First, brand value helps the consumer to interpret, process, and retrieve all the information related to products and brands. Then, it should be noted that brand value affects the consumer's confidence in the purchase decision, i.e., a consumer tends to choose a brand that he or she has already used and considers having the high quality or is familiar with. Finally, the author states that brand value, particularly perceived quality, and brand associations, provide value to the consumer, thereby increasing consumer satisfaction when the individual uses the product or brand.

18.7 Conclusions

To obtain a better understanding of CSR and how it may be important for companies, it is required to analyze its impact on several variables. This research aimed to understand which relevant consequents of CSR also impact customer satisfaction. According to the literature review, it could be said that Corporate social responsibility may have impact on consumers' perception towards the brand, especially on brand image and brand value. These important consequents may act as mediators on CSR impact on the brand. So, CSR may impact on consumer satisfaction towards the brand. Although some authors have developed research about this subject, further research may be developed to understand the relationship between CSR brand value and brand image and to understand if together they are able to explain consumer satisfaction.

This chapter aimed to be a theoretical one, so future research may focus on developing a model which tests the relationship between the suggested constructs, to verify if these can explain customer satisfaction.

References

Aaker, D. A. (1991). *Managing brand equity*. Wiley.

Aaker, D. A. (1992). The value of brand equity. *Journal of Business Strategy, 13*(4), 27–32.

Aaker, D. A. (1996). *Building strong brands*. Library of Congress Cataloging-in-Publication Data.

Aguinis, H., & Glavas, A. (2019). On corporate social responsibility, sensemaking, and the search for meaningfulness through work. *Journal of Management, 45*(3), 1057–1086.

Bansal, P. (2005). Evolving sustainably: A longitudinal study of corporate sustainable development. *Strategic Management Journal, 26*(3), 197–218.

Beig, F. A., & Nika, F. A. (2019). Brand experience and brand equity. *Vision, 23*(4), 410–417.

Bhattacharya, C. B., & Sen, S. (2004). Doing better at doing good: When, why, and how consumers respond to corporate social initiatives. *California Management Review, 47*(1), 9–24.

Bhattacharya, A., Good, V., & Sardashti, H. (2020). Doing good when times are bad: The impact of CSR on brands during recessions. *European Journal of Marketing, 54*(9), 2049–2077.

Bowen, H. R. (2013). *Social responsibilities of the businessman*. University of Iowa Press.

Brunk, K. H. (2010). Exploring origins of ethical company/brand perceptions – A consumer perspective of corporate ethics. *Journal of Business Research, 63*(3), 255–262.

Burmann, C., Jost-Benz, M., & Riley, N. (2009). Towards an identity-based brand equity model. *Journal of Business Research, 62*(3), 390–397.

Carroll, A. B. (1991). The pyramid of corporate social responsibility: Toward the moral management of organizational stakeholders. *Business Horizons, 34*(4), 39–48.

Carroll, A. B. (1999). Corporate social responsibility: Evolution of a definitional construct. *Business & Society, 38*(3), 268–295.

Carroll, A. B. (2000). Conceptual and consulting aspects of stake-holder theory, thinking, and management. *Public Administration and Public Policy, 81*, 169–182.

Carroll, A. B. (2015). Corporate social responsibility: The centerpiece of competing and complementary frameworks. *Organizational Dynamics, 44*(2), 87–96.

Carroll, A. B. (2021). Corporate social responsibility: Perspectives on the CSR construct's development and future. *Business and Society, 60*(6), 1258–1278.

Chaudary, S., Zahid, Z., Shahid, S., Khan, S. N., & Azar, S. (2016). Customer perception of CSR initiatives: Its antecedents and consequences. *Social Responsibility Journal, 12*(2), 263–279.

Cho, E., & Fiore, A. M. (2015). Conceptualization of a holistic brand image measure for fashion-related brands. *Journal of Consumer Marketing, 32*(4), 255–265.

Choi, S., & Ng, A. (2011). Environmental and economic dimensions of sustainability and price effects on consumer responses. *Journal of Business Ethics, 104*(2), 269–282.

Chwiłkowska-Kubala, A., Cyfert, S., Malewska, K., Mierzejewska, K., & Szumowski, W. (2021). The relationships among social, environmental, economic CSR practices and digitalization in polish energy companies. *Energies, 14*, 7666.

Cuong, D. T. (2020). The impact of customer satisfaction, brand image on brand love and brand loyalty. *Journal of Advanced Research in Dynamical and Control Systems, 12*(06), 3151–3159.

Cuong, D. T., & Long, N. T. (2020). The impact of service quality and brand image on customer satisfaction and behavioral intention in Vietnam fashion market. *Test Engineering and Management, 83*, 389–398.

Currás-Pérez, R., Dolz-Dolz, C., Miquel-Romero, M. J., & Sánchez-García, I. (2018). How social, environmental, and economic CSR affects consumer-perceived value: Does perceived consumer effectiveness make a difference? *Corporate Social Responsibility and Environmental Management, 25*(5), 733–747.

Dahlsrud, A. (2008). How corporate social responsibility is defined: An analysis of 37 definitions. *Corporate Social Responsibility and Environmental Management, 15*(1), 1–13.

Davis, K. (1973). The case for and against business assumption of social responsibilities. *Academy of Management Journal, 16*(2), 312–322.

Dong, Y. (2016). Whom your brand is more like? Analysis of brand image and brand personality. In *2016 international conference on contemporary education, social sciences and humanities* (pp. 370–374). Atlantis Press.

Eells, R., & Walton, C. (1974). *Conceptual foundations of business* (3rd ed.). Irwin.

Esa, E., Zahari, A. R., Rajadurai, R. J., & Tamyez, P. F. M. (2020). Corporate social responsibility and brand equity of Malaysian top 100 brand companies. In *9th international economics and business management conference* (pp. 48–56).

Giese, J. L., & Cote, J. A. (2002). Defining consumer satisfaction. *Academy of Marketing Science Review, 1*(1), 1–22.

Hsieh, S.-W., Lu, C.-C., & Lu, Y.-H. (2018). A study on the relationship among brand image, service quality, customer satisfaction, and customer loyalty – Taking 'the Bao Wei Zhen Catering Team' as an empirical study. *KnE Social Sciences, 3*(10), 1768–1781.

Husted, B. W., Allen, D. B., & Kock, N. (2015). Value creation through social strategy. *Business and Society, 54*(2), 147–186.

Keller, K. L. (1993). Conceptualizing, measuring, and managing customer-based brand equity. *Journal of Marketing, 57*, 1–22.

Keller, K. L. (2001). *Building customer-based brand equity: A blueprint for creating strong brands* (Working paper). Marketing Science Institute.

Kennedy, S. (1977), "Nurturing corporate images", *European Journal of Marketing, 11*(3), 120–1964

Khudhair, H. Y., Jusoh, A., Mardani, A., & Nor, K. M. (2019). Quality seekers as moderating effects between service quality and customer satisfaction in airline industry. *International Review of Management and Marketing, 9*(4), 74–79.

Lantos, G. P. (2001). The ethicality of altruistic corporate social responsibility. *Journal of Consumer Marketing, 19*(3), 205–232.

Leone, R. P., Rao, V. R., Keller, K. L., Luo, A. M., McAlister, L., & Srivastava, R. (2006). Linking brand equity to customer equity. *Journal of Service Research, 9*(2), 125–138.

Lin, M. S., & Chung, Y. K. (2019). Understanding the impacts of corporate social responsibility and brand attributes on brand equity in the restaurant industry. *Tourism Economics, 25*(4), 639–658.

Luo, X., & Bhattacharya, C. B. (2006). Corporate social responsibility, customer satisfaction, and market value. *Journal of Marketing, 70*(4), 1–18.

Maldonado-Guzman, G., Pinzón-Castro, S. Y., & Leana-Morales, C. (2017). Corporate social responsibility, brand image and firm reputation in Mexican small business. *Journal of Management and Sustainability, 7*(3), 38–47.

Malmelin, N., & Moisander, J. (2014). Brands and branding in media management-toward a research agenda. *International Journal on Media Management, 16*(1), 9–25.

Martínez, P., & Rodríguez del Bosque, I. (2013). CSR and customer loyalty: The roles of trust, customer identification with the company and satisfaction. *International Journal of Hospitality Management, 35*, 89–99.

Martínez, P., Pérez, A., & del Bosque, I. R. (2014). CSR influence on hotel brand image and loyalty. *Academia Revista Latinoamericana de Administracion, 27*(2), 267–283.

Mijatovic, I., Maricic, M., & Horvat, A. (2019). The factors affecting the environmental practices of companies: The case of Serbia. *Sustainability, 11*(21), 5960.

Mohammed, A., & Rashid, B. (2018). A conceptual model of corporate social responsibility dimensions, brand image, and customer satisfaction in Malaysian hotel industry. *Kasetsart Journal of Social Sciences, 39*(2), 358–364.

Naidoo, C., & Abratt, R. (2018). Brands that do good: Insight into social brand equity. *Journal of Brand Management, 25*(1), 3–13.

Okoye, A. (2009). Theorising corporate social responsibility as an essentially contested concept: Is a definition necessary? *Journal of Business Ethics, 89*, 613–627.

Oliver, R. L. (1980). A cognitive model of the antecedents and consequences of satisfaction decisions. *Journal of Marketing Research, 17*(4), 460–469.

Oliver, R. L. (1997). Satisfaction: A Behavioral Perspective on the Consumer, New York: Mc GrawHill. Peppers D and Rogers, M. 2004. Managing Customer Relationships. Hoboken.

Othman, N. Z., & Hemdi, M. A. (2015). Predictors of guest retention: Investigating the role of hotel's corporate social responsibility activities and brand image. In S. M. Radzi, C. T. Chick, M. S. M. Zahari, Z. Mohi, M. F. S. Bakhtiar, & F. I. Anuar (Eds.), *Theory and practice in hospitality and tourism research* (pp. 139–144). CRC Press.

Palací, F., Salcedo, A., & Topa, G. (2019). Cognitive and affective antecedents of consumers' satisfaction: A systematic review of two research approaches. *Sustainability, 11*(2), 431.

Porter, M. E., & Kramer, M. R. (2006). The link between competitive advantage and corporate social responsibility. *Harvard Business Review, 84*(12), 78–90.

Potočan, V., Mulej, M., & Nedelko, Z. (2021). Society 5.0: Balancing of Industry 4.0, economic advancement and social problems. *Kybernetes, 50*(3), 794–811.

Prayag, G., Hassibi, S., & Nunkoo, R. (2019). A systematic review of consumer satisfaction studies in hospitality journals: Conceptual development, research approaches and future prospects. *Journal of Hospitality Marketing and Management, 28*(1), 51–80.

Schmitt, B. (1999). *Experiential marketing. How to get customers to sense, feel, think, act, relate to your company and brands*. The Free Press.

Sen, S., Bhattacharya, C. B., & Korschun, D. (2006). The role of corporate social responsibility in strengthening multiple stakeholder relationships: A field experiment. *Journal of the Academy of Marketing Science, 34*(2), 158–166.

Stanaland, A. J. S., Lwin, M. O., & Murphy, P. E. (2011). Consumer perceptions of the antecedents and consequences of corporate social responsibility. *Journal of Business Ethics, 102*(1), 47–55.

Su, L., Swanson, S. R., & Chen, X. (2015). Social responsibility and reputation influences on the intentions of Chinese Huitang Village tourists: Mediating effects of satisfaction with lodging providers. *International Journal of Contemporary Hospitality Management, 27*(8), 1750–1771.

Sun, W., & Cui, K. (2014). Linking corporate social responsibility to firm default risk. *European Management Journal, 32*(2), 275–287.

Szymanski, D. M., & Henard, D. H. (2001). Customer satisfaction: A meta-analysis of the empirical evidence. *Journal of the Academy of Marketing Science, 29*(1), 16–35.

Tate, W. L., Dooley, K. J., & Ellram, L. M. (2011). Transaction cost and institutional drivers of supplier adoption of environmental practices. *Journal of Business Logistics, 32*(1), 6–16.

Torugsa, N. A., O'Donohue, W., & Hecker, R. (2013). Proactive CSR: An empirical analysis of the role of its economic, social and environmental dimensions on the association between capabilities and performance. *Journal of Business Ethics, 115*(2), 383–402.

Virvilaite, R., & Daubaraite, U. (2011). Socialinės atsakomybės raiška formuojant įmonės įvaizdį. *Engineering Economics, 22*(5), 534–543.

Votaw, D. (1972). Genius becomes rare: A comment on the doctrine of social responsibility Pt. I. *California Management Review, 15*(2), 25–31.

Waddock, S. A., & Graves, S. B. (1997). The corporate social performance-financial performance link. *Strategic Management Journal, 18*(4), 303–319.

Westbrook, R. A. (1987). Product/consumption-based affective responses and postpurchase processes. *Journal of Marketing Research, 24*(3), 258–270.

Zhao, Y., Abbas, M., Samma, M., Ozkut, T., Munir, M., & Rasool, S. F. (2021). Exploring the relationship between corporate social responsibility, trust, corporate reputation, and brand equity. *Frontiers in Psychology, 12*, 766422.

Part VII

Mechanics

Comparison of Numerical and Experimental Approaches to Determine Tensile Strength in Mechanical Components

19

Américo F. G. Costa, António T. K. B. Ferreira,
Fernando A. V. Figueiredo, and César M. A. Vasques

19.1 Introduction

To withstand a wide range of loads and operating scenarios, mechanical parts are typically conceived and designed with complex geometries that take advantage of available materials, while keeping production costs to a minimum. Virtual testing and numerical experiments using the finite element method (FEM) are typically employed during detailed design stages if these goals are to be met, precluding the necessity of a complex test campaign.

The FEM solves engineering and mathematical differential equations numerically, and structural analysis is nowadays a common area of application. In fact, the

A. F. G. Costa (✉)
ISPGAYA, Vila Nova de Gaia, Portugal

Centro de Formação Profissional da Indústria Metalúrgica e Metalomecânica (CENFIM), Lisboa, Portugal
e-mail: acosta@ispgaya.pt

António T. K. B. Ferreira
ISPGAYA, Vila Nova de Gaia, Portugal

Escola Profissional de Gaia (EPGaia), Vila Nova de Gaia, Portugal

F. A. V. Figueiredo
ISPGAYA, Vila Nova de Gaia, Portugal

Smile.Tech – Robótica, Vila Nova de Gaia, Portugal

proMetheus, Instituto Politécnico de Viana do Castelo (IPVC), Viana do Castelo, Portugal

C. M. A. Vasques
proMetheus, Instituto Politécnico de Viana do Castelo (IPVC), Viana do Castelo, Portugal

Escola Superior de Tecnologia e Gestão, Instituto Politécnico de Viana do Castelo (IPVC), Viana do Castelo, Portugal

© The Author(s), under exclusive license to Springer Nature Switzerland AG 2023
F. L. Almeida et al. (eds.), *Multidimensional Sustainability: Transitions and Convergences*, Springer Proceedings in Earth and Environmental Sciences,
https://doi.org/10.1007/978-3-031-24892-4_19

FEM originated from the need to solve complex elasticity and structural analysis problems in civil and aeronautical engineering. Its invention date is difficult to pinpoint but its origins can be traced back to the early 1940s, to the works of Hrennikoff (1941) and Courant (1943), where the concept of FE was conceived, albeit in a laborious and constrained manner. Clough was the first to coin the term "finite element method" in the early 1960s (Clough, 1960, 1990). In the 1960s and 1970s, the FEM was further developed by Argyris, Clough, Zienkiewicz, Ciarlet, their coworkers, and many others (Gupta & Meek, 1996; Sabat & Kundu, 2021). In these years, available open-source FE programs provided further impetus to the use of the method on the analysis of structural problems, one example being the original version of the FEM software NASTRAN initially supported by NASA and still in use these days.

Nowadays, the FEM theory (e.g., Zienkiewicz et al., 2013, 2014) is very well established and spread across several areas of application and is frequently available in conventional CAD (computer-aided design) software used by the industry. A list of FE software packages with a comparison of their main features can be found in Wikipedia ("List of finite element software packages," 2022). Review and comparative studies of FE software packages and applications can be found in the open literature (e.g., Fadiji et al., 2018a, b; Magomedov & Sebaeva, 2020). Expert systems for FE structural analysis were addressed by Naganarayana and Prathap (1992), such that the necessary expertise is available to a novice FE analyst to perform the same job even in the absence of trained experts. A review of FE analysis techniques, capabilities, and limitations was performed by Hatheway (1992). A review of useful guidelines for the FE analysis of steel structures was performed by Abambres and Arruda (2016). The stress analysis of a gear using photoelastic method and FE approach was reviewed by Ramachandra et al. (2022).

The FEM has demonstrated its value in simulating numerous design concepts and determining their behavior in virtually any environment. Consequently, it facilitates the evaluation of the viability of a new design after a model has been produced and validated, without the need to waste time and resources on a prototype. Some of the most noteworthy benefits of FEM include its capability to handle extremely complicated geometry and complex constraints; complicated loads, such as nodal loads, distributed element loads (e.g., inertia force, heat, and pressure), and time or frequency-dependent loading, can be managed; it is a dependable tool because it can do different analyses on the same model in different circumstances by modifying the loads, material attributes, or boundary conditions as the problem dictates; it can be used to solve a vast array of engineering problems, including solid mechanics, heat transfer, dynamics, fluid, and electrostatic ones, among others. Some downsides of the FEM include its approximation nature, yielding somewhat imprecise and mesh-dependent results; the requirement for pricey computer software and computational infrastructure; errors and blunders done by users can be disastrous; when using displacement-based formulations, stress analysis is typically less accurate than continuous displacement solution at the nodal locations.

When performing stress analysis with the FEM, the equivalent stresses of concern are the highest values reached during the various analyses conducted (Brnic et al., 2008; Cao et al., 2011; Chen et al., 2018, 2021; Golovanov et al., 2009; Miriţoiu & Ilincioiu, 2011; Nor et al., 2012; Yu & Fan, 2014).

This chapter aims to provide clarity regarding the appropriate manner in which the results of FEM analyses should be interpreted and compared. In order to accomplish this goal, two electrical connectors produced by the company SKELT are evaluated through the use of computational virtual testing and experimental methods, and the results obtained from these tests are critically analyzed and compared. The purpose is to evaluate the use of yield failure criteria in virtual FEM and actual tensile testing experiments for the strength design of mechanical components. In terms of methodology, the von Mises failure criterion is utilized in both experiments comparing the results of mechanical virtual FEM tensile tests on the same components to those obtained through actual experiments. Various mechanical components are subjected to a series of tensile tests to failure using a custom-built hydraulic press equipped with a 50-ton load cell manufactured by AEP. The same parts undergo FEM analysis using Autodesk Inventor 2022 for modeling and Inventor Stress Analysis for FEM simulations. Lastly, the results are critically assessed and compared.

19.2 Tensile Testing

Tensile testing, also known as tension testing, is a fundamental test well known in materials science and engineering in which a specimen is subjected to a controlled tension until failure. Properties that are directly measured via a tensile test are ultimate tensile strength, breaking strength, maximum elongation, and reduction in area. From these measurements the following properties can also be determined: Young's modulus, Poisson's ratio, yield strength, and strain-hardening characteristics ("Tensile testing," 2022). In this study, tensile tests are conducted on manufactured parts with function-specific geometries, rather than on tensile test pieces. This ensures that real-world tensile testing and simulation using the FEM are comparable.

Tensile Tests are performed with a custom-built press equipped with a 500KN Load cell with the reference CTC4165500KNI and manufactured by Italian company AEP. This load cell is calibrated once a year by an independent audit company and is classified as normalized accuracy 1 which means the max error it produces is 1% of the full scale, i.e., 5KN.

19.2.1 Component C14D41103-1

The first component to be analyzed is an electric linker with reference C14D41103-1, shown on the left side of Fig. 19.1; the part is designated as *Half-Fork* "R" and it is manufactured in Domex 460 steel (SSAB, 2022). Table 19.1 displays the relevant properties of the used Domex 460 steel while Table 19.2 enumerates the relevant data of the lot certificates from the supplier.

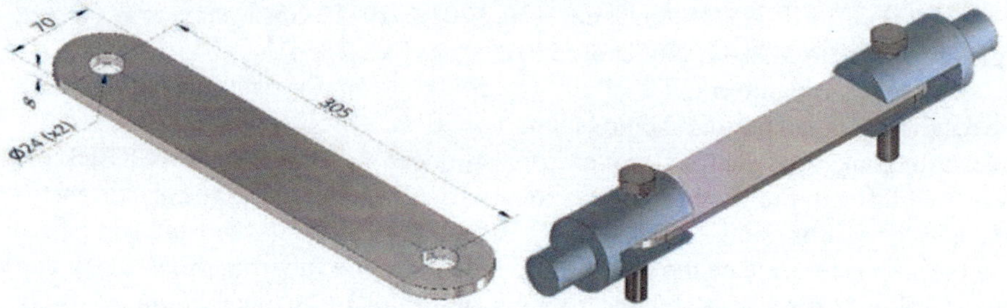

Fig. 19.1 Component C14D41103-1 (left); adaptor for component C14D41103-1 (right)

Table 19.1 Domex 460 steel properties (SSAB, 2022)

Parameter	Value	Unit
Yield strength	460	MPa
Ultimate tensile strength	520–670	MPa

Table 19.2 Domex 460 supplier certificate data

Parameter	Value	Unit
Certificate number	2010000369	
Raw material lot	2010LC110012	
State	Raw	
R_m (ultimate tensile strength)	573	MPa
Report numbers	47, 48, 49	

The component C14D41103-1 was tensile tested three times until failure; to adapt the component to the tensile testing machine, the mechanical device represented on the right side of Fig. 19.1 was designed.

The three tested samples are shown in Fig. 19.2, where the critical section where these components yield is clearly displayed.

The analysis of the results obtained in the real tests is presented in Table 19.3, where it can be seen that the failure of the components occurs at load values close to 170,000 N.

19.2.2 Component C1L27524-1

The second component to be analyzed is an electric linker with reference C1L27524-1, designated as *Triangular Rocker Arm*, produced with S350JO steel (British Steel, 2020) and with the dimensions indicated in Fig. 19.3. The properties of the S355JO steel, yield strength, and tensile strength are shown in Tables 19.4, 19.5 and Fig. 19.4.

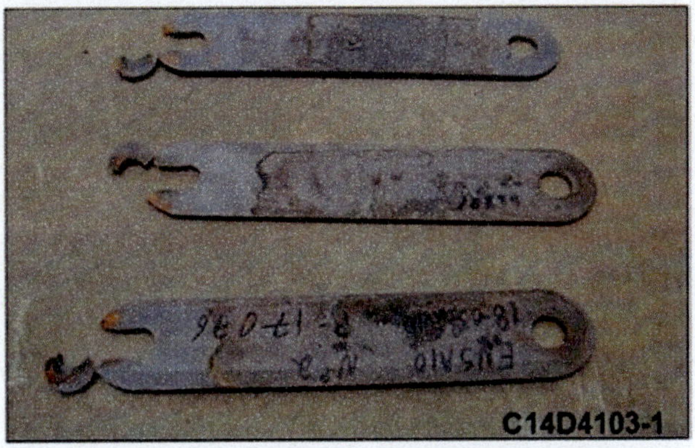

Fig. 19.2 Samples of component C14D41103-1 after the experimental tensile tests

Table 19.3 Results of failure tests of component C14D41103-1

Test	Force	Unit
1	169,380	N
2	170,760	N
3	168,890	N
Average	169,680	N

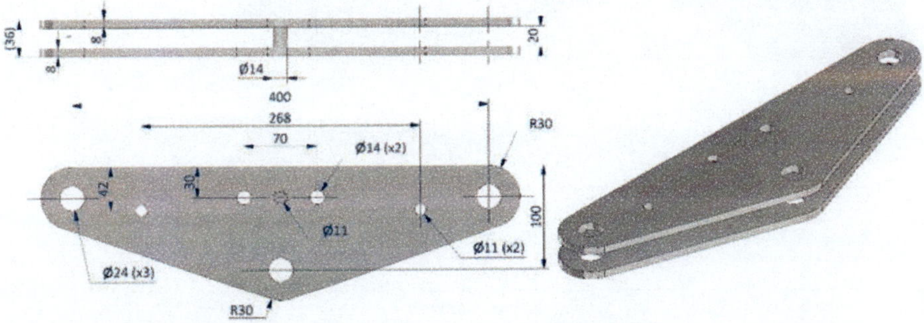

Fig. 19.3 Component C1L27524-1

Table 19.4 S355JO steel properties (British Steel, 2020)

Parameter	Value	Unit
Yield strength	355	MPa
Ultimate tensile strength	470–630	MPa

The component C1L27524-1 was tensile tested three times, until failure; Fig. 19.5 shows the samples after the test. The analysis of the results obtained is presented in Table 19.6. The failure of the components occurs at load values between 215,550 and 232,780 N. The average value of the applied breaking load is 226,610 N.

Table 19.5 S355JO supplier certificate data

Parameter	Value	Unit
Certificate number	2010000287	
Raw material lot	2010LC100012	
State	Raw	
R_m (ultimate tensile strength)	576	MPa
Report numbers	56, 57, 58	

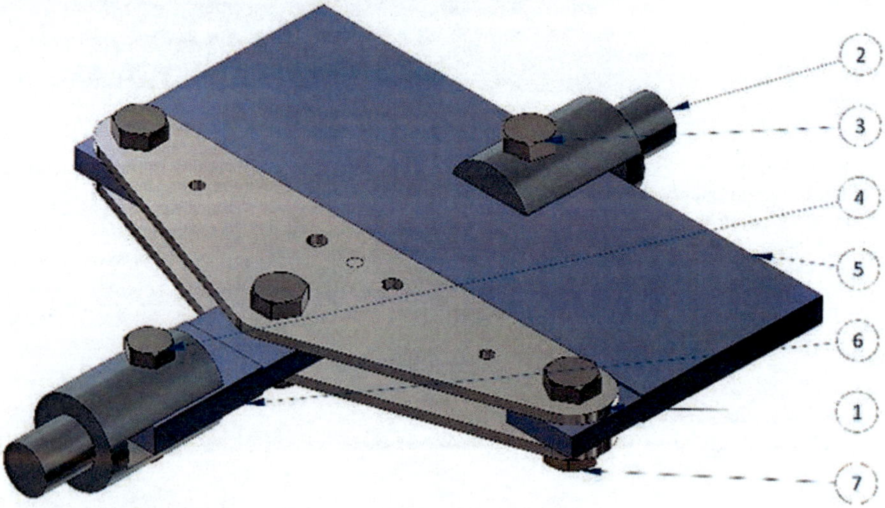

Número	Quantidade	Designação
1	1	Componente C1L27524-1
2	2	Bocal
3	4	Parafuso ISO 4017 - M22 x 60
4	1	Parafuso ISO 4014 - M20 x 80
5	1	Chapa
6	1	Chapa central
7	3	Porca ISO 4032 - M22

Fig. 19.4 C1L27524-1 component fastening device

19.2.3 Component B7419103-1

The component B7419103-1 shown in Fig. 19.6, designated as *Right Binder*, was manufactured with steel S275JR (OVAKO, 2022). In this component's analysis, the clamping device was not modeled in 3D; instead, constraint and load conditions were applied directly to the component. The purpose of this test is to validate the conclusions drawn previously. S275JR steel's mechanical characteristics, yield strength, and tensile strength are shown in Tables 19.7, 19.8, and 19.9.

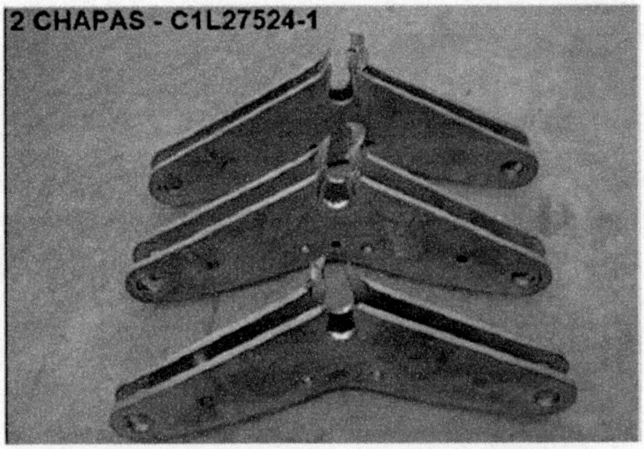

Fig. 19.5 Samples of component C1L27524-1 after the experimental tensile tests

Table 19.6 Results of failure tests of component C1L27524-1

Test	Force	Unit
1	215,550	N
2	231,510	N
3	232,780	N
Average	226,610	N

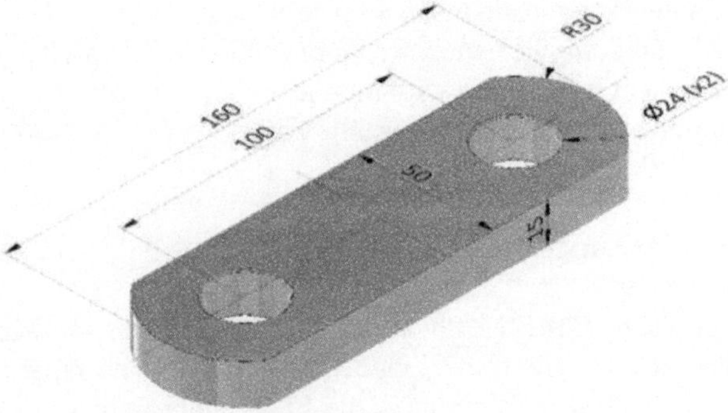

Fig. 19.6 Component B7419103-1

Table 19.7 S275JR steel properties (OVAKO, 2022)

Parameter	Value	Unit
Yield strength	275	MPa
Ultimate tensile strength	410–560	MPa

Table 19.8 S275JR supplier certificate data

Parameter	Value	Unit
Certificate number	2010000437	
Raw material lot	2010LC130007	
State	Raw	
R_m (ultimate tensile strength)	456	MPa
Report numbers	53, 54, 55	

Table 19.9 Results of failure tests of component B7419103-1

Test	Force	Unit
1	198,730	N
2	202,020	N
3	183,740	N
Average	194,830	N

The component B7419103-1 was tested three times, as above, and the physical results of these tests are presented in Fig. 19.7.

19.3 Finite Element Analysis

Following the tensile tests, a set of virtual testing and numerical experiments using FEM and computer analyses were performed, applying a varied set of loads. Each component is analyzed with increasing loads in order to evaluate the behavior of the component at different stress intensities.

19.3.1 Component C14D41103-1

Component C14D41103-1 and its adaptor were modeled in 3D, using Autodesk Inventor software, to reproduce in a computer the same conditions as in the tensile tests (Fig. 19.8). To simplify the FEM analysis, it was decided to exclude the two components labeled as *bocal*, element 3 on Fig. 19.8, fix one of the bolts, and apply variable loads to the other bolt.

The mesh generated in each of the components was finely defined according to the processing capacity of the available hardware. The mesh definition parameters are visible in Fig. 19.9.

The amplitude of the applied loads will vary between a minimum value of 60,000 N and a maximum of 175,000 N. Within this range are the three values obtained in the tensile tests, namely, 169,380, 170,760, and 168,890 N. The average load value achieved in these three tests is 169,680 N. When the load intensity is close to the values recorded in the tests, a finer analysis will be made.

Fig. 19.7 Samples of component B7419103-1 after the experimental tensile tests

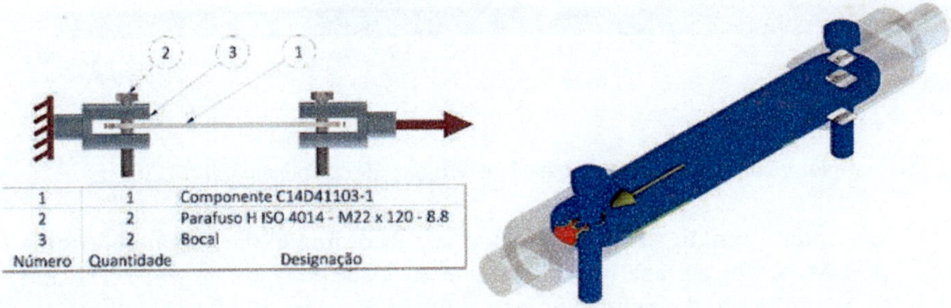

Fig. 19.8 Actual analysis conditions for component C14D41103-1 (left) and conditions under analysis defined in Autodesk Inventor software for component C14D41103-1 (right)

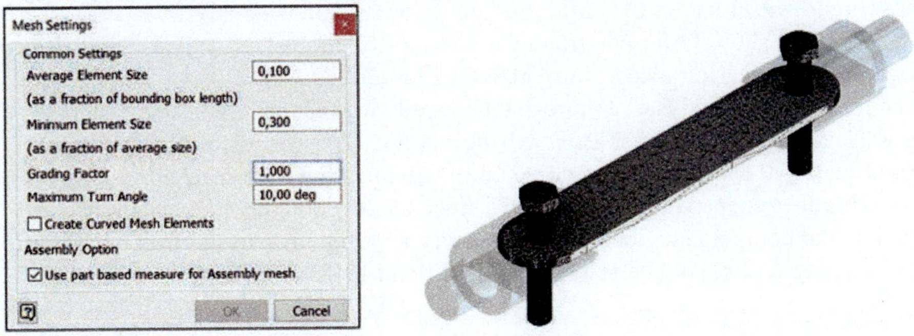

Fig. 19.9 Mesh definition for component C14D41103-1

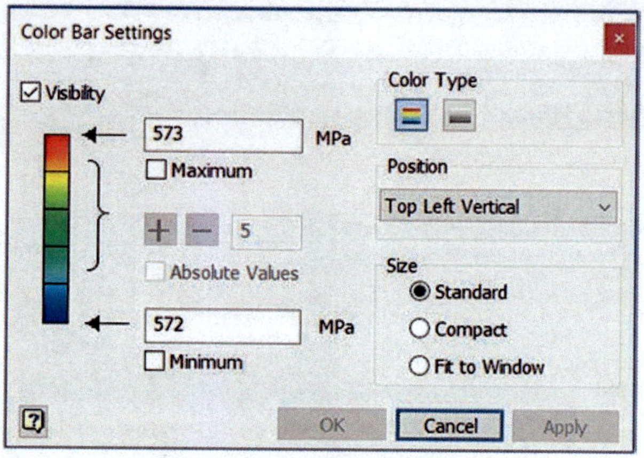

Fig. 19.10 Definition of the color map as a function of the equivalent stress for component C14D41103-1

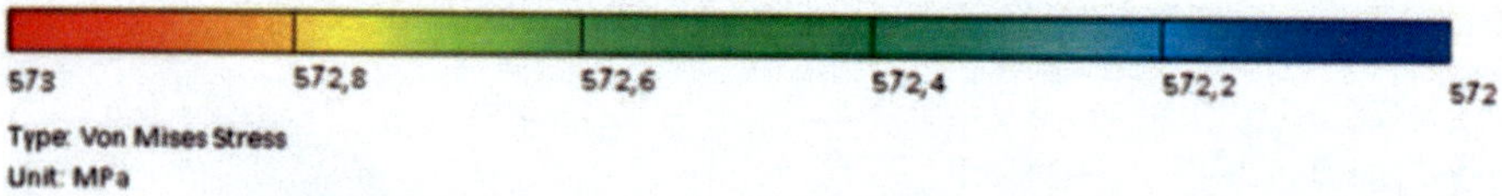

Fig. 19.11 Color gradient for different equivalent tensions for component C14D41103-1

The maximum strength value of the material, according to the supplier's certificate, is 573 MPa. So, the analysis will focus on a stress range between 572 and 573 MPa. According to the defined color configuration, Fig. 19.10, when the color shown is dark blue, it will have an equivalent stress below 572 MPa and in red equivalent stresses above 573 MPa.

In this analysis, the main interest is to analyze the equivalent stresses close to the failure stress, in the critical sections of the component, when subjected to different loads. Figure 19.11 shows the color gradient to be used in the analysis.

Table 19.10 shows different graphical analyses of equivalent stresses for different values of increasing loads, from 60,000 to 174,500 N.

In the following analyses, a more detailed study will be made, for which increasing loads with intervals of 500 N will be applied. A value of 500 N represents approximately 0.3% of the average value of the breaking load reached in the real tests. This increment value will allow a finer analysis of the equivalent stresses found in the component. Loads slightly lower and higher than the average value obtained in the tests will be used (Tables 19.11 and 19.12).

19.3.2 Component C1L27524-1

As was the case with the preceding component, the part and its adaptor device were modeled in 3D using Autodesk Inventor software in order to simulate the tensile test conditions in a computer (Fig. 19.12).

Table 19.10 Results of failure tests of component C14D41103-1 (part 1)

Load	Graphical analysis	Comments
60,000 N 35% of average tensile failure load		At a load of 60,000 N, red zones are already visible in the component whose calculated equivalent stresses are above 573 MPa. Despite these results obtained with a load of 60,000 N, the total rupture will only happen in real tests, close to a substantially higher value, 170,000 N.
70,000 N 41% of average tensile failure load		For a 70,000 N load, there is, as would be logical, an increase in the section of the component whose calculated equivalent stress is greater than the tensile strength of the material, 573 MPa.
80,000 N 47% of average tensile failure load	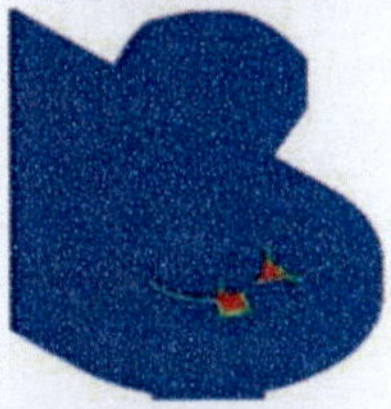	For load levels of 80,000 N and 90,000 N, the evolution of the areas whose equivalent stresses exceed the tensile strength increases gradually.
90,000 N 53% of average tensile failure load		
120,000 N 71% of average tensile failure load		For load steps of 120,000 N, the area whose equivalent stress is greater than the tensile strength of the material continues to spread, but still without crossing an entire resistant cross-section of the component.

The mesh generated in each of the components was finely defined according to the processing capacity of the available hardware. The mesh definition parameters are visible in Fig. 19.13.

The amplitude of the applied loads varies between a minimum of 180,000 and a maximum of 235,000 N. Within this range are the three values obtained in the real tests, namely, 215,550, 231,510, and 232,780 N. The value of the R_m of the material,

Table 19.11 Results of failure tests of component C14D41103-1 (part 2)

Load	Graphical analysis	Comments
163,500 N 96% of average load		For load steps of 163,500 N, it is already visible the appearance of a new area distinct from the previous one in which the calculated equivalent stress is higher than the failure stress, but still without crossing an entire resistant cross-section of the component.
164,000 N 96.6% of average load		At the 164,000 N load plateau there is a natural gradual increase in the areas where the equivalent stress exceeds the tensile strength, and the connection between the two previous areas is already visible. It is visible that this connection is slightly off-centered in relation to the piece, going against the photos observed in the real tests in Fig. 19.2. This load represents 96.6% of the average load, therefore very close to the failure zone verified in the real tests. For the first time in these analyses, it is verified that there is an entire resistant section whose equivalent stress exceeds the breaking stress of the material. We can suggest then that there is a connection between the total yielding of the component and the existence of an entire section with a stress equivalent to the breaking limit of the material.
164,500 N 96.9% of average load		At the load plateau of 164,500 N, about 96.9% of the average test load, two tenuous connections between the two most critical sections are already visible.
165,000 N 97.2% of average load		At the 165,000 N plateau and onwards the connection between the two areas becomes more and more pronounced. We are at 97.2% of the average load, so from here on it is possible that total failure in the real test could happen at any time. It should also be noted here that any change in the mesh definition causes slight changes in the graphical data obtained. These changes are very slight, but they go against what is expected, less element in the mesh, more definition in the calculation and heavier processing.
165,500 N 97.5% of average load		In the following load steps, it is verified, as expected, that the two critical sections will merge into a single section, totally in red, i.e., with an equivalent stress higher than the failure stress of 573 MPa. In the following analyses, the intervals between the load steps are not regular and can be higher than 500 N. It can be seen that with load increments of 500 N the graphical changes are not significant.

Table 19.12 Results of failure tests of component C14D41103-1 (part 3)

Load	Graphical analysis	Load	Graphical analysis
166,000 N 97.8% of average load		166,500 N 98.1% of average load	
168,000 N 99.0% of average load		168,500 N 99.3% of average load	
169,000 N 99.6% of average load		169,500 N 99.9% of average load	
173,500 N 102.2% of average load		174,500 N 102.8% of average load	

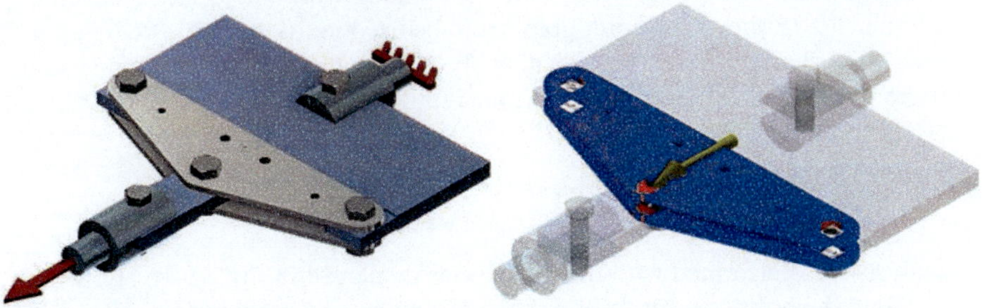

Fig. 19.12 Actual analysis conditions for component C1L27524-1 (left); conditions under analysis defined in Autodesk Inventor software for component C1L27524-1 (right)

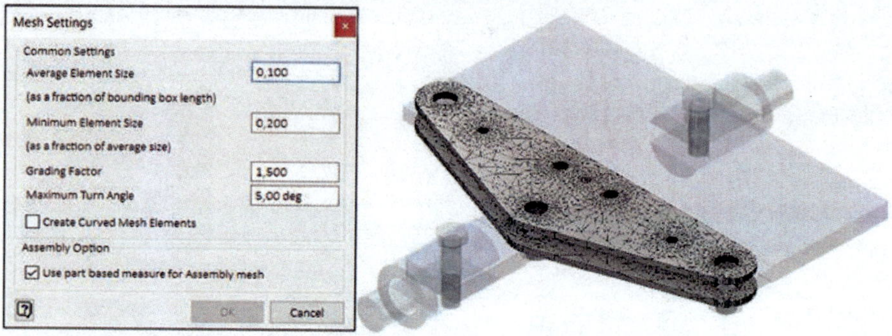

Fig. 19.13 Mesh definition for component C1L27524-1

Fig. 19.14 Mesh definition for component C1L27524-1

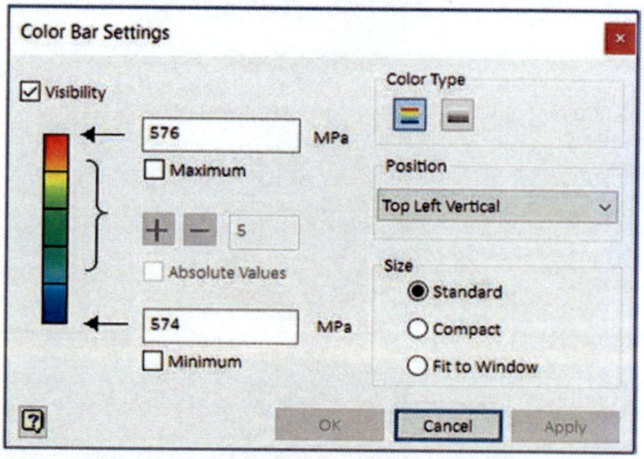

maximum strength according to the supplier's certificate, is 576 MPa. Therefore, we will concentrate on a stress range between 574 and 576 MPa (Fig. 19.14).

In this analysis, the main interest is to analyze the equivalent stresses close to the failure stress, in the critical sections of the component, when subjected to different loads. Figure 19.15 shows the color gradient to be used in the analysis.

Table 19.13 shows different graphical analyses of equivalent stresses calculated for different values of increasing loads, from 180,000 to 240,000 N.

19.3.3 Component B7419103-1

The analysis was performed with the restrictions displayed in Fig. 19.16 and without modeling the bolts and adaptors. For the execution of the analysis, the component was rigidly fixed by the face of one of the holes of 24 mm diameter, represented in yellow, and with a force applied perpendicularly to the face in red in the other hole, represented by an arrow.

Fig. 19.15 Color gradient for different equivalent tensions for component C1L27524-1

Table 19.13 Results of failure tests of component C1L27524-1

Load	Graphical analysis	Comments
180,000 N 79.4% of average load	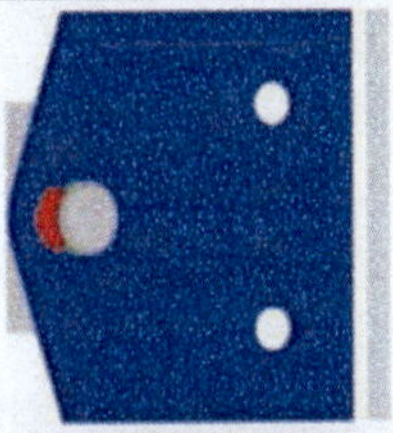	With a load of 180,000 N, it is already clearly visible that the component presents zones where the calculated equivalent stresses are above 576 MPa, zones in red. Despite these results obtained with a load of 180,000 N, this value represents only 79.4% of the average load found in the real tests and is still far beyond the load of 215,550 N, the minimum load obtained in the tests.
200,000 N 88.3% of average load	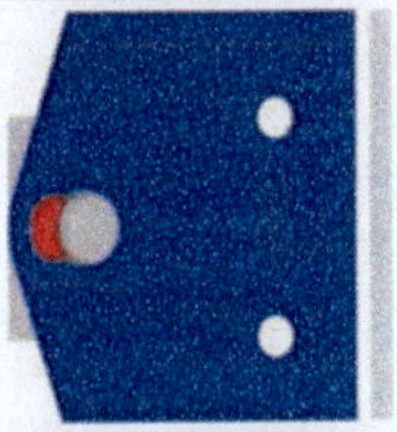	For a load of 200,000 N, the zone where the equivalent stress exceeds the tensile strength will widen even more and will not cross the whole resistant section of the component.
215,550 N 95.1% of average load	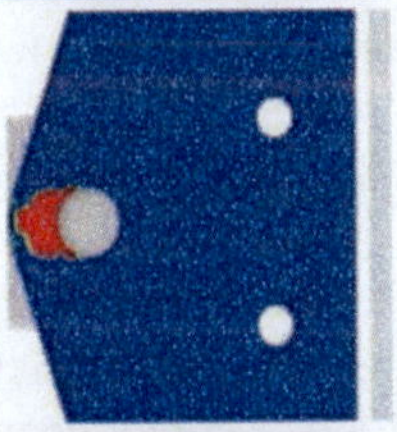	For a load of 215,550 N, failure load of one of the three tests, it is visible that there is a zone, whose stress exceeds the failure stress of the material, which almost crosses an entire resistant section of the component.
226,610 N 100% of average load	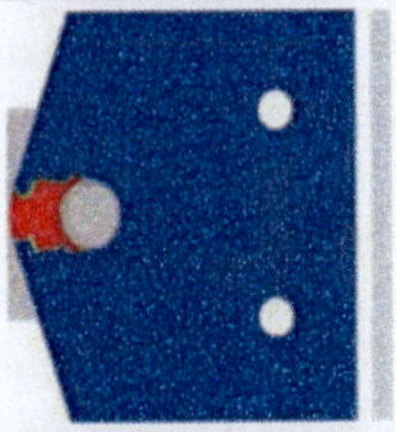	For a load of 226,610 N, the average load value of the tests, it is perfectly visible that there is an entire resistant section of the component that is crossed a stress higher than the admissible breaking stress of the material.
235,000 N 103.7% of average load	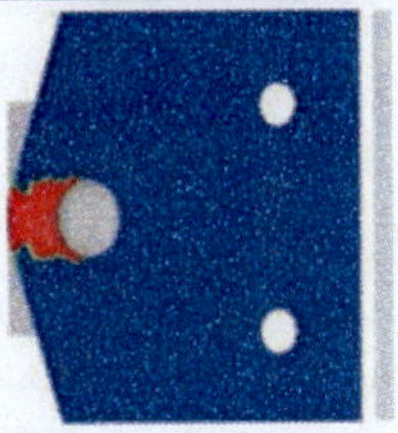	For a load of 235,000 N, higher than the maximum loads obtained in the tests, the red zone continues to propagate and becomes uniform in width across the critical section of the component.

Fig. 19.16 Conditions in Autodesk Inventor Analysis for Component B7419103-1

Table 19.14 presents an identical analysis to that performed in the previous component. Several analyses are presented with increasing load values and with shorter intervals at load values close to the average failure load obtained in the tensile tests.

19.4 Discussion of Results

In the FE analysis of the component C14D41103-1, it is visible, at 80,000 N of load, zones that exceed the admissible tensile strength of the material. However, when this same component is tested in the tensile testing machine, its total failure only comes to happen near 170,000 N. The final customer requires the real test of the component, and this is the value that will appear in the sheet of technical characteristics approved for the component.

The manufacturer must provide a report of a tensile test prior to the final validation of a new component. But, should a FEM analysis be required? In fact, to minimize the use of raw materials in the production of this type of component, it makes sense to perform previous analysis with FE software during the design phase. However, for these tests to have any logical correspondence with the reality of the tests, it is essential that professionals can interpret their results correctly; this means that it shouldn't be considered that a part will fail if it presents a node with an equivalent stress higher than the ultimate tensile strength of the material. Total failure usually occurs with a manifestly higher load. Since the safety factor is calculated as the quotient between the highest equivalent stress and the ultimate tensile stress of the material, it also means that this value should not be used as a result of the analysis.

This dissonance between the FE simulation and the real test occurs because the software for elemental analysis defines the safety coefficient by dividing the maximum admissible stress for the material by the maximum stress found, i.e., it validates the element with the highest stress, but the physical reality of the components when subjected to stresses does not work that way, but by the values of the average stresses. The maximum point values will always be greater than the average values,

Table 19.14 Results of failure tests of component B7419103-1

Load	Graphical analysis	Comments
100,000 N 51.3% of the average load		For a load of 100,000 N, red zones are already visible in the component whose equivalent stresses are above 456 MPa. Despite these results obtained for a load of 100,000 N, about 51.3% of the average load, the total rupture will only be verified in the real tests, close to a substantially higher average value, 194,830 N.
180,000 N 92.3% of the average load		For a load of 180,000 N, about 82.3% of the average load, there is clearly a substantial increase in the red area representing an equivalent stress in excess of 456 MPa, the tensile strength of the material, but there is still no visible band across an entire resistant section of the part where this stress is greater than the tensile strength.
185,000 N 95.1% of average load		With a 185,000 N load the area in red continues to increase, but still without crossing an entire resistant section of the component.
190,000 N 97.5% of average load		With a load of 190,000 N it is already visible, albeit faintly, that there is a band that crosses an entire resistant section of the component. This situation is more evident in the lower part of this table.
195,000 N 100.1% of average load	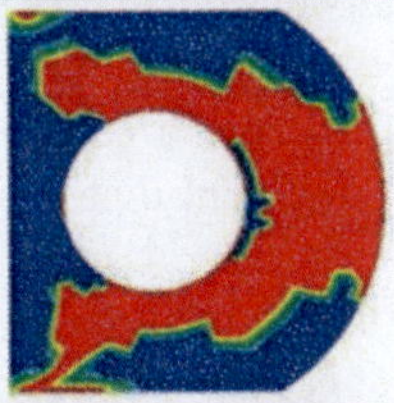	At the 195,000 N load plateau, 100.1% of the average load, it is already clearly visible that there is an entire section crossing the component where the calculated equivalent stress is higher than the yield strength of the material 456 MPa.
203,000 N 104.1% of average load		With a load of 203,000 N, which is a value slightly above one of the values obtained in the real test, a zone crossing the whole resistant section of the piece can be clearly seen, whose equivalent stress exceeds the breaking stress of the material, 456 MPa.

Fig. 19.17 Analysis of component C14D41103-1 made with a load of 168,500 N

so from this perspective, the maximum equivalent stress calculated by the software in a simulation will always be greater than the average stress calculated numerically.

Neglecting stress concentration and other complicating effects, the normal stress may be calculated by dividing the exerted force by the resisting section, but this calculated stress is the average value of that section and not its maximum value. This difference causes that the safety coefficients found by FE analysis software will always be clearly lower than real tests show.

When using FE analysis software, a qualitative analysis must be made; stresses must be evaluated and their significance in terms of location and extent of their area must be assessed. In the case of component C14D41103-1, the effective total failure happens if there is an entire resistant section of the component that is crossed by a band in which the stresses found are always higher than the admissible limit for the material. This is clearly seen in the image obtained from the analysis performed for a load of 168,500 N in Fig. 19.17. In addition to this image, we can find other identical ones for loads immediately above or below this load. In these images, the two most critical areas, red in the image, are connected by bands in which the stress is also higher than the material's limit, also represented in red. As a conclusion we could say that the designer, if he/she wants to predict with some accuracy the value obtained in real tensile testing, should look for an analysis in which the whole resistant section of the part was traversed by a stress higher than the failure stress.

Components C1L27524-1 and B7419103-1 present similar results. As a final analysis of component C1L27524-1, it can be said that the component yielded in the same way as the previous component, that is, the final yielding of the component to failure occurred when an entire resistant section exceeded the admissible tensile strength for the material over its entire length. With respect to component B7419103-1, it can be said that the component yielded in the same way as the previous two components, i.e., the final yielding of the component to failure occurred when an entire resistant section exceeded the admissible tensile strength for the material over its entire length.

19.5 Conclusions

The metalworking and machine design industry represents an enormous economic opportunity for Portugal and other nations, but it exhibits some competitive weaknesses. The ability to accomplish more and better with fewer resources is crucial for its survival. With this study, it is hoped to optimize the use of the FEM tools available in CAD software, but which are underutilized due to users' lack of understanding of how to interpret the results.

One conclusion that can be drawn from the analyses presented here is that the behavior of all components is comparable and that the tensile tests confirm that the software-suggested most critical sections are the effective failure sections identified in real tensile testing. One additional conclusion that can be drawn is that qualitative analysis is necessary for correctly interpreting the results. As evidenced by the real tensile test results, the presence of small, localized areas with equivalent stresses greater than the failure stress does not indicate that the component will fail at that load level; failure will typically occur at much higher loads, particularly when one of the component's resistant sections is traversed by a stress greater than the tensile strength. However, it should be noted that this study assumes that the applied load is static and not cyclic, as in that case, the fatigue effect would be dominant and the previous conclusions would be inadequate in terms of the mechanical resistance of the components over the long term.

For the success of these analyses, it is also essential to precisely define the mechanical properties of each of the component materials and the constraint and loading conditions for each of them. Regarding the analysis performed by the FE software, one must always keep in mind that the safety factor is calculated using the maximum value of the equivalent stress found, which is clearly a higher value than the average stress value. Therefore, in static situations in which the components are rigidly fixed and the load intensity is constant, structural designs in which reduced sections of certain components exceed the yield strength or tensile strength can be validated as long as this section does not cross an entire resistant section of the part. As this study evolves, it is planned in the future to evolve this simplistic approach to more complex geometries, stress-complicating effects, nonlinear material behavior, and continuous damage models to determine if the pattern repeats itself, i.e., that the total failure of the component occurs whenever a resistant section has an equivalent stress greater than the material's allowable limits.

Acknowledgments The third and fourth authors gratefully acknowledge the support provided by the Foundation for Science and Technology (FCT) of Portugal, within the scope of the project of the Research Unit on Materials, Energy and Environment for Sustainability (proMetheus), Ref. UID/05975/2020, financed by national funds through the FCT/MCTES.

References

Abambres, M., & Arruda, M. R. (2016). Finite element analysis of steel structures – A review of useful guidelines. *International Journal of Structural Integrity, 7*(4), 490–515. https://doi.org/10.1108/ijsi-07-2015-0020

British Steel. (2020). *Declaration of performance of BS section S355J0*. British Steel Limited. https://britishsteel.co.uk/media/323488/s355j0.pdf

Brnic, J., Canadija, M., Turkalj, G., & Lanc, D. (2008). Finite-element modelling and shear stress analysis of engineering structural elements. *Proceedings of the Institution of Mechanical Engineers, Part G: Journal of Aerospace Engineering, 222*(6), 861–872. https://doi.org/10.1243/09544100jaero296

Cao, L. M., Zeng, Q. L., Xiao, X. Y., & Cui, J. (2011). Finite element stress analysis on structure of hydraulic support. *Advanced Materials Research, 321*, 84–87. https://doi.org/10.4028/www.scientific.net/AMR.321.84

Chen, L. H., Wang, S., & Kondo, Y. (2018). Stress analysis and design of main reducer housing using finite element method. *MATEC Web of Conferences, 221*. https://doi.org/10.1051/matecconf/201822104008

Chen, L., Li, H., Yu, J., Luo, Y., Wang, Z., Zhu, X., et al. (2021). Stress analysis of screw connection of key structural components in pump turbine. *IOP Conference Series: Earth and Environmental Science, 804*(3), 032037. https://doi.org/10.1088/1755-1315/804/3/032037

Clough, R. W. (1960, September 8–10). The finite element method in plane stress analysis. In *2nd ASCE conference on electronic computation*, Pittsburgh, PA, US.

Clough, R. W. (1990). Original formulation of the finite element method. *Finite Elements in Analysis and Design, 7*(2), 89–101. https://doi.org/10.1016/0168-874x(90)90001-u

Courant, R. (1943). Variational methods for the solution of problems of equilibrium and vibrations. *Bulletin of the American Mathematical Society, 49*(1), 1–23. https://doi.org/10.1090/s0002-9904-1943-07818-4

Fadiji, T., Berry, T. M., Ambaw, A., Coetzee, C., & Opara, U. L. (2018a). Finite element analysis (FEA) – An effective and efficient design tool in food packaging industries: A review. *Acta Horticulturae, 1201*, 245–252. https://doi.org/10.17660/ActaHortic.2018.1201.33

Fadiji, T., Coetzee, C. J., Berry, T. M., Ambaw, A., & Opara, U. L. (2018b). The efficacy of finite element analysis (FEA) as a design tool for food packaging: A review. *Biosystems Engineering, 174*, 20–40. https://doi.org/10.1016/j.biosystemseng.2018.06.015

Golovanov, A. I., Mitryaikin, V. I., & Shuvalov, V. A. (2009). Stress analysis of helicopter main rotor torsion bar. *Russian Aeronautics (Iz VUZ), 52*(1), 104–109. https://doi.org/10.3103/s1068799809010188

Gupta, K. K., & Meek, J. L. (1996). A brief history of the beginning of the finite element method. *International Journal for Numerical Methods in Engineering, 39*(22), 3761–3774. https://doi.org/10.1002/(SICI)1097-0207(19961130)39:22<3761::AID-NME22>3.0.CO;2-5

Hatheway, A. E. (1992). A review of finite element analysis techniques: Capabilities and limitations. *Proceedings of SPIE, 10265*, 367–397. https://doi.org/10.1117/12.61113

Hrennikoff, A. (1941). Solution of problems of elasticity by the framework method. *Journal of Applied Mechanics, 8*(4), A169–A175. https://doi.org/10.1115/1.4009129

List of finite element software packages. (2022). In *Wikipedia*. https://en.wikipedia.org/wiki/List_of_finite_element_software_packages. Accessed 7 Oct.

Magomedov, I. A., & Sebaeva, Z. S. (2020). Comparative study of finite element analysis software packages. *Journal of Physics: Conference Series, 1515*(3), 032073. https://doi.org/10.1088/1742-6596/1515/3/032073

Miriţoiu, C., & Ilincioiu, D. (2011). A stress analysis of a metallic structure using finite element method. *World Academy of Science, Engineering and Technology, 52*, 761–765. https://citeseerx.ist.psu.edu/viewdoc/download?doi=10.1.1.298.9000&rep=rep1&type=pdf

Naganarayana, B. P., & Prathap, G. (1992). Expert systems and finite element structural analysis – A review. *Sadhana, 17*(2), 275–298. https://doi.org/10.1007/bf02812054

Nor, M. A. M., Rashid, H., Mahyuddin, W. M. F. W., Azlan, M. A. M., & Mahmud, J. (2012). Stress analysis of a low loader chassis. *Procedia Engineering, 41*, 995–1001. https://doi.org/10.1016/j.proeng.2012.07.274

OVAKO. (2022). *Material datasheet steel grade S27JR*. OVAKO AB. https://steelnavigator.ovako.com/steel-grades/s275jr/pdf

Ramachandra, P. M., Sutar, S., & Kumara, G. C. M. (2022). Stress analysis of a gear using photoelastic method and finite element method: Review. *Materials Today: Proceedings, 65*, 3820–3828. https://doi.org/10.1016/j.matpr.2022.06.579

Sabat, L., & Kundu, C. K. (2021). History of finite element method: A review. In *Recent developments in sustainable infrastructure* (pp. 395–404). https://doi.org/10.1007/978-981-15-4577-1_32

SSAB. (2022). *Data sheet 2278 SSAB Domex 460M*. https://www.ssab.com/api/sitecore/Datasheet/Get?key=eae213522eb94938b4f4cd29dc908b02_en

Tensile testing. (2022). In *Wikipedia*. https://en.wikipedia.org/wiki/Tensile_testing. Accessed 7 Oct.

Yu, N. H., & Fan, J. J. (2014). Stress analysis of the press cage in screw press by finite element method. *Applied Mechanics and Materials, 543–547*, 164–167. https://doi.org/10.4028/www.scientific.net/AMM.543-547.164

Zienkiewicz, O. C., Taylor, R. L., & Zhu, J. Z. (2013). *Finite element method: Its basis and fundamentals* (7th ed.). Butterworth-Heinemann.

Zienkiewicz, O. C., Taylor, R. L., & Fox, D. D. (2014). *The finite element method for solid and structural mechanics* (7th ed.). Butterworth-Heinemann.

Indoor Navigation of an Autonomous Guided Vehicle Using ArUco Markers

20

Fernando A. V. Figueiredo, Eduardo G. C. Pereira, and César M. A. Vasques

20.1 Introduction

The evolution of technology and the emergence of computer systems and digitalization applied to industry have led to an important new technology development, which increasingly tries to reduce manual work and produce more complex equipment aimed at saving manpower and simultaneously increasing productivity. In the past, in factories and warehouses, loads were transported using manually guided carts, such as a pallet truck. Nowadays, with the help of mobile robotics, these are being replaced by mobile autonomous guided vehicles (AGVs), as autonomous moving robots capable of transporting loads as intended, safely and without colliding with other AGVs or humans that may be working together. As a result, complex networks of AGVs working together, known as AGV systems (AGVSs), are becoming more prevalent in the manufacturing and logistics industries, where battery-powered and computer-controlled AGVs have become a reality.

According to Ullrich (2015), since the invention of the AGVs in the 1950s, their 60-year existence can be divided into four distinct eras. These eras can be

F. A. V. Figueiredo (✉)
ISPGAYA, Vila Nova de Gaia, Portugal

Smile.Tech – Robótica, Vila Nova de Gaia, Portugal

proMetheus, Instituto Politécnico de Viana do Castelo (IPVC), Viana do Castelo, Portugal
e-mail: ffigueiredo@ispgaya.pt

E. G. C. Pereira
ISPGAYA, Vila Nova de Gaia, Portugal

C. M. A. Vasques
proMetheus, Instituto Politécnico de Viana do Castelo (IPVC), Viana do Castelo, Portugal

Department of Mechanical Engineering, University of Aveiro, Aveiro, Portugal

© The Author(s), under exclusive license to Springer Nature Switzerland AG 2023

F. L. Almeida et al. (eds.), *Multidimensional Sustainability: Transitions and Convergences*, Springer Proceedings in Earth and Environmental Sciences,
https://doi.org/10.1007/978-3-031-24892-4_20

characterized by the available technology at the time and the emotional attitude toward the systems. They can also be viewed as evolutionary stages during which there were little technological advances, but which transitioned relatively suddenly into the subsequent era.

The first era of AGVS began in the United States in 1953 with the invention of a driverless, horizontally moving, AGV which was simply a tow truck that followed an electrically conductive wire in the floor instead of a rail (Wikipedia, 2022). This principle is now referred to as *inductive track guidance*, in which the AGV orients itself while driving along an induced magnetic field. For safety, tactile "sensors" such as bumpers and emergency stop handle with mechanical switches were used at that time, and the stations at which loads (goods) were to be transferred from/to, were coded with magnets embedded in the floor and detected by sensors in the vehicles. After its invention, the technology was subsequently followed by Europe a few years later. This period spanned more than two decades without significant technological advancements.

The second era started in the 1970s and lasted until the early 1990s, and it was marked by the introduction of electronics, basic onboard computers, and control cabinets. Active induction track guidance utilizing a wire on the floor, or stainless steel strips (Hollingum, 1998), became the norm, and data transfer utilized the same wire, infrared, or radio signals. By then, the classic AGVS was finally developed, combining the ever-increasing productivity with the use of manned transport systems created a need for a greater degree of automation to reduce production costs over the long term. Several technical innovations were introduced by then. High-performance electronics and microprocessors entered the scene by that time, enabling faster computing and consequently more complex use scenarios. Powerful new batteries were introduced, and battery charging became fully automatic. Intelligent and increasingly automated load handling occurred, and the mobility of vehicles increased. The production processes were streamlined to accommodate AGVs, which could then serve as mobile workbenches (assembly lines).

During the third era, which lasted from the mid-1990s to roughly 2010, technological standards and new markets were established. Electronic guidance and contactless sensors controlled by standard personal computers began to appear on AGVs containing powerful microprocessors. Conductive cable guidance has been discontinued. Magnetic and laser navigation have been adopted as traditional "free" navigation technologies, while WLAN has become a standard method of data transfer. AGVs have benefited from advances in materials flow and warehouse technology, mechanical engineering production methods, and assembly technology. Improvements in computing and sensory technology lead to key advances in vehicle and controlling technology. Improved sensory technology allowed faster driving, maneuvering, and load handling. Vehicles have become simpler and less expensive. Inductive energy transmission and new navigation methods (e.g., magnetic, laser-based, transponders, and building navigation) in AGVs controlled by personal computers have been adopted. Data transfer through WLAN and new areas of functionality, such as serving bulk-storage warehouses or hospitals. Further details and an extensive list of AGVSs companies and key developments involving current

AGVSs manufacturers during these three eras can be found in Ullrich (2015, pp. 11–13).

Since the 2010s, the world of AGVSs has entered a new evolutionary phase with a profound effect on both technology and new applications. This new era will not completely replace the last one, but rather expand upon its successes. This means that the applications and technological solutions of the third era will continue to be utilized! The technological foundation for the ensuing changes is supported by new, low-cost, and intelligent sensor systems, internet-based sensing software developments, and the ongoing industry digitalization.

A large variety of laser, radar, LIDAR, infrared, ultrasound, and video systems are used in today's AGVSs designs, and these will continue to advance in tandem with advancements driven by the automotive industry. It is not yet entirely known, however, which technologies or technology combinations will prove most useful. The problems that exist in the present day in terms of functionality are a direct result of the use of existing systems. In short, these may include (Ullrich, 2015): truly autonomous driving, avoiding obstacles, recognizing disruptions, pallet finder, truck loading, learning new tasks quickly, moving traffic in outdoor usage, reacting quickly, speech guidance, energy use, and battery disposal. As a result of the fourth era's functional challenges and technological advancements, a wide variety of new application scenarios and end users will emerge, complementing and expanding upon today's well-established markets.

The technology associated with AGVSs and guidelines for when and how to employ it has been provided by Feledy and Luttenberger (2017). Developments on traffic management of AGVSs in flexible manufacturing systems were addressed by Olmi (2011). The state-of-the-art and comprehensive studies of AGVs have been reported in the open literature (Ali & Khan, 2010; De Ryck et al., 2020; Fragapane et al., 2021; Hoff & Sarker, 1998; Liang et al., 2021; Lu et al., 2022; Lynch et al., 2018; Mohammadpour et al., 2021; Oyekanlu et al., 2020; Reis et al., 2022; Vijayaram, 2006; Vis, 2006). In short, the available technology, recent and ongoing studies have addressed:
Types and categories of AGVs (Egemin, 2014; Ullrich, 2015, §3.3.1);

- Guidance control systems (Bao et al., 2019; Cai et al., 2011; Chen et al., 2004; Djordjevich et al., 1999; Feng & Jiao, 2017; Hu et al., 2020; Iñigo & Alley, 1991; Rajagopalan et al., 1992a, b; Turpin, 1986; Yadav et al., 2018);
- Navigation methods (Ai et al., 2021; Cui et al., 2021; Culley & Baldur, 1988; Iñigo & Alley, 1991; Isrofi et al., 2021; Kusche, 1996; Lin et al., 2021; Liu et al., 2020; Man et al., 2010; Pan, 2021; Ray et al., 2008; Song et al., 2016; Tang et al., 2021; Wang et al., 2020; Yang et al., 2021; Yao & Zeng, 2012; Zhou et al., 2021);
- Obstacle avoidance (Ball, 1996; Bosnak & Skrjanc, 2021; Chen, 2010; Cucchiara et al., 2007; Digani et al., 2014; Li et al., 2021; Norton et al., 2019; Praserttaweelap et al., 2019; Shirai, 1990; Teso-Fz-Betoño et al., 2019; Wu et al., 2017; Xie, 1995);

- Safety (Cheshire, 1985; Kang et al., 2010; Nanthavanij et al., 1995; Praserttaweelap et al., 2019; Sabattini et al., 2016; Stimming et al., 2015; Tracey, 1985; Yang et al., 2017);
- Batteries and charging (Dehnavi-Arani et al., 2019; Pian et al., 2019; Schlieck, 1986; Schmidt et al., 2015);
- Environmental and sustainability concerns (Aguiar et al., 2019; Małopolski, 2018; Stefanini & Vignali, 2022);
- Operational, functionality, and optimization aspects (Arques Corrales & Aznar Gregori, 2020; Cao & Peng, 2020; Ganesharajah et al., 1998; Jiang et al., 2020; Lin et al., 2023; Sauer et al., 2022; Sun et al., 2021; Wang et al., 2021; Wu et al., 2022; Xie et al., 2020; Zou et al., 2022); and
- Fields of application (Ullrich, 2015).

The AGV can navigate guided or free, so it is necessary to plan the route. To help in this navigation and guarantee safety, it is equipped with sensors that stop it in case of an obstacle. The problem of autonomous navigation of mobile robots is divided into three main areas: localization, mapping, and path planning (Stachniss, 2009). Localization consists in determining in an exact manner the current pose of the robot in an environment (Where am I?). Mapping integrates the partial observations of the surroundings into a single consistent model (What does the world look like?) and path planning determines the best route in the map to navigate through the environment (How can I reach a given location?) (Fuentes-Pacheco et al., 2015).

Nowadays, as presented in Fig. 20.1, there are many AGV navigation options with diverse features that can be divided essentially into two groups (BlueBotics, 2022a, b):

- Navigating along physical lines (a.k.a. line following); this could be magnetic tape, painted lines, inductive wire, or (less often) rails; for simplicity, we can also include tags.
- Navigation technologies that guide vehicles along digital (software-based) paths.

Over the years, navigation technology has become more sophisticated, and today, automated vehicles are mainly laser navigated, but computer vision systems are taking their place alongside the development of autonomous automobile navigation. This chapter is focused on AGV localization and, more particularly, on the use of ArUco markers to enable it using only one camera. The ArUco is a square-based fiducial marker system which, in turn, is composed of a set of valid markers and an algorithm that performs their detection and possible correction (Garrido-Jurado et al., 2014). The square-based fiducial markers system's main advantage is that the presence of four prominent points can be employed to obtain the pose, while the inner region is used for identification (either using a binary code or an arbitrary pattern such as an image).

The different parts of the project of the AGV reported in this chapter were carried out simultaneously. In order to reduce costs, some components that were offered were used, which, in a way, determined some of the choices made in this project.

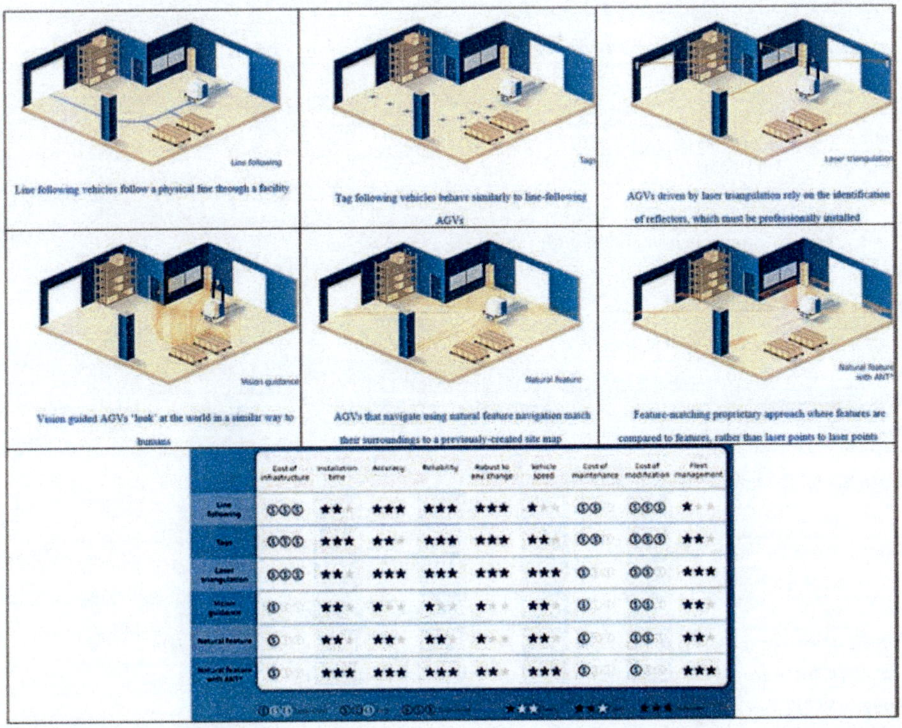

Fig. 20.1 AGV navigation technologies and features comparison. (Source: Adapted from BlueBotics, 2022a, b)

There are four main aspects of the development of the AGV that have been considered in the organization of the chapter, namely, locomotion, structure, control, and location. The locomotion consists of the selection of the motors to be used and the components that will give movement to the AGV. The structure concerns the chassis of the AGV as well as all armor. The control part, whose function is to command and control the motors of the AGV, was developed simultaneously with the chassis and the part that refers to the location, i.e., control of the position and orientation of the AGV, was the one where the work started. This chapter introduces a novel approach and contributes to advance the state of the art by simplifying the design of an AGV's indoor navigation system by using ArUco markers.

20.2 Specification and Requirements

The specifications of the developed AGV should enable it to transport a payload of at least 50 kg with a tare weight not exceeding 20 kg (Table 20.1).

The project is however constricted by the technical data of components that were kindly offered by companies that sponsored the project (Table 20.2).

Table 20.1 AGV requirements

Parameter	Value	Unit
Payload	> 50	kg
Tare weight	< 20	kg
Inclination	<= 5	%
Emergency stop distance	< 50	mm

Table 20.2 Limitations associated with "sponsored" components

Parameter	Value	Unit
Motor with reducer output torque	12.5	N.m
Motor with reducer output velocity	65	rpm
Wheel friction ratio	0.4	
Wheel diameter	125	m

Table 20.3 Dynamic capabilities of the AGV

Parameter	Value	Unit
Output force for each wheel	200	N
Friction for each wheel[a]	137.2	N
Linear velocity	< 425.4	mm/s
Acceleration for emergency stop	< 1.8	m/s^2
Required force for emergency stop	< 80.47	N
Time to achieve a full stop	0.23	s
Payload height to length ratio for AGV stability	2.5	mm/mm

[a] This means that the full motor output torque is not available

Considering the selection from the available components, calculations were made to confirm the dynamic capabilities of the vehicle; Table 20.3 depicts the dynamic capabilities of the AGV.

20.3 Mechanical Architecture and Motion

Wheeled locomotion is one of humanity's great innovations. The wheel was invented around 3000 BCE and the two-wheeled cart around 2000 BCE. Today four-wheeled vehicles are ubiquitous, and the total automobile population of the planet is over one billion. The effectiveness of cars, and our familiarity with them, makes them a natural choice for robot platforms that move across the ground (Corke, 2017).

There are multiple options to perform motion, positioning, and orientation of such platforms such as steerable wheels, omnidirectional wheels, and differential steering in conjunction with passive casters. The latter was the chosen solution for our AGV. Figure 20.2 shows the proposed concept with two independent motorized

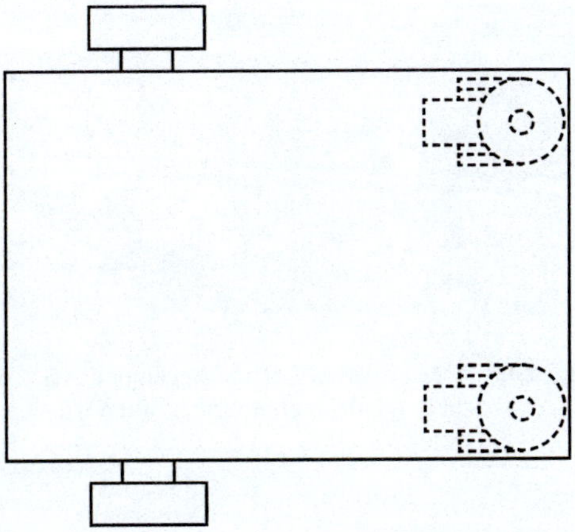

Fig. 20.2 Differential steering structure with passive casters

Fig. 20.3 Developed AGV: isometric representation (left); photograph with detail of the interior structure and electronics (center); photograph of the complete vehicle with the camera (right)

wheels that perform the differential steering on the rear side and two passive casters on the front of the vehicle.

Figure 20.3 shows the isometric rendering of the structure of the vehicle and the picture of the constructed device.

Considering the lack of information about the bearings of the motor's reducer, a transmission shaft, and the corresponding bearings, were sized and designed to guarantee the correct function under the specified loads. Figure 20.4 shows details of the complete assembly.

20.4 Computer Vision and Pose Estimation

As mentioned in the introduction, this project is focused on localization, particularly, the use of ArUco markers to enable it using only one camera. With a well-calibrated camera, after detecting an ArUco marker, knowing the position of the marker's four corners, it is possible to determine its pose relative to the camera and the camera's pose relative to the marker. The feature included in OpenCV was used both to calibrate the camera and to detect and calculate the marker's position and

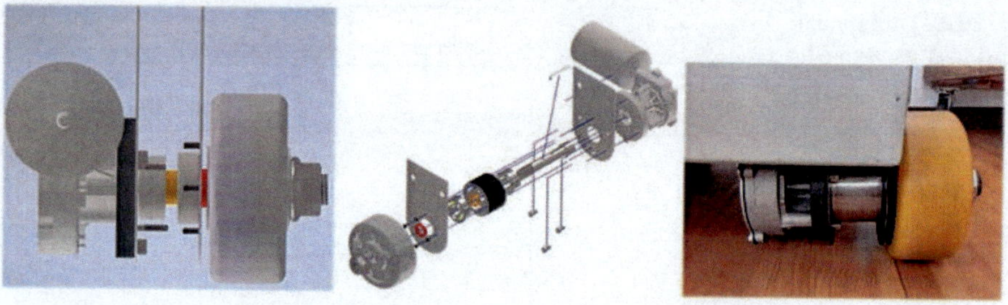

Fig. 20.4 Design details of the developed AGV: wheel assembly (left); exploded view of the wheel assembly (center); photograph of the wheel assembly (right)

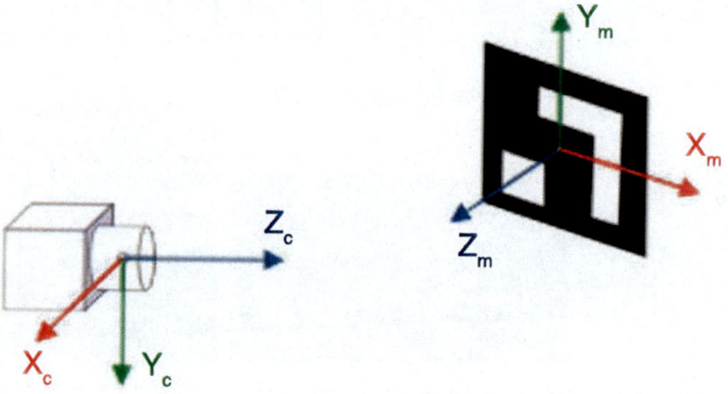

Fig. 20.5 Camera and marker with their referentials

distance, in order to have available the ArUco libraries with Python 3 the *opencv-contrib-python* was installed over the OpenCV library.

Figure 20.5 shows the relative pose of a marker relative to a camera, where X_c, Y_c and Z_c are the referential coordinates of the camera position considering its origin on the focal point of the camera, and X_m, Y_m and Z_m represent a referential linked to the pose of the marker.

The function *aruco.detectmarkers* provides a list of read markers and a list of their corresponding outer corners. After this step, the function *aruco.estimatePoseSingleMarkers* provides a list of rotation and translation vectors from camera to each detected marker (*rvecs* and *tvecs*). Figure 20.6 shows the meaning of the *tvec* coordinates.

The *rvecs* represent the rotation from the camera referential to each detected marker's pose; the *cv2.Rodrigues* method easily converts a *rvec* to a rotation matrix which can then be converted to Euler angles or to a quaternion. This rotation matrix **R** can be used to transform the *tvec* from camera coordinates to the marker coordinate system, which can be useful for navigation calculations:

$$\mathbf{R} = \text{cv2.Rodrigues}\left(\text{rvec}\right) \text{and tvec}_{\text{m}} = R \bullet \text{tvec}.$$

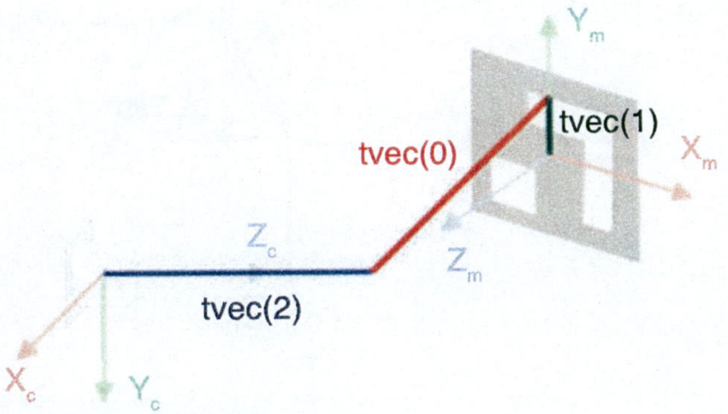

Fig. 20.6 Position of the marker in camera reference frame

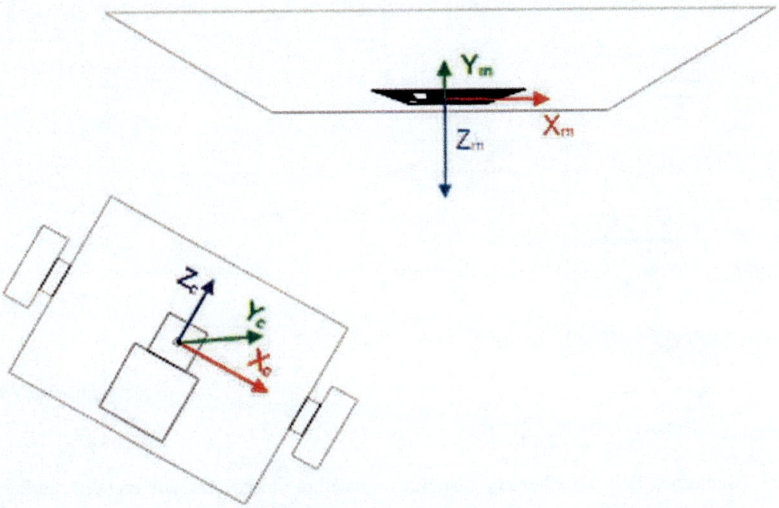

Fig. 20.7 Vehicle with camera facing a wall with a marker

With this information, it is possible to calculate both the relative position of the marker to the vehicle and the relative position of the vehicle relative to the marker which are both useful, for instance, to position the vehicle parallel to a wall where the marker is placed. Consider Fig. 20.7 where the vehicle is facing the marker at a random position). To place the vehicle parallel to the wall, Z_c must be aligned with X_m; to do this, a rotation around Y_c is required. Figure 20.8 shows the required rotation.

Once the rotation is performed the marker will probably fall out of the camera's view. However, the location of the vehicle with respect to the marker is still known. Figure 20.9 shows the vehicle parallel to the wall and the distance it needs to move in order to position itself in front of the marker.

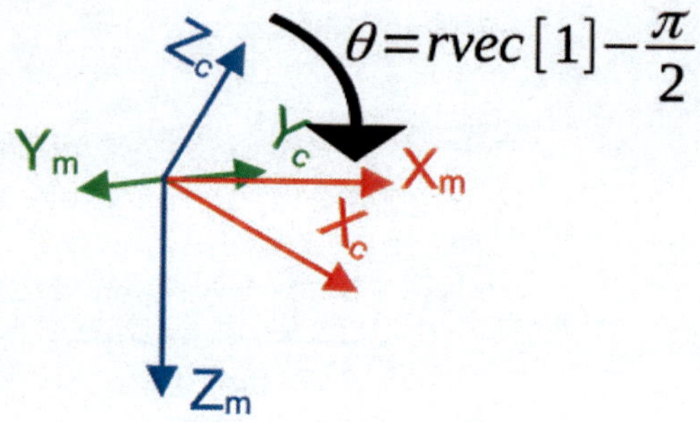

Fig. 20.8 Required rotation to place the vehicle parallel to marker

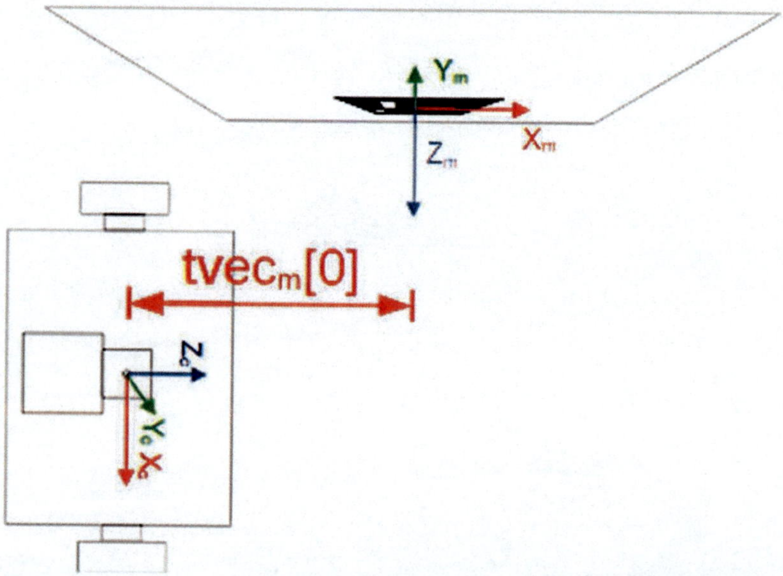

Fig. 20.9 Vehicle parallel to marker and distance required to place vehicle in front of marker

Once the vehicle has advanced the required distance to position itself directly in front of the marker, it should rotate 90° and after this movement it should once again have the marker in the camera's vision field; in fact, the marker should be very close to the center of the camera's image, where distortions are minimal, and this can be used to correct the pose estimation for the rest of the maneuver.

The vehicle can now move orthogonally toward the marker in a perfect maneuver for docking on a loading/unloading, a charging station, or a position for the performance of any kind of task by a machine or manipulation robot.

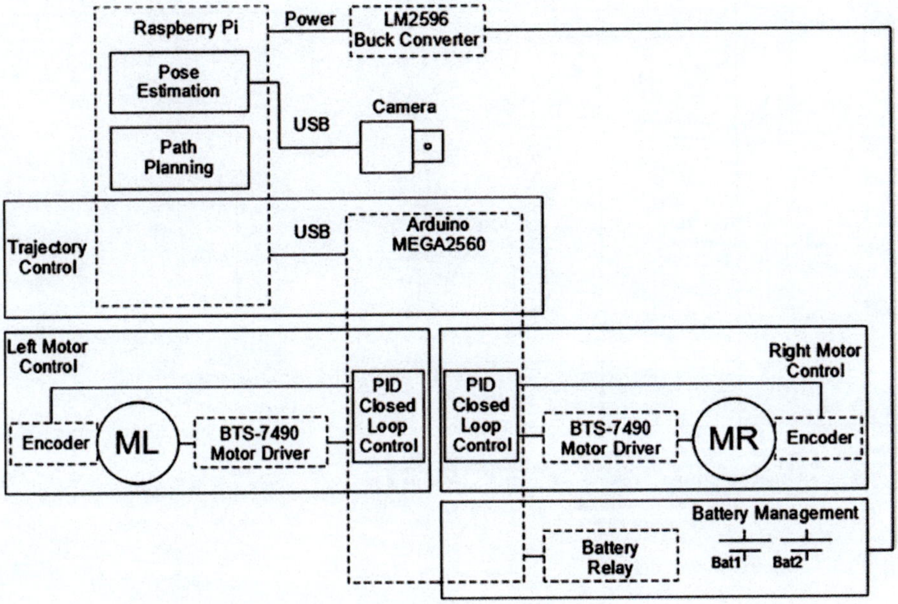

Fig. 20.10 Functional diagram of the control system

20.5 Electronics and Control

The control system performs motor and trajectory control, sensor feedback, battery management, pose estimation with computer vision, and path planning. Some of these features are implemented with lightweight algorithms that require cycle time reliability. These are run in and Arduino Mega2560; other algorithms, such as pose estimation and path planning require higher processing power but don't really need to be performed in real time, these are run in a Raspberry Pi. Figure 20.10 presents the functional diagram of the control system.

The motors and the encoder require a 24VDC power supply, all other components are supplied with 5VDC. Two 12VDC batteries are used to supply the vehicle, one of the batteries supplies the electronic circuitry through a Buck converter. The other battery is connected to a two-contact inverting relay that connects it in series with the first battery for vehicle operation and in parallel for simultaneous charging. The rest position of the relay places the circuit in charging mode; this feature guarantees the possibility of charging both batteries even when they are dead flat. The detailed control circuit is presented in Figs. 20.11 and 20.12.

The Raspberry Pi is running the control application in Python 3 with OpenCV and the *opencv-contrib-python* extension that includes the ArUco library. Raspberry Pi OS includes the required drivers to interface with camera and Arduino USB making the connection to both seamless.

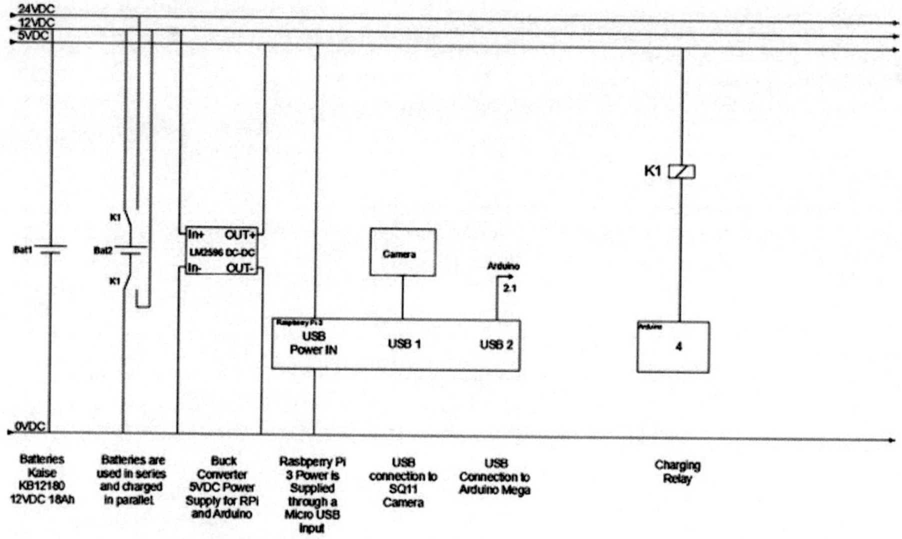

Fig. 20.11 Control circuit diagram 1

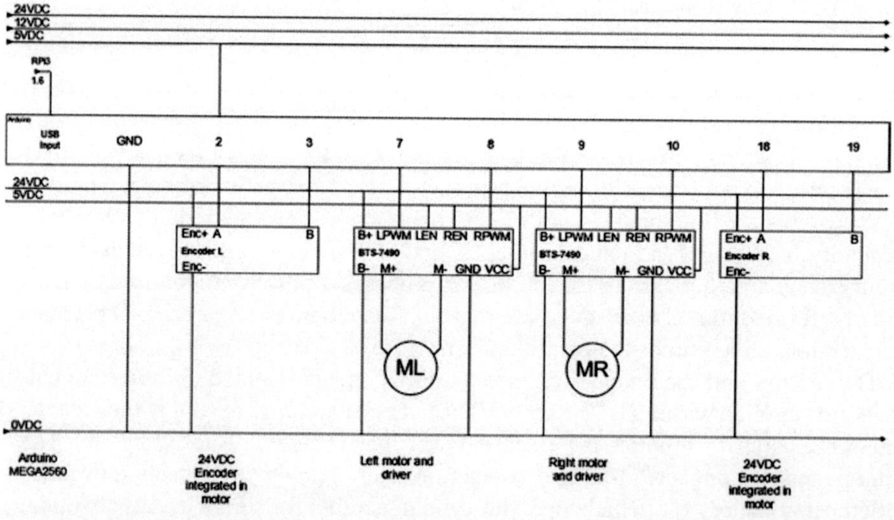

Fig. 20.12 Control circuit diagram 2

The communication between the Raspberry and the Arduino is done through an emulated serial port that runs over a USB connection. A simple communication protocol was developed for the first tests and proof of concept. Three commands are possible: “r”, “s” and “ “(space); “r”(run) turns on the motors and activates the closed-loop control to follow the position reference, “s”stops the motors and sets

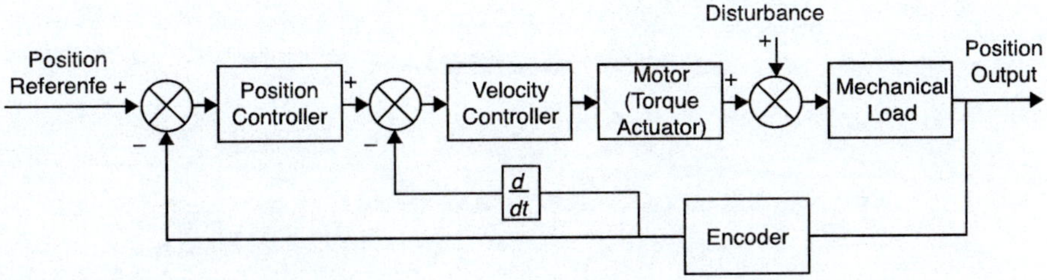

Fig. 20.13 Implemented cascade control loop for each wheel

the reference to current position, and " " (space) turns off the motors and resets the closed loop controllers.

If the closed loop controllers are active, the client (Raspberry Pi) may cyclically send a message with the required positions of each wheel (left and right), in decimal format and separated by a coma, e.g., (200.5,-52.25). The server (Arduino) always replies with the current position of both wheels in the same format. The cascade PID controller attempts to follow the reference positions sent by the Raspberry Pi.

A closed-loop control algorithm controls each wheel independently. Initially, a simple PID control was implemented, however, it was very hard to find correct settings that produced stable control whether the vehicle is loaded or unloaded and dealing with floor irregularities such as tile joints. Not willing to make the control of the wheels the core subject of this project, autotuning or treating this control as a nonlinear system was not desired. The implemented solution is well known in industrial environments as a PIV controller; in academia, this type of arrangement is called a cascade controller and it provides higher immunity to disturbances.

Cascade control is of great value where high performance is needed in the face of random disturbances, or where the secondary part of a process contains a significant time lag or has nonlinearity (Altmann & Macdonald, 2005). This arrangement also has a good tolerance to nonoptimal control parameters, meaning the optimal control parameters for the vehicle with half the payload, will work well whether the vehicle is fully loaded or with no load at all. Figure 20.13 is a graphical model of the implemented controller on each wheel.

Note that the installation has a velocity controller, but the motor shaft has no tachometer or a velocity sensor of any kind. To produce feedback for this controller, the position attained from the encoder is differentiated, this is also a source of disturbance that the control setup deals with well. The position controller is a complete PID controller as is the velocity controller, however, the latter integral and derivative gains are set to zero rendering it a proportional controller.

20.6 Tests and Validation

It was found that with the two models of camera used and with the tests performed, there is always some error. The cameras were calibrated using pictures of a chessboard in many poses and distances with the function *cv2.calibrateCamera*; then

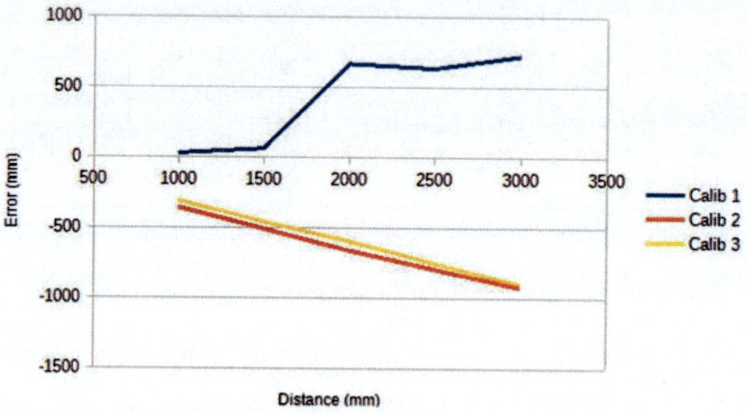

Fig. 20.14 Distance estimation error with three different camera calibrations

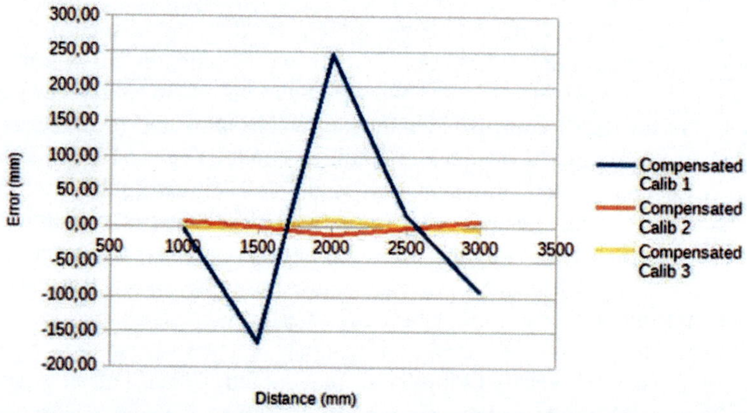

Fig. 20.15 Distance estimation error with three different camera calibrations after compensation with linear regression rule

they were tested with an ArUco marker with 100 × 100 mm size. Figure 20.14 shows the measured errors with each calibration.

Calibration 1 was performed with the chessboard at distances between 1000 and 2000 mm while calibrations 2 and 3 were performed with the chessboard between 1000 and 3000 mm; a linear regression was used for each calibration and an error compensation was used to obtain more accurate values; the application of this compensation resulted in what is shown in Fig. 20.15.

Calibration 2 was chosen resulting in a max error of 11.6 mm in distance calculation. Table 20.4 shows the distance estimation error after calibration and compensation with a linear regression rule.

The closed loop controller was tuned to a point where the final positioning error was under 2°, which results in a position error for each wheel under 2.18 mm;

Table 20.4 Marker distance measurements (mm)

Distance	Calculated distance	Error	Error with linear compensation
1000	644	– 356	+ 7.8
1500	994	– 506	– 2.4
2000	1341	– 659	– 11.6
2500	1708	– 792	– 0.8
3000	2076	– 924	+ 7.0

Fig. 20.16 Sequential steps of the AGV searching and approaching the marker (time advancing from left-to-right and top-to-bottom)

considering that it was possible to maneuver the vehicle in order to make a first-course approach to the marker and afterwards make a second and more accurate movement, the combined position error achieved was under 10 mm (Fig. 20.16).

20.7 Conclusion

Computer vision combined with a fiducial marker system is said to be an accurate way to combine features in the real world with features generated in virtual models. The ArUco system is a variant of such a system that is optimized for applications with low computing resources; yet it provides fast and accurate results that can be used for augmented reality but also for robotics applications. This project proves that the ArUco can be used for position and pose estimation and can be used with off-the-shelf components to guide a vehicle with a payload of 50 kg with an accuracy close to 1 cm. Further work can be done and this technology reveals good potential for application both in the field of AGVs and in the field of manipulation robotics.

References

Aguiar, G. T., Oliveira, G. A., Tan, K. H., Kazantsev, N., & Setti, D. (2019). Sustainable implementation success factors of AGVs in the Brazilian industry supply chain management [Conference paper]. *Procedia Manufacturing, 39*, 1577–1586. https://doi.org/10.1016/j.promfg.2020.01.284

Ai, C., Geng, D., Qi, Z., Zheng, L., & Feng, Z. (2021). Research on AGV navigation system based on binocular vision [Conference paper]. In *IEEE international conference on real-time computing and robotics (RCAR 2021)* (pp. 851–856). https://doi.org/10.1109/RCAR52367.2021.9517359

Ali, M., & Khan, W. U. (2010). Implementation issues of AGVs in flexible manufacturing system: A review [Review]. *Global Journal of Flexible Systems Management, 11*(1–2), 55–62. https://doi.org/10.1007/bf03396578

Altmann, W., & Macdonald, D. (2005). *Practical process control for engineers and technicians*. Elsevier/Newnes.

Arques Corrales, P., & Aznar Gregori, F. (2020). Swarm AGV optimization using deep reinforcement learning [Conference paper]. In *3rd international conference on machine learning and machine intelligence (MLMI 2020)* (pp. 65–69). https://doi.org/10.1145/3426826.3426839

Ball, N. R. (1996). Application of a neural network based classifier system to AGV obstacle avoidance. Mathematics and Computers in Simulation, 41(3–4), 285–296. https://doi.org/10.1016/0378-4754(95)00078-x.

Bao, J., Luo, Z., & Li, D. (2019). Research on omnidirectional motion of AGV based on vision guidance [Conference paper]. In *10th international symposium on precision engineering measurements and instrumentation (ISPEMI 2018)* (Vol. 11053, p. 1105335). https://doi.org/10.1117/12.2512102

BlueBotics. (2022a). *AGV navigation methods 1: Line following and tags*. Retrieved 2022, August 12 from https://bluebotics.com/agv-navigation-line-following-tags/

BlueBotics (2022b). *AGV navigation methods 2: Virtual path following*. Retrieved 2022, August 12 from https://bluebotics.com/agv-navigation-methods-virtual-path-following/

Bosnak, M., & Skrjanc, I. (2021). Obstacle avoidance for line-following AGV with local maps [Conference paper]. In *IEEE 15th international symposium on applied computational intel-*

ligence and informatics, proceedings (SACI 2021) (pp. 193–198). https://doi.org/10.1109/SACI51354.2021.9465550

Cai, J., Cai, H., Zhang, G., & Zhou, Z. (2011). Guidance line identification algorithm for AGV in complex environment. *Nongye Jixie Xuebao/Transactions of the Chinese Society of Agricultural Machinery, 42*(10), 20–24+89.

Cao, M. L., & Peng, Z. (2020). Research on loading and unloading path optimization for AGV at automatic container terminal based on improved particle swarm algorithm [Conference paper]. In *Proceedings of the 4th annual international conference on data science and business analytics (ICDSBA 2020)* (pp. 1–5). https://doi.org/10.1109/ICDSBA51020.2020.00010

Chen, T. D. (2010). Fast computing scheme for AGV obstacle distance measure and road recognition [Conference paper]. *Advanced Materials Research, 108–111*, 500–506. https://doi.org/10.4028/www.scientific.net/AMR.108-111.500

Chen, C., Wang, B., & Ye, Q. T. (2004). Application of automated guided vehicle (AGV) based on inductive guidance for newsprint rolls transportation system. *Journal of Dong Hua University (English Edition), 21*(2), 88–92.

Cheshire, K. (1985). Some safety aspects of AGVs [Conference paper]. In *Proceedings of the 3rd international conference on automated guided vehicle systems* (pp. 309–315).

Corke, P. (2017). *Robotics, vision and control: Fundamental algorithms in MATLAB* (2nd ed.). Springer.

Cucchiara, R., Perini, E., & Pistoni, G. (2007). Efficient stereo vision for obstacle detection and AGV navigation [Conference paper]. In *14th international conference on image analysis and processing (ICIAP 2007)* (pp. 291–296, 4362794). https://doi.org/10.1109/ICIAP.2007.4362794

Cui, G., Bai, Y., & Li, S. (2021). AGV research based on inertial navigation and vision fusion [Conference paper]. In *5th CAA international conference on vehicular control and intelligence (CVCI 2021)*. https://doi.org/10.1109/CVCI54083.2021.9661175

Culley, G., & Baldur, R. (1988). Free wheel approach to AGV navigation [Conference paper]. *Proceedings of Computers in Engineering, 1988*, 267–274.

De Ryck, M., Versteyhe, M., & Debrouwere, F. (2020). Automated guided vehicle systems, state-of-the-art control algorithms and techniques [Review]. *Journal of Manufacturing Systems, 54*, 152–173. https://doi.org/10.1016/j.jmsy.2019.12.002

Dehnavi-Arani, S., Sabaghian, A., & Fazli, M. (2019). A job shop scheduling and location of battery charging storage for the automated guided vehicles (AGVs). *Journal of Optimization in Industrial Engineering, 12*(2), 121–129. https://doi.org/10.22094/JOIE.2018.543203.1511

Digani, V., Caramaschi, F., Sabattini, L., Secchi, C., & Fantuzzi, C. (2014). Obstacle avoidance for industrial AGVs [Conference paper]. In *Proceedings of the IEEE 10th international conference on intelligent computer communication and processing (ICCP 2014)* (pp. 227–232). https://doi.org/10.1109/ICCP.2014.6937001

Djordjevich, A., Tso, S. K., Zhu, H. Y., & Pjevalica, V. (1999). AGV guidance by fiberoptic tactility [Conference paper]. In *Proceedings of the 1999 sensors and controls for intelligent machining and manufacturing mechatronics* (Vol. *3832*, pp. 206–211).

Egemin. (2014). *Automatic guided vehicles (AGVs): Types*. Retrieved 2022, August 12 from https://web.archive.org/web/20140329035738/http://www.egeminusa.com/pages/agvs/agv_types.html

Feledy, C., & Luttenberger, M. S. (2017). *A state of the art map of the AGVS technology and a guideline for how and where to use it*. MSc, Lund University.

Feng, T., & Jiao, B. (2017). The vision guidance and image processing of AGV [Conference paper]. *Journal of Physics: Conference Series, 887*(1), 012093. https://doi.org/10.1088/1742-6596/887/1/012093

Fragapane, G., de Koster, R., Sgarbossa, F., & Strandhagen, J. O. (2021). Planning and control of autonomous mobile robots for intralogistics: Literature review and research agenda [review]. *European Journal of Operational Research, 294*(2), 405–426. https://doi.org/10.1016/j.ejor.2021.01.019

Fuentes-Pacheco, J., Ruiz-Ascencio, J., & Rendón-Mancha, J. M. (2015). Visual simultaneous localization and mapping: A survey. *Artificial Intelligence Review, 43*(1), 55–81. https://doi.org/10.1007/s10462-012-9365-8

Ganesharajah, T., Hall, N. G., & Sriskandarajah, C. (1998). Design and operational issues in AGV-served manufacturing systems. *Annals of Operations Research, 76*, 109–154. https://doi.org/10.1023/a:1018936219150

Garrido-Jurado, S., Muñoz-Salinas, R., Madrid-Cuevas, F. J., & Marín-Jiménez, M. J. (2014). Automatic generation and detection of highly reliable fiducial markers under occlusion. *Pattern Recognition, 47*(6), 2280–2292. https://doi.org/10.1016/j.patcog.2014.01.005

Hoff, E. B., & Sarker, B. R. (1998). An overview of path design and dispatching methods for automated guided vehicles [Review]. *Integrated Manufacturing Systems, 9*(5), 296–307. https://doi.org/10.1108/09576069810230400

Hollingum, J. (1998). SmartCart is low-cost AGV. *Assembly Automation, 18*(3), 213–214. https://doi.org/10.1108/01445159810224824

Hu, X., Luo, Z., & Jiang, W. (2020). AGV localization system based on ultra-wideband and vision guidance. *Electronics, 9*(3), 448. https://doi.org/10.3390/electronics9030448

Iñigo, R. M., & Alley, D. (1991). Algorithms for path planning, navigation and guidance of an AGV. *Robotics and Autonomous Systems, 7*(4), 309–326. https://doi.org/10.1016/0921-8890(91)90062-P

Isrofi, A., Setiawan, B., & Wibowo, S. (2021). Automated guided vehicle (AGV) navigation control using matrix method applying radio frequency identification (RFID) point address [Conference paper]. In *Proceedings of the 1st international conference on electrical and information technology (IEIT 2021)* (pp. 139–146). https://doi.org/10.1109/IEIT53149.2021.9587440.

Jiang, J., Zhang, S., & Zhu, W. (2020). Structural optimization of multi-functional heavy-load AGV based on distribution method of bearing forces. *Jisuanji Jicheng Zhizao Xitong/Computer Integrated Manufacturing Systems, 26*(8), 2083–2091. https://doi.org/10.13196/j.cims.2020.08.008

Kang, W., Lee, E., Park, J., Heo, M., & Sim, E. (2010). Safety improvement test of a GPS based AGV system using a dead reckoning algorithm [Conference paper]. In *Proceedings of 2010 Asia-Pacific international symposium on aerospace technology (APISAT 2010)* (pp. 656–659).

Kusche, J. (1996). On AGV's navigation in 3-dimensional space. *Journal of Geodesy, 70*(4), 183–187. https://doi.org/10.1007/BF00873699

Li, B., Zhang, Y., Ouyang, Y., Liu, Y., Zhong, X., Cen, H., & Kong, Q. (2021). Fast trajectory planning for AGV in the presence of moving obstacles: A combination of 3-dim A∗ search and QCQP [Conference paper]. In *Proceedings of the 33rd Chinese control and decision conference (CCDC 2021)* (pp. 7549–7554). https://doi.org/10.1109/CCDC52312.2021.9602686

Liang, Y., Wang, H., Ji, M., & Wang, Y. (2021). Quantized preview control of networked AGV systems with data dropout [Conference paper]. In *2021 China automation congress (CAC 2021)* (pp. 2675–2680). https://doi.org/10.1109/CAC53003.2021.9727494

Lin, P. T., Liao, C., & Liang, S. (2021). Probabilistic indoor positioning and navigation (PIPN) of autonomous ground vehicle (AGV) based on wireless measurements. *IEEE Access, 9*, 25200–25207. https://doi.org/10.1109/ACCESS.2021.3057415

Lin, Y., Xu, Y., Zhu, J., Wang, X., Wang, L., & Hu, G. (2023). MLATSO: A method for task scheduling optimization in multi-load AGVs-based systems. *Robotics and Computer-Integrated Manufacturing, 79*, 102397. https://doi.org/10.1016/j.rcim.2022.102397

Liu, J., Wang, Y., Sheng, J., Zhang, Y., Qi, J., & Yu, L. Q. (2020). Design of AGV positioning navigation control system based on vision and RFID [Conference paper]. *Advances in Intelligent Systems and Computing, 1063*, 16–24. https://doi.org/10.1007/978-3-030-31967-0_2

Lu, W., Guo, S., Song, T., & Li, Y. (2022). Analysis of multi-AGVs management system and key issues: A review [Review]. *CMES – Computer Modeling in Engineering and Sciences, 131*(3), 1197–1227. https://doi.org/10.32604/cmes.2022.019770

Lynch, L., Newe, T., Clifford, J., Coleman, J., Walsh, J., & Toal, D. (2018). Automated ground vehicle (AGV) and sensor technologies – A review. In *12th international conference on sensing technology (ICST)* (pp. 347–352). https://doi.org/10.1109/ICSensT.2018.8603640

Małopolski, W. (2018). A sustainable and conflict-free operation of AGVs in a square topology. *Computers and Industrial Engineering, 126*, 472–481. https://doi.org/10.1016/j.cie.2018.10.002

Man, Z. G., Ye, W. H., Zhao, P., Lou, P. H., & Wu, T. J. (2010). Research on RFID and vision-based AGV navigation [Conference paper]. *Advanced Materials Research, 136*, 298–302. https://doi.org/10.4028/www.scientific.net/AMR.136.298

Mohammadpour, M., Zeghmi, L., Kelouwani, S., Gaudreau, M. A., Amamou, A., & Graba, M. (2021). An investigation into the energy-efficient motion of autonomous wheeled mobile robots [Review]. *Energies, 14*(12), 3517. https://doi.org/10.3390/en14123517

Nanthavanij, S., Yenradee, P., & Techapichetvanich, K. (1995). An intelligent safety feature for AGV's economic operations: A simulation analysis. *Journal of Human Ergology, 24*(1), 73–80.

Norton, A., Gavriel, P., & Yanco, H. (2019). A standard test method for evaluating navigation and obstacle avoidance capabilities of AGVs and AMRs. *Smart and Sustainable Manufacturing Systems, 3*(2), 106–126. https://doi.org/10.1520/SSMS20190028

Olmi, R. (2011). *Traffic management of automated guided vehicles in flexible manufacturing systems* PhD, Università degli Studi di Ferrara.

Oyekanlu, E. A., Smith, A. C., Thomas, W. P., Mulroy, G., Hitesh, D., Ramsey, M., et al. (2020). A review of recent advances in automated guided vehicle technologies: Integration challenges and research areas for 5G-based smart manufacturing applications [Review]. *IEEE Access, 8*, 202312–202353. https://doi.org/10.1109/ACCESS.2020.3035729

Pan, Y. (2021). Challenges in visual navigation of AGV and comparison study of potential solutions [Conference paper]. In *2021 international conference on signal processing and machine learning (CONF-SPML 2021)* (pp. 265–270). https://doi.org/10.1109/CONF-SPML54095.2021.00058

Pian, C. Y., Liu, J., & Zhang, L. (2019). Estimation and simulation of charged state of battery for AGV. *International Journal of Mechatronics and Applied Mechanics, 2*(6), 120–126.

Praserttaweelap, R., Kaitwanidvilai, S., & Aoyama, H. (2019). Safety path planning with obstacle avoidance using particle swarm optimization for agv in manufacturing layout. *International Journal of Innovative Computing, Information and Control, 15*(1), 351–368. https://doi.org/10.24507/ijicic.15.01.351

Rajagopalan, R., Cheng, R. M. H., & Lequoc, S. (1992a). Guidance control scheme employing knowledge-base for AGV navigation [Conference paper]. In *Proceedings of the American control conference* (Vol. 2, pp. 942–947). https://doi.org/10.23919/acc.1992.4792224

Rajagopalan, R., Cheng, R. M. H., & Lequoc, S. (1992b). A guidance control scheme for accurate track following of AGVs [Conference paper]. In *IEEE international conference on robotics and automation* (Vol. 1, pp. 188–193). IEEE.

Ray, A. K., Gupta, M., Behera, L., & Jamshidi, M. (2008). Sonar based Autonomous Automatic Guided Vehicle (AGV) navigation. In *IEEE International Conference on System of Systems Engineering, SoSE 2008* (pp. 1–6). IEEE. https://doi.org/10.1109/SYSOSE.2008.4724179

Reis, W. P. N., Couto, G. E., & Junior, O. M. (2022). Automated guided vehicles position control: A systematic literature review [Review]. *Journal of Intelligent Manufacturing, 34*, 1483–1545. https://doi.org/10.1007/s10845-021-01893-x

Sabattini, L., Cardarelli, E., Digani, V., Secchi, C., & Fantuzzi, C. (2016). Multi-AGV systems in shared industrial environments: Advanced sensing and control techniques for enhanced safety and improved efficiency [Conference paper]. *Symposium on Autonomous Industrial Vehicles: From the Laboratory to the Factory Floor, ASTM Special Technical Publication STP, 1594*, 57–81. https://doi.org/10.1520/STP159420150052

Sauer, C., Lyczkowski, E., Schmidt, M., Nüchter, A., & Hoßfeld, T. (2022). Testing AGV mobility control method for MANET coverage optimization using procedural simulation. *Computer Communications, 194*, 189–201. https://doi.org/10.1016/j.comcom.2022.07.033

Schlieck, D. (1986). AGVs with unlimited battery power supply [conference paper]. In *Proceedings of the 3rd international conference on automated materials handling* (pp. 141–149).

Schmidt, J., Meyer-Barlag, C., Eisel, M., Kolbe, L. M., & Appelrath, H. J. (2015). Using battery-electric AGVs in container terminals – Assessing the potential and optimizing the economic viability. *Research in Transportation Business and Management, 17*, 99–111. https://doi.org/10.1016/j.rtbm.2015.09.002

Shirai, Y. (1990). Flexible path selection and obstacle detection for AGV [Conference Paper]. In *Proceedings of the IEEE international workshop on intelligent motion control (IMC 1990)* (Vol. 1, IP-39-IP-44, 687278). https://doi.org/10.1109/IMC.1990.687278

Song, Z., Wu, X., Xu, T., Sun, J., Gao, Q., & He, Y. (2016). A new method of AGV navigation based on Kalman Filter and a magnetic nail localization [Conference Paper]. In *IEEE international conference on robotics and biomimetics (ROBIO 2016)* (Vol. 952–957, p. 7866447). https://doi.org/10.1109/ROBIO.2016.7866447

Stachniss, C. (2009). *Robotic mapping and exploration*. Springer.

Stefanini, R., & Vignali, G. (2022). Environmental and economic sustainability assessment of an industry 4.0 application: The AGV implementation in a food industry. *International Journal of Advanced Manufacturing Technology, 120*(5–6), 2937–2959. https://doi.org/10.1007/s00170-022-08950-6

Stimming, C., Krengel, A., Boehning, M., Vatavu, A., Mandici, S., & Nedevschi, S. (2015). Multi-level on-board data fusion for 2D safety enhanced by 3D perception for AGVs [Conference paper]. In *Proceedings of the IEEE 11th international conference on intelligent computer communication and processing (ICCP 2015)* (Vol. 239–244, p. 7312636). https://doi.org/10.1109/ICCP.2015.7312636

Sun, J., Yu, Y., & Xin, L. (2021). Research on Path Planning of AGV Based on Improved Ant Colony Optimization Algorithm [Conference paper]. In *Proceedings of the 33rd Chinese Control and Decision Conference (CCDC 2021)* (pp. 7567–7572). https://doi.org/10.1109/CCDC52312.2021.9601807

Tang, Y., Hao, H., Zhou, J., Lin, Y., & Dong, Z. (2021). Research on outdoor AGV precise navigation based on BDS/INS data fusion. *Journal of Intelligent and Fuzzy Systems, 41*(3), 4295–4306. https://doi.org/10.3233/JIFS-189690

Teso-Fz-Betoño, D., Zulueta, E., Fernandez-Gamiz, U., Aramendia, I., & Uriarte, I. (2019). A free navigation of an AGV to a non-static target with obstacle avoidance. *Electronics, 8*(2), 159. https://doi.org/10.3390/electronics8020159

Tracey, P. M. (1985). AGV safety, new developments to meet changing needs [Conference paper]. In *Proceedings of the 3rd international conference on automated guided vehicle systems* (pp. 47–55).

Turpin, D. R. (1986). Inertial guidance: Is it a viable guidance system for AGVS? [Conference paper]. In *Proceedings of the 4th international conference on automated guided vehicle systems: AGVS-4* (pp. 301–320).

Ullrich, G. (2015). *Automated guided vehicle systems: A primer with practical applications* (2nd ed.). Springer.

Vijayaram, T. R. (2006). Materials handling technology and significance of expert systems to select appropriate handling equipments in engineering industries: A review. *Journal of Scientific and Industrial Research, 65*(8), 619–624.

Vis, I. F. A. (2006). Survey of research in the design and control of automated guided vehicle systems. *European Journal of Operational Research, 170*(3), 677–709. https://doi.org/10.1016/j.ejor.2004.09.020

Wang, T. C., Tong, C. S., & Xu, B. L. (2020). AGV navigation analysis based on multi-sensor data fusion. *Multimedia Tools and Applications, 79*(7–8), 5109–5124. https://doi.org/10.1007/s11042-018-6336-3

Wang, J., Pan, J., Huo, J., Wang, R., Li, L., & Nian, T. (2021). Research on optimization of multi-AGV path based on genetic algorithm considering charge utilization [Conference paper]. *Journal of Physics: Conference Series, 1769*(1), 012052. https://doi.org/10.1088/1742-6596/1769/1/012052

Wikipedia. (2022, July 20). *Automated guided vehicle*. Wikipedia. Acessed in 2022, August 11. https://en.wikipedia.org/wiki/Automated_guided_vehicle

Wu, H., Luo, Y., Zhou, Q., & Li, J. (2017). Obstacle avoidance strategy of AGV formation based on time efficiency first [Conference Paper]. In *Proceedings of the IEEE 2nd advanced informa-*

tion technology, electronic and automation control conference (IAEAC 2017) (pp. 772–777). https://doi.org/10.1109/IAEAC.2017.8054120

Wu, X., Zhai, J., Xiao, H., Hu, Y., & Lou, P. (2022). Deadlock-free intersection-access sequence optimization and collision avoidance for multi-load AGV system. *Jisuanji Jicheng Zhizao Xitong/Computer Integrated Manufacturing Systems, 28*(4), 979–989. https://doi.org/10.13196/j.cims.2022.04.002

Xie, M. (1995). Trinocular vision for AGV guidance: Path localization and obstacle detection. *Computers and Electrical Engineering, 21*(6), 441–452. https://doi.org/10.1016/0045-7906(95)00021-L

Xie, D., Guo, D., Ji, Y., & Gao, Z. (2020). Simulation research on optimization of AGV charging strategy for automated terminal. *Xitong Fangzhen Xuebao/Journal of System Simulation, 32*(10), 1927–1935. https://doi.org/10.16182/j.issn1004731x.joss.20-FZ0299E

Yadav, A., Gaur, A., Jain, S. M., Chaturvedi, D. K., & Sharma, R. (2018). Development navigation, guidance & control program for GPS based autonomous ground vehicle (AGV) using soft computing techniques [Conference paper]. *Materials Today: Proceedings, 29*, 530–535. https://doi.org/10.1016/j.matpr.2020.07.309

Yang, X., Zhao, X., & Huang, C. (2017). Safety issues and requirements of AGV [Conference paper]. In *49th international symposium on robotics (ISR 2017).*

Yang, Y., Quan, Y., & He, Y. (2021). Research on multi-AGV management system of autonomous navigation AGVs for manufacturing environment [Conference paper]. *Journal of Physics: Conference Series, 1910*(1), 012025. https://doi.org/10.1088/1742-6596/1910/1/012025

Yao, X., & Zeng, L. (2012). Experiment study of AGV navigation based on multi-sensor [Conference paper]. *Advanced Materials Research, 472–475*, 484–487. https://doi.org/10.4028/www.scientific.net/amr.472-475.484

Zhou, X., Jin, L., & Liu, Y. (2021). Modeling and simulation research of heavy-duty AGV tracking control system based on magnetic navigation [Conference paper]. *Journal of Physics: Conference Series, 1746*(1), 012021. https://doi.org/10.1088/1742-6596/1746/1/012021

Zou, W. Q., Pan, Q. K., Wang, L., Miao, Z. H., & Peng, C. (2022). Efficient multiobjective optimization for an AGV energy-efficient scheduling problem with release time. *Knowledge-Based Systems, 242*, 108334. https://doi.org/10.1016/j.knosys.2022.108334

Part VIII

Sustainability & Innovation

21 Plastic and Copper Reduction in Residential Building, an Empirical Analysis on Materials Usage and Energy Efficiency

Nelson Castro Neves and José Castro Oliveira

21.1 Introduction

From food packaging to surgery and construction, plastic is everywhere nowadays. The synthetic plastic we use today started its life as Bakelite at the beginning of the twentieth century and was further developed for multiple applications after World War II and is currently derived from fossil hydrocarbons, having been used as one of the most man-made materials to date (Geyer et al., 2017). The amount of plastic used in domestic and industrial usage generates large amounts of waste, making it one of the world's biggest environmental concerns.

Recycling plastic is a complicated task due to the diversity of applications, density, and other properties, even with a straight recovery of plastic, to be economically feasible it requires separation and processing by plastic type (Corbitt, 1999). Being so, and to keep up with the plastic waste generated from the global demand, landfill is one of the possibilities, since plastic is very stable on decomposition (Akah et al., 2015). On the other hand, the lack of available space for landfill and its operational costs are leading to recycling options to avoid pollution and recover raw materials which can be reused (Aguado & Serrano, 1999).

Plastic pollution has become one of the most pressing environmental issues as the rapid increase in the production of single-use plastic items strains the world's ability to deal with them.

N. C. Neves (✉)
ISPGAYA, Vila Nova de Gaia, Portugal
e-mail: nneves@ispgaya.pt

J. C. Oliveira
REMIT – Research on Economics, Management and Information Technologies,
Vila Nova de Gaia, Portugal

© The Author(s), under exclusive license to Springer Nature Switzerland AG 2023

F. L. Almeida et al. (eds.), *Multidimensional Sustainability: Transitions and Convergences*, Springer Proceedings in Earth and Environmental Sciences,
https://doi.org/10.1007/978-3-031-24892-4_21

Plastic pollution is most visible in developing countries in Asia and Africa, where waste collection systems are often inefficient or nonexistent. However, in the developed world, especially in countries with low recycling rates, there are also problems in collecting and discarding plastics properly (National Geographic, 2019).

In addition to plastic reduction, it is important to introduce the concept of a Circular Economy. There are several ways to define Circular Economy (Kirchherr et al., 2017), and some fall within the scope of this project. In the production of goods, significant volumes of materials are commonly lost in the chain between mining and final manufacturing. For instance, the Sustainable Europe Research Institute (SERI) estimates that each year the manufacturing of products in OECD countries consumes over 21 billion tons of materials that aren't physically incorporated into the products themselves (Ellen MacArthur Foundation, 2013).

One of the sources of waste is the general obsolescence of products. This, when not programmed, occurs due to variations in the supply and demand relationship. In any case, the obsolescence of products causes the early appearance of waste, namely, plastic (Sandborn, et al., 2007). Thus, as we will see, one of the project's stakeholders defines itself as capable of guaranteeing the value chain to delay as much as possible the obsolescence of lighting fixtures, guaranteeing the supply of consumables, in this case, LED lamps, extending the warranty period of the products supplied.

The aforementioned value chain is supported by the ability of lighting fixtures to connect to a Wi-Fi network, which can signal the occurrence of faults. With small parameterizations, the occurrence of the fault can be notified to the registered supplier of the same equipment, with the prior authorization of the user. Thus, the solution falls within the scope of the Internet of Things (IoT) concept (Guo, et al., 2020). This aspect increases the efficiency of the lighting solution, increasing the overall sustainability of the solution (Casini, 2014).

The present conceptual research has two main objectives: (1) Design and apply intelligent design in domestic lighting aiming the reduction of plastic, and other metals and (2) Contribute to the literature enhancement, as there is no available literature on plastics reduction for domestic lighting.

This chapter is organized as follows: It starts with an introduction to general plastics and lighting, and in Sect. 21.2, a brief literature review addressing the general use of plastic. Section 21.3 focuses on the methodology used in the reduction of plastic and metals in a real application project. Section 21.4 discusses the initial results obtained in the project. Finally, Sect. 21.5 presents the general conclusions, contributions, limitations, and clues for future projects.

21.2 Literature Review

According to an OECD report (2018), "Plastics are widely used materials that deliver a range of important benefits to society. Their global production and use are expected to increase fourfold to 2050." Being so, a constant rise is expected in

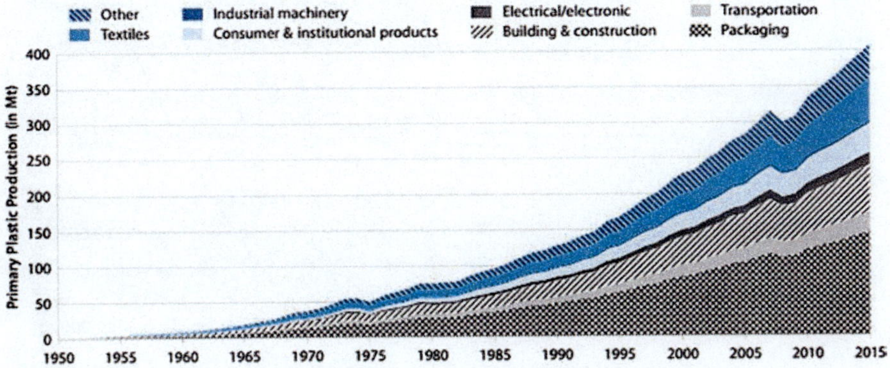

Fig. 21.1 Global primary plastics production. (Source: Geyer et al., 2017)

plastic production and consumption in all areas and sectors; one may ask, is there any sector in particular that stands out in plastic production? According to Geyer et al. (2017), domestic and industrial packaging leads the production rank, followed by building and construction (Fig. 21.1).

On the other hand, another question is posed, what to do with plastic waste after its lifetime? According to Geyer et al. (2017), the majority of the produced plastic is either discarded or incinerated, the remaining part is recycled (Fig. 21.2).

Since recycling plastic may delay the process instead of avoiding its final disposal, if the current production and waste management continue within this trend, it is expected that by 2050, 12,000 Mt. will be deposited in landfills or spread over the environment (Geyer et al., 2017).

Taking the above problems into consideration, it makes more sense to think of solutions to avoid reaching critical values of plastic waste, this can be (1) intelligent use of available technologies and (2) considering the clever design to avoid plastic production and use (Geyer et al., 2017).

21.3 Methodology

Assuming the second above option (intelligent use of available technologies), the authors designed a feasible project together with the most relevant stakeholders (academia, industry, and municipality) to install intelligent lighting instead of normal domestic lighting in a real application project of construction of two apartments of T2 configuration (two bedrooms, kitchen, living room, one WC, and hall).

The apartments are municipality property. The first apartment is being built using traditional domestic lighting (using copper cabling, switches, and Led lights). The second apartment is being built using intelligent lightning (using intelligent Led lights) and reducing as much as possible copper cabling, switches, and other electrical components using plastic.

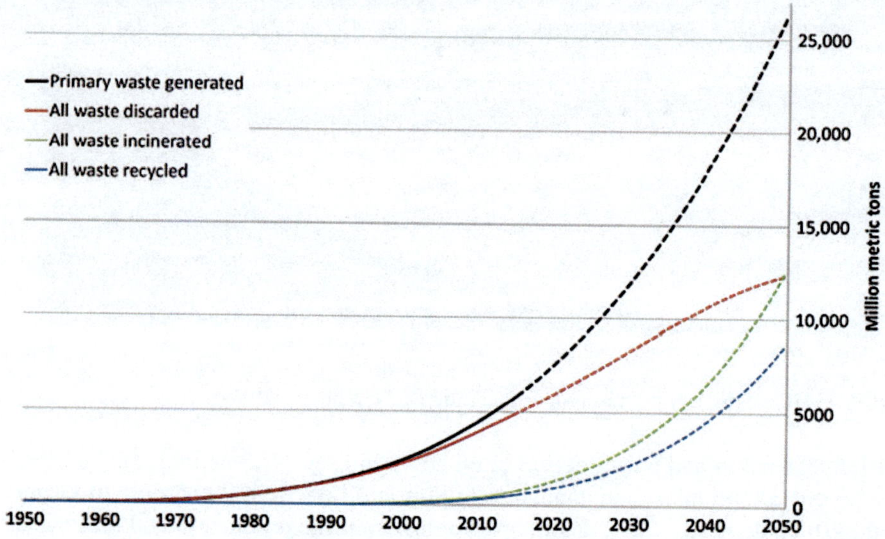

Fig. 21.2 Cumulative plastic waste generation and disposal (in million metric tons). Solid lines show historical data from 1950 to 2015; dashed lines show projections of historical trends to 2050. (Source: Geyer et al., 2017)

At the end of the construction, a certified electric and independent company will provide the final figures and comparison between the two apartments in terms of plastic, copper, and other metals reduction.

As for this chapter, the results obtained are real laboratory figures, using mathematical figures from samples of the material being used in the current construction of the apartments.

This joint project aims to reduce the use of plastic, as well as copper, in new housing construction, or reconstruction by placing intelligent lighting controlled by wireless communication in on/off modes or dimming, not using different components such as switches, piping, and electrical wires used in a classic assembly between control points and junction boxes.

The adoption of this lighting solution makes the concepts of energy efficiency and IoT (Internet of Things) connectivity appear in the project, which promotes efficiency and sustainability. An ordinary smartphone through an application controls the luminaires that will be used in the smart installation.

Another aspect that should be highlighted is the concept of circular economy underlying the project, as the lighting is built using 3D printing by additive manufacturing with recycled plastic. In this way, we have a circular economy in parallel with the reduction of plastic through its recycling in the production of filament for 3D printers (Despeisse et al., 2017).

21.3.1 Stakeholders

This chapter reflects a real project between three relevant stakeholders (academia, industry, and municipality) to achieve plastic and metal reduction in the area of construction.

As for the stakeholders' names, and due to the confidentiality statements required, the involved entities will be described as A, B, and C.

21.3.1.1 Entity A (Academia)

The academia is represented by a Higher Education Politecnic, which exists in Vila Nova de Gaia for over 30 years. It develops higher education courses such as CTeSP (level 4), Bachelors (level 5), and Masters (level 6). The competence for this project comes mainly from the CTeSP courses in Electronics and Industrial Automation and Renewable Energies and Energy Efficiency of the Degree in Electronics and Automation Engineering. In addition to higher education. Moreover, it develops R&D projects applied to the provision of services to the community. In this project, it does all the electric and electronic calculations, laboratory simulations and tests, and project monitoring as part of the monitoring committee.

21.3.1.2 Entity B (Industry)

The industry is represented by a multinational lighting corporation founded at the end of the nineteenth century. It is a world leader in lighting for professionals, domestic consumers, and lighting for the Internet of Things. Its energy-efficient lighting products, systems, and services enable its customers to enjoy the superior light quality and make people's lives safer and more comfortable, making businesses more productive and cities more livable. In this project, it advised and supplies all the intelligent lighting (and its applications) for apartment 2. Furthermore, it provides project monitoring as part of the monitoring committee.

21.3.1.3 Entity C (Municipality)

The municipality chosen for this project is amongst the biggest municipalities in Portugal and offered its services and construction site as part of their Urban Planning, Social Housing, and Urban Rehabilitation department. Its mission is to perform with rigor, competence, and excellence the attributions delegated by the municipality administration. It intends to be a reference of excellence in the services provided to citizens, nationally and internationally, being an open, intelligent, ecological, and innovative municipality. This municipality is part of the European Smart Cities. In this project, it has provided access to their construction site (Apartments 1 and 2). Moreover, it provides project monitoring as part of the monitoring committee.

21.3.2 Presentation of the Object of Study Application

The prototype installation and monitoring (apartments 1 and 2) will be carried out in a building under reconstruction located in Largo Senhora do Monte, parish of Pedroso, Vila Nova de Gaia (Figs. 21.3, 21.4, and 21.5).

The building is intended for multifamily occupancy with one T3 apartment, two T2 apartments, and one T1 apartment. Each fraction consists of a kitchen, living room, one, two, or three bedrooms, and complete sanitary installation. As mentioned above, our research will be carried out on two completely identical apartments (T2 apartments) (Fig. 21.6).

Fig. 21.3 Building location. (Source: GAIURB, 2022)

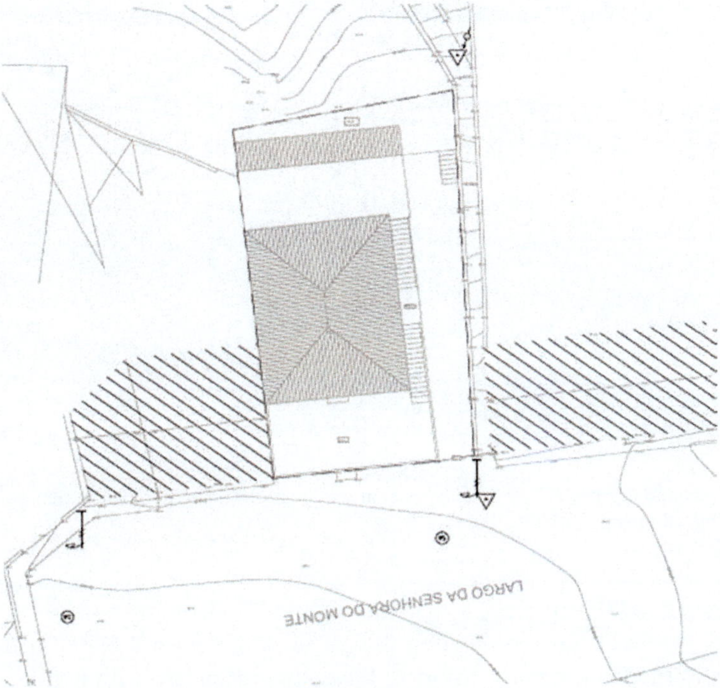

Fig. 21.4 Building detail. (Source: GAIURB, 2022)

Fig. 21.5 An aerial view of the building (to be started). (Source: Google Maps, 2022)

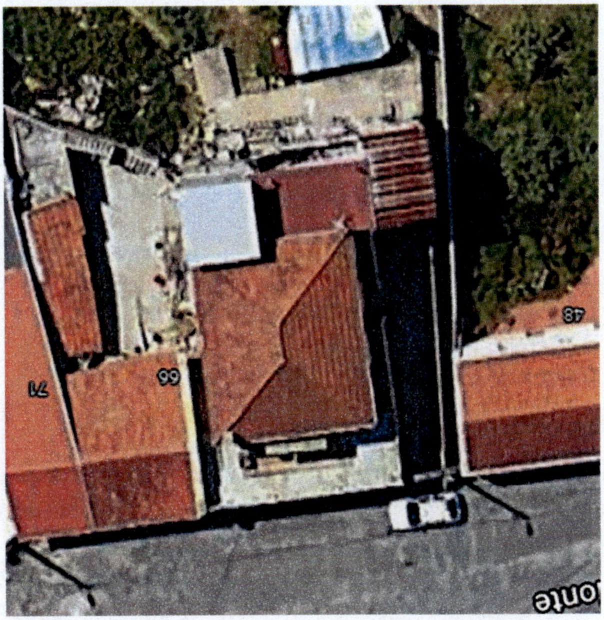

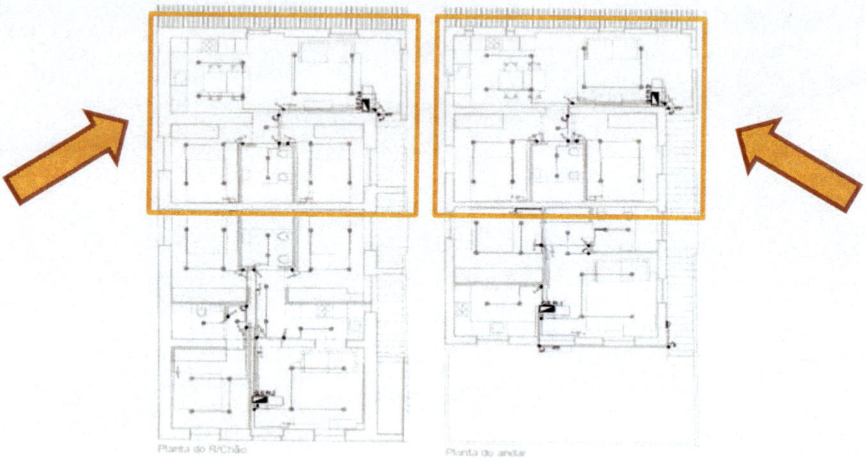

Fig. 21.6 Indication of the T2 apartments, one on the lower floor and another on the upper floor of the building. (Source: GAIURB, 2022)

21.4 Results Discussion

As the reconstruction has not yet started, the entire comparative study of the solutions is based only on the Autocad file plans provided by the municipality services.

According to the reconstruction project engineer, the T2 apartments will have 20 light points, 2 "switches," and 6 "two-way switches." In addition to these materials, we have inventoried a set of plastic or plastic components with metal that make up the electrical circuits, guaranteeing safety according to electrical regulations, e.g., electrical wires, electrical junction boxes, electrical wire routing tubes, switchgear boxes, connectors, and unions.

To estimate the quantities of material to be used in a classic installation, it is necessary to remember how to implement each of the circuits:

- Simple switching: Always used to control a single circuit, with one or more lamps, from a single place. For example, bathroom, kitchen, and pantry.
- Two-way switching: Used to control a single electrical circuit from two different locations. For example, stairs, hallways, and rooms.

Figures 21.7 and 21.8 show the single-line and multiline schemes for each case.

As one of the objectives of the research is to quantify the reduction of plastic and metals through the adoption of wireless technology that allows the switching of light points, it is important to quantify which components are left to exist with the adoption of this same technology.

As can be seen in Figs. 21.7 and 21.8, the elimination of switches has the impact of eliminating electrical wires and pipes in different numbers and distances between

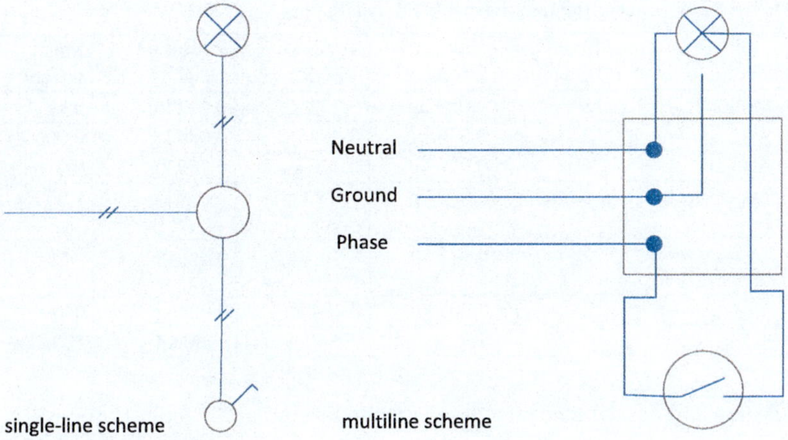

Fig. 21.7 Simple switching connection. (Source: Authors)

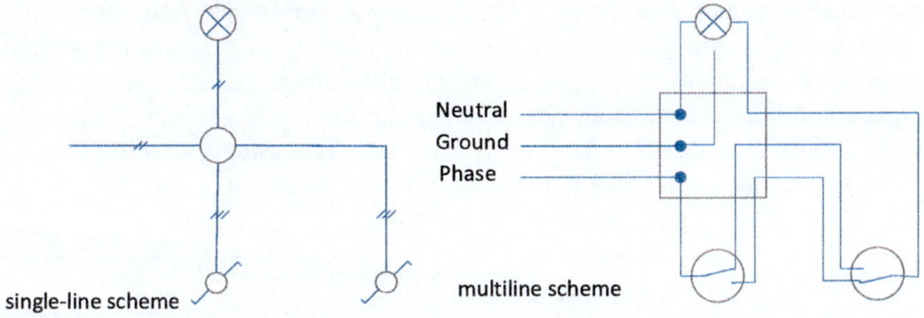

Fig. 21.8 Stair or room switching connection. (Source: Authors)

the switchgear box and the junction box. Thus, the phase and neutral installation between the junction box and the light points remain intact.

Thus, it is clear that the material that is discarded in an electrical installation for wireless-controlled lighting is the following:

- Switchgear boxes.
- Switch or two-way switch.
- Electrical wires, from the Switchgear boxes to the Electrical junction box.
- Electrical wire routing tubes, from the Switchgear boxes to the Electrical junction box.
- Unions.

A unit of each component was purchased, out of which the global weight and the partial weight were estimated considering the plastic and the material.

Table 21.1 summarizes the weight and quantities for each of the components of the lighting electrical circuit considering the electrical diagrams consulted in

Table 21.1 Components of the lighting electrical installation

Electrical component	Quantity	Unit weight	Total weight	Plastic weight	Metal weight
Switchgear boxes	8	0,029 kg	0,232 kg	0,232 kg	0 kg
Switch	2	0,046 kg	0,092 kg	0,086 kg	0,006 kg
Two-way switch	6	0,046 kg	0,276 kg	0,258 kg	0,018 kg
Simple copper electrical wires	53 m	0,030 kg	1914 kg	0,854 kg	1,06 kg
Electrical wire routing tubes	53 m	0,122 kg	6466 kg	6466 kg	0 kg
Unions	28	0,003 kg	0,084 kg	0,084 kg	0 kg
		Totals	9064 kg	7980 kg	1084 kg

Autocad files and descriptive memories attached to the architectural process of the building's recovery project. It also disaggregates between the plastic and metal weight of the analyzed components.

As can be concluded from Table 21.1, in an intelligent lighting installation in a T2 apartment, a reduction of about 9 kg of material can be achieved. Out of this reduction, almost 8 kg is plastic and just over 1 kg is metal. Apart from the positive environmental impact, this type of solutions will slow down the increment in plastic production and usage Geyer et al. (2017). Furthermore, it will encourage a circular economy approach, entities can be seen as part of the ecosystem and will influence the decisions of future consumer purchasing (Castro Oliveira et al., 2022).

21.5 Final Considerations

21.5.1 General Conclusions

Although being in a construction phase, the authors could simulate in a laboratory the expected reduction in plastics and metal for two similar apartments, one using traditional lighting and the other using intelligent lighting.

The results obtained (circa 8Kg reduction in plastic, and about 1Kg in metal) are very promising, and set new approaches in domestic lighting, having in mind plastic and metal reduction, besides encouraging companies to a circular economy approach (Castro Oliveira et al., 2022).

21.5.2 Contributions

This research has contributed to knowledge in two forms: (1) Literature enrichment, and (2) Managerial advice.

From the literature point of view, this research opened a new area in the literature, as from the authors' knowledge the current literature is a generalist and does not focus on this particular area (plastic reduction in building and construction).

As for managerial advice, this research (although with laboratory results) can contribute to new ways to project and build domestic buildings, having in mind a sustainable manner focused on plastic and metals reduction, sustainability, and customer satisfaction.

21.5.3 Limitations

This chapter faced some limitations, first on the literature, as mentioned earlier the available literature is generalist and not focused on the area of research. Being so, some parts of this paper had to be brief (i.e., Literature review).

On the other hand, and since the project is in the building phase, some of the results were obtained through laboratory experimentation and calculation.

21.5.4 Clues for Future Projects

Since this is a new area, having in mind construction and materials reduction, this research can be applied to other types of buildings.

Furthermore, when applied to civil and electrical engineering, the actual results may lead to new formulas to calculate material reduction (plastic and others).

References

Aguado, J., & Serrano, D. P. (1999). *Feedstock recycling of plastic wastes*. RSC Clean Tech/Monographs.

Akah, A., Hernandez-Martinezb, J., Rallan, C., & Garforth, A. (2015). Enhanced feedstock recycling of post consumer plastic waste. *Chemical Engineering Transactions, 43*, 2395–2400. https://doi.org/10.3303/CET1543400

Casini, M. (2014). Internet of things for energy efficiency of buildings. *International Scientific Journal Architecture and Engineering, 2*(1), 24–28. http://architecture.scientific-journal.com/articles/2/4.pdf

Castro Oliveira, J., Lopes, J. M., Farinha, L., Silva, S., & Luízio, M. (2022). Orchestrating entrepreneurial ecosystems in circular economy: The new paradigm of sustainable competitiveness. *Management of Environmental Quality, 33*(1), 103–123. https://doi.org/10.1108/MEQ-11-2020-0271

Corbitt, R., & A. (1999). *Standard handbook of environmental engineering* (2nd ed.). McGRAW-HILL. https://www.accessengineeringlibrary.com/content/book/9780070131606

Despeisse, M., Baumers, M., Brown, P., Charnley, F., Ford, S. J., Garmulewicz, A., et al. (2017). Unlocking value for a circular economy through 3D printing: A research agenda. *Technological Forecasting and Social Change, 115*, 75–84. https://doi.org/10.1016/j.techfore.2016.09.021

Ellen MacArthur Foundation. (2013). Towards the circular economy. *Journal of Industrial Ecology, 2*(1), 23–44. https://www.werktrends.nl/app/uploads/2015/06/Rapport_McKinsey-Towards_A_Circular_Economy.pdf

Geyer, R., Jambeck, J. R., & Law, K. L. (2017). Production, use, and fate of all plastics ever made. *American Association for the Advancement of Science, Science Advances, 3*, 1700782. https://doi.org/10.1126/sciadv.1700782

Guo, Y., Wang, N., Xu, Z. Y., & Wu, K. (2020). The internet of things-based decision support system for information processing in intelligent manufacturing using data mining technology. *Mechanical Systems and Signal Processing, 142*, 106630. https://doi.org/10.1016/j.ymssp.2020.106630

Kirchherr, J., Reike, D., & Hekkert, M. (2017). Conceptualizing the circular economy: An analysis of 114 definitions. *Resources Conservation and Recycling, 127*, 221–232. https://doi.org/10.1016/j.resconrec.2017.09.005

National Geographic. (2019). https://www.nationalgeographic.com/environment/article/plastic-pollution

OECD. (2018). *Policy highlights improving markets for recycled plastics – Trends, prospects and policy responses*. Available at: https://search.oecd.org/environment/waste/Policy-Highlights-Improving-Markets-for-Recycled-Plastics-Trends-Prospects-and-Policy-Response.pdf

Sandborn, P. A., Mauro, F., & Knox, R. (2007). A data mining based approach to electronic part obsolescence forecasting. *IEEE Transactions on Components and Packaging Technologies, 30*(3), 397–401. https://doi.org/10.1109/TCAPT.2007.900058

Sustainability in Society 5.0 Perspective

22

Zehranur Sanioğlu Tanış

22.1 Introduction

We live in a socially challenging, uncertain, and complex age. Increasing world population, industrialization and excessive consumption habits of people have increased the need for energy. In order to meet this energy need, people have recklessly used natural resources, leading to their depletion, environmental degradation, and global disasters. People are starting to understand that they need to take action to prevent all these problems that the world is facing. In order to be protected from the destructive effects of the digital transformation experienced with Industry 4.0 and to benefit from its constructive effects at the highest level, the Society 5.0 approach, which puts people and society at the center, has emerged. The Society 5.0 philosophy is to transform states, institutions, and society into a smart society. It paves the way for energy, cities, regions, disaster prevention and mitigation, agriculture and food, logistics, manufacturing service, finance, healthcare, and public services. While realizing these, it is also to ensure that the Sustainable Development Goals of the United Nations can be achieved. In this direction, in this study, we first tried to give information about the philosophy of Society 5.0 and the sustainable development approach. Then, we examined and summarized the results of different studies dealing with Society 5.0 practices that mediate the achievement of Sustainable Development Goals.

Z. Sanioğlu Tanış (✉)
Akşehir Faculty of Economics and Administrative Sciences, Department of Business Administration, Selcuk University, Konya, Turkey
e-mail: zsanioglu@selcuk.edu.tr

© The Author(s), under exclusive license to Springer Nature Switzerland AG 2023

F. L. Almeida et al. (eds.), *Multidimensional Sustainability: Transitions and Convergences*, Springer Proceedings in Earth and Environmental Sciences,
https://doi.org/10.1007/978-3-031-24892-4_22

22.2 Society 5.0 and Sustainable Development

Society 5.0 has emerged as a philosophy that focuses on society by providing solutions with new technologies in all sectors and social activities to ensure sustainable development (Higashihara, 2018; Keidanren Policy and Action, 2016; Nakanishi, 2019). The Society 5.0 model aims to create a human-centered society and uses technology and innovation to achieve this (Higashihara, 2018). In 2016, the Japanese government launched what it calls a "Super Intelligent Society" or "Society 5.0" to tackle the problems it faces such as the declining birthrate, aging population, and deteriorating infrastructure (Council of Science, Technology and Innovation, 2016). Society 5.0 follows Industry 4.0 to some extent. However, it aims to contribute to the quality of life, social welfare, sustainability, and economic growth by taking the transformation realized with Industry 4.0 beyond the industry. In order to achieve this, it tries to create "Super Smart Societies" where new knowledge and values are revealed by making more use of the results of Industry 4.0 and technology (Ferreira & Serpa, 2018; Fukuda, 2020).

One of the focal points of Society 5.0 is to combine cyberspace and physical space to provide appropriate infrastructures for a sustainable life (Higashihara, 2018; Keidanren, 2019). In other words, the Society 5.0 model predicts the creation of a sustainable society where various values are interconnected by cyber-physical systems and where people can live under safe and comfortable conditions (Shiroishi et al., 2018). In the report prepared by Keidanren, the Japanese Federation of Economic Organizations, the main objectives of Society 5.0 are: developing solutions against the aging world population, making the virtual world and the real-world work together, making use of the Internet of Things by considering the interests of the society, and producing solutions for environmental pollution and natural disasters. Again, in the same report, the obstacles to be faced in the creation of Society 5.0 are listed as follows (Keidanren, 2016):

- Barriers in the legal system
- Scientific gaps in the digitization of objects
- Lack of qualified personnel
- Sociopolitical prejudices
- Social resistance

In order to overcome all these obstacles, five steps have been determined. First of all, it was stated that national strategies should be created by *Ministries and Agencies* and the state incentive system should be integrated. It was emphasized that the next step was to develop laws suitable for new technologies by emphasizing the *Legal System*. In the third step, serious research and development studies on technologies such as cyber security, bio, nano, and robot were made and the importance of *Technology* development and use was explained. In the fourth step, it was stated that *Human Resources* should be developed by making educational reforms such as information technology literacy and equipping existing manpower with advanced numerical skills. In the last step, it was understood that Society 5.0 is also relevant to citizens, governments, academic circles, and all other stakeholders, and its *Scope* has been revealed.

Fig. 22.1 Society 5.0 and the Sustainable Development Goals (Source: http://www.keidanren.or.jp/policy/2018/095.html)

Society 5.0 will bring about major changes in lifestyles and industry. In addition to the transformation of existing industries, close partnerships spanning borders and industries across many sectors will enable people to lead different lifestyles. The aim of Society 5.0 is to enable all people to pursue their own happiness and lifestyles and to achieve sustainable development in harmony with nature by solving social problems. In short, achieving Society 5.0 will contribute to solving all global problems facing not only Japan but also the world but also meeting the Sustainable Development Goals established by the United Nations (Cabinet Office, 2017) (Fig. 22.1).

We can examine it under seventeen headings: (1) eradication of poverty, (2) end of hunger, (3) health and care, (4) quality education, (5) gender equality, (6) clean water and hygiene, (7) renewable clean energy, (8) equitable and non-exploitative work and economic growth, (9) innovation economy, (10) waste reduction, (11) sustainable cities and communities, (12) responsible and resourceful production, (13) climate protection, (14) protecting seabed life, (15) protecting natural life, (16) peace, justice and strong institutionalization, and (17) unity and cooperation in goals (Fukuyama, 2018). These Sustainable Development Goals (SDGs) enumerated enable to reveal the creative perspective from various perspectives, and the necessary solutions to achieve these goals should be supported by digital transformation (Fukuyama, 2018).

In order to ensure sustainable development, the Japanese Businessmen's Federation Keidanren (2019) has gathered the above-mentioned goals under nine headings. In this report, where Society 5.0 is declared to the world, information about the areas where the creation of the new society, called "Super Smart Society", will benefit and what kind of innovations will come, cities and regions, energy, disaster prevention and reduction, health, agriculture and food, logistics, production, and service, finance and public services (Keidanren, 2019: 15–20) (Table 22.1).

Sustainable development is a process of economic transformation that optimizes existing economic and social benefits without compromising the possibility of obtaining the same benefits in the future (Goodland & Ledec, 1987). It requires a socioeconomic system that can support an increase in income, improved education, improved public health, and overall quality of life (Pearce et al., 1989). Society 5.0

Table 22.1 Community 5.0 practices to achieve Sustainable Development Goals

Cities and regions	In urban areas, energy, transportation, logistics, waste, etc. related data will be shared for smart solutions. It is aimed to create a comfortable life.
Energy	Microgrids integrating decentralized renewable energy, energy storage systems, and demand-side controls will be developed and be compatible with local conditions. Affordable, reliable energy will be available to all.
Disaster prevention	Digital technologies will be used to mitigate natural disasters by taking effective measures to prevent daily maintenance and infrastructure degradation. Maintenance and rapid restoration of water and sewer infrastructure will ensure the water supply will continue in the event of natural disasters and accidents.
Health	The concept of health will be transformed into lifelong health services and will cover the fields of health, nursing, and medicine. Treatment according to individual health will be provided at the preventive stage to stop the onset and exacerbation of the disease. Access to high-quality healthcare will use next-generation high-speed communication networks and artificial intelligence-based medical and telemedicine technologies.
Agriculture and foods	Cutting-edge technologies such as artificial intelligence, agricultural robots, and remote monitoring and control by autonomous drones for on-site agricultural work will be leveraged.
Logistics	New platforms will be created using various trading procedures to increase efficiency. Most jobs will be automated with autonomous driving, drones, and robots without the use of human power.
Production and services	Analyzing data and using it to produce useful goods and services requires a great deal of investment and professional knowledge. This situation will be facilitated by artificial intelligence modules, which will increase in efficiency thanks to digital transformation.
Finance	Digital transformation will provide a variety of ready-made financial services, including payment, finance, and insurance. Low-cost, convenient, fast, secure, and versatile payment methods will enable people to live anywhere without cash.
Government policy	Central and local governments will rebuild their systems based on digitization. By digitizing many of their tasks and quickly sharing data between various actors, they will deliver more effective public services.

Source: http://www.keidanren.or.jp/policy/2018/095.html

is a more integrated environment that facilitates the merging of the digital and real worlds. The society of the future will continually develop new values and services that make people's lives more affordable while ensuring sustainable growth (Budziewicz-Guzlecka & Drab-Kurowska, 2021: 401). Creating Super Smart Societies with Society 5.0 will have an important role in the realization of the Sustainable Development Goals.

22.3 Sustainable Development with Community 5.0 Practices in the World

The aim of this chapter is to analyze and summarize the results of different studies dealing with Society 5.0 practices that mediate the achievement of Sustainable Development Goals. For this reason, we have tried to make a summary of the previously published studies below. To this end, we reviewed studies that analyze world practices in the fields of *cities and regions, energy, disaster prevention and mitigation, health, agriculture and food, logistics, production and service, finance and public services* within the scope of Society 5.0, which will ensure the achievement of the United Nations' Sustainable Development Goals. Within the scope of Society 5.0, which aims to increase the welfare and peace of the society and to realize Sustainable Development Goals, there are application examples from various parts of the world, most of which are from Japan.

In 2017, Artificial Intelligence Technology Strategies was published, aiming at the development and implementation of artificial intelligence, which plays a key role in the creation of Japan Society 5.0. With the growing awareness of the importance of data sharing rules, Japan has started to work to create a global rule and in March 2019 the Principles of the human-centered Artificial Intelligence Society have been announced (Fukuda, 2020: 1). Japan, with its competence in technology, constitutes the elements of smart faktory, Internet of Things (IoT), machine-to-machine communication (M2M), industrial robots, sensors and semiconductors, robotic logistics solutions (such as transporter robots, conveyors), software programs and systems for industrial manufacturers, cloud It produces and sells advanced technology solutions in areas such as edge computing and edge computing (BTK, 2018: 23). At the end of Japan's G-20 Presidency, the Society 5.0 vision, whose basic principles were accepted by the G-20 and B20 Leaders in their Summit Declarations, is launched as a roadmap that also serves the United Nations Sustainable Development Goals (Kara, 2020).

Faced with increasing medical and social security expenses and demands, as a solution to this situation within the scope of Japan Society 5.0; It aims to (1) support people's independence through the provision of connectivity and information sharing, including check-up, treatment, and care records, with medical data users; (2) the implementation of remote medical care services; and (3) the use of AI and robots in medical care facilities. Thus, by combining and sharing the scattered medical data in different hospitals, a more effective treatment based on data will be possible. Thanks to remote medical care, it can be possible to measure health data such as heart rate and decrease in the number of elderly people coming to hospitals (WHO, 2021).

In Japan, resulting in the elimination of the distortions on behalf of public infrastructure, information and communication technologies (ICT), robots, and systems that require expertise by using new technologies that contain sensors for inspection and maintenance, repair areas that require early detection and diagnosis can be made and thus minimizing unexpected accidents and reducing the time spent on repairs (BTK, 2018: 24). A total of 229 smart city projects have been put into operation in 157 regions in Japan in public–private–academia partnership (Deguchi, 2020).

In addition to Japan, other countries are trying to implement Society 5.0 practices in order to achieve Sustainable Development Goals. There are many countries in the world trying to build "Smart Cities" that encourage and facilitate the participation of their citizens in physical and digital spaces, online and offline processes. For example, Tallinn, Singapore, and Amsterdam are trying to establish smart cities using digital programs (Puutio, 2018). Similarly, San Francisco, Silicon Valley, New York, and London, which are known as the most powerful technology centers, are some of the cities that are compatible with the future (Kelly, 2018).

Society 5.0 aims to use data to create efficient energy networks. For this purpose, we can give an example of the Pakistani government's plan that proposes the elimination of taxes on the production of solar and wind energy equipment in the country in 2019 in order to increase renewable energy production (Mukhtar, 2019). In the Rajasthan region of India, the use of solar energy is encouraged by the government (Banerji, 2018). Benban Solar Park in Egypt is one of the world's largest and most ambitious projects. This solar park is expected to meet 20% of Egypt's energy needs from renewable energy sources (Fleming, 2019). Mexico, on the other hand, has established 4000 Megawatts of new solar farms and more than 1000 new wind farms with the power generation reform (Ávila, 2019).

One of the fields in which advanced technologies are used with Community 5.0 applications is the health sector. The use of artificial intelligence, which is one of the most effective products of advanced technologies, in health services is quite common. For example, the UK has created a digital physician thanks to an artificial intelligence-based chatbot and has established an organization called Babylon Health, which enables patients to interact directly with this physician in a very short time through a mobile application. The biotechnology company Insilico Medicine, headquartered in Hong Kong, has started collaborative work with a team from Oxford University to design drugs with fewer side effects using artificial intelligence (Büyükgöze & Dereli, 2019).

22.4 Conclusions

In this study, we tried to deal with Society 5.0 practices, which are important in the realization of the Sustainable Development Goals. We have revealed the philosophy and practices of Society 5.0, which is thought to serve the world to overcome the ecological crisis it is in. The digital transformation that comes with Industry 4.0 has begun to change lives, cities, jobs, and behavior patterns, that is, society. The aim of Society 5.0 is to create societies that put society and people at the center by preventing the negative effects of technological advances. Thus, it will contribute to the adoption of practices that serve to bring about an economic transformation that optimizes current economic and social benefits without jeopardizing the possibility of obtaining the same benefits in the future. Many countries around the world, especially Japan, have started to implement these practices. These practices continue to increase in order to achieve the United Nations' Sustainable Development Goals in the fastest and most stable way. New applications will continue to appear every day. All these applications are still very new. For this reason, although it is difficult to determine how effective the practices are, it is possible to say that they are trying to remedy the ecological crisis that the world is in.

References

Ávila, J. M. (2019). *Can free trade deliver cheaper renewable energy?* Ask Mexico, https://www.weforum.org/agenda/authors/juan-manuel-avila

Banerji, A. (2018, November). Women in India are tackling the rural energy crisis with solar power. In *World Economic Forum* (Vol. 22). https://www.weforum.org/agenda/2018/11/girl-power-indias-solar-gal-pals-light-up-rural-homes/

BTK. (2018). *Toplum 5.0 Araştırma Rapor.* https://www.btk.gov.tr/uploads/pages/arastirma-raporlari/toplum-5-0-arastirma-raporu.pdf

Budziewicz-Guzlecka, A., & Drab-Kurowska, A. (2021). Good regulation in the telecommunications and postal services markets as a necessary tool in the aspect of sustainable development and society 5.0. *European Research Studies Journal, 24*(3B), 399–411. https://doi.org/10.3390/su13052682

Büyükgöze, S., & Dereli, E. (2019). Toplum 5.0 ve Dijital Sağlık. In *Uluslararası Bilimsel ve Mesleki Çalışmalar Kongresi-Fen ve Sağlık* (Vol. 7). Akademisyen Kitabev.

Cabinet Office. (2017). Society 5.0. http://www8.cao.go.jp/cstp/english/society5_0/index.html

Council of Science, Technology and Innovation. (2016). *Fifth science and technology basic plan.* Government of Japan.

Deguchi, A. (2020). *Society 5.0 and its application to American smart cities.* https://www.americancityandcounty.com/2020/06/10/society-5-0-and-its-application-to-american-smart-cities

Ferreira, C. M., & Serpa, S. (2018). Society 5.0 and social development. *Management and Organizational Studies, 5*(4), 26–31. https://doi.org/10.5430/mos.v5n4p26

Fleming, S. (2019). *Egypt is building one of the world's largest solar parks.* https://www.weforum.org/agenda/2019/01/egypt-is-building-one-of-the-worlds-largest-solar-parks/

Fukuda, K. (2020). Science, technology and innovation ecosystem transformation toward society 5.0. *International Journal of Production Economics, 220*, 107460. https://doi.org/10.1016/j.ijpe.2019.07.033

Fukuyama, M. (2018). Society 5.0: Aiming for a new human-centered society. *Japan Spotlight, 27*(5), 47–50.

Goodland, R., & Ledec, G. (1987). Neoclassical economics and principles of sustainable development. *Ecological Modelling, 38*(1–2), 19–46. https://doi.org/10.1016/0304-3800(87)90043-3

Higashihara, T. (2018). *A search for unicorns and the building of society 5.0.* World Economic Forum.

Kara, T. (2020). *Toplum 5.0, Akıllı Fabrikalar ve Japonya Örneği.* https://www.ekonomidoktorunuz.com/is-dunyasi/toplum-5-0-akilli-fabrikalar-ve-japonya-ornegi-20200417

Keidanren. (2016). *Toward realization of the new economy and society.* http://www.keidanren.or.jp/en/policy/2016/029_outline.pdf

Kelly, J. (2018). *These are the world's most future-proof cities.* https://www.weforum.org/agenda/2018/06/worlds-most-future-proof-cities-jll

Mukhtar, I. (2019). *Energy-short Pakistan moves to power up solar manufacturing.* https://www.weforum.org/agenda/2019/02/energy-short-pakistan-moves-to-power-up-solar-manufacturing/

Nakanishi, H. (2019). *Modern society has reached its limits – "Society 5.0" will liberate us.* World Economic Forum.

Pearce, D., Markandya, A., & Barbier, E. B. (1989). *Blueprint for a green economy Earthscan.* Publications Limited.

Puutio, T. A. (2018). Here are 5 predictions for the future of our cities. In *World Economic Forum.* https://www.weforum.org/agenda/2018/02/here-are-5-predictions-for-the-cities-of-the-future/

Shiroishi, Y., Uchiyama, K., & Suzuki, N. (2018). Society 5.0: For human security and well-being. *Computer, 51*(7), 91–95. https://doi.org/10.1109/MC.2018.3011041

WHO. (2021). *Ageing and health.* https://www.who.int/news-room/fact-sheets/detail/ageing-and-health

The Role of Mathematics on the Sustainability: A Systematic Review

23

Catarina Oliveira Lucas and Joana Becker Paulo

23.1 Introduction

How can Mathematics contribute to improve sustainability today? It is already known the need to take care of the future to avoid greater damage to nature, the shortage of raw materials, and the very life of the planet. Mathematics helps with its concepts and calculations to carry out analyses, which provide recommendations on decision-making in favor of company's objectives. Sustainability is also related to the economic issue, companies are currently concerned about becoming sustainable in terms of reducing costs and, simultaneity, of increasing their profits. Therefore, Mathematics can contribute considerably to greater economic, environmental, and ecological sustainability in the sense of using techniques and methods that lead to the optimization of companies' resources, to a better use of existing assets.

Sustainability has opened a precedent to try a balance where the environment is preserved without harming the companies in their business issues. Thus, a new strategic model had to be thought, so that nature was protected without harming business. According to Clarke et al. (1994), business and the environment can progress simultaneously in a win–win relationship. Being sustainable is no longer a cost when doing business; it is a catalyst for innovation, new market opportunities, wealth, and value creation.

This chapter aims to present a comparative analysis and synthesis of online information related to: Sustainability and Statistics, Sustainability and Mathematics, and Sustainability and Modelling. The generating research question was: "What is the role of Mathematics in the field of Sustainability?". To facilitate answering such a broad question, several more concrete derived questions were formulated.

C. O. Lucas (✉) · J. B. Paulo
ISPGAYA, Vila Nova de Gaia, Portugal
e-mail: clucas@ispgaya.pt

© The Author(s), under exclusive license to Springer Nature Switzerland AG 2023

F. L. Almeida et al. (eds.), *Multidimensional Sustainability: Transitions and Convergences*, Springer Proceedings in Earth and Environmental Sciences,
https://doi.org/10.1007/978-3-031-24892-4_23

23.2 Literature Review

For energy sustainability alone, mathematics has much to contribute to finding better and less polluting ways to explore new energy, in increasing combustion efficiency, in the development of alternative energy, in the management of energy grids and networks, and in minimizing the climate consequences of energy use. For sustainability of financial markets and economic systems, the role of mathematics is equally ubiquitous and essential (Levin, 2013).

Levin (2013) identified a set of mathematical challenges toward achieving sustainability: i) develop statistical mechanics of ecological communities, socioeconomic systems, and the biosphere; ii) model the emergence of an ecological pattern; iii) determine indicators of impending critical transitions between states; and iv) develop schemes for robust governance in these multiscale systems.

The greatest challenge is to achieve cooperation in dealing with problems of the Global Commons (Hardin, 1968; Levin, 1999; Skyrms, 1996), especially as regards public goods and common pool resources. This brings to the fore a different set of mathematical tools—control theory, game theory, voting theory, and mechanism design (Akcay et al., 2012; Barrett, 2007; Dixit et al., 2012; Maskin, 2008; Myerson, 2008; Saari, 2008a, b)—for identifying under what conditions cooperation is possible and how best to achieve it.

The problems of achieving sustainability are urgent and huge and will require complementary inputs of diverse disciplines (Levin & Clark, 2010; Rehmeyer, 2011). Obviously, Mathematics has a great deal to contribute to addressing these problems; but it is equally certain, as has always been the case, that new mathematics will be stimulated by the energy and freshness that comes from new applications and new challenges. The discipline of mathematics has much to contribute, and much to gain, from engagement (Levin, 2013).

Lafuente-Lechuga et al. (2021) have reviewed international research focused on sustainability, big data, and the mathematical techniques used for its analysis. According to the authors, the mathematical techniques used in the empirical work are mainly regression analysis, neural networks, and multi-criteria decision methods.

Sadhukhan et al. (2021) discussed a novel digital output using mathematical computation of life cycle sustainability assessment for design decisions on systemic holistic sustainability of technical systems.

23.3 Methodology

A qualitative methodology was used which consisted of a review based on a predefined systematic method to exhaustively search and identify the most relevant answers that Google provides to the research question.

According to Donato and Donato (2019), a Systematic Review is a less expensive scientific investigation, it is a research article with pre-defined systematic methods to systematically identify all relevant published and unpublished documents for a research question, assesses the quality of these articles, extracts the data, and synthesizes the results. When statistical methods are used to combine the results of two or more studies, it is called a meta-analysis (Donato & Donato, 2019). Also, MacLure et al. (2016) described the reasons for conducting a systematic review including to: (a) identify, evaluate, and interpret available research evidence relevant to a particular topic; (b) identify effective and ineffective interventions; (c) help inform practice and policy by providing integrated and unbiased evidence on which to base decisions; (d) and identify gaps in the literature to inform future studies (MacLure et al., 2016).

There are four essential *criteria* for a systematic review:

It must be comprehensive: all relevant literature in the field must be included.

A rigorous methodology must be followed—defining the research question, writing a protocol, searching the literature, collecting, screening, and analyzing the literature. The whole process should also be carefully documented.

A thorough literature search to find all relevant articles on the topic. Thus, it is important that the search strategy is rigorously developed with high sensitivity to find all potentially relevant articles and carry out this search in various databases and other resources.

At least two people should be involved, especially for article screening and data extraction.

Steps in the systematic review process (Donato & Donato, 2019):

1. Formulate the research question.
2. Produce a research protocol and record it (items 1 and 3–8 should be in the protocol for writing the systematic review).
3. Define the inclusion and exclusion criteria.
4. Develop a search strategy and search the literature—find the studies.
5. Selecting the studies.
6. Evaluating the quality of the studies.
7. Data extraction.
8. Synthesis of data and evaluation of the quality of the evidence.
9. Dissemination of results—Publication.

23.4 Protocol

Responses from published and unpublished documents (articles, reports, and blogs) were analyzed. The quality of the extracted data was assessed, the answers provided by the different documents were compared and the results were synthesized.

The following steps were followed in the process:

23.4.1 Research Question

Q_0: What is the role of Mathematics in the Sustainability scope?

To facilitate answering such a broad research question, more concrete derivative questions were formulated:

Q_1: What is the idea of sustainability?
Q_2: What is the meaning of suitable development?
Q_3: What is the difference between sustainability and sustainable development?
Q_4: What are the two key concepts of sustainable development?
Q_5: What are the three E's of sustainability?
Q_6: What are the three principles of sustainability?
Q_7: Why are the three pillars of sustainability important?
Q_8: What are the four scientific principles of sustainability?
Q_9: What is the most important pillar of sustainability?
Q_{10}: What are the four general indicators of sustainable development?
Q_{11}: What are the primary goals of sustainability?
Q_{12}: What are the seven dimensions of sustainable development?
Q_{13}: What is a good example of sustainability?
Q_{14}: Why is sustainability so important?
Q_{15}: What does sustainability mean in education?
Q_{16}: What is the role of students in sustainable development?
Q_{17}: What does sustainability mean in Maths?
Q_{18}: How does mathematics help our nation to be sustainable?
Q_{19}: How does mathematics relate to the environment?
Q_{20}: Does environmental science involve Maths?
Q_{21}: Does environmental studies use Maths?
Q_{22}: Why do environmental scientists use mathematical models?
Q_{23}: What kind of Maths is used in environmental science?
Q_{24}: How is Maths used in climate change?
Q_{25}: What kind of Maths is used in environmental engineering?
Q_{26}: Do environmental engineers do a lot of Maths?
Q_{27}: How do environmental engineers use Maths?
Q_{28}: Why do environmental scientists use mathematical models?

23.4.2 Research Protocol

This step will encompass all the other steps.

So, the work consisted in looking for answers to the questions formulated in Google, in different databases (articles, blogs, reports, etc.) about:

(a) Sustainability
(b) Sustainability and Statistics
(c) Sustainability and Mathematics
(d) Sustainability and Modelling

The data sources used are indicated in the references section.

23.4.3 Inclusion and Exclusion Criteria

Initially, works describing concepts related to Sustainability were included. Also included were those that directly related to Mathematics and Sustainability. Papers that presented the usefulness of Mathematics in a more general scope of Science that was not directly related to Sustainability were excluded. For example, answers from works such as this one were excluded:

> Answer: Conceptual and mathematical models are especially powerful because *they are more precise models.* They help explain an idea or relationship in a way that is easy to understand. (Environmental Science Flashcards | Quizlet)

23.4.4 Search Strategy

It used the frequently asked questions suggested by Google as a tool that allows the description of the most viewed information related to Mathematics and Sustainability, it means the most accessed data information by Google users: researchers, teachers, students, professionals, and the general population (Fig. 23.1).

23.4.5 Selection of Studies

The selection of each of the studies providing answers to the questions formulated was also performed automatically by Google itself. Question and answer pages are web pages that contain information in the format of a question followed by its answers. For content that represents a question and its answers, it is possible to mark the data with the *schema.org* types QAPage, Question, and Answer. Properly tagged pages qualify for *enhanced search* displayed on the search results page. This enhanced treatment helps sites reach the right users in search (Google Developers, 2022).

23.4.6 Evaluating the Quality of the Studies

To qualify for enhanced search, a question must have at least one answer, either "*acceptedAnswer*" or "*suggestedAnswer.*" However, questions may not have answers when they are first published. It used a URL linked directly to the answer. It is highly recommended to provide a URL for each answer because it improves the user experience when they click through to the site.

STEM (science, technology, engineering, and maths)-related educational Q&A pages where the focus is on providing a correct answer to a question submitted by

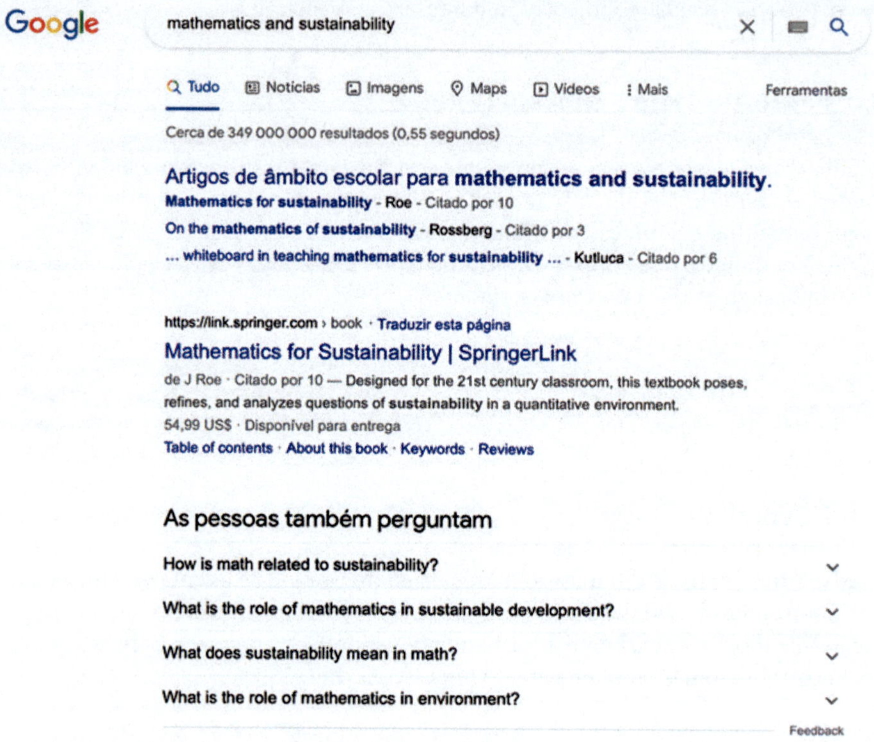

Fig. 23.1 Description of the search strategy on Google

users may qualify for a carousel experience. These pages can only have a single answer provided or selected by internal experts, not users. For example, an educational page where a user submits a question, and the best answer is selected by experts (Google Developers, 2022).

23.4.7 Data Extraction

After the evaluation referred to in Sect. 23.4.6, several answers to the 28 questions presented in Sect. 23.4.1 were obtained. So, the next step was to separate them into categories.

23.4.8 Synthesis of Data and Evaluation of the Quality of the Evidence

At this stage, a procedure will be described for combining questions-answers, excluding some repetitions, joining some that can be similar, and compiling the questions-answers into groups:

(A) Related to concepts/definition of Sustainability
(B) Related to Sustainability in Education
(C) Related to Maths and Sustainability
(D) Related to Mathematics' curricular units of Degrees of Environment Engineering or Maths in studies related to Environment

The results of this data aggregation and evaluation are presented in Table 23.1.

Table 23.1 Compilation of questions and answers in groups

Group	Questions	Answers
A—Sustainability	Q_1 Q_2 Q_3 Q_4 Q_5 Q_6 Q_7 Q_8 Q_9 Q_{10} Q_{11} Q_{12} Q_{13} Q_{14}	Sustainable development is development that meets the needs of the present without compromising the ability of future generations to meet their own needs. While sustainability is often thought of as a long-term goal (i.e., a more sustainable world), sustainable development refers to the many processes and pathways to achieve it (e.g., sustainable agriculture and forestry, sustainable production and consumption, good government, research and technology transfer, education...). Three community dynamics are particularly important to building healthy and prosperous communities over the long term: economy, ecology, and equity (the 3 E's). Sustainability is made up of four pillars: human, social, economic, and environmental. The four general indicators of sustainable development are: resources, biodiversity and habitat, productive natural resources, and sustainable energy. The 17 Sustainable Development Goals (SDGs) of the United Nations are: No Poverty; Zero Hunger; Good Health and Wellbeing; Quality Education; Gender Equality; Clean Water and Sanitation; Affordable and Clean Energy; Decent Work and Economic Growth; Industry, Innovation and Infrastructure; Reduced Inequality; Sustainable Cities and Communities; Responsible Consumption and Production; Climate Action; Life Below Water; Life on Land; Peace, Justice and Strong Institutions; Partnerships for the Goals The seven dimensions of sustainable development are: economy, community, occupational groups, government, environment, culture, and physiology. Renewable clean energy is probably the most obvious example of sustainability. Here are two examples. Solar energy: Once the sun's electromagnetic radiation is captured, it produces electricity and heat. Wind energy: Wind turbines convert the kinetic energy in the wind into mechanical power. Sustainability is important for many reasons, including: environmental quality—In order to have healthy communities, we need clean air, natural resources, and a nontoxic environment. It is required more resources such as energy, water, and space.

(continued)

Table 23.1 (continued)

Group	Questions	Answers
B—Sustainability in education	Q_{15} Q_{16}	"Education for Sustainable Development allows every human being to acquire the knowledge, skills, attitudes, and values necessary to shape a sustainable future. Students usually play a vital role as initiators, drivers, and contributors of the sustainability processes, but the institutionalization of this engagement is useful through structures that assign responsibilities and provide resources and funds.
C—Maths and sustainability	Q_{17} Q_{18} Q_{19}	Mathematics not only helps us to understand natural phenomena, it also allows us to sustain the majority of human activity on the planet. Transport networks, the Internet, and business transactions are all practical applications of research, graph theory, and number theory. Maths provides confidence in climate change models, and it helps to improve existing renewable technologies. Maths is also key in assessing renewables based on observations from the environment. For example, weather data helps to predict efficiency of solar cells
D—Maths in studies related to environment	Q_{20} Q_{21} Q_{22} Q_{23} Q_{24} Q_{25} Q_{26} Q_{27} Q_{28}	Core courses in the environmental science major include biology, chemistry, geology, and mathematics. Mathematics for environmental studies: To solve environmental problems, which cover ecological systems, weather, and ocean, it is effective to quantity and analyze these systems as mathematical models. For example, scientists use the models to relate the amount of energy reflected from objects to the objects' physical condition. By calculating averages, analyzing variance, and making diagrams, it can find out whether the climate has changed and how. Predicting future climate requires mathematical modeling with differential equations and stochastic methods. Maths plays an important role in the investigation of environmental science and climate studies (ESCS) issues. It is recommended that ESCS-focused maths programs include at least four applied courses such as mathematical modeling, numerical analysis, operations research, ordinary differential equations, partial differential equations, statistics, and probability. For example, Shawnee University's bachelor of science in environmental technology has 7 h of maths in the form of calculus, probability, statistics, algebra, trigonometry, and differential equations. Aspiring environmental engineers must complete plenty of maths, science, and engineering courses and gain practice applying those concepts to solving environmental problems. Environmental practitioners should master basic maths definitions and the formation of problems, as daily operations require calculation of percentages, averages, simple ratios, geometric dimensions, threshold odor numbers, and force, pressure, and head, as well as the use of dimensional analysis and advanced maths, etc. Mathematical models are recognized as effective tools that could help examine economic, environmental, and ecological impacts of alternative pollution-control and resources-conservation actions, and thus aid planners or decision-makers in formulating cost-effective management policies.

23.5 Discussion

The results presented essentially show: a description of the different terms related to Sustainability (Group A); the role of students and educational institutions to ensure a sustainable development that meets the needs of the present without compromising the ability of future generations to meet their own needs (Group B); the relevance of the presence of Mathematics curricular units in the Environmental Science or Environmental Engineering degrees (Group D); and the recognition of mathematical models as effective tools to plan, predict and assist decision-makers in solving sustainability problems (Groups C and D).

23.6 Conclusions

It was concluded that, on the one hand, the discipline of mathematics has much to contribute to solving the urgent and huge sustainability problems; on the other hand, mathematics has much to gain from its involvement in this field: new mathematics is appearing stimulated by the energy and freshness that comes from new applications and challenges. In particular, in the field of education, complementarity, and cooperation of various disciplines is essential to create solutions for sustainable development, in order to explore different processes and ways to achieve it, by contributing to: sustainable agriculture and forestry, sustainable production and consumption, good governance, sustainable research and technology transfer, sustainable education, and among others.

The research presents some limitations, such as the fact that it was carried out only for the English language and not for other languages. So, to extend this work in the future, it is intended to seek the answers to these same questions using the Spanish and Portuguese languages, i.e., by searching for the expressions described above ("Sustainability", "Sustainability and Mathematics", etc.) in Portuguese and Spanish, and observe what Google suggests as questions–answers. It is also intended to extend the work with results from other databases such as articles, reports, and blogs and from other data sources.

References

Akçay, E., Meirowitz, A., Ramsay, K. W., & Levin, S. A. (2012). Evolution of cooperation and skew under imperfect information. *Proceedings of the National Academy of Sciences of the United States of America, 109*(37), 14936–14941. https://doi.org/10.1073/pnas.1212925109

Barrett, S. (2007). *Why cooperate: The incentive to supply global public goods*. Oxford University Press.

Clarke, R., et al. (1994). The challenge of going green. In *Corporate social responsibility*. Harvard Business Review. https://hbr.org/1994/07/the-challenge-of-going-green

Dixit, A., Levin, S., & Rubenstein, D. (2012). Reciprocal insurance among Kenyan pastoralists. *Theoretical Ecology, 6*(2), 173–187.

Donato, H., & Donato, M. (2019). Stages for undertaking a systematic review. *Acta Médica Portuguesa, 32*(3), 227–235. https://doi.org/10.20344/amp.11923

Google Developers. (2022). *Central da Pesquisa Google: Perguntas e Respostas*. Updated 2022-08-04 UTC. https://developers.google.com/search/docs/advanced/structured-data/qapage?hl=pt-br#suggestedanswer

Hardin, G. (1968). The tragedy of the commons. *Science, 162*, 1243–1248.

Lafuente-Lechuga, M., Cifuentes-Faura, J., & Faura-Martínez, U. (2021). Sustainability, big data and mathematical techniques: A bibliometric review. *Mathematics, 9*(20), 2557. https://doi.org/10.3390/math9202557

Levin, S. (1999). *Fragile dominion: Complexity and the commons*. Perseus Books.

Levin, S. (2013). The mathematics of sustainability. *Notices of the AMS, 60*(4), 392–393. https://doi.org/10.1090/noti982

Levin, S., & Clark, W. (2010). *Toward a science of sustainability: Report from the NSF toward a science of sustainability conference*. Princeton University Printing and Mailing Services.

MacLure, K., Paudyal, V., & Stewart, D. (2016). Reviewing the literature, how systematic is systematic? *International Journal of Clinical Pharmacy, 38*, 685–694. https://doi.org/10.1007/s11096-016-0288-3

Maskin, E. (2008). Mechanism design: How to implement social goals. *American Economic Review, 98*(3), 567–576. https://doi.org/10.1257/Aer.98.1.567

Myerson, R. (2008). Perspectives on mechanism design in economic theory. *American Economic Review, 98*(3), 586–603. https://doi.org/10.1257/Aer.98.3.586

Rehmeyer, J. (2011). *Mathematical and statistical challenges for sustainability*. American Mathematical Society.

Saari, D. (2008a). Complexity and the geometry of voting. *Mathematical and Computer Modelling, 48*(9–10), 1335–1356. https://doi.org/10.1016/J.Mcm.2008.05.033

Saari, D. (2008b). Disposing dictators, demystifying voting paradoxes. *Mathematical and Computer Modelling, 48*(9–10), 1671–1673. https://doi.org/10.1016/J.Mcm.2008.05.016

Sadhukhan, J., Sen, S., & Gadkari, S. (2021). The mathematics of life cycle sustainability assessment. *Journal of Cleaner Production, 309*, 127457, ISSN 0959-6526. https://doi.org/10.1016/j.jclepro.2021.127457

Skyrms, B. (1996). *Evolution of the social contract*. Cambridge University Press.

24 Effects on Students and Teachers of a Course on Strategies for Well-Being

Adelinda Candeias, António Portelada, Adriana Félix, and Edgar Galindo

24.1 Introduction

Despite scientific advances, humanity faces environmental challenges that lead to the unsustainable degradation of human resources, conditioning their well-being (Ronen & Kerret, 2020). In a society where consumption prevails, it is increasingly important to find ways to reconcile values that involve environmental sustainability (Lengyel et al., 2019). Sustainability education is a concept derived from the early efforts to give in schools an education more adequate to the problems of the present world, especially, environmental issues. Consequently, a main objective has been to help students to acquire knowledge, skills, values and competences to understand and give a response to the above-mentioned issues (Grange, 2017), including topics like injustice in sharing the world's resources and problems regarding economical development, justice, peace and conflict, and human rights and dignity, along with the defense of ecosystems (Wals & Jickling, 2002). Consequently, it has been a growing interest to study the development of the corresponding skills and qualities among educators and students (Bürgener & Barth, 2018) at different educational levels. In short, it has been concluded that in order to transform the reality, teachers and students need competences and qualities, i.e., a new pedagogy emphasizing self-directed learning, participation and collaboration, problem orientation, and emotional management, based in trans disciplinarity, and able to build bridges between formal and informal education (Rieckman, 2018). In this context,

A. Candeias (✉) · A. Félix · E. Galindo
Comprehensive Health Research Center (CHRC), Évora University, Évora, Portugal
e-mail: aac@uevora.pt

A. Portelada
ESE Santarém, Santarém, Portugal

© The Author(s), under exclusive license to Springer Nature Switzerland AG 2023
F. L. Almeida et al. (eds.), *Multidimensional Sustainability: Transitions and Convergences*, Springer Proceedings in Earth and Environmental Sciences,
https://doi.org/10.1007/978-3-031-24892-4_24

researchers have been exploring models of educator competences in sustainability education (Mulà et al., 2017).

In this context, psychologists and educational scientists have been giving increasing attention to social and emotional learning (SEL). Probably, as a result of the work on emotional intelligence of Daniel Goleman and the contributions of researchers such as James Comer on positive school climates (Comer, 2009), scientists have discovered the importance of social environment and emotional states of students and teachers after many decades studying learning at school as a cognitive issue. SEL has been defined as the process by which children and adults learn to understand and manage emotions, maintain positive relationships, and make responsible decisions (O'Conner et al., 2017). Current SEL approaches have been influenced by multiple perspectives. Nevertheless, the main topics have been the way children and adults understand themselves (self-awareness), the way they regulate their emotions and behaviors (self-management), the way they empathize with others (social awareness) and/or interact with others (Social Relationships), and the way they make decisions affecting themselves and other people (responsible decision-making). Currently, it is accepted that SEL is significantly correlated with (a) school outcomes, including academic achievement, and behavior, (b) workplace satisfaction, (c) a reduction in stress and anxiety, (d) measures of well-being, and (e) healthy relationships (Haynes, 2021). Here, it is important to mention the work developed by the Collaborative Consortium for Academic, Social and Emotional Learning (CASEL, 2003) created to improve the quality of socioemotional learning based on empirical evidence and to promote the inclusion of SEL as an integral part of school-based education from preschool through secondary. This approach (CASEL, 2012, 2016) posits that socioemotional skills are cognitive, affective, and behavioral in nature and that interventions should develop five main domains: (1) self-awareness skills, (2) self-management skills, (3) social awareness skill, (4) relationship skills, and (5) skills in responsible decision-making. Social-emotional well-being is associated with the way each individual interprets himself/herself and his/her social context (Manning & Fleming, 2019). Several studies have been carried out on the well-being of different actors in the school (OECD, 2023), but a deep knowledge is still lacking, since most investigations are limited to quantitative studies or qualitative reports of student emotions (Scott et al., 2021).

A sustainable approach emerged in the last three decades to refer to an interconnected system of durable human and natural values, where the survival and prosperity of the species depend on maintaining the value of biodiverse environments (Hargreaves, 2002) abandoning the individual conception, which focuses on strengthening personal skills such as self-esteem and optimism in detriment of collective skills such as cooperation (Ecclestone & Rawdin, 2016). The collective approach of well-being proposes a representation of well-being linked to the community (Ascorra et al., 2021). It implies strengthening the school community's collective well-being fostering the interactions between the different school actors (López et al., 2022). In this context, teachers are a key player in the socioemotional development of students and, consequently, the well-being in the school context. Teachers are important mediators of students' self-esteem and motivation processes

(Choi, 2018). Their positive relationship with students leads to the positive development of students' academic, physical, and emotional needs (Jones & Kessler, 2020), promoting overall well-being (Johnston et al., 2022). This positive dynamic has been proven through several studies (Denston et al., 2022; Durlak et al., 2011; Zembylas, 2007).

Consequently, the development of teaching skills that promote strategies for well-being, socioemotional learning, and executive functioning is an essential pillar for a sustainable and well-being-oriented education. Teachers' perception of socioemotional well-being determines their performance and engagement with students in order to develop socioemotional skills (Denston et al., 2022). Teachers who see themselves as agents of change feel more motivated to develop programs with a socioemotional scope; at the same time, teachers who are more aware of their impact on the emotions of others have greater socioemotional skills (Martinsone & Damberga, 2017).

Several studies have shown that the emotions and cognitions developed in the student-teacher relationship are related to students' adaptation, the school environment, and academic results (Milatz et al., 2015; Roorda et al., 2011). Consequently, the positive modification of the school environment can lead to improved results in terms of health and well-being of students and can have important implications for the resilience of students who are in a situation of emotional and academic disadvantage (Saab & Klinger, 2010). In the same line of thought, Velasco (2021) adds that the principles of inclusion, equity, and sustainability are also inherent to improving education and health in our school contexts as a way to face the challenges of the twenty-first century. Also Schleicher (2018) proposes that the educational community's greatest dilemma is that cognitive skills should be only one part of school aims, because schooling today needs to be more focalized on "ways of thinking (involving creativity, critical thinking, problem solving and judgement), ways of working (including communication and collaboration), tools for working (including the capacity to recognize and exploit the potential of new technologies) and the capacity to live in a multi-faceted world as active and responsible citizens" (p. 31).

Based on these guidelines, in this study well-being is understood from a sustainability perspective (Shirley et al., 2020) and a collective perspective (Ascorra et al., 2021), and the role of SEL in well-being is taken into account. In the current study, a proposal of intervention course has been developed, based on the paradigm of psychological well-being of teachers as the starting point to improve students well-being and a more sustainable teaching, It follows the model of reflexive development of teachers based on the "Teacher Centered Coaching" Model (Wang, 2017) to improve intervention in a comprehensive context that included five main topics (Fig. 24.1):

1. Learning for well-being in the classroom – concepts and strategies
2. Executive functioning and self-regulation of learning
3. Social awareness, interpersonal relationships, and empathy
4. Learning environments for well-being
5. Learning for community well-being

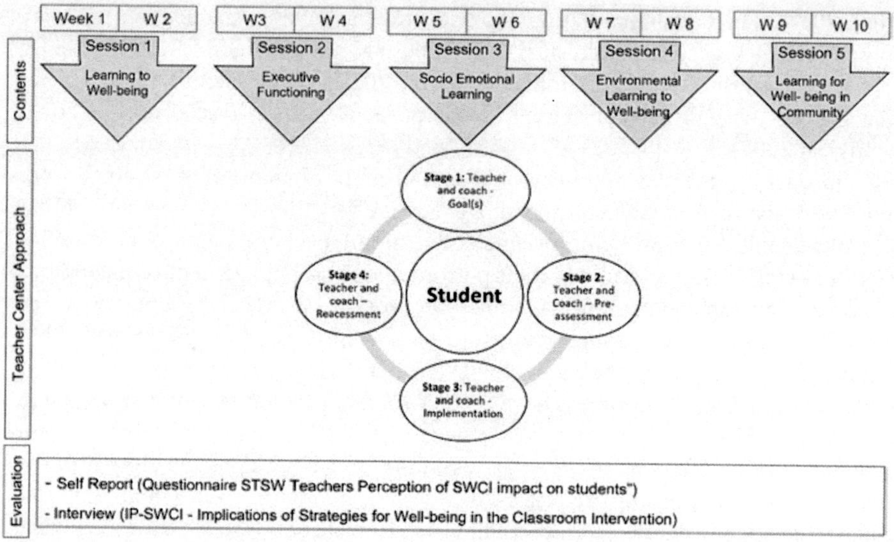

Fig. 24.1 Diagram showing the application of the course of strategies for well-being in the classroom (SWCI)

Based on the assumption that the development of teaching skills that promote strategies for well-being, socioemotional learning, and executive functioning is an essential pillar for a sustainable and well-being-oriented education, in the current study, a proposal of intervention course has been developed as the starting point to improve students well-being and a more sustainable teaching, namely how that intervention contributes to change: (i) perceptions of teachers about impacts on student well-being and (ii) perceptions of teachers about impacts on themselves, on students, and on classroom context, as we present in next topics.

24.1.1 Goals

The purpose of this study is to describe how teachers that apply a training intervention focalized in a multidimensional approach to Strategies for Well-Being in the Classroom perceive themselves as mediators to generate the well-being of their students in the classroom. The main aims are as follows:

1. To describe teacher's perceptions of the impact of implementing Strategies for Well-Being in the Classroom on students' well-being, socioemotional learning, executive functioning, and classroom environment
2. To describe teacher's representations about the implications of the intervention on themselves as mediators to generate the well-being of students in the classroom

24.2 Method

24.2.1 Design

A phenomenological-hermeneutic qualitative design (Fuster, 2019) was applied to improve the understanding of the meanings and practices in a specific group of participants. The four phases recommended by Fuster (2019) have been carried out, namely, allowing access to an in-depth understanding of the meanings surrounding the training experiences: elucidation of the perspectives of the research team, collection of the experience of participants in relation to well-being in classroom, reflections on the data reported by participants, and writing about this lived experience.

24.2.2 Participants

Participants were 30 basic schoolteachers, 26 women and 4 men, from all regions of Portugal, who were inscribed online and voluntarily, in the teachers training intervention. The data used for developing SWC came from 8 participants (6 women and 2 men). Twenty-two participants were excluded from the data analysis for different reasons, who were working with a population of 8- to 15-year-old students ($n = 423$). Of the eight teachers included, six were women (75%) and two men (25%); these percentages aligned with the Portuguese proportions, where the percentage of female teachers is 78.1% (PORDATA, 2021). Three teachers were aged between 30 and 45 years, three between 45 and 55, and two between 60 and 65. All of them had more than 8 and less than 40 years of service. The majority managed classes at different levels of education (92%), only one (8%) worked with a full-time class (first cycle).

24.2.3 Intervention

This study was carried out in several schools in Portugal, after the pandemic COVID-19 lockdown. Data were collected via a 10-week online course that took 25 h in total (3 h for face-to face sessions during uneven weeks, devoted to the development of contents, and 2 h for online coaching during even weeks). This course was supported by the Portuguese Scientific Council of In-Service Training for Teachers and took place at the University of Evora in the frame of a project called REFLECT[1] – *Raising awareness and stimulating Executive Functioning and Socioemotional Learning by integrating Evidence-based strategies in the Classroom to empower pupils, teachers and parents* (see https://my.reflectproject.eu). The name of the course is "Strategies for Well-Being in the Classroom Intervention" (SWCI), whose aims are as follows:

[1] Projeto ERASMUS+: REFLECT/2019–1-BE02-KA201–060353

1. Identify aspects of learning for well-being, socioemotional learning, and self-regulation in practice
2. Enable participants to consolidate their learning, reflect on practice, and set specific goals to implement strategies in the classroom
3. Create a safe and trusted learning environment that enables active learning
4. Explore responsible decision-making as a component of socioemotional learning
5. Increase understanding of what are the essential elements of friendly environments (safe spaces for learning) and their implications for individual and community well-being

The course is organized into five main topics:

1. Learning for well-being in the classroom – concepts and strategies
2. Executive functioning and self-regulation of learning
3. Social awareness, interpersonal relationships, and empathy
4. Learning environments for well-being
5. Learning for community well-being

The face-to-face sessions had (a) an introductory component of content presentation, (b) a component of analysis and discussion of practical situations, (c) a component of planning the implementation of exercises and practical strategies in the classroom, (d) a component for reflection (make my day), and (e) an evaluative component. The reflective component was followed in 2 h online sessions. In each face-to face session, the participants had the autonomy to apply the exercises and strategies that they would develop in the classroom during the 2 weeks that elapsed between each face-to-face session. The work carried out was recorded and analyzed in the following sessions.

24.2.4 Instruments

Two qualitative research instruments were used:

1. Self-report questionnaire about "Teachers Perception of SWCI Impact on Students" (STSW). This tool aims to operationalize teacher's perceptions of the impact of implementing Strategies for Well-being in the Classroom on students' well-being, socioemotional learning, executive functioning, and behavior.

   This tool comprises three open questions:

   1. Please remember the strategies/activities you have implemented with your students related to Strategies for Well-Being in the Classroom during the period of the course and list your perceptions about:

      1.1. Students impacts on social emotional competences, executive functioning, and well-being.

2. Please remember the strategies you have implemented with your students related to Strategies for Well-Being in the Classroom during the period of the course and list your perceptions about students:
   2.1. Most liked strategies
   2.2. Most positive impact
3. Please remember the strategies/activities you have implemented with your students related to Strategies for Well-Being in the Classroom during the course and list your perceptions about:
   3.1. Impact on Students' Behavior in Classroom

2. The teachers have been interviewed in order to characterize their representations about the implications of Strategies for Well-Being in the Classroom Intervention (IP-SWCI), on themselves as mediators to generate the well-being in their classrooms with their students. This protocol comprises four open questions:
   1. The main topics that you consider critical to improve well-being in the classroom
   2. The main methods that you consider critical to improve well-being in the classroom
   3. The main changings in your classroom practices during and after your participation in this intervention about Strategies for Well-Being in the Classroom
   4. The main relevance for you from your participation in this intervention about Strategies for Well-Being in the Classroom

24.2.5 Procedures

The instruments STSW and IP-SWCI were applied after the intervention by a member of the team through the Zoom virtual communication platforms. Interviewers agreed previously on the conditions and confidentiality protection. This agreement was supported by a written declaration sent by Zoom platform (chat) of the researcher and a letter of consent following the regulations of the Ethics Committee of first author's institution.

24.2.6 Data Analysis

Data were analyzed using a qualitative approach. The data collected from the STSW and IP-SWCI were analyzed using content analysis and responded to the phases of content analysis of the phenomenological-hermeneutic method. First, successive readings were made of the entire corpus and then preliminary coding was done for each actor in order to have a manageable subset of data (Fuster, 2019). To do this, the information given by the teachers in the interviews and in the self-reported questionnaire has been divided in themes, and a code has been given to each theme.

The methodology is shown as the steps to designing the SWC in Fig. 24.2.

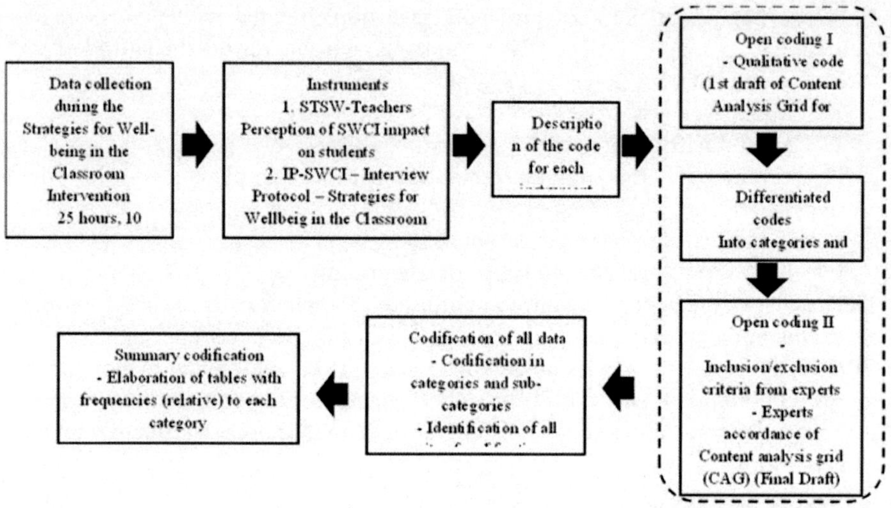

Fig. 24.2 Diagram of the methodology of the SWCI

24.2.7 Research Process of Content Analysis

To answer research questions, the researchers accomplished the following research process.

24.2.7.1 What Are the Codes and Their Components in the CAG-SWC?

- The theoretical framework of SWC was applied to code the raw data collected from IP-SWCs.
- IP-SWCIs used at the end of the course was applied.
- The redundant codes in both instruments have been eliminated.

24.2.7.2 What Is the Quality of the CAGs?

- Three experts on educations were asked to check the correspondence between the codes and components of implications on pupils and teachers.
- CAG-STSW and CAG-IP-SWCI were revised according to the experts' comments with a concordance of 85% and the final grids were reached (CAG-STSW – Appendix A; CAG-IP-SWCI – Appendix B).
  1. The perception of the teachers about the impact of the applied strategies on the well-being, the socioemotional learning and the executive functioning of the students, as well as on the classroom environment.
  2. The representations of the teachers about the effects of the intervention on them as mediators for the generation of well-being of the students.

24.3 Results

In this research, two instruments were used to collect data: STSW and SWCI. The answers of participant teachers to both instruments were analyzed through content analysis, as previously explained. The tables below were drawn up to illustrate the results with categories and subcategories, with their respective partial and absolute frequencies of the number of mentions or Recording Units RU), and are organized into the following two main topics.

Perceptions of Teachers About Impacts on Student Well-Being

Teachers' perceptions were organized into three parts:

1. Students impact of implementing Strategies for Well-Being in the Classroom
2. Most liked/most positive impact Strategies for Well-Being in the Classroom
3. Students' most positive impact Strategies for Well-Being in the Classroom

Perception of Teachers About Impacts on teachers, students, and classroom context.

We will present the results and its discussion in the following two topics.

24.3.1 Perceptions of Teachers About Impacts on Student Well-Being

24.3.1.1 Students Impact of Implementing Strategies for Well-Being in the Classroom

The content analysis of this first part is shown in Table 24.1, with a total of 60 recording units divided into five categories: socioemotional learning, executive functions, well-being, classroom context, and school context. The socioemotional learning category, with 26 recording units, is divided into 8 subcategories: personal relationship, self-knowledge, acceptance of others, respect, empathy, behavior in the classroom, responsibility, and participation. The executive functions category, with 20 recording units, is divided into 6 subcategories: attention ability, active listening, concentration, mental flexibility, curiosity, and motivation. The well-being category, with six recording units, is divided into four subcategories: being more conscious of how everyday moments make them feel; handle better your emotions; establishing and maintaining healthy relationships; and handle frustration and stress. The classroom context category, with four recording units is divided into three subcategories: focus your attention; organize; and listen. The school context categories, with four recording units, is divided into three subcategories: improving your social interactions; use of verbal regulation; and controlling your impulses. The category with the most recording units is the socioemotional learning category, with 43.33% of the recording units for this section. Next is the executive functions category with 33.33% of the recording units.

Table 24.1 Students impact of implementing Strategies for Well-Being in the Classroom

Categories	Subcategories	*F*	*F* %
Well-being	Being more conscious of how everyday moments make you feel	2	6 (10%)
	Handler better your emotions	2	
	Establishing and maintaining healthy relationships	1	
	Handle frustration and stress	1	
Socio emotional learning	Personal relationship	5	26 (43.33%)
	Self-knowledge	2	
	Acceptance of the other	5	
	Respect	4	
	Empathy	2	
	Behavior in the classroom	2	
	Responsibility	2	
	Participation	4	
Executive functions	Attention ability	5	20 (33.33%)
	Active listening	2	
	Concentration	5	
	Mental flexibility	3	
	Curiosity	2	
	Motivation	3	
Classroom context	Focus your attention	1	4 (6.67%)
	Organize	2	
	Listen	1	
School context	Improving your social interactions	2	4 (6.67%)
	Use of verbal regulation	1	
	Control your impulses	1	
Total			60 (100%)

24.3.1.2 Most Liked Strategies for Well-Being in the Classroom

The content analysis of this part can be seen in Table 24.2, with a total of 227 recording units divided into 3 categories: well-being, socioemotional competencies, and executive Functioning. The well-being category, with 71 recording units, is divided into 4 subcategories: make my day, optimism, positive and realistic image, and motivation. The socioemotional competencies category, with 107 recording units, is divided into 5 subcategories: empathy, being able to accept yourself/others, affinity, communication, and cooperation. The executive functioning category, with 49 recording units, is divided into 6 subcategories: attention, planning abilities, cognitive flexibility, inhibition, organization, and working memory. The category with the most recording units is the socioemotional competencies category, with 47.14% of the recording units for this section.

24.3.1.3 Students' Most Positive Impact Strategies for Well-Being in the Classroom

These data are shown in Table 24.3, with 149 recording units divided into 3 categories: socioemotional competencies, well-being, and executive functioning. The

Table 24.2 Students' most liked/most positive impact Strategies for Well-Being in the Classroom

Categories	Subcategories	*F*	*F* %
Well-being	Make my day	31	71 (31.28%)
	Optimism	18	
	Positive and realistic image	4	
	Motivation	18	
Socioemotional competences	Empathy	31	107 (47.14%)
	Being able to accept yourself/others	18	
	Affinity	9	
	Communication	22	
	Cooperation	27	
Executive functioning	Attention	12	49 (21.59%)
	Planning abilities	9	
	Cognitive flexibility	5	
	Respond inhibition	9	
	Organization	12	
	Working memory	2	
Total			227 (100%)

Table 24.3 Students' most positive impact strategies for well-being in the classroom

Categories	Subcategories	*F*	*F* %
Well-being	Motivation	25	43 (28.86%)
	Self-regulated behavior	18	
Socioemotional competences	Self-consciousness	20	77 (51.68%)
	Self-regulation of emotions	14	
	Social consciousness	17	
	Capacity to interact with others	17	
	Making decisions	9	
Executive functions	Active listening	15	29 (19.46%)
	Planning homework and other school activities at home	6	
	Informal learning through games	8	
Total			149 (100%)

socioemotional competencies category, with 77 recording units, is divided into 5 subcategories: self-consciousness, self-regulation of emotions, social consciousness, capacity to interact with others, and making decisions. The well-being category, with 43 recording units, is divided into 2 subcategories: motivation and self-regulated behavior. The executive functioning category, with 29 recording units, is divided into 3 subcategories: active listening, planning homework and other school activities at home, and informal learning through games. The category with the most recording units is the socioemotional competencies category, with 51.68% of the recording units for this section.

24.3.2 Perception of Teachers About Impacts of Strategies for Well-Being in the Classroom Intervention on Teachers, Students, and Classroom Context

As we could observe, Table 24.4 shows the impacts of Strategies for Well-Being in the Classroom Intervention with 257 units of content, organized into 4 categories: impact on the approach to well-being (24.12%), impacts on the methods to improve well-being in the classroom (21.01%), impacts on the practices in the classroom (28.79%), and impacts on the teacher training models (26.07%). The category with

Table 24.4 Impacts of strategies for well-being in the classroom intervention

Categories	Subcategories	*F*	*F* %
1. Impacts on the approach to well-being in the classroom	1.1 Learning for personal and professional well-being/self-reflection (make my day/ coaching)	10	62 (24.12%)
	1.2. Executive functioning and self-regulation	13	
	1.3 SEL/social awareness, relationship building, empathy, and responsible decision-making	20	
	1.4 Friendly learning environments (classroom)	7	
	1.5 Promoting well-being in the classroom in community and (in)formal learning	12	
2. Impacts on the methods to improve well-being in the classroom	2.1. Distance training	6	54 (21.01%)
	2.2. Sharing and active discussion based on experiences	8	
	2.3. Sharing instruments/material	16	
	2.4. Reflective and self-knowledge methodologies	16	
	2.5. Practice based	8	
3. Impacts on the practices in the classroom	3.1. SEL	18	74 (28.79%)
	3.2. EF (self-regulation)	7	
	3.3. Student Well-being	11	
	3.4. School-student-family relationship	7	
	3.5. Student-centered practice	10	
	3.6. Classroom environment	10	
	3.7. Teacher well-being	11	
4. Impacts on the teacher training models	4.1. Potential training model	15	67 (26.07%)
	4.2. Training model/limitations	9	
	4.3. Well-being in the educational community	7	
	4.4. Atmosphere in the classroom	6	
	4.5. Strategies for different classroom contexts	6	
	4.6. Guidelines for the development of self-reflection	9	
	4.7. Socioemotional learning	4	
	4.8. Executive functioning	4	
	4.9. Student learning and Well-being	7	
Total			257 (100%)

the most recording units is the impacts on the practices in the classroom with 28.79% of the recording units for this section. Next is the impact on the teacher training models with 26.07% of the recording units. An analysis of the data obtained for each category is presented in the following part.

24.4 Discussion

Our objective was to describe the perception of teachers about school well-being as a collective and sustainable construct. It was analyzed whether teachers see themselves as mediators to generate well-being in their classrooms with their groups of students. It was observed that teachers perceive that wellness strategies, especially those that promote socioemotional learning and executive functioning, improve the personal development of teachers and students. The skills observed were awareness, empathy, attention, concentration, organization, and self-regulation. Teachers showed themselves as more motivated, optimistic, and emotionally empowered people. All of this produced concrete improvements in the teaching and learning process, as well as in the classroom and school atmosphere. According to the data, the intervention generated a positive impact on the personal and professional well-being of teachers.

This study must be considered as part of a movement to study the development of new emotional skills and qualities among educators and students (Bürgener & Barth, 2018) with similar positive outcomes, in this case at the level of the primary schools. As it has been mentioned in the introduction, it seems that both students and teachers need skills in order to attain the goal of creating a new pedagogy that takes into account not only cognitive aspects of learning, but also emotional and social aspects, as it is emphasized in SEL (Rieckman, 2018). This study adds one more evidence showing that the development of teaching skills that promote strategies for well-being, socioemotional learning, and executive functioning should be an essential part for a sustainable education (Denston et al., 2022). It adds evidence showing that teachers who see themselves as agents of change feel more motivated to develop new pedagogies (Martinsone, & Damberga, 2017) and seems to show once again that emotions and cognitions developed between students and teachers are related to school environment and academic results (Milatz et al., 2015; Roorda et al., 2011). Furthermore, this study adds further evidence on the importance of SEL at school, this time not only reporting students' emotions (Scott et al., 2021), but in a controlled intervention. Once again, these results show that the social environment and the emotional states of students and teachers are essential in the process of learning and teaching (O'Conner et al., 2017), crating conditions for a positive and more sustainable climate at schools.

Thus, it shows the need to develop new training approaches, centered on a sustainable vision of the well-being of teachers and students, as well as oriented toward the development of socioemotional learning and executive functions. Teacher training should include strategies such as sharing and active discussion based on experiences, sharing instruments/materials, and reflective methodologies, as well as models based on the well-being of the educational community. These results agree with the above-cited studies on SEL, sustainability education, and socioemotional well-being, which recognize the relationship between education and health and promotes principles of equity, sustainability, and inclusion (Velasco, 2021). In summary, all aspects of the school have an impact on the health of students and the school must provide much more than academic learning: The school should focus on promoting physical and mental health, as well as social well-being.

24.5 Conclusions

This study seems to demonstrate that teachers who understand the importance of well-being in education promote improvements in themselves and in the students in the teaching-learning process and in the educational context. According to the data, the intervention carried out had a positive impact on the personal and professional well-being of the participating teachers and subsequently led to a positive impact on the sustainable well-being of their students. The importance of teacher training in themes related to well-being that integrated the five stages of the SWCI course must be emphasized.

Teachers seem to have positively experienced this reflective training model – teacher-centered coaching. This indicates the importance of adopting training models centered on the well-being of the teacher, the student, and the educational community. Last but not least, this study highlights once more the importance of SEL in the modern schools, showing that the development of teaching skills that promote strategies for well-being, socioemotional learning, and executive functioning should be an essential part for a sustainable education and more positive involvement of students and teachers in the process of learning and development. These characteristics of teaching and learning create the conditions for a more accurate learning for well-being and sustainability.

Finally, this study has certainly some limitations. The sampling method was restricted to a sample of convenience in order to explore the studied fields, i.e., it is difficult to generalize these results. As usual in a natural environment, many variables can intervene in producing a result, i.e., these results could have been influenced by other unknown factors. Additionally, the participants were teachers, who were all motivated to improve well-being in the classroom, and thus comparisons and generalizations are difficult. In future research, more diversified samples of the changes in students' well-being might yield data suitable to look for causal relationships and predictive capabilities. Various socioeconomic groups and ethnic minorities should be included. It will also be important to evaluate families as well as other members of the school staff. It is then important to promote the replication of this study in other contexts to promote health, for example, other teaching centers. Finally, this study highlights the possibilities of a multidimensional approach to the well-being of students and teachers in educational contexts and the usefulness of opting for teacher-centered coaching models, in order to make the teacher a mediator for the well-being of students.

Acknowledgments We are grateful to the psychologists and teachers who contributed to the Strategies for Well-Being in the Classroom Intervention.

Funding This study was funded by the Erasmus + Program of the European Union through the REFLECT Project "REFLECT – Raising awareness and stimulating Executive Functioning and Socioemotional Learning by integrating Evidence-based strategies in the Classroom to empower pupils, teachers and parents (Ref. 2019-1-BE02-KA201-060353)". The support of the European Commission for the production of this publication does not constitute an endorsement of its content, which reflects the views only of the authors, and the Commission cannot be held responsible for eventual uses that could be made with the information contained therein.

References

Ascorra, P., Morales, M., Cuadros, O., & Leal, F. (2021). *Bienestar sostenible y colectivo: de la comprensión individual a la comprensión colectiva* (pp. 14–26). Revista Docencia.

Bürgener, L., & Barth, M. (2018). Sustainability competencies in teacher education: Making teacher education count in everyday school practice. *Journal of Cleaner Production, 174*, 821–826. https://doi.org/10.1016/j.jclepro.2017.10.263

Choi, A. (2018). *Emotional well-being of children and adolescents: Recent trends and relevant factors*. OECD education working papers (No. 169). OECD Publishing. https://doi.org/10.1787/41576fb2-en.

Collaborative for Academic Social and Emotional Learning (CASEL). (2003). *Safe and sound: An educational leader's guide to evidence-based social and emotional learning (SEL) programs*. CASEL.

Collaborative for Academic Social and Emotional Learning (CASEL). (2012). *Effective social and emotional learning programs: Preschool and elementary school edition*. http://casel.org/wp-content/uploads/2021/~01/

Collaborative for Academic Social and Emotional Learning (CASEL). (2016). *SEL impact*. http://www.casel.org/impact/

Comer, J. (2009). *What I learned in school. Reflections on race, child development and school reform*. Jossey-Bass.

Denston, A., Martin, R., Fickel, L., & O'Toole, V. (2022). Teachers' perspectives of social-emotional learning: Informing the development of a linguistically and culturally responsive framework for social-emotional well-being in Aotearoa New Zealand. *Teaching and Teacher Education, 117*, 103813. https://doi.org/10.1016/j.tate.2022.103813

Durlak, J. A., Weissberg, R. P., Dymnicki, A. B., Taylor, R. D., & Schellinger, K. B. (2011). The impact of enhancing students' social and emotional learning: A metaanalysis of school-based universal interventions. *Child Development, 82*(1), 405–432. https://doi.org/10.1111/j.1467-8624.2010.01564.x

Ecclestone, K., & Rawdin, C. (2016). Reinforcing the 'diminished' subject? The implications of the 'vulnerability zeitgeist' for well-being in educational settings. *Cambridge Journal of Education, 46*(3), 377–393. https://doi.org/10.1080/0305764X.2015.1120707

Fuster, D. (2019). Investigación cualitativa: método fenomenológico hermenéutico. *Propósitos y Representaciones, 7*(1), 201–229. https://doi.org/10.20511/pyr2019.v7n1.267

Grange, L. L. (2017). Environmental education after sustainability. In B. Jickling & S. Sterling (Eds.), *Post -sustainability and environmental education* (1st ed., pp. 93–107). Palgrave Macmillan.

Hargreaves, A. (2002). Sustainability of educational change: The role of social geographies. *Journal of Educational Change, 3*, 189–214. https://doi.org/10.1023/A:1021218711015

Haynes, N. M. (2021). *Toward a grounded theory of social and emotional learning (SEL)* (p. 3618). Academia Letters. https://doi.org/10.20935/AL3618

Johnston, O., Wildy, H., & Shand, J. (2022). That teacher really likes me' – Student-teacher interactions that initiate teacher expectation effects by developing caring relationships. *Learning and Instruction, 80*, 101580. https://doi.org/10.1016/j.learninstruc.2022.101580

Jones, A. L., & Kessler, M. A. (2020). Teachers' emotion and identity work during a pandemic. *Frontiers in Education, 5*, 583775. https://doi.org/10.3389/feduc.2020.583775

Lengyel, A., Kovács, S., Müller, A., Dávid, L., Szőke, S., & Bácsné Bába, É. (2019). Sustainability and subjective well-being: How students weigh dimensions. *Sustainability, 11*(23), 6627. https://doi.org/10.3390/su11236627

López, V., Ramírez, L., López-Concha, R., Ascorra, P., Álvarez, J. P., Carrasco-Aguilar, C., Jervis, P., Squicciarini, A. M., Simonsohn, A., Contreras, T., & Opazo, H. (2022). Well-being of school communities in the context of COVID-19 pandemic: A qualitative study in Chilean low-SES schools. *Frontiers in Psychology, 13*, 853057. https://doi.org/10.3389/fpsyg.2022.853057

Manning, M., & Fleming, C. (2019). Understanding well-being. In C. Fleming & M. Manning (Eds.), *Routledge handbook of indigenous well-being* (pp. 3–10). Routledge. https://doi.org/10.4324/9781351051262

Martinsone, B., & Damberga, I. (2017). Qualitative analysis of teachers' written self-reflections after implementation of a social-emotional learning program in Latvia. *International Journal of School & Educational Psychology, 5*(4), 215–225. https://doi.org/10.1080/21683603.2016.1225236

Milatz, A., Lüftenegger, M., & Schober, B. (2015). Teachers' relationship closeness with students as a resource for teacher well-being: A response surface analytical approach. *Frontiers in Psychology, 6*, 1949. https://doi.org/10.3389/fpsyg.2015.01949

Mulà, I., Tilbury, D., Ryan, A., Mader, M., Dlouhá, J., Mader, C., Benayas, J., Dlouhý, J., & Alba, D. (2017). Catalysing change in higher education for sustainable development: A review of professional development initiative for university educators. *International Journal of Sustainability in Higher Education, 18*, 798–820. https://doi.org/10.1108/IJSHE-03-2017-0043

O'Conner, R., De Feyter, J., Carr, A., Luo, J. L., & Romm, H. (2017). *A review of the literature on social and emotional learning for students ages 3–8*. Department of Education. http://ies.ed.gov/ncee/edlabs

OECD. (2023). *Schools as hubs for social and emotional learning: Are schools and teachers ready?* OECD Education Spotlights, No. 4, Paris: OECD Publishing. https://doi.org/10.1787/f6d12db7-en

PORDATA. (2021). *Statistics about Portugal and Europe*. Fundação Francisco Manuel dos Santos. https://www.pordata.pt/en/Subtheme/Portugal/Teaching+Staff-43

Rieckman, M. (2018). Key themes in education for sustainable development. In A. Leicht, J. Heiss, & W. J. Byun (Eds.), *Issues and trends in education for sustainable development* (pp. 39–59). UNESCO.

Ronen, T., & Kerret, D. (2020). Promoting sustainable well-being: Integrating positive psychology and environmental sustainability in education. *International Journal of Environmental Research and Public Health, 17*(19), 6968. https://doi.org/10.3390/ijerph17196968

Roorda, D. L., Koomen, H. M. Y., Spilt, J. L., & Oort, F. J. (2011). The influence of affective teacher–student relationships on students's chool engagement and achievement a meta-analytic approach. *Review of Educational Research, 81*(4), 493–529. https://doi.org/10.3102/0034654311421793

Saab, H., & Klinger, D. (2010). School differences in adolescent health and well-being: Findings from the Canadian health behaviour in school-aged children study. *Social Science & Medicine, 70*(6), 850–858. https://doi.org/10.1016/j.socscimed.2009.11.012

Schleicher, A. (2018). *World class: How to build a 21st-century school system, strong performers and successful reformers in education*. OECD Publishing.

Scott, S. R., Rivera, K. M., Rushing, E., Manczak, E. M., Rozek, C. S., & Doom, J. R. (2021). "I hate this": A qualitative analysis of Adolescents' self-reported challenges during the COVID-19 pandemic. *The Journal of Adolescent Health, 68*(2), 262–269. https://doi.org/10.1016/j.jadohealth.2020.11.010

Shirley, D., Hargreaves, A., & Washington-Wangia, S. (2020). The sustainability and unsustainability of teachers' and leaders' well-being. *Teaching and Teacher Education., 92*, 102987. https://doi.org/10.1016/j.tate.2019.102987

Velasco, V. (2021). *Health-promoting schools in the time of the COVID-19 pandemic*. School for Health in Europe (SHE). https://www.schoolsforhealth.org/sites/default/files/SHE-HPS-in-time-of-pandemic.pdf

Wals, A. E. J., & Jickling, B. (2002). Sustainability in higher education: From doublethink and newspeak to critical thinking and meaningful learning. *International Journal of Sustainability in Higher Education, 3*, 221–232. https://doi.org/10.1016/S0952-8733(02)00003-X

Wang, S. (2017). "Teacher centered coaching": An instructional coaching model. *Mid-Western Educational Researcher, 29*(1), 20–39. https://www.mwera.org/mwer/volumes/v29/issue1/v29n1-wang-voices-from-the-classroom.pdf

Zembylas, M. (2007). Emotional ecology: The intersection of emotional knowledge and pedagogical content knowledge in teaching. *Teaching and Teacher Education, 23*(4), 355–367. https://doi.org/10.1016/j.tate.2006.12.002

Peace as an Essential Condition for Sustainable Development

25

Candido Alberto Gomes and Susana de Oliveira e Sá

25.1 Introduction

Although they have been side-lined to the background due to the pandemics and the war in Europe, the United Nations has established interconnected sustainable development goals expected to be in force by 2030. However, literature portrays wars as sources of relevant damage to the environment, in opposition to such goals (United Nations, 2015). During the interwar period, Freud (2011) studied the continuous struggle between Eros and Thanatos, constructive and destructive impulses, life and death. Could this account for the perenniality of warfare, such as World War II itself, the Cold War, and successive intra-national and international conflicts? Wars obey an instinctual destructive force, the destruction of living beings in general, human beings, their environment, and their things, which bring significant social, economic, political, and cultural impacts. In such conflicts, where one breathes an atmosphere of fear and anomie, ethics and the power of law fade away, as whatever means to defeat the enemy with efficiency and effectiveness are put into practice.

In fact, literature concerning the effects of wars on the environment has been established. This chapter aims to briefly summarize research findings, in an attempt to measure its series of effects, concluding with considerations on the unfulfilled norms of public international law, despite the fact that its authors are no jurists.

Time and again land is interdicted and rendered unusable due to contamination. Nowadays, relatively often rail lines and roads in Europe are interrupted to defuse bombs, sometimes even bombs from the First World War period. However,

C. A. Gomes
CIDI' Research Center, Fafe, Portugal

S. de Oliveira e Sá (✉)
CIDI' Research Center and Joint Director of its Research Center, Fafe, Portugal

Post-Doctoral in Catholic University Brasília, Education, Brasília, Brazil

© The Author(s), under exclusive license to Springer Nature Switzerland AG 2023

F. L. Almeida et al. (eds.), *Multidimensional Sustainability: Transitions and Convergences*, Springer Proceedings in Earth and Environmental Sciences,
https://doi.org/10.1007/978-3-031-24892-4_25

destruction and risks are not limited to Europe, impairing countries in the Southern Hemisphere as well. Although there are theories about the historical end of colonialism, power relations of "more developed" countries over "less developed" ones prevail, pursuant to globalization and the evolution of contemporary capitalism.

25.2 Wars in the Northern Hemisphere

Focusing on industrial age wars, let us first discuss the environmental damage that took place in the Northern Hemisphere, and then what happened in the Southern Hemisphere. The most prominent fact is that rebuilding and restoring costs are much higher and more time-consuming than warfare destruction. The first far-reaching war after the Industrial Revolutions, World War I recorded a great deal of contribution from the chemical and mechanical industries. In the case of France, once peace was established, the army and civilian companies developed techniques for the destruction of chemical weapons near Verdun, the decisive battle site, where arsenic bombs, perchlorate explosives, and others were located. It was imperative to destroy them due to the high risks they posed to the civilian population, including risks to crops and drinking water contamination. In addition to French armaments, the German Army left behind unexploded ammunition even in houses and railroad cars. The French Army also accumulated weapons and did not warn the civilian population about them. Industrial weapon destruction plants had to be erected, but they also led to soil and water contamination, something almost inevitable given the technology available back then (Lin, 2021).

In World War II, when atomic bombs were eventually used, the massive bombings in Italy alone created an impact on the landscape, resulting in risks to personal safety, soil, water, and archaeological sites. Often archaeologists find unexploded bombs in excavations. Comparison of aerial photographs before and after attacks is employed at great expense. The use of GPR (ground penetrating radar) and other techniques also help detecting them. Estimates indicate that 10–15% of the dropped bombs did not explode, something between 80,000 and 100,000 tons (Barone, 2019). Such high number do not even consider mines. Johnston (2007) estimates that worldwide 70% of mines are located in the territories of native peoples, that is, the most vulnerable persons.

From economic and ecological perspectives, war intensifies the use of non-renewable natural resources and thus increases the demand for mining. According to Cairns (2004), war is an unsustainable practice on a finite planet with finite resources. Less recent estimates indicate that in the post-Cold War period, mining for military purposes, often with polluting effects, accounted for 11% of the world's copper, 9% of iron, and 8% of lead, aside from rare metals (Biswas, 2001). Biswas (2001) presented a global estimate of land use for military purposes, and it ranged from 750,000 to 1.5 million square kilometers, an area larger than France and the United Kingdom combined.

There are also armed conflicts over resources in developing countries carried by groups linked to and sponsored by multinational companies based in the Northern

Hemisphere, i.e., conflicts that are "exported" to poor countries. Billon (2001) exemplifies conflicts between Israel and Palestine over water; over coffee in El Salvador; involving Nigeria and Biafra and Sudan over oil, over marijuana and timber in the Philippines. Peace depends, then, on external forces. Even in peaceful times, mining companies cut costs by using outdated technologies in the Southern Hemisphere, in countries like Brazil. The history of dam disasters in Brazil alone includes the tragedies of Mariana and Brumadinho, with the spreading of tailings that are highly damaging to both the environment and the public health (Freitas et al., 2019).

Still during the Cold War, experiments with atomic weapons were transferred to tropical islands controlled by great powers. Far from major centers, they were in areas inhabited by first peoples, tribal groups, and other ethnic minorities, i.e., populations of low socioeconomic status and little power (Baum, & Barrett, 2018; Johnston, 2007). Given the inherent secrecy of military activities, it is difficult to fully identify their effects on human health and the environment. Some 504 nuclear weapons have been exploded in the atmosphere at 13 primary sites, emitting radioisotopes and dangerous heavy metals. Thermal radiation, electromagnetic pulse, as well as physical and mental impacts on the population are all part of a long list of side effects.

However, the Cold War became hot in selected locations and non-world conflicts, not always involving or reflecting clashes between the Western and Eastern blocs. In the Persian Gulf War, with the confrontation between Iraq and the Coalition, widespread oil spills took place near destroyed facilities and on the Saudi Arabian border to prevent Coalition troops from moving in. These spills reached the Red Sea, causing marine life to become unbalanced. Oil wells were torched in Kuwait, leading to great environmental damage, including to the atmosphere. The objective of war is to defeat the enemy in the shortest amount of time, without considering means or consequences. It is estimated that more than a million mines were laid in Kuwait, not counting the unexploded bombs. As for the Coalition, at least a third of its bombs and missiles failed to explode. Around 1250 civilians and over 50 mine deactivation experts died because of these mines after the conflict. Cairns (2004) stated that wars over finite resources will continue until the human species learns about sustainability.

25.3 Wars in the Southern Hemisphere

The Southern Hemisphere, with its vast majority of developing countries, endowed with more abundant natural resources and weaker resistance to pollution, becomes a target for the greed of post-industrial economies. The disregard for sustainability translates into their consumerist economic system, demand for raw materials, energy, and their own industrial products, particularly from polluting companies transferred to developing countries. It is like a voracious machine, aiming at the continuous growth of the gross domestic product (GDP) while ignoring externalities. Ignoring externalities is closing one's eyes to the consequences, thus coating oneself with a deep ideological bias, which is far from innocent. Thus, the report by Stiglitz et al. (2009), which among other propositions advised that GDP was calculated in such a way as to subtract non-renewable natural resources used for production, was abandoned. According to this method, of course, the continuous growth dogma would fade away and the post-growth society would be later achieved (Cassiers et al., 2018).

Economic globalization has connected developed countries with developing countries, including the latter in the process in a subordinate position as to satisfy its expansionism, with its hunger for consumption of materials and energy. However, it is rather an exclusionary inclusion, in the sense that it keeps the new "colonies" in subordinate conditions in many ways. In general terms, after the imperialism and colonialism of the nineteenth and twentieth centuries, the great powers have found in globalization and network societies less costly forms of domination and spoliation than before. Once the non-renewable resources are exhausted, these countries can be abandoned with all their woes, struggling with the weakness of their institutions and difficulty or impossibility of fulfilling their basic functions. They then seek another export "economic cycle" or some parallel economic activity, even drug trafficking, to achieve temporary prosperity, which will be concentrated in a few wealthy groups. Therefore, such country is excluded from the international system, as well as the majority of its population, since income tends to become more and more concentrated (Chancel et al., 2022), excluding the very members of society, who only marginally get a share of this wealth. Even worse, what remains for the less well-off parts of society is equally distributed to the detriment of those of lower status, such as minorities in general, especially women, ethnic minorities, and all those who, like pets, live off the crumbs that fall from their owners' tables (Piketty, 2014; Vitaud, 2022).

Still with the framework of the Cold War, the long Vietnam War had serious ecological consequences for that country and also for Cambodia. Fumigations, especially using the so-called "agent orange", and 1.8 million tons of bombs dropped on rice fields in Vietnam alone got intertwined with poverty and famine. Defeating the enemy justifies any means. In Cambodia, bombings led farmers in contaminated areas to produce 50–60% less rice, the main source of nutrition of the local population, lowering incomes by 60% (Lin, 2020).

After the Cold War, traditional sources of funding for armed insurgent groups dried up. Many then sought alternative resources outside the state, in the shadows of smuggling and drug trafficking. In mining areas, such as in sub-Saharan Africa, such groups can be financed by rival companies interested in capturing mining operations. Environmental ecological terrorism has been practiced in several areas, with deliberate use of force to affect normal use of the environment and natural resources (Lavaux, 2004). Since the Middle Ages, fresh water has been a target and an instrument of warfare. It is therefore not surprising that the intentional pollution of its sources and courses are military targets, as well as dams.

One of the most enduring, half-century-long, conflicts has been that of the Revolutionary Armed Forces of Colombia (FARC) (1966–2016). At first an informal arm of the Communist Party, in the framework of the Cold War, it developed most of its activities in the Amazon jungle, an ecologically vulnerable territory where the presence of the state was rare. It was a response to the horrible living standards of the population, to which economic groups were continually insensitive. After the Cold War, there was a lack of financial resources to sustain and expand the movement. To compensate for this, just like other armed groups (Billon, 2001), they resorted to kidnappings, forced contributions, and other criminal activities, especially growing coca and processing it for sale. In addition to destroying the environment, they have engaged in a tangle of illegal acts, provoked a population exodus, with a high number of internal refugees, and the death of countless people. The reaction of the Colombian state, with US support, manifested itself in the form of combating the production of coca, although the largest consumer markets are in developed countries, among populations with high purchasing power. Aerial fumigation has had disastrous effects on the jungle, contaminating soil and water, particularly in the Amazonian state of Putumayo, as if the cultivation and refining of coca per se were not enough, bringing along the use of fertilizers, pesticides, and other substances to increase productivity, as in any capitalist activity. Deforestation and defoliation have affected the subtle ecological balance of the tropical jungle. The annual rate of forest loss, which was 218.1 square kilometers in 2000–2010, grew in the middle of the second decade of this century to no less than 295.9 square kilometers (Aquino et al., 2017), apart from forest fragmentation. Therefore, the number of human and animal victims, migrants, refugees, and damage to ecosystems is immense.

These war tactics, also due to the extent of the damage caused to the planet, heritage of all people, have worried armed forces. For example, the Peruvian Army aims to establish a commitment to environmental protection by proposing in its military instruction the Military Program of Environmental Education and Control (Adolfo, 2020). India has developed ecological intervention units for the first time in the world, dedicated to natural preservation missions such as pasture and soil care, water purification, and stabilization of dunes. China has taken the same initiative (Keucheyan, 2018). On the other hand, great powers also begin to worry about such problems. The prevailing capitalism is not interested in ecological crises and political-military instability caused by damage to ecosystems (Keucheyan, 2018). An ancient and illustrative historical example lies in the Kingdom of Mari, which developed in Mesopotamia, on the banks of the Euphrates, as the most important city-state during part of the second millennium before Christ. No records of military conflicts were found, as the basis of its economy was trade. Therefore, the kingdom's elite were vehemently opposed to wars, once these disrupted their business (Wright, 1975).

25.4 Conclusions

Literature shows that wars are harmful to the environment, to natural and human life, and to human beings and the context in which they live. Damage has existed ever since, but with the increasing technification of conflicts, the negative consequences tend to increase. Why?

This is no legal work since it is not our specialty. However, we must examine the issue, even if only from a bird's eye view. Distance and contradiction in the face of law are not only specific to military conflicts but are also found in public policies and individually, through the actions of individuals or groups, who often take refuge in the formal minutiae rather than in the spirit of laws, in other words, hiding behind details to escape public criticism.

First, there *is* a complicated *pas de deux* between the legal and the social fact. Particularly when writing a constitution and dealing with human and social rights, one of the dilemmas is to find the balance point between advancing them and the possibilities for society to fulfil them. If an advancement risk being massively disrespected by state bodies and civil society, the norm may lose its legitimacy. In case of retreat, a historical window may be lost and will not be reopened any time soon. So, what to do? For example, how far is it possible to go in terms of gender equality in a typically macho society? How will legal and social facts collide? Acts can be criminalized, but what twists and subterfuges can lead to the perception of injustice and the consequent revolt? Another example is the participation of homosexuals in the armed forces, which varies from country to country. In Portugal, the army presented homosexuality as a "contraindication", later abolished. Higher standards, such as the EU Declaration of Human Rights, do not justify discrimination. Consequently, the expectation is that regulations follow the constitutional and legal hierarchy. However, the persistence of dissonance due to sociocultural tradition cannot be ignored. Regulations, therefore, must be enforced and deviations sanctioned. Otherwise, they become illegitimate and considered as prescriptions that do not need to be fulfilled, falling into oblivion.

Legislation, its enforcement, and judgments can be compared to an arena where customs and laws clash, as well as organized groups, the media, public opinion, and risks of losing or winning votes in a democratic state (Allison & Zelikow, 1999). Media itself can create parallel realities or imitations of reality, according to Baudrillard (1981). It is pretty much an arena, in the sense of the Roman circus, that is, a ground that absorbs the blood of fighters, both human and animal. In this place, forces and strategies, beliefs and values, attitudes and behaviors are measured. Bureaucracies—from *kratós*, power— themselves can deviate from an ideal type of neutrality (that's why it is an ideal type; Weber, 1968), because they are susceptible to becoming politicized, partisan, ideologized, affecting their legitimacy in a more or less serious way.

Legitimacy and legitimation refer to Habermas' great work, *Legitimation crisis*. What is this immaterial and abstract process that proves so important in society and in law? Its absence creates intangible, yet widely noticeable voids in social life, as in contemporary society. The author, a member of the Frankfurt School, considers that capitalism is in crisis and Western society has reached anomie, in simple words, the lack of norms, a Durkheimian concept. Having passed the stage of liberal capitalism analyzed by Marx, it would need to rely on a new form of analysis. Capitalism has changed its features and content, so that it has gradually moved away from pre-capitalist moral values. Why? Habermas attributes this situation to the failures of the state while correcting contradictions of recent capitalism. According to Piketty

(2021), the concentration of income continues to grow, with a tendency to continue this way, which means for the German philosopher a promise breach regarding results distribution for all. Instead, there is impoverishment of large parts of the population. Consequently, the economic-political power to obtain laws favorable to the status quo's interests grows. Symptomatic cases are the loosening of social and labor rights and attempts for flexibility of environmental norms, in order to obey the "invisible hand" within certain moralistic and conservative ideological guidelines, such as the "right to life". In truth, this right becomes a folding screen of contradictions: the right to life is practically denied by the fragile livelihood of families, scarcity of decent juvenile employment, and environmental degradation, the latter manifested by health indicators, including mortality. In other words, it is life exchanged for death. What is the logic of defending the "right to life" through religion if the population around a cement factory, especially children, have their survival limited by respiratory diseases? The sense of justice expects norms and acts to be internally and externally coherent, otherwise delegitimization and the void of anomie will ensue.

Weber (1968) drew attention to the need for legitimation and justification of any kind of domination and authority. After all, force and coercion are not enough to guarantee acceptance and fitting into a different legal and social order. Thus, feelings of injustice emerge, which delegitimize justice and prompt the "law" of the strongest. In our view, this gap helps explain the increase of authoritarian and liberal states in the world, at the same time the number of really Democratic States under the Rule of Law is shrinking. As a result, the Law's aspiration to provide protection and security to everyone is undermined. Would it be a mere coincidence that the growing belief in men concentrating power, able to solve "everything", as in a huge company, and the resurgence of imperial daydreams, similar to those of Napoleon and Hitler? Like the current of a river, particularly the river of Heraclitus, history does not turn back.

In these impasses Habermas (1976) resorts to a Hegelian concept, that of *Sittlichkeit*, one of those German words not very prone to translation. It could be understood as ethical life or ethical order. Lacking a rational consensus, a new *Sittlichkeit* must be woven, to coherently address the new order.

At this point we return. What drives the clear disobedience of wars towards the Hague and Geneva conventions on public international law? One of the answers lies in the crisis of legitimacy. Another one is the acceptance of any means to achieve desired ends, as codified (not created) by Machiavel (1980). This classic author lived in Florence, and the events he witnessed as a child were vivid illustrations of the ambience of the time, largely reflected in his work. War constitutes a situation of anomie, where crimes against humanity stand out—as history details it. In the case of the war in Ukraine, it is widely documented by reliable media. What does it matter, in the face of a highest goal, to defeat the enemy with the greatest efficiency and cause him dread, to instill in everyone the feeling that the attacker may hide in any space or time, ready to erode defenses and render defensive efforts useless? What is the problem with killing through starvation if the cost of it is not high? Ethics and law are ignored. Punishment, strictly speaking, emerged with the International

Court in the Hague, starting in Nuremberg. Then, the sense of impunity prevails. It will take a defeat for states to hand over those responsible for such crimes, if those responsible do not die beforehand. Furthermore, the coercive power to apply sanctions to a state that disobeys the norms of public international law is largely lacking.

In conclusion, these are dilemmatic situations that humanity must overcome. Certainly, the Hegelian *Sittlichkeit* applied by Habermas (1976) is one of the fundamental keys here. A concerted worldwide action may be likely to contribute to reversing environmental damage, for environmental solutions need to be global to become effective. Then it is likely that chemical and biological weapons, including those stored and used to deter attacks, will be outlawed. And that environment, people, and things will no longer drift toward a horizon of collective suicide, whether slowly or suddenly. If the ship sails through space carrying all of us, we all either float or we all sink with the boat. According to Morin (1999), social diversity is as essential as biodiversity.

References

Adolfo, G. H. (2020). Educación ambiental en la instrucción militar del Ejército Peruano sobre la historia de los conflictos de guerra y sus efectos. In *Trabajo de suficiencia profesional*. Escuela Militar de Chorrillos.

Aquino, A. R., Paletta, F. C., & Almeida, J. R. (2017). Vulnerabilidade ambiental. São Paulo, Blucher.

Allison, G., & Zelicow, P. (1999). *Essence of decision: Explaining the Cuban missile crisis* (2nd ed.). Longman.

Barone, P. M. (2019). Bombed archaeology: Towards a precise identification and a safe management of WWII's dangerous unexploded bombs. *Heritage, 4*, 2704–2711.

Baudrillard, J. (1981). *Simulacres et simulations*. Galilée.

Baum, S., & Barrett, A. (2018). *A model for the impacts of nuclear war.* N.p., Global Catasthrophic Risk Institute Working Paper 18-2. https://www.researchgate.net/publication/342612113

Billon, P. (2001). The political ecology of natural resources and armed conflicts. *Political Geography, 20*(5), 561–584.

Biswas, A. K. (2001). Scientific assessment of the long-term environmental consequences of war. In J. Austin & C. Brush (Eds.), *The environmental consequences of war: Legal, economic, and scientific perspectives* (pp. 303–315). Cambridge University Press.

Cairns, J. (2004). Allocating finite resources on a finite planet. *Ethics of Science and Environmental Politics., 4*, 25. https://doi.org/10.3990/heritage2040167

Cassiers, I., Maréchal, K., & Méda, D. (2018). *Vers une société post-croissance: Intégrer les défis écologiques, économiques et sociaux*. Éds. De l'Aube.

Chancel, L. (leading author), Piketty, T., Saez, E., & Zuckman, G. (2022). *World inequality report 2022*. World Inequality Lab, n.np., 0098-21_WIL_RIM_RAPPORT_A4.pdf (wid.world).

Freitas, C., Barcellos, C., Asmus Silva, M., & Xavier, D. (2019). Da Samarco em Mariana à Vale em Brumadinho: Desastres em barragens de mineração e saúde coletiva. *Cadernos de Saúde Pública, 35*(5), 1–7, e00052519. https://doi.org/10.1590/0102-311X00052519

Freud, S. (2011). *O mal-estar na civilização*. Penguin & Companhia das Letras. "Exército português apresentava homossexualidade como uma 'contra-indicação'".

Habermas, J. (1976). *Legitimation Crisis*. Polity.

Johnston, B. (2007). *Half-lives, half-truths, and other radioactive legacies of the Cold War*. School for Advanced Research. www.sarpress.sarweb.org

Keucheyan, R. (2018). La nature est un champ de bataille. In *Essai d'écologie politique*.

Lavaux, S. (2004). *Degradación ambiental y conflictos armados: Las conexiones*. Centro Editorial Universidad del Rosario.

Lin, E. (2020). How war changes land: Soil fertility, unexploded bombs, and the underdevelopment of Cambodia. *American Journal of Political Science, 66*(02), 1–16. https://doi.org/10.1111/ajps.12577.

Lin, E. (2021). How war changes land: Soil fertility, unexploded bombs, and the underdevelopment of Cambodia. *American Journal of Political Science, 00*(00), 1–16. https://doi.org/10.1111/ajps12577

Machiavel, N. (1980). *Le Prince*. Flammarion.

Morin, E. (1999). *Les sept savoir nécessaires à l'éducation du futur*. UNESCO.

Piketty, T. (2014). *Capital in the twenty-first century*. The Belknap Harvard University Press.

Piketty, T. (2021). *Une brève histoire de l'égalité*. Éds. du Seuil.

Stiglitz, J., Sen, A, & Fitoussi, J.-P. (2009). *Report by the commission on measurement of economic performance and social progress*. http:///www.stiglitz-sen.fitoussi.fr/documents/rapport_anglais.pdf.

United Nations. (2015). *Transforming our world: The 2030 agenda for sustainable development*. Transforming our world: the 2030 Agenda for Sustainable Development | Department of Economic and Social Affairs. un.org

Vitaud, L. (2022). *En finir avec la productivité: Critique féministe d'une notion phare de l'economie et du travail*. Payot.

Weber, M. (1968). *Economy and society: An outline of interpretive sociology*. Bedminster.

Wright, G. (1975). *Arqueología bíblica*. Eds. Cristiandad.

Check for updates

The Sustainability of the Cork Sector Through a Second Life of Retired Wine Bottle Corks: An Empirical Study

26

Catarina Ferreira, Marta Cardoso, Marta Moreira, Pedro Sousa, and Tânia Moura

26.1 Introduction

Cork has unique properties, which gives it the possibility of being used for a variety of purposes. It is 100% recyclable and extracted from cork oak without ever harming the normal development of the species. Also, this material resists friction and wear and is highly durable (APCOR, 2018).

The diversification and complexity of the different products that result from cork lead the producers to carry out intense quality control and permanent research into both the continuous improvement of existing products and new and innovative techniques of application.

Research facilities and laboratories in the most advanced production units play a fundamental role and ensure the quality demanded by the end clients, as well as being responsible for validating/certifying new products.

After being transformed from planks into corks, it can re-enter the production process as it is fully recyclable. Corks can be recycled by grinding, and the granulate resulting from this process can be used in other products.

The use of cork and respect for the maintenance of the tree and its life cycles are an example of excellence and a mark of ecological standards.

Being a key producer of cork, Portugal has a responsibility to give a second destination to used cork, thus enabling a circular economy and contributing to society well-being. Therefore, it is necessary to alert all stakeholders to the sector's shortcomings and weaknesses and should include industry, government, education, research, and agriculture.

C. Ferreira (✉) · M. Cardoso · M. Moreira · P. Sousa · T. Moura
ISPGAYA, Vila Nova de Gaia, Portugal
e-mail: ispg2018102855@ispgaya.pt

© The Author(s), under exclusive license to Springer Nature Switzerland AG 2023

F. L. Almeida et al. (eds.), *Multidimensional Sustainability: Transitions and Convergences*, Springer Proceedings in Earth and Environmental Sciences,
https://doi.org/10.1007/978-3-031-24892-4_26

Corks are light in weight, totally impermeable to liquids and practically impermeable to gases, flexible, and provide thermal and acoustic insulation. Additionally, it is a natural fire-retardant material, which doesn't cause flames or release toxic gases and prevents the appearance of mites (contributing to protection against allergies).

The European Union is the largest producer of cork (more than 80%), and in the western Mediterranean countries, Portugal stands out (with more than 50%) and is the world's largest producer and processor of cork and cork products (Gil, 2007).

The cork sector is represented in Portugal by APCOR (Portuguese Cork Association), whose objectives are to promote, publicize, and research the Portuguese cork industry. This association has its headquarters in Santa Maria de Lamas, in Santa Maria da Feira municipality, and has 238 associated companies that represent 85% of exports. Santa Maria da Feira is also the region where most of the companies are based in Portugal.

In Portugal, the cork industry includes cork preparation activities (CAE 16293), the manufacture of corks (CAE 16294), and the manufacture of other cork products (CAE 16295).

Regarding circular economy and sustainability, cork sector is in an infancy stage. Considering the potential of cork infinite reuse, the adoption of measures becomes necessary to face a possible scarcity of resources in the future and minimize the environmental impacts, nonetheless little is being done to recycle or reuse cork. Such need becomes increasingly urgent due to global warming, which causes harmful effects to the environment and humanity.

The main objectives of this research are to (1) improve the sustainability of the sector by promoting the reuse of used corks in other (non-food) products, given that this is a waste product whose recycling is environmentally "friendly," considering its investment/profitability factor and (2) contribute to the literature as this subject isn't covered sufficiently by the actual literature.

26.1.1 Gap

Initially, the authors faced a lack of awareness regarding the recycling of corks in a general context. After investigation, we discovered that despite the existence of various models implemented in the national circuit some years ago, the result of the gathering of used products still shows very low values when compared to the quantity placed on the market (new corks). This is because, according to statistical data published by APCOR (2020a), on average, 40 million corks are produced per day in Portugal, of which 95% are exported. Thus, around two million corks are produced daily for the domestic market.

Within the scope of corks recycling, we've found the following projects nationally: Green Cork; "Rolhão"; Mobile Ecocentre; Corkscrew Action; Maiambiente Collection APCOR (2020b). Among the above-mentioned projects, Green Cork stands out: a project initiated in 2008, which has enabled the collection, to date, of around 295 tonnes of corks and, in return, the planting of 476,000 trees in 2012

(2016 data). 2012 was the year in which the largest collection of corks was observed in Portugal (89 tonnes), while of the approximately 320 million corks produced annually, the collection carried out by Green Cork reaches only 6%. The latest records show that 98,414,732 corks have been collected (since 2009), resulting in the plantation of 1,225,811 trees.

By analyzing the campaigns established to this date, we can see that the programs implemented haven't sorted the desired effects, highlighting the difficulty in raising awareness of the collection of used cork material and the difficulty in obtaining enough collected product to achieve the viability of these projects.

Some inconsistencies were found which can deteriorate the projects underway:

- Collection points in "Continente" supermarkets/hypermarkets aren't very visible/existent
- Lack of positioning and dissemination strategy to the community
- Lack of awareness campaign regarding cork collection and recycling
- Lack of "incentives" for consumer participation

Such gaps, when properly addressed, could enable an increase in the recycling quota. Additionally, it creates the opportunity for the development of new business models and job opportunities.

This chapter is organized as follows: It starts with the current introduction, where the theme of the study and the gaps found in the literature are debated. In Sect. 26.2, a review of the existing literature about the theme of cork is reflected. In Sect. 26.3, the applied methodology is discussed. In Sects. 26.4 and 26.5, respectively, the obtained results are presented and discussed. Finally, in Sect. 26.6, the main conclusions, suggestions for managers, limitations, paper originality, and clues for future investigations are discussed.

26.2 Literature Review

26.2.1 Companies Within the Sector

Portugal is the leading cork-producing country in the world, being responsible for 46% of the world's cork production (APCOR, 2021).

According to Caldeira (2020), the cork industry creates more than 9000 jobs, direct and indirect, throughout the country, with greater incidence in the northern region, where most of the cork manufacturing companies are located (about 83% of the total companies in the sector), and is mostly composed of small- and medium-sized companies (99.4%).

Concerning the report "Information Bureau 2019 – Cork in Numbers" by APCOR (2019), cork production is oriented almost exclusively to the wine industry sector, which retains about 72% of all production, followed by the construction sector with 25% and, finally, with almost 3% for other products.

To analyze the companies in the sector, statistical data from Banco de Portugal (2022) was used. Regarding companies manufacturing other cork products, between 2019 and 2020, there is a decrease in companies and, consequently, in the number of employees. On the other hand, the volume of sales and services showed an increase. Although an overall analysis of the sector shows a decreasing trend in all indicators, in 2020, a greater evidence in sales and services was observed.

These figures reveal a stronger presence of companies manufacturing other cork products, which highlights the importance of investing in improving the circular economy of these companies.

The second largest world producer of cork is Spain (33%), followed by Morocco (6%), Algeria (5%), Tunisia (4%), Italy (3%), and France (3%). In some of these countries, circular economy projects have been implemented as a way of making use of this material. It should be noted the case of France with the lowest production of cork and one of the largest consumers of corks, where there's already a commitment to the use of this material - with one of the best collection results (APCOR, 2021).

26.2.2 The Sustainability of Cork Industry and Circular Economy

Regarding the transformation of cork, only 30% is used in the manufacture of natural corks. However, nothing is lost in the transformation of this material, since the waste is ground and converted into granules, returning to the production process for technical transformation.

According to Corticeira Amorim (2022), the world's largest cork transformation group, "Corks are biodegradable and absorb CO_2, but their life doesn't end once a bottle is opened. Recycling and reusing materials allow the life cycle of the corks to extend, contributing to a better world. Although recycled cork is never used to make new corks, it can be used in various applications" by being reintegrated into the production cycle. In terms of recycling, Corticeira Amorim adds that one should "reuse the by-products of the industrial process, to ensure that nothing is lost and everything is transformed."

Corticeira Amorim (2022) also explains that the surplus from the transformation is converted into biomass (cork dust), which is used to produce energy that covers more than 60% of the group's needs. As a result of the above, Amorim states that "from planks to cork dust, a perfect circular economy model becomes reality."

During the research, with the aim of obtaining valid scientific data on the recycling of corks, we came across a clear lack of scientific information related to the recycling process: "what happens from the collection of corks to obtaining new products?" However, we did obtain information on several projects, both national and international, of corks recycling initiatives, the main driver being Corticeira Amorim (Portugal). In this field, there's only evidence that Amorim Cork Composities, with its sustainability program, shows results from the implementation of this type of program. In the year 2019, Amorim Cork Composities recycled 485 tons of corks, with this initiative increasing in the following years, registering, in 2021, 801 tons of corks. With the implementation of this project, Amorim Cork

Composities achieves "only" the recycling of 3% of the total number of corks produced annually, and the company intends to find ways to increase this figure (Ernst & Young Global Limited, 2022).

This shows that the sustainability of the sector is efficient and effective regarding the use of the raw material and its production cycle. The cork sector in Portugal is well positioned at this level, since almost all resources are used to their maximum potential and waste is minimized.

In addition, a great demand for ground cork is visible, especially on the part of SMEs, with the intention of incorporating it into new products.

João Rui Ferreira, President of APCOR, argues that "it's not difficult to recycle cork, what's difficult is to collect the cork that exists in the world and transform it, especially in Portugal, where a large part of our production is located" (Porto Business School, 2017).

The reuse of cork allows for an "almost" infinite life cycle when the means of collection and recycling are effective and efficient, making the circular economy issue impossible to ignore. In this way, it's possible to improve the participation of the public, given the high recycling/reuse potential of this material. For this, the collection/treatment program(s) must be created/formulated to obtain a higher percentage of product.

In this way, an awareness-raising strategy must be created for existing initiatives and collection points, with the aim of improving the amount of product collected.

26.3 Methodology

This chapter is organized as follows (Fig. 26.1).

In this study, a combination of qualitative and quantitative methodology was used.

The analysis of the cork sector begins by researching the characteristics of the product and the evolution of the sector, such as production, export, recycling/reuse, and transformation of the product.

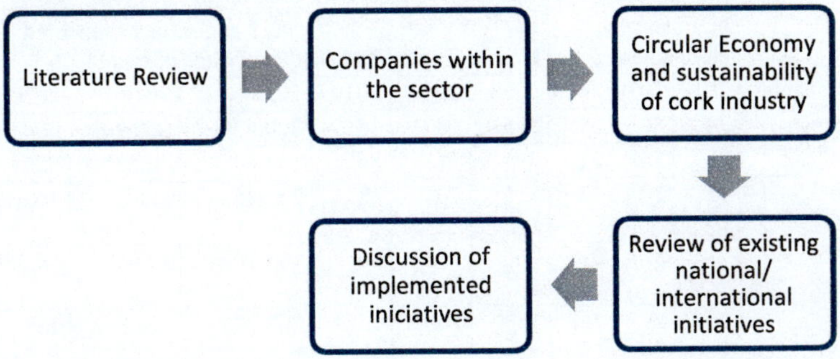

Fig. 26.1 Methodology

This study is based on various statistical data of the cork sector, conclusions/dispositions concerning reports, and the analysis of collection projects already implemented at national and international levels.

For the development of this study, information obtained from books, reports, yearbooks, and articles published on websites was used, as well as field research regarding cork recycling/reuse initiatives already implemented. To complement this information, interviews were conducted with representatives of APCOR (Portuguese Cork Association) and CTCOR (Technological Cork Centre).

26.4 Results

Statistical data concerning the evolution of the cork sector in Portugal was analyzed, with special attention to the manufacture of other cork products. To complement the analysis of the sector, initiatives for the collection of used material with a view to subsequent reuse/recycling were studied.

On the first part, we analyzed statistical data from Banco de Portugal (2022) about the number of companies, employees, and sales and services provided (€) (Table 26.1).

Regarding the companies manufacturing other cork products, in the transition from 2019 to 2020, there was a decrease in companies and in the number of employees, in general. In contrast, the volume of sales and services provided increased by approximately 1.23%, which leads us to the need to invest in the circular economy of the sector, both to drive new business opportunities (without the use of virgin raw material) and to leverage the economic sustainability of the sector.

Analyzing the table of all companies in the sector, in 2020 all indicators show a decrease, with greater evidence in sales and services rendered, in the order of 7.57%. This fall may be due, totally or partially, to the restrictions and, consequently, to the economic crisis caused by the COVID-19 pandemic. This is because it was a period in which there was a limitation on the movement/conviviality of people, which

Table 26.1 Manufacture of other cork products

Economic activity sector: 16295 – Manufacture of other cork products									
Distribution by age of companies	Number of companies			Number of employees			Sales and services rendered (Thousands of euros)		
	2018	2019	2020	2018	2019	2020	2018	2019	2020
All companies	74	82	77	1852	1891	1849	320,429	300,909	304,530
Until 5 years	28	35	31	89	111	86	14,034	16,821	5254
From 6 to 10 years	13	10	11	115	70	83	28,545	5257	16,812
From 11 to 20 years	7	8	9	76	103	96	10,610	13,963	14,338
More than 20 years	26	29	26	1572	1607	1584	267,240	264,868	268,126

Source: Adapted from Banco de Portugal (2022)

Table 26.2 Total companies in the sector

Total CAEs: 16293 – cork preparation industry; 16,294 – manufacture of corks; 16,295 – manufacture of other cork products

Distribution by age of companies	Number of companies			Number of employees			Sales and services rendered (thousands of euros)		
	2018	2019	2020	2018	2019	2020	2018	2019	2020
All companies	768	775	751	8412	8545	8222	1,773,808	1,742,421	1,619,978
Until 5 years	220	219	182	963	1010	853	196,234	206,165	169,308
From 6 to 10 years	140	143	148	1161	1153	1061	263,237	253,130	232,627
From 11 to 20 years	163	163	168	991	968	956	232,010	220,782	184,459
More than 20 years	245	250	253	5387	5414	5351	1,082,328	1,062,345	1,033,585

Source: Adapted from Banco de Portugal (2022)

mainly affected the HORECA sector (hotels, restaurants, and coffee shops) (Table 26.2).

At national level, some initiatives for the collection of used material (corks) have been implemented, some more recently and others have been discontinued.

The main cork recycling initiatives in Portugal were as follows:

Green Cork – developed by Quercus, in collaboration with Amorim, Continente, Dolce Vita, schools, scouts, municipalities, waste collection companies, wineries, and wine producers. It began in 2008 and has enabled the collection, to date, of 295 tonnes of corks and, in return, the planting of 476,000 trees in 2012 (Green Cork, 2020).

Rolhão – created by the São Brás de Alportel municipality (Algarve) at the beginning of 2005, encouraging its citizens to place corks in containers provided for that purpose, next to the ecopoints (APCOR, 2020b).

Ecocentro Móvel – in 2021, Lipor inaugurated the mobile ecopoint in various cities in the metropolitan area of Porto, with the aim of collecting waste that could be recycled, such as corks (APCOR, 2020b).

Ação Saca-Rolhas – this is a project created by the Guias de Portugal Association (AGP) in 2005, with the aim of promoting the collection of used corks, reusing them for new products/projects. By December 2012, AGP had collected 20,000 kg of corks (APCOR, 2020b).

Recolha Maiambiente – LIPOR, Quercus and Maiambiente have created a door-to-door collection channel for corks from non-residential customers (restaurants, coffee shops, canteens, etc.) in Maia municipality. This project is supported by Green Cork, BA Glass, Corticeira Amorim, and Extruplás (Lipor, 2022).

Internationally, there are also some initiatives for collecting used products (corks), among which the following stand out:

EcoBouchon (France) – project established by Amorim Cork France in 2009, with the aim of strengthening customer relations by combining environmental protection with social solidarity. "Today it is the world's largest contributor to corks recycling" (Corticeira Amorim, 2022).

EtiCo – established in 2011 and counts with the participation of associations and institutions that mobilize around a thousand volunteers and manage more than 5000 collection points throughout Italy. The collection of corks is transformed into donations to charitable institutions and, simultaneously, favors the circular economy (Corticeira Amorim, 2022).

Cork2Cork – created in 2011, it is a partnership between Amorim Cork Flooring and the NH Hotel Group to promote a more sustainable world. In the first stage (2011), 68 containers for corks were delivered, in hotels in Spain, Belgium, Italy, Germany, France, and the Netherlands. In the second stage (2019), 74 containers were provided in hotels in Spain and Italy (Corticeira Amorim, 2022).

Amorim Cork Life – project started in 2013, in South Africa, whose aim is to collect used/surplus corks to give them a new use, creating jobs (Corticeira Amorim, 2022).

Recork – corks recycling program established in North America in 2008 by the Canadian footwear company SOLE. The project has more than three thousand partners, who send the collected corks to other manufacturers responsible for transformation, for future incorporation into the production of SOLE's shoes (Corticeira Amorim, 2022).

Worldwide, 550 million corks have been collected since 2018. France is the leader in this process, with the annual delivery of approximately 45 million corks. However, this quantity is only a very low percentage of the number of corks used per year (12 billion), thus not reaching 0.4% (100% Cork, 2022).

According to Per & Britt Karlsson (2021), only 2–3% of all corks were recycled in 2021 worldwide. Corroborating this data, the Ernst & Young Global Limited (2022), in the sustainability report of Amorim Cork Composites, mentions the collection of "only" 3% of the total corks produced annually.

Through sustainable cork collection and recycling, the cork sector is participating in the fight against climate change, preserving forests.

26.5 Discussion

From the analysis of the literature and the results obtained, it's necessary to answer some important questions: What is the evolution of the initiatives already implemented in Portugal? What's the result obtained in cork collection at an international level? What can be improved in terms of collection of used material?

At national level, Green Cork is the project that stands out. According to Green Cork's latest report (2019), the results presented were as follows: from 2009 to 2019, 408 tonnes of corks were collected, corresponding to 91,004,000 corks, which enabled the planting of 1,075,688 trees.

The remaining national initiatives show little expressiveness, either due to the prematurity of the project or due to the low adhesion of the public leading to their stagnation/extinction.

Internationally, the project with the best results is the one in France, with around 45 million cork stoppers collected. Another project worth highlighting is that of the NH Hotels group, Cork2Cork. With the product collected through this initiative, implemented in hotels in Spain and Italy, 300 rooms were insulated, improving comfort and energy efficiency.

"That way our guests will be able to witness how the corks are set aside and collaborate with the initiative directly" – NH Hotel Group (2019).

As the public is a key part of the success of such projects, it's also important to let them know what is done/improved after their input.

26.6 Conclusions

26.6.1 General Conclusions

Cork is a material increasingly used in a wide range of areas, as it has unique characteristics and strong potential both in terms of sustainability and circular economy.

As it's an eco-efficient natural product with an unlimited life cycle, this requires a sufficiently strong contribution from the consumer/buyer to enable its reintegration into a new production cycle.

Portugal is one of the largest producers, processors, and exporters of cork in the world, with the greatest share of this production going to the wine sector, through the manufacture of corks for bottles. On average, around 40 million corks are produced daily.

Regarding sustainability, the cork sector needs to adopt additional measures for its growth. This is a gradual process whose effects aren't seen immediately. However, given the increasingly pressing need to save resources and minimize the environmental footprint, it's important to find alternatives that enable a growing improvement in said sustainability. This improvement may be obtained from synergies both in recycling and in reusing this material.

However, by analyzing the initiatives already implemented, the consumer/buyer isn't aware of the impact that not reusing/recycling causes on the environment since it's a natural product. On the other hand, we are faced with a high demand for this material already used for incorporation into new products by some recent companies in the market from different areas of activity. These seek, through the cork, to highlight their products as ecologically sustainable and improve their contribution to the circular economy in favor of a better planet. Given the small volume of collected products available for processing, this objective is not always possible to achieve.

Another obstacle is the bureaucracy involved in the whole process of reusing already used cork, not only in terms of licensing but also in terms of integration into other products.

Concerning the facts presented above, we believe that, when properly dealt with, improvements in the recycling quota can be obtained. On the other hand, it may open doors to new business models and job opportunities.

We verify that the approached subject has a strong probability of success regarding the level of improvement of recycling and reuse. To this end, it's imperative that the activity sector in question takes initiatives/measures to encourage the participation of the community and thus make the circular economy and sustainability model a reality for the common benefit.

To take advantage of the synergies that cork allows and to meet the growing needs of this raw material, given its application in a wide range of areas, it's becoming increasingly necessary to invest in ways to recycle and reuse this product, adopting new measures that allow cork that has already been used to be used for purposes other than the production of corks, since corks cannot "become corks" again.

Moreover, to achieve the purpose of improving the recycling and reuse of cork, based on its sustainability, we propose to promote the recycling of this product with the collaboration of humanitarian associations/institutions which are very close to the community and which, at the same time, can encourage both morally and socially participation and improvement in practices such as this. Based on this incentive, it will be possible not only to continue to invest in cork recycling but also to increase this practice, achieving an improvement in cork recycling, a result which will enhance the increasing sustainability of the product.

The main shortcoming of existing initiatives is related to the fact that some collection/delivery points aren't located at strategic points, which makes it difficult to obtain the product for recycling in large quantities. Thus, there is a need for investment in the promotion and dissemination of existing campaigns and/or the creation of new initiatives that awaken the interest/need for the active participation of all.

We present some examples of initiatives that could be put into practice to increase the recycling quota of cork:

- Creation of a box that would be fixed to the glass recycling containers to allow easier access to recycling points for this waste.
- Promotion of a nationwide initiative with the Portuguese Firefighters League, a corks collection campaign among the hotel sector and individuals in each region.
  - Through knowledge and proximity to the community, it will help and be possible to reach values higher than the current ones registered.
  - For the above idea to be feasible, we suggest an incentive to the corporations; for example, in the total sum of tons at the end of each period, we offer equipment or fuel vouchers.
- To improve the amount of collection on the commercial surfaces, we suggest a new advertising and incentive strategy for customers visiting these spaces.
  - Place information regarding the reuse of cork and its benefit and repeatable life cycle next to the wine cellar area, as well as well signposted collection containers for this purpose.

– Create an incentive for the return of corks, for example, a simple device whereby the customer places the corks and a pre-established amount per kilogram is returned as a discount on purchases in the shop.

26.6.2 Suggestions for Managers of Companies Participating in the Sector

Recycling already used corks makes it possible to take advantage of synergies for recycling this material as a way of obtaining a larger quantity of raw material at a reduced "environmental cost." Given that the main shortcoming lies both in the incentive and in the volume of the collection, the essential effort for improvement should focus on the "massification" of collection points and collaboration with entities close to the community.

Thus, the promotion of initiatives to deliver/collect used corks should be intensified in the channels closest to the main buyers/consumers/partners as a way of increasing the amount of product for recycling/reuse.

Small companies, being one of the main stakeholders in the reuse of cork products, should be given easier access to waste and should be enabled to easily transform the product and, above all, place the final product on the market.

26.6.3 Limitations

One of the main limitations of this research is related to the difficulty in accessing information since companies in the sector refer in their sustainability and circular economy models to the total use of raw materials, but don't provide information regarding the collection of material for recycling/reuse, which leads us to question the extent to which these models are being implemented.

In addition to this difficulty, there is strong resistance to the dissemination of data and forms of action in the sector, which hinders research work and suggestions for improvement.

The issue of recycling after corks use is a subject that needs some work, as it's perceived with some lack of interest by the sector because it's a process with low profitability.

On the other hand, the product used isn't considered a waste by the transformers, as it's a 100% biodegradable material, but, by having this type of ideology, the potential of cork to be reintroduced in the production of new products is being lost.

The recycling/reuse of this product is considered unprofitable by the sector, since the recycling process often requires the selective separation of other elements, such as plastic and metal, which are an integral part of some structures (port and sparkling wine corks). The cost of the current separation process is a demotivating factor for the use of the product under consideration.

Furthermore, the literature on cork and cork reuse/recycle is very scarce.

26.6.4 Originality of the Work

The sustainability of the cork sector through the transformation of used corks is an original theme, given that there aren't many efforts to find processes to improve sustainability in this way. With this contribution, it would be possible to promote greater efficiency in the reuse of cork, at the same time avoiding its waste and harmful effects on the environment.

Given the urgency of combating climate change, which has harmful effects on the environment and native species, we emphasize the need to reuse all kinds of materials that make this possible and to take advantage of the characteristics of products that allow this, reducing the negative impact of the transformation industry on the environment.

It's therefore an original challenge, due to the lack of bibliographical content about sustainability/reuse of cork, which makes it a differentiating and necessary work for today and the future.

Once the GAP had been identified, we began a survey of existing initiatives and their results. Based on that, we carried out an analysis, case by case, and checked each program to develop its results, and useful life. In this way, we obtained information that allowed us to use the theme and carry out different and original work.

This is, therefore, an enormous opportunity to explore the subject to try to understand its impact and to help in the creation of new goals and incentive programs, the main purpose is to raise the awareness of the largest number of participants in the process and to awaken the need to create/update collection/reuse methods.

By performing this research, we intend to create influence on what exists and, at the same time, propagate the need for improvement in reuse and achieve a more sustainable future in the recovery of this material.

The new concepts that can be created for the new goals to be achieved should involve those responsible for the sector as well as the entire population and contribute to enrich the current literature.

26.6.5 Clues for Future Work

This topic, although already worked on a few years ago, may be explored in the future to find a more beneficial solution for the environment, the consumer, and the business sector.

Following are some of the points that could be updated or improved in the future:

- Study and analyze the recycling of used cork and the companies that carry out the transformation process for reintegration.
- To analyze the demand for recycled cork granules and the type of transformation process in the various industries that incorporate them into their final products.
- To continue the collection awareness project.
- To obtain sponsors for the collection campaign under consideration.

- To analyze possibility of new partnerships for product collection and/or transformation with the following companies/organizations:
  - SUMA (at the level of used product collection)
  - LIPOR (at the level of used product collection)
  - Municipal councils at the national level (with the support of the government in terms of both transport and communication)
  - Galp Energia
  - EDP
  - Portuguese Fire Brigade Association
  - Recycled glass processing companies (BA Glass)
- To develop a cork collection campaign that brings together fire brigades close to local society and that enables the use of the same to intervene with the catering sector and the community.
- To provide benefits per tonne collected over a given period, thus valuing the work carried out and encouraging them to actively collect the material. The aid could be, among others, fuel vouchers and firefighting equipment.

References

100% Cork. (2022, January 28). *How recycling corks can help combat global warming*. 100% Cork. https://100percentcork.org/how-recycling-corks-can-help-combat-global-warming/

APCOR. (2018). *APCOR – Cortiça: O que é?* https://www.apcor.pt/cortica/o-que-e/

APCOR. (2019). *Information Bureau 2019 – Cortiça em Números*. APCOR. https://www.apcor.pt/wp-content/uploads/2019/02/CORTI%C3%87A-em-n%C3%BAmeros_PT.pdf

APCOR. (2020a). *A fileira da cortiça: da floresta ao consumidor*.

APCOR. (2020b). *Reciclagem*. APCOR. https://www.apcor.pt/cortica/reciclagem/

APCOR. (2021). *Cork 2020 (anuário de cortiça)*. APCOR.

Banco de Portugal. (2022). *Quadros do Setor*. Banco de Portugal – Eurosistema. https://www.bportugal.pt/QS/qsweb/Dashboards

Caldeira, C. (2020). Portugal produz 50% da cortiça de todo o planeta. In *Sector tem tendência a crescer*. Agricultura & Mar. https://agriculturaemar.com/portugal-produz-50-da-cortica-de-todo-o-planeta-sector-tem-tendencia-a-crescer/

Corticeira Amorim. (2022). *Reciclagem: 5 continentes com programas de reciclagem*. Corticeira Amorim.

Ernst & Young Global Limited. (2022). *Relatório de Sustentabilidade 2021*. https://amorimcork-composites.com/media/8350/pt_-_relatorio_de_sustentabilidade.pdf

Gil, L. (2007). *A cortiça como material de construção*. https://www.apcor.pt/wp-content/uploads/2015/07/Caderno_Tecnico_F_PT.pdf

Green Cork. (2020, March 19). *Green Cork Projeto de reciclagem de rolhas de cortiça*. O Projeto. https://www.greencork.org/o-projecto/relatorios/

Karlsson, P., & Karlsson, B. (2021, December 29). *Cork and the environment: Recycling, biodiversity and the contribution to fight global warming*. Forbes. https://www.forbes.com/sites/karlsson/2021/12/29/cork-and-the-environment-recycling-biodiversity-and-the-contribution-to-fight-global-warming/?sh=6c9696501da0

Lipor. (2022, January 20). *LIPOR, Maiambiente e Quercus encetam projeto piloto de recolha de rolhas de cortiça*. Lipor. https://www.lipor.pt/pt/comunicados-de-imprensa/lipor-maiambiente-e-quercus-encetam-projeto-piloto-de-recolha-de-rolhas-de-cortica/

NH Hotel Group. (2019, December 4). *New life for cork stoppers: NH is recycling its bottle corks and turning them into construction materials*. NH Hotel Group. https://www.nh-hotels.com/corporate/press-room/news/new-life-cork-stoppers-nh-recycling-its-bottle-corks-and-turning-them-construction

Porto Business School. (2017, December 14). *A cortiça é um paradigma da economia circular*. Porto Business School. https://www.pbs.up.pt/pt/artigos-e-eventos/artigos/a-cortica-e-um-paradigma-da-economia-circular/

Part IX

Tourism & Leisure

27 A Grounded Theory of Historic Housing Tourism: A Methodological Path

José Luís Braga and Óscar Silva

27.1 Introduction

This chapter aims to account for the methodological path that has been taken in the concretion of an empirical study on historic housing tourism (henceforth HHT) (Braga, 2016).

Hence, the fieldwork started by seeking to contact a gatekeeper who could provide access to the substantive population of the study (Hammersley & Atkinson, 2019). Being aware of the important activity carried out by the Housing Tourism Association (TURIHAB) – an entity that has been striving for the recovery of heritage for more than three decades as well as for the promotion and qualification of the tourist offer in stately homes – the choice proved peaceful. Therefore, it seemed clear that the appropriate person to make the entry possible into the field would be the marketing director for TURIHAB. Thus, we held a preliminary interview with this TURIHAB collaborator to get into the problem under investigation.

The first technique we used to obtain relevant data to the ongoing investigation was participant observation. Nevertheless, we soon realized that an ethnographic method would produce a vast set of data that would only allow a descriptive study of the substantive area, without the possibility of generalizing, nor of establishing, a theory.

Therefore, by undertaking a literary review in manuals of research methods in tourism, the authors realized that classical grounded theory (henceforth CGT)

J. L. Braga (✉)
CIDI – Instituto Europeu de Estudos Superiores, Vila Nova de Gaia, Portugal
e-mail: josebraga@iesfafe.pt

Ó. Silva
Centro de Investigação Techn&Art – Instituto Politécnico de Tomar,
Vila Nova de Gaia, Portugal

© The Author(s), under exclusive license to Springer Nature Switzerland AG 2023
F. L. Almeida et al. (eds.), *Multidimensional Sustainability: Transitions and Convergences*, Springer Proceedings in Earth and Environmental Sciences,
https://doi.org/10.1007/978-3-031-24892-4_27

would best suit our project. So, the open coding of the field notes related to participant observation started by using the computer program *Atlas.ti*™ for the analysis of qualitative data. Nevertheless, quickly this system of storage, codification, and analysis of data was abandoned, and the authors opted for manual analysis. Actually, this decision was taken in accordance with what Barney Glaser, the co-creator of the methodology, recommends. The American sociologist advises against the use of computers to manage data in CGT: "as yet, computers block the abstract meaning making of good GT. The wonders of computers are great but not yet for GT" (Glaser, 2003, p. 17).

On the other hand, after reading Birks and Mills' handbook on grounded theory (2011), the researchers realized that their investigative work would benefit if they had online methodological mentoring. By *surfing* the internet, they found a portal, *Grounded Theory Online* (2022), which offered methodological counselling to young researchers using the methodology, what Glaser names "minus mentorees" (Glaser, 2014a).

So, the researchers started to adapt their data collection conforming it to the CGT principles, without disregarding the already significant corpus from the participant observation sessions that they had previously undertaken. However, these data were only subject to reasoned analysis when they finished the iterate process of collecting and analyzing data relating to the interviews they conducted with the owners, relevant to the theoretical sampling. In summary, contrary to the data from participant observation, the information extracted from the interviews was in line with the CGT methodology.

For almost 2 years, the authors held periodic meetings online (the minimum frequency was fortnightly and, as they began to adjust to the method, the meetings became sparser). These online seminars aimed to clarify doubts or look for ways to overcome the challenges arising from the use and sustainability of the CGT methodology capable of pointing out and providing answers/predictions to all stakeholders.

27.2 Data Collection

In this study, we collected considerable data that were gathered mainly in view of two techniques of social research: participant observation and unstructured interview. The details of how we collected the data are set out below.

27.2.1 Participant Observation

The brief period in which the authors dedicated themselves to participant observation included five sessions that took place in October, November, and December 2011 (then low season), on Saturday and on a public holiday (busy days). These sessions took place in the morning (approximately from 9:00 to 13:00). There were four research scenario houses, all located in Minho: the *Paço de Calheiros*, in Ponte de Lima; *Quinta de Santa Comba*, in Barcelos (which I observed on two occasions); the *House of Sezim*, in Guimarães, and the *Casa do Campo*, in Cabeceiras de Basto.

In these houses, we adopted an observer position as a participant (Cole, 2008). In fact, in the half dozen social situations, we tried to make descriptive observations which, as Spradley (1980, p. 78) points out, aim to guide the social scientist when he is still ignorant of the culture under study. According to what the American anthropologist prescribed, we undertook the task of identifying the nine most important dimensions of a given social situation:

1. *Space*: the physical place or places
2. *Actor*: the people involved
3. *Activity*: a set of related acts people do
4. *Object*: the physical things that are present
5. *Act*: single actions that people do
6. *Event*: a set of related activities that people carry out
7. *Time*: the sequencing that takes place over time
8. *Goal*: the things people are trying to accomplish
9. *Feeling*: the emotions felt and expressed (Ibidem)

Therefore, in the four houses which were studied, focus was placed on different scenarios and social situations. In the *Paço de Calheiros*, scene of our first session, we started the observation in the breakfast room; we then went to the kitchen and circulated around the house. So, we observed the interaction between host and guests at breakfast; we talked to members of the staff in the kitchen, noticed the interaction between host and guests outside the property, and witnessed a checkout that took place on the balcony. We also observed the apartments adjacent to the house. Then, we went up to the living room, where we talked to the owner/host and again watched the contact they had with guests.

In the first session, we held at the *Quinta de Santa Comba*, we observed the vicinity of the house, we visited its rooms and held an informal conversation with the owner's wife, as he was absent. She led us on a guided tour also along the chapel, the farm, and the kennel. On the second visit to the same setting, we were received by the owner who was in the breakfast room chatting with guests. We paid a close attention to this social situation. We went up to the top floor of the house. The owner showed us the room for the exclusive use of the family.

Days later we visited the *House of Sezim*. We saw the office. The host showed us the interior of the house, the balcony, and the guests' amenities. We also paid attention to the breakfast service and the interaction between staff, host, and guests. Then we talked to the kitchen owner and, finally, we went about the space surrounding the house and returned to the office.

The last participant observation session took place in the Casa do Campo's room, where we witnessed breakfast and the interaction between a group of guests and the owner's daughter. Then we talked informally with both the hosts, the mother and the daughter.

27.2.2 Characterization of the Sample

Throughout the study, 53 interviews were conducted with hosts/owners of stately homes. Of the more than half a hundred houses, only 10 were not classified as HHT. HHT houses are small-sized facilities. From the sample that we collected, the house with the smallest number of rooms had two and the one with the highest accommodation capacity had 15 rooms.

As for the type of interviews we conducted, they were mainly face-to-face (48), but also by telephone (3) and e-mail (2). Regarding the geographical origin of the houses, the vast majority are located in the former province of Minho (34). A residual number is found in Douro Litoral (4), Trás-os-Montes and Alto Douro (3), Beira Litoral (4), Beira Alta (1), Ribatejo (1), Alto Alentejo (2), Baixo Alentejo (3), Algarve (1) and Açores (2) – let it be stressed that the total number of houses is 55 since one of the interviewees owns three houses of HHT, two in the Azores islands and one in the Algarve region.

If we consider the sampling by district, the largest number of houses (21) are in Viana do Castelo, followed by Braga (12), Porto (5), Aveiro (4), Beja (3), Vila Real (2), Évora (2), Ponta Delgada (1), Horta (1), Faro (1), Viseu (1), Santarém (1) and Bragança (1) – the same distortion factor mentioned above maintains its effectiveness in this case.

Regarding the municipalities where the houses are located, Ponte de Lima (15) stands out – which is justified by the fact that TURIHAB is based there and because it was in this municipality that HHT had its beginning – and mainly in the parishes of Arca and Ponte de Lima (3) and Arcozelo (3).

27.2.3 Interviews to Generate Theory and Theoretical Sampling

For the first interview, we prepared four questions as neutral as possible, of a "grand tour" type (Spradley, 2016, p. 86):

1. *Can you please talk about your experiences as owner and host of house X?*
2. *What were the main things that happened in House X from the start of your management of tourism to the present?*
3. *How do you feel as a host?*
4. *Can you tell me how this activity has changed over time?*

In subsequent interviews, we tried to have a conversation with the interviewee, so we asked only questions 1 and 3. Our first goal was to ask questions that led to a verbal "regurgitation" ("instill the spill") by the interlocutor (Glaser, 2001, p. 175).

On the other hand, the authors tried to exercise ourselves in the practice of unstructured interviews by listening to news broadcasts on YouTube and registering the keywords on a sheet of paper. Thus, they avoided writing down everything so as to prevent descriptivism and promote conceptualization.

Similarly, at the end of the research interviews, the researchers expanded the field notes from what we had kept in our memory and from what we had written on paper.

Sometimes the researchers were confronted with owners who provided us with proper line data (Glaser, 1998, p. 9). But they soon came to the conclusion that all kinds of data were valuable and should be carefully analyzed, thereby preventing speculation. The authors considered that these data could prove useful at a later stage of the analysis.

In other circumstances they were so successful in the interviews that they could conduct them in full, using only two open questions or even just one. The way the interviews were conducted resembled the "Freudian" interview, since the researchers' asked questions and probed the information they thought was most relevant – as in the therapeutic interview where the therapist reflects and reformulates the patients' statements, emphasizing their emotional aspects, letting the patient say what s/he likes without interrupting him/her except when the climax is depleted (Kvale, 2011, p. 18).

However, not all interviews proved to be easy. There were times when the interviewer faltered when formulating the questions or when, despite the empathy of the interlocutor, the researcher could not properly probe, nor did he divide patterns of behavior in the discourse and sought in vain for substantial information in what was said. There were also cases in which the interviewees' statements were erratic. Our methodological memo illustrates well these vicissitudes:

> The interview with this owner didn't start in the best way. I wavered in the first question. The owner, however, spoke quite at ease. Initially – and in the final phase of the interview – his wife was there and sometimes intervened to introduce some topics she thought relevant. Both were courteous. At one point, the owner asked if I didn't have any more questions. I was waiting for him to give me additional information regarding certain points that he thought were relevant and that he had already suggested. But this time, I had a hard time finding substantial information. It seemed I could not find any relevant patterns that could be of use.
>
> Adding to this difficulty, the owner repeatedly asked me not to write information he meant to keep confidential. Therefore, large parts of the interview were not recorded on paper and my conscious memory discarded some of the confidential information he gave.

The authors' goal at this stage was to interview as many hosts as possible, and they came to the conclusion that, in this first stage of analysis, they needed 25 interviews. In fact, they needed to collect more data to refrain from being descriptivist and actually start conceptualizing. In addition, they tried to interview using other means, such as e-mail and telephone. Nevertheless, the scarce conceptual output garnered led them to focus only on the face-to-face interview mode, since the so important observational aspect was absent from the two other means.

However, the researchers' analysis of the interviews and the ideas they had in the memos, concerning the concepts and their interrelations, led them to infer who they would interview next. Indeed, Glaser (2014b) maintains that writing memoranda is a vital research procedure for CGT, so he recommends that memos should be continuously written, as they ensure the quality of the emerging theory.

On the other hand, their questions, which were previously descriptive, became more and more specific to collect data that could fill in the missing categories. One of the strategies they used to confront the interviewees with sensitive information was to remove illustrations of testimonies given by other hosts and to interpret them. Below the authors highlight how they put these questions together:

1. *Some people have told me that...*
2. *What do you think of...?*
3. *Can you tell me about...?*

Although at some point of the research they had also interviewed hosts of stately homes not belonging to the HHT category, their methodologist advised them to continue to interview hosts of this type of tourist facilities first.

As they improved their theoretical sampling, they interviewed hosts that had close connections to the founders of the house, hosts who had purchased the house but had no connection to the original family, as well as hosts who had no connection to the house other than being employees there. Their effort to theoretically saturate the categories continued when they interviewed newly open to HHT home owners; owners of houses that had shut down their activity not long ago; owners of houses that had been the object of sharing, as well as owners of highly reputed HHT facilities in the market.

27.3 Data Analysis

Data analysis in CGT is a meticulous work involving what follows: open coding; constant comparative analysis; theoretical sampling; the identification of the core category; selective coding; sorting of memos; theoretical coding; and the writing of memos in the form of theory. Below we try to clarify the various methods and praxis of the analysis in CGT.

27.3.1 Open Coding and Constant Comparative Analysis

The field notes which were took during the interview were expanded in the word processor and line-by-line coding was first implemented, by questioning the incidents as follows:

What category does this incident indicate?
What property of which category does this incident indicate?
What is the main concern of the participant? (Glaser, 1998, p. 140)
The codes have been assigned using the word processor's "comment" feature. The procedure is illustrated in Fig. 27.1.

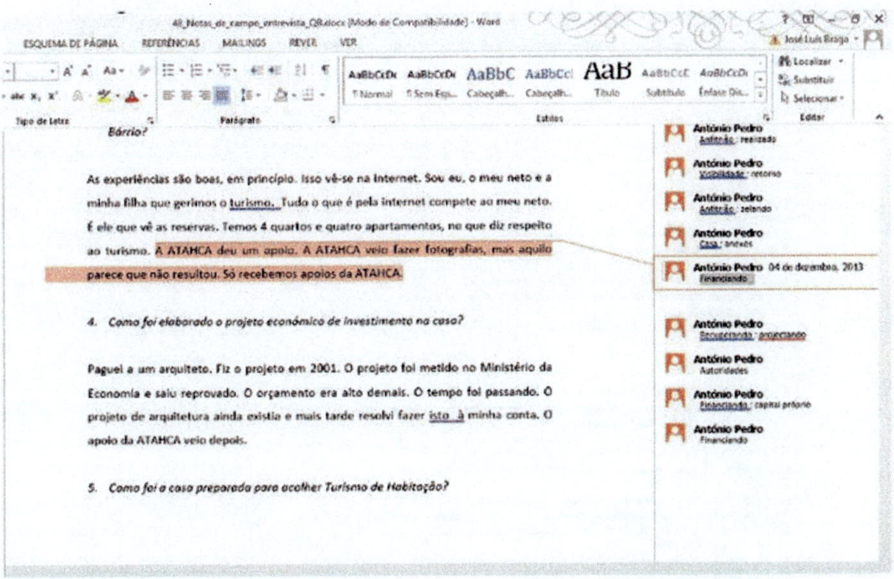

Fig. 27.1 Open coding process in the word processor. (Source: Authors)

After the open coding process, the identified incidents were then "copied" and "pasted" into another new word document under the appropriate code title, as shown in the example below:

Code 1
Incident a
Incident b
Etc.
Code 2
Incident c
Incident d
Etc.

Then the researchers used the "highlight" feature of the same word processor to be able to suppress incidents and let the codes alone to remain visible, so that they could compare codes with codes. Later, they expanded the codes so that they could see the incidents and enable the comparison of incident to incident within a code and immediately after they collated incidents through codes. The process is shown in Figs. 27.2 and 27.3.

They then wrote memos about the codes and their relationships with other codes by creating a separate document as shown in Fig. 27.3.

In other words, the researchers grouped incidents under their properties or categories while simultaneously writing memos, for example, concerning the "host"

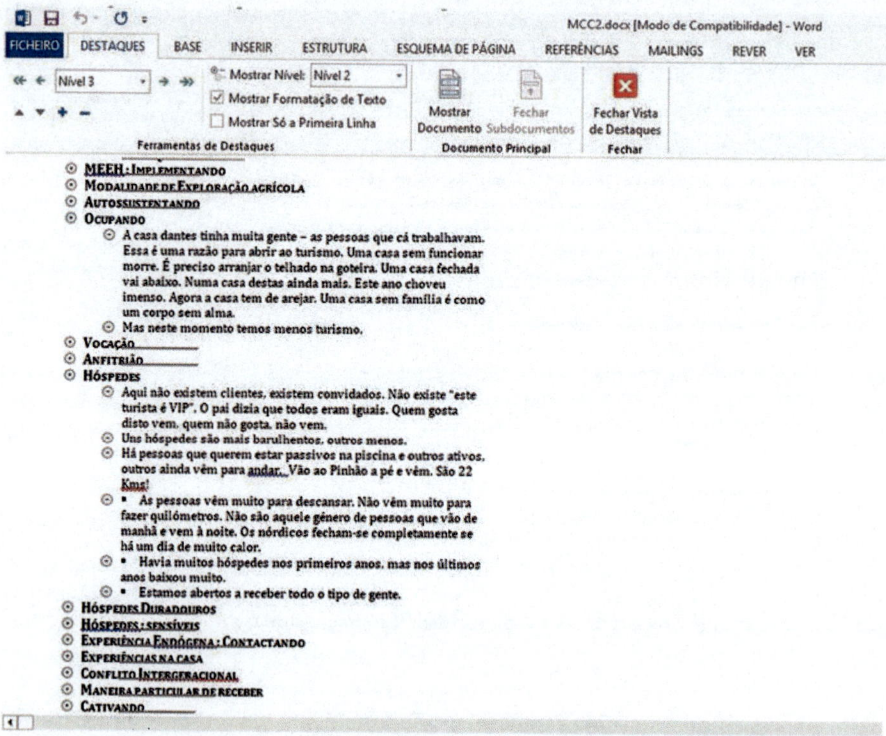

Fig. 27.2 Constant comparison method in the word processor. (Source: Authors)

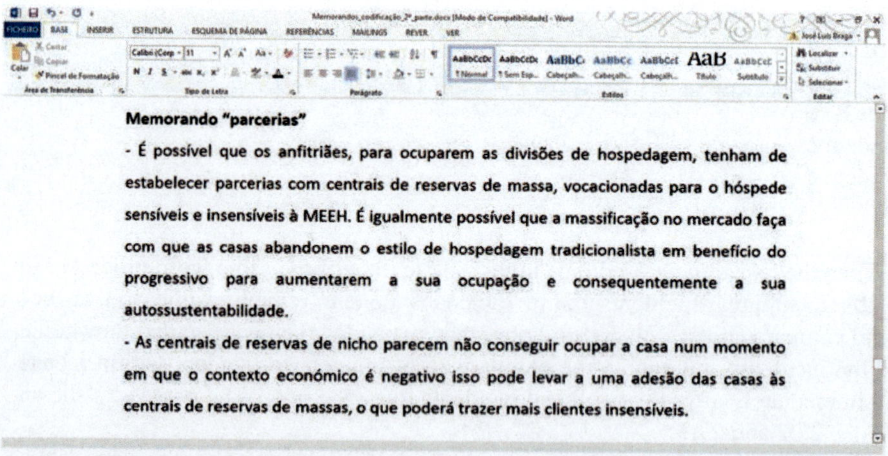

Fig. 27.3 Writing of memos in the word processor. (Source: Authors)

category. As they became more experienced in the coding process, they found that the names they gave to the incidents were conceptualizing them (rather than describing them). Whereas in the memos they began to unravel whether the concepts they were discovering were properties of another concept or if they were autonomous concepts. As have been stated, CGT is an interactive process, so they had to analyze (i.e., code and write memos) each interview before they started the next one. Moreover, this research instrument must be written in the present, abstracting us in its writing of people, time, and places.

Thus, they begun to focus their analytical attention on behaviors, namely on the conduct of the host in response to the actions of the guests, in order to discern the properties of the guest and the host. They also set up a typology of hosts. The next moment they tried to raise our level of conceptualization, putting details apart to uncover patterns. In addition, they peered their interviews to extract all incidents that concerned the "host" concept. Indeed, they have found that some incidents defined "host"; some others concerned their behavior, others focused on the way the host behaved in response to guests. Soon from our analysis emerged also a distinction between "refunctionalization styles" (i.e., classic; hybrid; modern). In addition, for each interview, the researchers tried to find out what was happening to the participants and what solutions they were adopting for the problems they were facing.

With regard to the writing of memos, they started to look for "gray" areas of analysis and to record all the hindrances found during research, thus gaining awareness of the difficulties. However, they did not abandon the constant comparative method, as they used it to continue the comparison of the most relevant codes, since they still needed to ensure that the concepts that came out of the research were well grounded. In fact, Glaser and Strauss (1967) divide the method of constant comparison into four phases:

> (1) Comparing incidents applicable to each category, (2) integrating categories and their properties, (3) delimiting the theory, and (4) writing the theory. Although this method of generating theory is a continuously growing process – each stage after a time is transformed into the next – earlier stages do remain in operation simultaneously throughout the analysis and each provides continuous development to its successive stage until analysis is terminated.

At this stage, they raised the conceptual level by distinguishing properties that would specify or define the concepts, dimensions that shaped a concept, and degrees that would give them the extension of a concept in a continuum – e.g., category: "host"; property: "attachment"; dimension: degrees of "attachment" (too much; too little, etc.) (see Glaser, 1978). Thus, looking for the properties of the concept "host" and determining its variation, they were able to define its typologies. On the other hand, that enabled them to highlight the structural conditions in which hosts operate (e.g., category "political structure").

27.3.2 Main Concern

The first goal of CGT is to identify the main concern of the participants and how to solve it. In fact, in this methodological approach, the research problem emerges from the participants' statements, rather than being preconceived by the researcher. Therefore, in order to identify the main concern of the participants and the process by which to solve it, the researchers asked the above-mentioned questions to the incidents, compared incident to incident, code to code and, later, category to category. This process led to the emergence of the main concern and the development of preliminary concepts and categories (cf. Andrews et al., 2012).

Nevertheless, the procedure was gradual. Initially they made sure that the main concern of the hosts when buying the estates could be different from those that inherited them. Wanting to know the motivations of this last type of hosts, they found out that the change of hands of the property explained their behavior. They noticed that lineage hosts, in some cases, created a hosting exploration modality that included a family contract detailing how the exploration of the estates must be carried out and that these guidelines determine their future behaviors. In fact, there remains the need to preserve the family ties the house stands for. The following statement from an owner explains it:

> However, people must be first. I have children. If I had to feed them and didn't have any money, I'd sell the house. However, there is a sentimental value that is priceless. I'm not that much of a well-to-do person, but if you gave me 50 million euros for this house, I still wouldn't sell it, because I know I wouldn't come back here to see it.
>
> But if you told me you'd have this house, but your kids won't be healthy, then I'd sell it. Now, for money, I wouldn't sell it for 50 million euros, I wouldn't want that money. What I feel for it comes first; feelings; people. I have this emotional connection because every corner here is a trace of my great-great-grandfather, whom I never saw, but who looks like my father. (Interview No. 8)

- Thus, what ensures the continuity of the family heritage is the handing of the house in the best possible conditions to the next generation. However, it was not possible to know whether this concern was also common to the hosts by acquisition (i.e., those who do not have a kinship connection to the founder of the house). It may also happen that these participants intend to improve the house's reputation. To reach a proper conclusion, the authors interviewed both types of owners, starting with the following question:

  What does this house mean to you?

After listening to the relevant participants for theoretical sampling, they concluded that both types of hosts (lineage and acquisition) shared the same main concern, as summed up in the following extract from an interview to a host by acquisition:

> This estate will surely pass to the next generation. The house was bought in 1974. This activity started 21 years ago and therefore the next generation is already there. They will go on. When there are generational conflicts – and there always are – they will always be greater if the older generation do not accept the innovations of the younger generations. (Interview No. 30)

The researchers then wrote a working paper. In fact, they wrote it when they completed the first theoretical sorting of the memos, which could then be written. The purpose of the working paper is to put the theory into paper through a draft. This was the way to describe the main concern of the participants: "*after more than four dozen interviews, I concluded that the main concern of the participants is the preservation of the house's history (and family bonds). The way to address this concern is to refunctionalize it!*" (Braga, 2016, p. 318).

27.3.3 Core Category Identification and Selective Coding

There is a time when the researcher starts to selectively code for a core variable and when open coding ceases. Glaser (1978) suggests that the theory is delimited to a single core category (which is often a basic social process). This does not mean that the other variables become extinct, but the concentration of the analysis in a single core variable only lowers other potential core categories to a subordination function.

In fact, at this stage of the analysis, the researchers were under pressure, since, after 20 interviews, the memo bank and the constant comparative analysis had not yet led them to a definitive core category. In other words, they needed to raise the level of conceptualization in order to unveil a core category and possibly four subcategories and delimit their study to the way their participants approach their capital problem. Glaser (1998, p. 150) refers to the decision of initiating selective coding for a core category in the following terms:

> Some people find this a brave move, since they are not sure that they have found the core category or that it correct to pursue. They must make this selection as it is one of the prime delimiters of grounded theory. By selective coding the theory is boiled down and codified, by saturation, more focused memos, selective theoretical sampling and the shift to a more focused theoretical perspective.

Finally, after about 2 years of research, the authors reached the core category: "*refunctionalizing*". This is a variable that seems to interconnect all the main concerns of the participants. From this moment they stopped coding neutrally and started doing it only for the code "*refunctionalizing*". On the other hand, they have come to define this category in relation to its properties. They also realize that this refunctionalization is formed by a structure. In this sense, some behaviors of the hosts that did not fall into the core category were left out. This was the case of the attitudes that dated back to the setting up of the house. Therefore, at this stage we favored the analysis of behavior in the present.

27.3.4 Sorting of Memos and Theoretical Coding

Before starting the memo sorting, the researcher has already written in paper the conceptual generalizations – so hard to obtain – concerning how a core category continuously solves the main concern inserted in the memos. Therefore, the analyst now has a bank of mature memos that claim to be sorted in the form of a theory.

Thus, before the researchers entered the memo sorting stage, they read Glaser's compendium (2005) on theoretical codes, which allowed them to integrate them conceptually better. The same author distinguishes theoretical codes from substantive codes: "*there are basically two types of codes to generate: substantive and theoretical. Substantive codes conceptualize the empirical substance of the area of research. Theoretical codes conceptualize how substantive codes may relate to each other as hypotheses to be integrated into the theory*" (Glaser, 1978, p. 55).

From our classification effort what came about was a combination of theoretical codes: a basic social process (for a definition of this term, see Glaser and Holton, 2005; Carrero Planes et al., 2006); "refunctionalizing" made of two stages, "improvising" and "professionalizing"; a typology of hosts ("dedicated host"; "undedicated host"; "lineage host"; "host by acquisition"; "host initiator"; "continuing host"; "professional host"; "manipulative host"); and three refunctionalization styles ("classic", "hybrid," and "modern"). On the other hand, we developed an amplifying casual looping model in a vicious cycle ("less dedication" → "less personalization" → "obtaining negative evaluations of the accommodation" → "negative reputation of the house" → "reduction of sustainability" → "lower equity" → "lower pace/intensity of home recovery" → "less certainty of home continuity") and, by contrast, a virtuous cycle ("greater dedication" → "greater personalization" → "getting positive reviews of the accommodation" → "positive reputation of the house" → "increased sustainability" → "higher equity" → "higher pace/intensity of home recovery" → "greater certainty of home continuity").

The researchers made a preliminary classification of the memos by code names to see what "came about". Then they also classified the interrelations between concepts. For example, they tried to determine the variations of hosts' behavior when facing more severe, softer, or even absent legal frameworks. The sorting of memos was made by conceptual ideas. The researchers have therefore sorted the ideas contained in our memos which referred to the concepts and concepts interrelations. The mechanics of memo preparation comprised the printing of A4 sheets of paper and cut them with a boxcutter to highlight the most relevant ideas concerning the emerging theory. Then they collected the first piece of paper from the stack of memos, and if they could see more than one idea in it, they would cut it into sections. On the other hand, they let the kinetic relationship emerge from the pieces of paper they had removed from the pile. Then they would put together the pieces of paper that could be interrelated. On top of each pile, they wrote the name of the idea that put them together or kept them isolated waiting for an idea to emerge. The relationship between the pieces of paper led to conceptualization (an example of such a relationship may be the dimension "performing more"/"performing less"). Later the researchers wrote the aforementioned working paper of more than half-hundred pages where the outcomes of their sorting work were. They needed no grammar, spelling, punctuation, syntax or formatting. This exercise helped them raise their level of conceptualization. Afterward, they made another memo sorting, seeking to highlight patterns of behavior and structures. In fact, they reiterated the analysis of memos referring to the "host" category, in order to understand whether these were defined by their properties.

In addition, they classified the remaining memos, for example, in order to understand what the hosting exploration modality offered guests. At this point, they tried to intensify the pace of sorting by trying to capture memos that denoted patterns. However, they were left with a large number of memos (150 A4 pages), so they excluded those that were too close to the data. When they finished the second memo sorting session, they wrote the following memo:

> I now end the second sorting with several concepts that refer to behaviour: hosting; creating a reputation; dedicating; recovering; sustaining; valuing; contacting; Economic Model of the House; Personalizing; Integrating; Professionalizing; Broadcasting; Engaging; Enjoying; House; Political Structure; Financing; Economic Model of the House. Next, I will classify each of these concepts, trying to highlight the properties, dimensions and degrees of the categories.

In the subsequent session, the third one, the authors framed the sorting by connecting the memos with the "refunctionalizing" core category. At this time, they also ceased our methodological supervision, since they were convinced that they already had a solid theory, despite their methodologist being of the opinion that they could still raise our level of conceptualization further and delimit their study to the way our participants solve their main concern.

In fact, this stage of the process has proved troublesome because they have accumulated a multitude of memos. For this reason, their classification of memos was time-consuming and tedious (it could take several days and even weeks). They tried to sort the memos in a fluid and spontaneous way. To solve the problem, they restricted themselves only to the new memos and the "best" memos of the earlier stages of their study to come up with conceptual ideas. Nevertheless, it proved impossible for them to summarize the sorting effort in 2–3 h, as their methodologist requested.

Although they have not delimited their substantive theory to the fullest, which would eventually lead to a further reduction in the number of categories and a theory with a smaller set of high-level concepts, they believe that the final result is solid and that it fits the criteria of adherence to the data, relevance, work, and modifiability that are indispensable to a CGT (see Tarozzi, 2008).

27.4 Conclusions

In these lines, the authors have tried to present the essential methodological aspects of the empirical research that they have been developing and which seeks to clarify how the owners of stately homes solve the main concern with which they are faced in the pursuit of their task of HHT hosts. Thus, the authors' aim was to demonstrate the potential of CGT. So, from a social unit of analysis (the manor house of HHT), we can establish a theory applicable to a number of everyday situations within that substantive area, and not just a single type of scenario.

Similarly, throughout this chapter, we highlighted the way data were collected through participant observation and unstructured interviews, as well as clarifying the methods of data analysis used to generate a grounded theory of historic housing tourism: open coding; constant comparative method; theoretical sampling; memo writing; selective coding; memo sorting; theoretical coding; and theoretical writing.

References

Andrews, L., Higgins, A., Andrews, M. W., & Lalor, J. G. (2012). Classic grounded theory to analyze secondary data: Reality and reflections. *The Grounded Theory Review, 11*(1), 12–26. http://groundedtheoryreview.com/2012/06/01/classic-grounded-theory-to-analyse-secondary-data-reality-and-reflections/

Birks, M., & Mills, J. (2011). *Grounded theory: A practical guide*. Sage Publications.

Braga, J. L. (2016). *Refuncionalizando a Casa Solarenga: uma Grounded Theory* PhD Thesis. Universidad de Santiago de Compostela. https://minerva.usc.es/xmlui/handle/10347/14805

Carrero Planes, V., Soriano Miras, R., & Trindad Requena, A. (2006). *Teoría Fundamentada GroundedTheory: El desarrollo de teoría desde la generalización conceptual*. Centro de Investigaciones Sociológicas. Coleção "Cuadernos Metodológicos" (p. 37).

Cole, S. (2008). Action ethnography: Using participant observation. In B. W. Ritchie, P. Burns, & C. Palmer (Eds.), *Tourism research methods: Integrating theory with practice* (pp. 64–72). CABI.

Glaser, B. (1978). *Theoretical sensitivity: Advances in the methodology of grounded theory*. The Sociology Press.

Glaser, B. G. (1998). *Doing grounded theory: Issues and discussions*. Sociology Press.

Glaser, B. G. (2001). *The grounded theory perspective: Conceptualization contrasted with description*. Sociology Press.

Glaser, B. G. (2003). *The grounded theory perspective II: Description's remodeling of grounded theory methodology*. Sociology Press.

Glaser, B. G. (2005). *The grounded theory perspective III: Theoretical coding*. Sociology Press. Glaser.

Glaser, B. (2014a). Choosing grounded theory. *The Grounded Theory Review, 13*(2), 3–19. http://groundedtheoryreview.com/2014/12/19/choosing-grounded-theory

Glaser, B. G. (2014b). *Memoing: A vital grounded theory procedure*. Sociology Press.

Glaser, B., & Holton, J. (2005). Basic social processes. *The Grounded Theory Review: An international journal, 4*(3), 1–27. http://groundedtheoryreview.com/2005/06/22/1533/

Glaser, B. G., & Strauss, A. L. ([1967] 2008). The discovery of grounded theory: Strategies for qualitative research. Aldine Transaction.

Grounded Theory Online. (2022). *About us*. https://www.groundedtheoryonline.com/who-we-are/

Hammersley, M., & Atkinson, P. (2019). *Ethnography: Principles in practice* (4th ed.). Routledge.

Kvale, S. (2011). *Doing interviews*. Sage Publications Ltd..

Spradley, J. P. (1980). *Participant observation*. Cengage Learning. Inc.

Spradley, J. P. (2016). *The ethnographic interview*. Waveland Press.

Tarozzi, M. ([2008] 2012). *Che cos'è la grounded theory?* (3rd Reprint). Carocci editore S.p.A.

28 Attitudes of Internal Stakeholders About Museums in Vojvodina and Their Offer in Tourism

Snežana Besermenji, Jelica Ilić, Staša Arsenović Koprivica, and Radovan Sremac

28.1 Introduction

Stakeholders are "groups or individuals, who can be or are influenced by realization of the mission of the organizational system" (Freeman, 1984, p. 46). In a museum, a stakeholder is a person who has an interest in the politics of programs of the museum (Kotler, 1998, p. 68). Stakeholders can be internal or external. Internal are the owners, managers, and other employees, so the internal stakeholders in the museum are represented by managers, curators, IT specialists, restorers, art historians, and all other employees. Employees who have an interest in museum programs aimed at tourism and tourists will be presented in this chapter under the term internal stakeholders.

The aim of the work is to analyze the attitudes of employees in the museums of Vojvodina (province of the Republic of Serbia) and examine to what extent the museums are focused on tourism and participate in the formation of the tourist offer of the region, in addition to their primary task of preserving and protecting the existing cultural heritage within their framework. In the existing literature, museums are often viewed as a tourist product and usually a static opinion is given about the importance of tourism for museums, and museums, due to their special role, should be distinguished from other tourist attractions such as monuments or theme parks (Culley, 2010).

The relationship between cultural tourism and museums should be observed through their contrasts, as well as similarities. Their difference is mainly based on the non-profit nature of museums and the profit nature of cultural tourism. Their

S. Besermenji (✉) · J. Ilić · S. A. Koprivica · R. Sremac
Department of Geography, Tourism and Hotel Management, Faculty of Sciences, University of Novi Sad, Novi Sad, Serbia
e-mail: snezana.besermenji@dgt.uns.ac.rs

© The Author(s), under exclusive license to Springer Nature Switzerland AG 2023

F. L. Almeida et al. (eds.), *Multidimensional Sustainability: Transitions and Convergences*, Springer Proceedings in Earth and Environmental Sciences,
https://doi.org/10.1007/978-3-031-24892-4_28

similarities are based on providing opportunities for improving experience, education, as well as cross-cultural acquaintance and understanding. The public dimension has been for a long time the function of the museum, so the issue of visitors has been a frequent topic of discussion, and the public dimension is extremely important for tourism, because it deals with the perspectives of users, i.e., consumers (Benediktsson, 2004). The museum audience is strongly connected with tourism, because tourists are part of that audience, and for many museums the largest numbers of visitors are tourists (Kotler et al., 2008). The presence of a museum is most often seen as the presence of something valuable and relevant, which should be shared with the public. The main task of a museum is to educate its visitors about the history, culture, and nature of a city, region, or country, as well as about a specific object of special interest, while at the same time to preserve these elements for future generations. The main role of the museum is to protect cultural heritage and attract as many tourists as possible (Pekarik, 2003). According to the research of Allan and Altal (2016), the main motivational factor for visiting the museums is the exploration and the major emotional dimension is the pleasure. Nowadays museums include a very wide range of objects, such as aquariums, vivarium's, botanical gardens, planetariums, and similar, and they are becoming more and more interesting attractions for tourists. They also become destination visitor and interpretive centers where people can find information and learn about the place (Krivosejev, 2012). The tourism industry is based on diversity. Tourists in general, and cultural tourists in particular, are interested in topics they do not have at home, which are different customs, a different way of life, manifestations, cultural heritage in general, as well as museums with different settings. This is the reason museums should participate in the formation of the identity of a destination (Gajski et al., 2011).

In their empirical research, Conti et al. (2017) identified clusters of museum visitors: occasionals, aestheticals, routine-breakers, global experientials, and aesthetic learners. "Some clusters may be considered the "classic" museum visitors who are mainly interested in the aesthetic and learning dimensions of the museum experience, while other clusters represent the "new targets" who are interested in other dimensions of experience, such as entertainment, evasion, and/or socialization" (Conti et al., 2017, p. 4).

In order to attract as many tourists as possible and achieve economic effects, it is necessary to include museums in joint tourist packages. However, it is necessary to take care that the number of visitors does not endanger the collections or their values in any way (McManus, 2010). For museums to be successfully involved in tourism, it is necessary that they become open to an entrepreneurial approach, while at the same time to achieve their basic task of heritage preservation. Museums are increasingly using the Internet to promote themselves and attract visitors (Bauer et al., 2019). Digital technology serves as an important tool to innovate things and processes in all areas of the museum institution, not only in promotion activities (Navarrete, 2019). Augmented reality is increasing engagement in visitors' experience in museum, and technological advancements are expanding the innovative potential and creative opportunities of businesses (Serravalle et al., 2019). Also, tourism entities, such as hotels, travel agencies, entertainment attractions, and others, need to be ready to accept new ideas (Silberberg, 1995). In modern world, museums face many challenges such as numerous competitions and the need of increasingly sophisticated visitors (Lynch et al., 2000). Nowadays, the focus of the museum is increasingly moving toward the overall experience, instead of toward the objects (replicas, simulations, performances, and electronic media) (Ames, 2004). Museums today devote more resources to social and participatory experiences than earlier (Kotler, 2001). Some experts believe that the process of popularization and commercialization can reduce the value of cultural experiences, which questions the definition of traditional museums (Tufts & Milne, 1999). However, due to financial reality, museums have also undertaken commercial services, so research and conservation take a back seat to market issues (Genoways & Ireland, 2003). The traditional view of the museum becomes outdated, because today many museums are based on social events, i.e., the opportunity to spend time with friends and family (Thyne, 2001).

28.2 Research Methodology

The research was conducted on the territory of Autonomous Province of Vojvodina in the period from October to December 2020. The research included internal stakeholders of museums in Vojvodina. For the purpose of the research, a questionnaire was created, which mostly contained closed questions. In order to investigate the perception of internal stakeholders, a five-point Likert scale was used, and the points were expressed on a scale from 1 to 5, where point 1 expresses absolute disagreement or the lowest value and point 5 expresses absolute agreement and the highest value. The questionnaire was created in accordance with Sara Kali's research, and it was adapted for the research needs discussed in this chapter. In the questionnaire, in relation to the set topics, different statistical methods were applied. Statistical processing of the collected data was performed using the Statistical Package for Social Sciences (SPSS 17.0). The questionnaire consisted of four parts. The first part consisted of the socio-demographic characteristics of the respondents, the second part related to their views on the most important goals and the activities of the museum, the third part related to their views on the presence of tourists, and the fourth part dealt with possibilities for improving the museum's offer. The questionnaire was answered anonymously by all interested internal stakeholders. Applied statistical methods are descriptive, reliability of the measuring instrument, and factor analysis.

Descriptive statistical method included the methods of data collection, arrangement, and presentation, and the grouping of data was carried out according to the values or modalities of observed features. The reliability of the measuring instrument was tested using Cronbach alpha coefficient. In order to reduce a large set of items, factor analysis was applied, and this was achieved by finding groups of closely related items. To determine the validity of the factor analysis, Bartlett's reliability test (Bartlet, 1954) and Kaiser-Meyer-Olkin indicator were applied (Gorsuch, 1983). The factors explained 70% of the total variance. The number of factors was determined by using the rotation method.

28.2.1 Socio-demographic Characteristics of Internal Stakeholders

In this research, internal stakeholders are employees of museums, who participate in the creation and presentation of programs for the public. For the purpose of the research, 77 respondents answered the questionnaire, and the questionnaire was distributed via e-mail to all museums in Vojvodina (34). Table 28.1 shows the socio-demographic characteristics of the sample (internal stakeholders).

In the total sample, men make 28% and women 49%. The age groups from 31 to 40 with 26% and from 41 to 50 with 27% are the most represented. According to education, the largest number of respondents completed college or university, i.e., 31%, and master and post-graduate studies, i.e., 33%. In the occupational structure of internal stakeholders, curators are the most numerous, namely 2 curators, 1 graduate ethnologist anthropologist senior curator, 3 archaeologist curators, 4 ethnologist curators, 3 historian curators, 1 art historian curator, 4 pedagogue curators, 1 anthropologist curator, and 2 ethnologist senior curators. In this research were also included 10 archaeologists, 11 historians, and 29 of others.

Table 28.1 Socio-demographic characteristics of the sample (internal stakeholders)

Characteristics		Frequency (%)
Sex	Men	+-28 (36.4)
	Women	49 (63.6)
Age	Less than 20	1 (1.3)
	From 21 to 30	1 (1.3)
	From 31 to 40	26 (33.8)
	From 41 to 50	27 (35.1)
	From 51 to 60	18 (23.4)
	More than 61	4 (5.2)
Education	Primary school	–
	Secondary school	5 (6.5)
	College or university	31 (40.3)
	Master and post-graduate studies	33 (42.9)
	PhD studies	8 (10.4)
Occupation	Curators	27 (33.8)
	Archaeologists	10 (13.0)
	Historians	11 (14.3)

28.3 Results

28.3.1 Perception of Internal Stakeholders About the Importance of Certain Museum Activities

In modern tourism, museums represent a significant part of the tourist offer. In Vojvodina, tourism in most museums is not clearly emphasized in museum business. However, the surveyed internal stakeholders mentioned the protection of local heritage as the most important function of museums (92.2%), followed by the interpretation of local history or culture for tourists, which indicates their awareness that museums should be more represented in the tourist offer in Vojvodina (Table 28.2).

28.3.2 Perception of Individual Items of Internal Stakeholders About the Presence of Tourists

This analysis includes the perception of internal stakeholders on the impact of the presence of tourists on the work of the museum. Calculating the overall mean value of perception was preceded by the recording of items that had a negative meaning, more precisely recording of items that show a negative attitude toward the impact of the presence of tourists.

According to data from Table 28.3, it can be concluded that the attitude of internal stakeholders toward tourists is positive, because they provide new sources of income (46.8% absolutely agree and 28.6% agree). They absolutely disagree (59.7%) with the attitude that tourists disrupt conservation efforts. They also absolutely disagree with the attitude that the presence of tourists causes more damage than the local population (58.4%) and that tourists (51.9%) interfere with the proper preservation of artifacts.

Table 28.2 The most important museum activities for internal stakeholders

Items	Valid percentages (%)					Mean value (M)	Std. deviation (q)
	1	2	3	4	5		
Interpretation of local history or culture for tourists	0.0	0.0	6.5	5.2	88.3	4.82	0.53
Interpretation of local history or culture for local visitors	0.0	0.0	7.8	10.4	81.8	4.74	0.59
Protection of local heritage	0.0	0.0	2.6	5.2	92.2	4.90	0.38
Education of local visitors	0.0	1.3	6.5	14.3	77.9	4.69	0.65
Education of tourists	0.0	0.0	6.5	10.4	83.1	4.77	0.56

Table 28.3 Basic characteristics of individual items of the perception of internal stakeholders about the presence of tourists ($N = 47$)

Items	Valid percentage (%)					Mean value (M)	Std. deviation
	1	2	3	4	5		
Provides new sources of income	3.9	2.6	18.2	28.6	46.8	4.1	1.05
Puts more pressure on available resources	15.6	20.8	40.3	13.0	10.4	2.8	1.16
Restricts access to other visitors	55.8	23.4	16.9	0.0	3.9	1.7	1.00
Interferes proper conservation of artifacts	51.9	22.1	16.9	5.2	3.9	1.9	1.11
Disrupts conservation efforts	59.7	20.8	15.6	0.0	3.9	1.7	1.00
Causes more damage to the locality than the presence of local visitors	58.4	16.9	19.5	1.3	3.9	1.8	1.06
Local visitors show more respect for provided information than tourists	9.1	24.7	37.7	7.8	20.8	2.6	1.06

Table 28.4 Basic characteristics of individual items for the analysis of internal stakeholders' reflection on the needs of tourists when performing business

Items	Valid percentage (%)					Mean value (M)	Std. deviations
	1	2	3	4	5		
Exhibition planning	3.9	2.6	27.3	20.8	45.5	4.01	1.9
Designing exhibitions	3.9	2.6	23.4	20.8	49.4	4.09	1.9
Development of educational programs	3.9	6.5	20.8	19.5	49.4	4.03	1.15
Text and information about the exhibits	1.3	0.0	11.7	26.0	61.0	4.46	0.75
Advertising promotion	1.3	1.3	14.3	27.3	55.8	4.35	0.87

28.3.3 Review of Internal Stakeholders on the Needs of Tourists When Performing Their Business Tasks

The reflection of internal stakeholders on the needs of tourists when performing their business tasks was examined using a five-point Likert scale. A detailed analysis showed that there is a significant difference between the perceptions of business tasks by internal stakeholders. They absolutely pay attention to the needs of tourists when creating text and information about exhibits (61%), followed by advertising promotions (55,8%), designing (49.4%), and developing educational programs (49.4) (Table 28.4).

28.3.4 Stakeholders' Attitudes on the Possibilities for Improving the Museum Offer

In order to determine whether there are statistically significant differences in the perception of the possibilities for improving the museum offer among internal

Table 28.5 Suitability of the scale for performing the factor analysis procedure of the perception of possibilities for improving the museum offer

Kaiser-Meyer-Olkin measure of sampling adequacy	0.83
Bartlett's test of sphericity approx. chi-square	1220.245
df	231
Significance	0.000

Table 28.6 Extraction sums of squared loadings

Cumulative %	
1	44.678
2	54.524
3	60.183
4	65.773
5	70.730

stakeholders, a factor analysis was performed on the total sample of internal stakeholders for 22 questions. The value of the KMO indicator of the adequate sample was 0.83, which exceeds the threshold value (it should be greater than 0.6). Bartlett's test of sphericity reached the required statistical significance Sig. 0.000, which means that it is less than the threshold value 0.05. This confirms the justification of the application of factor analysis (Table 28.5).

These five factors determine 70.7% of the data, i.e., the variance. Table 28.6 shows the communality of individual motives, i.e., to what degree each one describes a given phenomenon. All items have a value greater than 0.5, so they are all included in further analysis. By analyzing the values and the curve diagram, it is concluded that it is meaningful to keep five dimensions, which describe the perception of the improvement of the museum offer (Table 28.7).

Principal components analysis revealed the presence of five components, which explain a total of 70.7% of the variance. The value of the Cronbach coefficient for all factors (from F1 to F5) exceeds the recommended value of 0.7. Cronbach coefficient alpha for the entire scale of 22 questions is F1–F5 = 0.94. The above data indicate that the set model is reliable (Nunnally, 1978) and that the obtained results are scientifically supported.

The first factor, F1 – modernization of the museum setting, combines five items, with factor loadings ranging from 0.43 (for the statement "computer room") to 0.76 (for the statement "audio accompaniment of the exhibition").

The second factor, F2 – animation, combines five items, with factor loadings ranging from 0.32 (with the statement "costumed animators" to 0.87 (with the statement "children's animations").

The third factor, F3 – complementary activity, combines three items, with factor loadings ranging from 0.44 (with the statement "demonstration of making replica souvenirs") to 0.76 (with the statement "treasure hunt").

Table 28.7 Results of factor analysis

Selected factors and attributes	Factorial workload	Eigenvalue	Variance explained	Cronbach's
F1 – modernization of the museum exhibition		9.829	44.678	0.875
Audio accompaniment of the exhibition (music in certain parts of the museum).	0.762			
Video views (films)	0.660			
Portable computers	0.629			
Use of QR codes	0.546			
Computer room-research center for visitors	0.430			
F2 – animation		2.166	9.846	0.840
Children's playgrounds	0.871			
Thematic restaurant in accordance with the museum offer	0.793			
Children's workshops	0.733			
Prize quizzes	0.711			
Costumed animators, who bring historical events to life	0.323			
F3 – complementary activity		1.245	5.659	0.774
The possibility of engaging in some traditional activity	0.715			
Demonstrations of making replicas, souvenirs	0.447			
"Treasure hunt" (a game in which, in each subsequent phase, the visitor chooses the next exhibit he should see)	0.761			
F4 – information and shopping		1.230	5.590	0.751
Possibility of buying replica souvenirs	0.861			
Free accompanying brochure/flyer	0.747			
Panels with texts and photos about the exhibit	0.539			
Possibility to buy books, CDs with a detailed description of the exhibitions	0.518			
F5 – way of visiting the museum		1.091	4.957	0.784
Virtual tours	0.818			
Audio tours/audio guides	0.777			
Fixed (touch screen) computers next to the exhibits	0.668			
Possibility of visiting the museum/gallery accompanied by a guide	0.581			
Possibility of visiting the museum/gallery independently	0.391			

The fourth factor, F4 – information and shopping, combines four items, with factor loadings ranging from 0.1 (with the statement "possibility of buying books and CDs") to 0.86 (with the statement "possibility of buying replicas and souvenirs").

The fifth factor, F5 – the way of visiting museums, combines five items, with factor loadings ranging from 0.39 (with the statement "possibility of visiting independently") to 0.81 (with the statement "virtual tours").

28.4 Conclusions

Museums are attractions that tourists like to visit in a destination. For this reason, museums should be oriented toward their visitors, both local residents and tourists. Through this research, it was learned that internal stakeholders in museums have a positive attitude toward tourism because they are aware that its development creates additional sources of income and that proper operations do not disturb the efforts for conservation and preservation of artifacts. When doing business, the needs of tourists are taken into account, especially when it comes to the formation and provision of information about exhibits, promotion, and other activities. Although emphasis is placed on the basic function of the museum, which is the protection and preservation of cultural heritage, the employees believe that the interpretation of local history and culture and the education of tourists are of great importance.

If it turned out that museum employees are open to the development of tourism and presence in the tourist offer at the destination, the question arose as to why museums in Vojvodina record a low number of visitors. The answer should be sought in the possibilities for improving the museum offer. During the research, a factor analysis model was created to measure the possibilities for improving the museum offer, where 22 items were grouped into five factors: (1) modernization of the museum setting, (2) animation, (3) additional activities, (4) information and shopping, and (5) the way to visit the museum.

The modernization of the museum setting would be realized by the introduction of audio accompaniment for exhibitions, video displays, portable computers, and the use of QR codes and the creation of a computer room, that is, a research center for visitors. Animation would be realized through children's playrooms, children's workshops, prize quizzes, themed restaurants, and costumed animators. Supplementary activities represent the possibility of engaging in some traditional activities, a demonstration of making replicas and souvenirs, as well as the introduction of the "treasure hunt" game. The way to visit the museum should be performed through virtual tours, audio tours, a fixed "touch screen", accompanied by a guide or independently. Information and shopping should provide the purchase of souvenirs or replicas, free accompanying flyers, panels with texts and photos, as well as the possibility of buying books or CDs.

The results of this work have theoretical and practical significance, because knowing the views of internal stakeholders, which are based on experience from practice, indicates that employees in museums are aware of the role of tourism in the work of museums and that they do not need additional education in that direction, but help in education how to write and apply projects and find financial and other resources for modernizing the museum offer. In the theoretical sense, a new model was created for measuring the possibilities for improving the museum offer, which can be the basis for some subsequent research.

References

Allan, M., & Altal, Y. (2016). Museums and tourism: Visitors motivations and emotional involvement. *Mediterranean Archaeology and Archaeometry, 16*(3), 43–50.

Ames, M. M. (2004). Museums in the age of deconstruction. In *Reinventing the museum: Historical and contemporary perspectives on the paradigm shift* (p. 80).

Bartlet, M. (1954). A note on multiplying factors for various chi-squared approximations. *Journal of the Royal Statistical Society, 16*, 296–298.

Bauer, J. E., Sohn, A. P. L., & Oliveira, B. S. D. (2019). Cultural Tourism: A study on museums and the Internet. *Turismo: Visão e Ação, 21*, 291–308. https://doi.org/10.14210/rtva.v21n3.p291-308

Benediktsson, G. (2004). Museums and tourism: Stakeholders, resource and sustainable development. In *International Museum Studies*. Museion/Groteborg University.

Culley, S. (2010). *Museums and tourists: A quantitative look at curator perceptions of tourism*. Master Thesis, http://uwspace.uwaterloo.ca/bitstream/handle/10012/5075/Culley_Sarah.pdf?sequence=1

Conti, E., Pencarelli, T., & Vesci, M. (2017, September). Museum visitors' profiling in the experiential perspective, value co-creation and implications for museums and destinations: An exploratory study from Italy. In *Heritage, Tourism and Hospitality International Conference HTHIC* (pp. 21–34).

Freeman, R. E. (1984). *Strategic management: A stakeholder approach*. Pitman Publishing.

Gajski, A., Klarić, V., Laszlo, Ž., Nevidal, R., & Pintarić, S. (2011). *Muzeji i turizam: Djelovanje muzeja kao dionika kulturnog turizma*. Muzej suvremene umjetnosti.

Genoways, H. H., & Ireland, L. M. (2003). *Museum administration: An introduction*. Altamira Press.

Gorsuch, R. L. (1983). *Factor Analysis*. Lawrence Earlbaum Associates.

Kotler, N. (2001). New ways of experiencing culture: The role of museums and marketing implications. *Museum management and curatorship, 19*(4), 417–425.

Kotler, P., & K. N. (1998). *Museum strategy and marketing: Designing missions, building audiences*. Generating Revenue and Resources.

Kotler, N. G., Kotler, P., & Kotler, W. I. (2008). *Museum marketing and strategy: Designing missions, building audiences, generating revenue and resources*. Wiley.

Krivošejev, V. (2012). *Muzeji, menadžment, turizam: ka savremenom muzeju, od teorije do prakse*. Narodni muzej.

Lynch, R., Burton, C., Scott, C., Wilson, P., & Smith, P. (2000). *Leisure and Change: Iimplications for museums in the 21st century*. University of Technology and Powerhouse Museum.

McManus, G. (2010). *Community facility or tourism product: Managing the multiple identities of a small museum*. Available on internet: www.intercom.museum/documents2-2McManus.pdf. Accessed on 03.06.2021.

Navarrete, T. (2019). Digital heritage tourism: Innovations in museums. *World Leisure Journal, 61*(3), 200–214. https://doi.org/10.1080/16078055.2019.1639920

Pekarik, A. J. (2003). Another view museums as symbols. *Curator The Museum Journal, 46*(2), 132–135.

Serravalle, F., Ferraris, A., Vrontis, D., Thrassou, A., & Christofi, M. (2019). Augmented reality in the tourism industry: A multi-stakeholder analysis of museums. *Tourism Management Perspectives, 32*, 100549. https://doi.org/10.1016/j.tmp.2019.07.002

Silberberg, T. (1995). Cultural tourism and business opportunities for museums and heritage sites. *Tourism Management, 16*(5), 361–365. https://doi.org/10.1016/0261-5177(95)00039-Q

Thyne, M. (2001). The importance of values research for nonprofit organisations: The motivation-based values of museum visitors. *International Journal of Nonprofit and Voluntary Sector Marketing, 6*(2), 116–130. https://doi.org/10.1002/nvsm.140

Tufts, S., & Milne, S. (1999). Museums: A supply-side perspective. *Annals of Tourism Research, 26*(3), 613–631. https://doi.org/10.1016/S0160-7383(99)00024-9

29 Perception and Expectations of an Employability Fair as a Generator of Sustainable Wealth in the Tourism Sector

Óscar Silva, Catarina Lucas, and José Oliveira

29.1 Introduction

In economically developed countries, a remarkable trend of better pay for intellectual and skilled workers has begun, meeting the expectations of knowledge-based companies and having characteristics such as initiative, creativity, problem-solving capacity, or flexibility (OECD, 1996). The "moment of truth" or the contact between tourism sector employees and consumers (Carlzon, 1987) will determine the extent to which the service fails, meets, or exceeds the expectations of tourists. These roles played by the workforce will determine the very survival of the tourism industry (Amoah & Baum, 1997). This concept is supported by Swarbrooke (1995) that, in the very nature of commercial tourist operations, human resources often make up the majority of the budget. As a result, the tourism industry faces more challenges than ever before in attracting skilled and motivated personnel than emerging sectors of the economy (Kelley-Patterson & George, 2001).

Thus career/employability fairs are, first of all, an attractive form and channel of marketing communication, as well as an experienced and tested way of stimulating

Ó. Silva (✉)
Centro de Investigação Techn&Art – Instituto Politécnico de Tomar,
Vila Nova de Gaia, Portugal
e-mail: osilva@ispgaya.pt

C. Lucas
ISPGAYA, Vila Nova de Gaia, Portugal

J. Oliveira
REMIT – Research on Economics, Management, and Information Technologies,
Vila Nova de Gaia, Portugal

© The Author(s), under exclusive license to Springer Nature Switzerland AG 2023

F. L. Almeida et al. (eds.), *Multidimensional Sustainability: Transitions and Convergences*, Springer Proceedings in Earth and Environmental Sciences,
https://doi.org/10.1007/978-3-031-24892-4_29

commercial exchange. For entrepreneurs, the fairs are also an important place for the acquisition of information about a specific trade or business (Gębarowski, 2010). Companies that attend career/employability fairs benefit from being able to connect with a large group of potential employees, recruit young talent, and reinforce awareness of the company's brand (Greening & Turban, 2000). Literature is abundant on definitions for fairs and exihibitions, that highlights the difference between both, stating that: "Fairs are market events of a specific duration, held at intervals, in which a large number of companies present the main range of products/services of one or more sectors of industry and sell it mainly based on samples" (UFI, 2011). The fairs predominantly attract visitors from commerce and companies, while "Exhibitions are market events of a specific duration, held at intervals, in which a large number of companies present a representative range of products/services from one or more sectors of the industry and sell it or provide information about it for sales promotion purposes" (UFI, 2011). Career/employability fairs are a way for companies to meet recruitment goals in an increasingly competitive context to attract the best talent (Silkes et al., 2010). Career/employability fairs also allow visitors to learn about organizations and their recruitment needs (Brennan et al., 2004). This research aims to study the value of career fairs as a recruitment tool, as perceived by their visitors.

A worker's perception of any industry will undoubtedly be determined by his commitment, attitude, and perception of work in the industry, as well as by the employability available in the sector. It is argued that this is a situation particularly relevant for the tourism and hospitality sector, since the message disseminated by the media, and generally present in the imagination of potential recruits, has a negative image of working in the industry (Aksu & Köksal 2005; Brien, 2004; Getz, 1994; Kusluvan & Kusluvan, 2000). Negative attitudes toward tourism work can result in the industry's inability to capture and retain the most qualified tourism students.

This research presents preliminary results to respond to the research question: How employability fairs for professional and high education students (in terms of motivation, satisfaction, and recommendation to further editions) contribute to the interdimensional sustainability of employers as generators of economic/social and environmental wealth?

This research attempted to reveal and point out some changes, which visitors consider would strengthen the career fair/employability experience, as well as the commitment of tourism students to the industry, to assess the status/future commitments of tourism jobs in the human resources planning/sustainability (HR) process for the tourism sector. This chapter is organized as follows: Section 29.1, is an introduction to the subject of tourism and fairs, in Sect. 29.2 and extensive literature is carried out on the theme under study. In Sect. 29.3, the methodology used and data collection is described. Section 29.4 presents and discusses the obtained results, and finally, Sect. 29.5 presents the major conclusions, contributions, limitations, and clues for future investigations.

29.2 Literature Review

Traditionally, business organizations use career/employability fairs to recruit highly qualified candidates, thus reducing the time it takes to fill vacancies (Hansen, 2006). In turn, career/employability fairs allow visitors/students to learn about different business organizations present, evaluate their profiles according to the profile required by the organization (Silkes et al., 2010), and learn about careers and career paths within an industry, and also for those who did not know that they existed before the fair (Payne & Sumter, 2005). Visitors/students benefit from a large number of business organizations that meet in the same place and businesses benefit by accessing a large number of visitors/students during a visit (Payne & Sumter, 2005). Being so, this type of fairs are a form to invest in a career or possible employability, as it allows to find the "fittest" candidates is reasonable. Interacting with a recruiter at a career/employability fair is also a way for students to get to know the recruiter before the interview and hopefully make a more meaningful impression. Recruiters can be newly licensed human resources who can better relate to students, managers, and experienced human resources directors to whom new hires will report in the future (Harris et al., 2006). These recruiters typically work in human resources or operations (Silkes et al., 2010). Sending employees to career/employability fairs not only boosts workers' morale but also transforms other employees into recruiters throughout the year (Hansen, 2006). Another advantage of career/employability fairs is the ability of employers to observe the competition, especially when attending industry-specific fairs.

Career fairs not only provide recruitment opportunities but also allow companies the opportunity to raise brand awareness (Hansen, 2006). This is an important opportunity because some visitors/students only know the company through the recruiter (Silkes et al., 2010). According to Silkes et al. (2010), the visitors/students who attended career/employability fairs had greater contact and knowledge of their chosen careers. Rynes and Cable (2003) have compiled research on the recruitment process and on several factors that make a company attractive to a candidate, such as their recruiter impression, the opportunity of the selection process, and the opinion of personal friends or family.

Candidates are thus influenced by "(I) how kind, empathetic, caring, concerned by the recruiter's knowledge of the company, (II) their positions, and (III) their candidates" (Silkes et al., 2010). In addition, the impressions that candidates have on recruiters affect how candidates approach the recruitment process, including their decision on job selection (Connerley & Rynes, 1997). Another very effective strategy is to use graduated students, who like to give career advice and feel more comfortable talking to a young student compared to the recruiter (Taylor, 2009). Some companies have found that paying a subsidy to a former trainee to become a representative of their business organization on their return to school is also an effective way to raise awareness of the company (Hansen, 2006). An alternative method companies use is to inform former interns of when they will be at the fair, invite them to a casual meeting, and ask them to bring a friend with them as a way to reach more students (Taylor, 2009). In addition, business organizations have found that the association between fairs and technology reinforces their recruitment strategies since fairs allow them to face time with potential quality candidates (Hansen, 2006). To better connect with Generation Y candidates, Sodexo recruiters use Twitter, Facebook, and the company's career blog, as well as attending career fairs (Frauenheim, 2009).

In other approaches, it is understood that the individual perception of a service landscape creates internal responses in the form of reactions (cognitive, emotional, and physiological), which lead to different behaviors (Bitner, 1992). Still, Bitner (1990) states that a customer's perception of the quality of the product/service affects their overview of the service company and its behavioral intentions. The approach behaviors occur when the client remains and enjoys the space during the encounter with the product/service and avoids behaviors when the client leaves and can feel feelings of disappointment. Approach behaviors can lead to affiliation or membership and commitment to a service company (Bitner, 1992). Still, and according to the same author, "membership implies that the customer becomes a regular or loyal user, to revisit, who may become a brand supporter, and that encourages others to engage with the service company." What a candidate thinks about the host/promoter of the employability/career fair reflects that their representatives need to be carefully selected (Turban et al., 1995), and the same care should be taken with recruiters attending a fair. The signaling theory provides a plausible explanation for the candidates' employment decisions since experiences with recruiters can serve as important, but not observable, signs of organizational characteristics (Connerley & Rynes, 1997). For example, "respondents interpret the interpersonal effectiveness of recruiters (i.e., nonverbal cues, symbols, sympathy, organizational skills, and verbal behavior) as signs of organizational effectiveness" (Wildes & Tepeci, 2004).

Previous studies have also shown that the amount of information provided to candidates positively affects the attractiveness of the company as a workplace (Turban et al., 1995). According to Boyd and Boyd (2020) "consider space and configuration in a way that incentivizes natural conversation and interaction and removes barriers – real and perceived – should facilitate better engagement". However, environmental conditions are background characteristics, which affect both customers and employees, temperature, lighting, aroma, ambient color, noise, music, etc., becoming more noticeable when they are extreme, when the customer spends a lot of time in the service landscape, or when these conditions conflict with their expectations (Bitner, 1990). Regarding the configuration of spatial layout and spatial functionality, it is the size and shape of furniture, equipment, machinery, and how all these items are positioned in space, and how these items facilitate "performance and achievement of objectives" (Bitner, 1992). All these signs, symbols, and artifacts, direct or implicit, communicate information about a service landscape to their customers. Bitner notes that this dimension is especially important for first impressions. In the case of visitors/students present at these employability/career fairs, they may perceive the lack of information as a sign that the business organization (host) does not "care" enough to provide them with accurate information.

Since research has shown that there is a strong relationship between the impressions of candidates during the early stages of recruitment and final decisions, the impressions recruiters make on candidates are of paramount importance (Connerley & Rynes, 1997). Job expectations are constantly changing and recruiters are the link between education and industry (Miranda, 1999).

A possibility for visitors/students to benefit from the relationship between recruiters and educators is that recruiters can advise on which courses and training students should take to better prepare for the area in which they specialize (Payne & Sumter, 2005). Educators, on the other hand, must monitor the changes in the sector to ensure that students are well prepared for future expectations and positions (Miranda, 1999). One of the ways to do this is through career fairs/employability and through relationships that are built between universities and recruiters. In addition, internship and job opportunities for students reinforce the recruiter-educator relationship. It often leads to other partnerships, such as research, as well as the increase in the profile of the school and the profile of the company in the minds of students (Chi & Gursoy, 2009). Other studies show that the results showed a gap between the perceived importance of the event among students and the performance of the event (Lee et al., 2019). Discussion of how to address this gap is included, so students can reap the full benefits of career fair. These opportunities lead to long-term relationships that will continue to optimize benefits for students in the future.

According to the studies developed by Boyd & Boyd (2020), "Careers fairs continue to play a significant role in facilitating the transition from university to graduate employment, but as part of a more involved process rather than providing the actual hiring moment itself." For Milman & Whitney (2014) and Ferreira et al. (2021), "the level of satisfaction with the career fair was attributed to current availability of jobs that appeal to them, willingness of employers' representatives to take their printed resume, and sufficient industry representation of the segment that the students were interested to pursue their careers. Implications for college career fair event planners, students, and recruiters are discussed."

29.3 Methodology

The methodology used for this research is qualitative, and this research is a case study, whereby questionnaires were applied. A case study should be performed when investigating contemporary events (Lopes et al., 2018; Whitehead, 2003). Therefore, a case study is found to be the most suitable method for this research as it allows to gather interpretations through observations (Ferreira et al., 2021; Lopes et al., 2022).

29.3.1 Data Collection

This research presents preliminary results to respond to the research question, "How employability fairs for high education students contribute to the interdimensional sustainability of employers as generators of economic/social and environmental wealth?" To support this goal, a questionnaire was delivered to a sample of visitors who visited the First Tourism Employability Fair that took place on April 6th in Porto in 2022 (after the COVID-19 pandemic). This data collection instrument is one of the most used methods among tourism researchers, being quite effective in the systematic collection of information from a large number of people (Altinay et al., 2015). The questionnaire survey was based on an author/study. The fair organizers had a mandatory script for the questionnaire that the authors had to follow. Given the interest in obtaining answers about satisfaction and recommendation of the event described, visitors were invited to answer a questionnaire about said satisfaction after having spent some time at the fair and during the event.

The population under study consists of all visitors to the tourism employability fair, on which dimension we do have not sufficient information. A convenience non-random sampling was adopted including all visitors who accepted to participate in the study, with an exclusion criterion of age under 16. Of all the visitors to the First tourism employability fair, 106 respondents were selected. A face-to-face survey was conducted at the venue with 23 questions which enabled data related to reasons that brought participants to the event, satisfaction with the date and venue, accessibility to the venue, opening and closing times, venue conditions, available places, registration, cleanliness, toilets, and event organization.

To assess the degree of satisfaction with the event, several items were answered on a 5-point Likert scale, ranging from 1 = "Very Dissatisfied" to 5 = "Very Satisfied." Questions were also asked to investigate how the participants evaluated the association between this event and socio-cultural events, tourist visits, and entertainment activities.

To assess the association between this event and other types of events (socio-cultural, tourist visits, and entertainment activities), a 3-point response scale was used, ranging from 1 = "Very Unsatisfactory" to 3 = "Suitable." The participants were asked the following questions: whether they would recommend the Tourism Employability Fair to their circle of friends (1 = "Yes"; 2 = "No"; 3 = "NA"); what would they suggest to improve this event; what other initiatives they would like the event to promote; and whether they would visit the fair if organized again (1 = "Yes"; 2 = "No"; 0 = "NA").

The following control variables were used: age, gender, qualification level, profession, and study interest area. Ethical approval was obtained from the event hosting institution. All participants were informed about the study and the anonymity of the research process.

29.3.2 Data Analysis

In terms of statistical software, Microsoft Excel, IBM SPSS Statistics, R version 4.1.2 (2021-11-01), and JASP were used. Given some gaps in the characteristics of the sample (the absence of gender proportionality, the sampling technique, the lack of knowledge of the population size, and consequent non-calculation of the response rate, etc.), at this point, inferential statistics were not used, but only descriptive statistics of the different variables were used. In order to predict the characteristics of possible future participants in the next edition of the event, proportions (%) of responses for the different groups (categories) of the control variables and some variables of interest were compared.

Although only preliminary results of a broader and comparative study of the sustainability of events within the scope of employability in the tourism sector are presented here, an attempt was also made to verify the existence/absence of a dependency relationship between satisfaction and loyalty to the event by comparing proportions.

29.4 Results and Discussion

In total, 81% of the surveyed individuals were female, aged between 17 and 62 years (26.45 ± 10.68). Although 13% of missing values were observed for the age variable, approximately 49% of the respondents were aged between 20 and 22 years (inclusive).

It was noted that 56% of the individuals surveyed learned about the event at school/work, 12% through e-mail, 12% through the internet in general, and 4% through social media. The data obtained from the survey concerns essentially nominal and ordinal variables, whose descriptive statistics are presented in Table 29.1.

It is noteworthy that regarding the variables measuring the degree of satisfaction, a median and mode of 5 ("Very satisfied") was observed for all variables (date,

Table 29.1 Descriptive statistics of nominal and ordinal variables

	Valid	Missing	Mode	Median	Minimum	Maximum
Reasons to attend the event	102	4	1	–	1	5
Knowledge of the event	100	6	2	–	1	8
Event date	105	1	5	5	1	5
Venue	105	1	5	5	1	5
Accessibility	105	1	5	5	1	5
Schedules	105	1	5	5	1	5
Venue conditions	105	1	5	5	1	5
Vacancies	104	2	5	5	1	5
Registration	104	2	5	5	1	5
Cleanliness	105	1	5	5	1	5
Toilets	98	8	5	5	1	5
Organization	104	2	5	5	1	5
Recommend to friends	105	1	1	–	1	3
Association with sociocultural events	100	6	3	3	1	3
Association with tourist visits	100	6	3	3	1	3
Assoc. with entertainment activities	102	4	3	3	1	3
Intends to repeat	106	0	1	1	0	2

venue, accessibility, schedules, venue conditions, vacancies, registration, cleanliness, toilets, and organization), which reveals a high satisfaction of respondents regarding the event. It was observed that the majority of respondents are "Very satisfied" (items ranging between 54% and 90%).

Moreover, the degree of Overall Satisfaction (average) of the individuals with the event corroborates these results (4.588 ± 0.573). As can be observed in Table 29.2, an absence of answers for the categories 2 = "Dissatisfied" and 3 = "Neither very nor somewhat satisfied" concerning the degree of Overall Satisfaction was noted.

According to the results, it is observed that individuals with Elementary or Secondary Education ($p = .010$, Kruskal-Wallis test) and individuals who were very satisfied with this event ($p = .028$, Kruskal-Wallis test) intend to return for the next edition of the event. The latter result may lead the study to the formulation and testing of the hypothesis that visitors' loyalty and visitors' satisfaction with the event are dependent variables.

The results presented in Fig. 29.1 also reveal a high level of satisfaction among respondents regarding various features of the venue and event organization.

The frequencies of answers of category 3 ("Neither very nor somewhat satisfied") were divided into two categories: "neutral positive" and "neutral negative."

Also, it was observed that around 68% of the individuals surveyed participated in the event in search of internship/work opportunities and around 22% went to the fair because it was suggested by teachers. It is known that 55 companies attended the event. It should be noted that no individual indicated that they would not recommend the event to their circle of friends.

According to some characteristics of the sample (such as: lack of gender proportionality, sampling technique, population size unknown and, consequently, absence of the response rate), inferential statistics were not used in the analyses, but only the proportions (%) of responses for the different groups (categories) were compared. In particular, the fact that a non-random convenience sampling was used, in a face-to-face survey at the venue of the event, may lead to a non-representative sample of the population and consequent bias in the results.

However, although only preliminary results of a broader and comparative study of the sustainability of events within the scope of employability in the tourism sector are presented here, an attempt was also made to verify the existence/absence of a dependency relationship between satisfaction and loyalty to the event by comparing proportions.

The results reveal a high level of satisfaction among respondents regarding this first fair event. Since a large part of the individuals that have participated in the event were looking for an internship/work opportunity, we can estimate that there is a possibility to generate employment in the tourism sector, consequently generating sustainable wealth in the local economy.

Table 29.2 Characterization of the public who intend to return in the next edition of the event

Variable	Categories	Code	Intends to repeat the event	
			Yes	NA
Gender	Male	1	16 (16.8%)	3 (37.5%)
	Female	2	79 (83.2%)	5 (62.5%)
Age	<21	1	25 (30.5%)	2 (28.6%)
	[21; 23]	2	26 (31.7%)	2 (28.6%)
	[23; 30]	3	10 (12.2%)	2 (28.6%)
	≥30	4	21 (25.6%)	1 (14.3%)
School level*	Elementary/Secondary	1	46 (50.0%)	0 (0.0%)
	Bachelors/Degree	2	34 (37.0%)	7 (100.0%)
	MSc/PhD	3	12 (13.0%)	0 (0.0%)
Reasons to attend the event ($p = .334$)	Internship/job opportunities	1	64 (69.6%)	5 (62.5%)
	Teachers' suggestions	2	18 (19.6%)	2 (25.0%)
	Fun/socializing	3	2 (2.2%)	0 (0.0%)
	Friends' suggestions	4	4 (4.4%)	1 (12.5%)
	Invitation offer	5	4 (4.4%)	0 (0.0%)
Knowledge of the event ($p = .917$)	E-mail	1	11 (12.4%)	1 (12.5%)
	School/work	2	49 (55.1%)	4 (50.0%)
	Internet	3	10 (11.2%)	2 (25.0%)
	Word of mouth recommendation	4	6 (6.7%)	0 (0.0%)
	Friends recommendation	5	9 (10.1%)	1 (12.5%)
	Social networking	8	4 (4.5%)	0 (0.0%)
Recommend to friends ($p = .898$)	Yes	1	**93 (98.0%)**	7 (100.0%)
	NA	3	2 (2.1%)	0 (0.0%)
Association with sociocultural events	Very inadequate	1	4 (4.5%)	2 (25.0%)
	Neither adequate nor inadequate	2	25 (28.1%)	1 (12.5%)
	Adequate	3	60 (67.4%)	5 (62.5%)
Association with tourist visits	Very inadequate	1	6 (6.7%)	2 (25.0%)
	Neither adequate nor inadequate	2	27 (30.3%)	1 (12.5%)
	Adequate	3	56 (62.9%)	5 (62.5%)
Association with entertainment activities	Very inadequate	1	10 (11.0%)	3 (37.5%)
	Neither adequate nor inadequate	2	29 (31.9%)	1 (12.5%)
	Adequate	3	52 (57.1%)	4 (50.0%)
Overall Satisfaction Score[a,*]	Very unsatisfied	1	1 (1.1%)	1 (12.5%)
	Satisfied	4	19 (20.2%)	0 (0.0%)
	Very satisfied	5	74 (78.7%)	7 (87.5%)

* $p < .05$, ** $p < .01$, *** $p < .001$

[a] The Overall Satisfaction Score was obtained as a unit-rounded average of the responses to all items regarding satisfaction with the event for different criteria

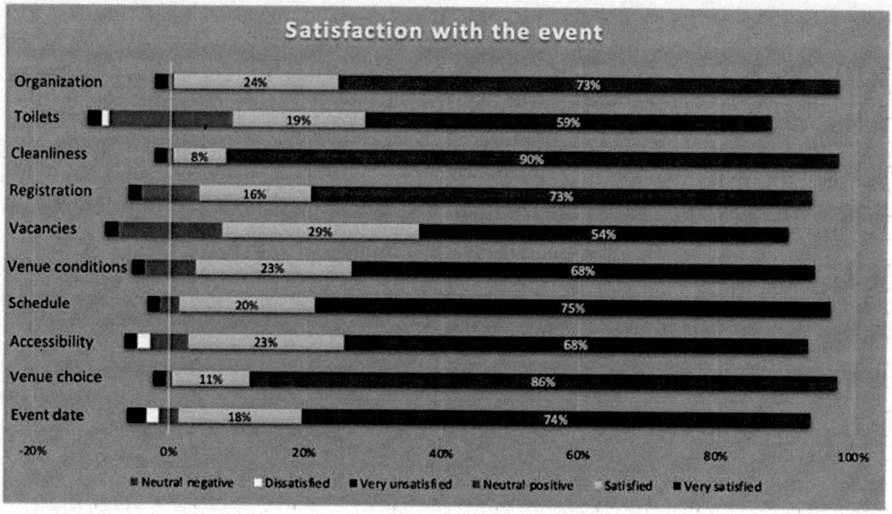

Fig. 29.1 Satisfaction with the event according to diverse items

29.5 Conclusions

This chapter has contributed to understanding the expectations of employability fairs as generators of sustainable wealth in the tourism sector and contributed to the success of both students and recruiting organizations. After the COVID-19 pandemic, employability fairs are one of the options to gather possible candidates, mainly graduates from higher education in tourism. On the other hand, it serves to detect and attract possible talents (Silkes et al., 2010).

This type of event is also important to enlighten students mainly from higher education organizations for the tourism market, especially for recruitment and selection, and this is a significant accomplishment in diminishing graduate labor markets, further debilitated by the COVID-19 pandemic, especially among private sector companies (Jackson et al., 2022).

Furthermore, this research, based on the First Tourism Employability Fair, has contributed to knowledge in two ways: first, it has enlightened the literature by adding new topics on tourism and the impact of fairs on employability. Secondly, due to the fact that a large part of the individuals who participated in the event were searching for an internship/work opportunity, we can estimate that there is a possibility to generate employment in the tourism sector, consequently generating sustainable wealth in the local economy.

As such, this may serve to advise companies and managers on how to take the most of employability fairs to capture new candidates just graduating mainly from high education (especially in the area of tourism) and detect possible new talents.

29.5.1 Limitations

One of the main limitations of this research was the sample size, which was mainly due to the survey setup (respondents were approached inside the fair). Future studies have to be performed on respondents at the entrance and before they diverge into the fair.

29.5.2 Clues for Future Work

This research can be used as a base for future investigations about employability fairs on tourism, either in Portugal or in EU countries. Furthermore, longitudinal studies can be performed yearly to compare the evolution. Moreover, future research should use a stratified sampling technique with gender proportion that allows for obtaining a representative sample of the population. This technique has a lot of advantages, such as the data collected is more homogeneous within each subset than in the total population and it is possible to obtain separate estimates of the population parameters for each stratum without choosing another sample, thus avoiding extra costs.

References

Aksu, A. A., & Köksal, C. D. (2005). Perceptions and attitudes of tourism students in Turkey. *International Journal of Contemporary Hospitality Management, 17*(5), 436–447. https://doi.org/10.1108/09596110510604869

Altinay, L., Paraskevas, A., & Jang, S. S. (2015). *Planning research in hospitality and tourism.* Routledge.

Amoah, V. A., & Baum, T. (1997). Tourism education: policy versus practice. *International Journal of Contemporary Hospitality Management, 9*(1), 5–12. https://doi.org/10.1108/09596119710157531

Bitner, M. J. (1990). Evaluating service encounters: the effects of physical surroundings and employee responses. *Journal of Marketing, 54*(2), 69–82. https://doi.org/10.1177/002224299005400206

Bitner, M. J. (1992). Servicescapes: The impact of physical surroundings on customers and employees. *Journal of Marketing, 56*(2), 57–71. https://doi.org/10.1177/002224299205600

Boyd, J., & Boyd, S. (2020, April). *Expectation and interaction: Student and employer approaches to university careers fairs*. p. Issue 44.

Brennan, C., Daly, M., Fitzpatrick, E., & Sweeney, E. (2004). Tradition versus technology: Careers fairs in the 21st century. *International Journal for Educational and Vocational Guidance, 4*(1), 23–41. https://doi.org/10.1023/B:IJVO.0000021056.29294.06

Brien, A. (2004). Do I want a job in hospitality? Only till I get a real job! In K. A. Smith & C. Schott (Eds.), *Proceedings of the New Zealand tourism and hospitality research conference, NZ.*

Carlzon, J. (1987). Moments of truth. *Industry Week, 234*, 40–42.

Chi, C. G., & Gursoy, D. (2009). How to help your graduates secure better jobs? An industry perspective. *International Journal of Contemporary Hospitality Management, 21*(3), 308–322. https://doi.org/10.1108/09596110910948314

Connerley, M. L., & Rynes, S. L. (1997). The influence of recruiter characteristics and organizational recruitment support on perceived recruiter effectiveness: Views from applicants and recruiters. *Human Relations, 50*(12), 1563–1586. https://doi.org/10.1023/A:1016923732255

Ferreira, C., Cardoso, C., Travassos, M., Paiva, M., Pestana, M., Lopes, J. M., & Oliveira, M. (2021). Disorders, vulnerabilities and resilience in the supply chain in pandemic times. *Logistics, 5*, 48. https://doi.org/10.3390/logistics5030048

Frauenheim, E. (2009). Sodexo. *Workforce Management, 88*(13), 23. https://scholarworks.umass.edu/cgi/viewcontent.cgi?article=1050&context=gradconf_hospitality

Gębarowski, M. (2010). *Współczesne targi.: Skuteczne narzędzie komunikacji marketingowej*. Regan Press.

Getz, D. (1994). Students' work experiences, perceptions and attitudes towards careers in hospitality and tourism: A longitudinal case study in Spey Valley, Scotland. *International Journal of Hospitality Management, 13*(1), 25–37. https://doi.org/10.1016/0278-4319(94)90057-4

Greening, D. W., & Turban, D. B. (2000). Corporate social performance as a competitive advantage in attracting a quality workforce. *Business & Society, 39*(3), 254–280. https://doi.org/10.1177/000765030003900302

Hansen, F. (2006). Far from obsolete, career fairs are a low-cost way for firms to hire quickly and fill pipelines. *Workforce Management, 85*(4), 46–47. Available on: https://workforce.com/news/far-from-obsolete-career-fairs-are-a-low-cost-way-for-firms-to-hire-quickly-and-fill-pipelines

Harris, K. J., Kwansa, F. A., & Lattuca, F. P., Jr. (2006). Recruiter opinion of hospitality programs: An assessment of selected programs. *Journal of Human Resources in Hospitality & Tourism, 5*(1), 17–33. https://doi.org/10.1300/J171v05n01_02

Jackson, D., Riebe, L., & Macau, F. (2022). Determining factors in graduate recruitment and preparing students for success. *Education+ Training*, (ahead-of-print). https://doi.org/10.1108/ET-11-2020-0348.

Kelley-Patterson, D., & George, C. (2001). Securing graduate commitment: an exploration of the comparative expectations of placement students, graduate recruits and human resource managers within the hospitality, leisure and tourism industries. *International Journal of Hospitality Management, 20*(4), 311–323. https://doi.org/10.1016/S0278-4319(01)00019-6

Kusluvan, S., & Kusluvan, Z. (2000). Perceptions and attitudes of undergraduate tourism students towards working in the tourism industry in Turkey. *Tourism Management, 21*(3), 251–269. https://doi.org/10.1016/S0261-5177(99)00057-6

Lee, M. J., Lee, P., & De Villa-Lopez, B. (2019). Hospitality and tourism career fairs: how important are they and how well do they work? *Journal of Teaching in Travel & Tourism, 19*(4), 326–340. https://doi.org/10.1080/15313220.2019.1592061

Lopes, J. N., Farinha, L. M., Ferreira, J. J., & Ferreira, F. A. (2018). Peeking beyond the wall: Analysing university technology transfer and commercialisation processes. *International Journal of Technology Management, 78*(1–2), 107–132. https://doi.org/10.1504/IJTM.2018.093936

Lopes, J. M., Gomes, S., & Mané, L. (2022). Developing knowledge of supply chain resilience in less-developed countries in the pandemic age. *Logistics, 6*(1), 3. https://doi.org/10.3390/logistics6010003

Milman, A., & Whitney, A. P. (2014). Evaluating students' experience and satisfaction at a hospitality and tourism college career fair. *Journal of Human Resources in Hospitality & Tourism, 13*(2), 173–189. https://doi.org/10.1080/15332845.2014.847304

Miranda, P. (1999). *The Level of technical skills and management competency demanded by the hospitality industry as perceived by hospitality recruiters*. Unpublished Master's thesis. University of Wisconsin-Stout.

OECD. (1996). *The knowledge-based economy, Paris*.

Payne, B., & Sumter, M. (2005). College students' perceptions about career fairs: what they like, what they gain, and what they want to see. *College Student Journal, 39*, 2269–2276.

Rynes, S. L., & Cable, D. M. (2003). Recruitment research in the twenty-first century. In W. C. Borman, D. R. Ilgen, & R. J. Klimoski (Eds.), *Handbook of psychology: Industrial and organizational psychology* (Vol. 12, pp. 55–76). Wiley.

Silkes, C., Adler, H., & Phillips, P. S. (2010). Hospitality career fairs: Student perceptions of value and usefulness. *Journal of Human Resources in Hospitality & Tourism, 9*(2), 117–130. https://doi.org/10.1080/15332840903383806

Swarbrooke, J. (1995). *The management and development of visitor attractions*. Butterworth-Heinemann.

Taylor, S. (2009). Capitalizing on college ties. *HR Magazine, 54*(12), 48–51.

Turban, D. B., Campion, J. E., & Eyring, A. R. (1995). Factors related to job acceptance decisions of college recruits. *Journal of Vocational Behavior, 47*(2), 193–213. https://doi.org/10.1006/jvbe.1995.1035

UFI (The Global Association of the Exhibition Industry) (2011). *The role of exhibitions in the marketing mix*. Ravensburg, Germany.

Whitehead, D. (2003). Case study research design and methods, 3rd edition. *Journal of Advanced Nursing, 44*, 108. https://doi.org/10.1046/j.1365-2648.2003.02790_1.x

Wildes, V. J., & Tepeci, M. (2004). Influences of campus recruiting on applicant attraction to hospitality companies. *Journal of Human Resources in Hospitality & Tourism, 2*(1), 39–51. https://doi.org/10.1300/J171v02n01_03

30 Rail Tourism – Emergencies and Possibilities on National Territory

Luísa Fernanda Ferreira Rodrigues

30.1 Introduction

Innovation and sustainability in tourism involve the creation of new ways of approaching the practices of using the resources available in the territory. Railroads, with their long history, are synonymous with technical and scientific evolution in the world of transport. Tourism is indelibly linked to transport because, in practice, they are the ones that enable the movement of people in space and the train has always played an important role in the way we have another perspective of the territory, revolutionizing the way of seeing landscapes, economies and cultures and the idea of progress has marked lasting local communities that directly coexisted with the presence of railway infrastructures.

The appeal of this places lies in their significance for the local communities and in the safeguarding and use of the cultural, material and immaterial heritage related to them, being places of departure and arrival, places of good or bad memories, places of fiction and places that democratized the discovery of the world. In an attempt to better understand these relationships between the old infrastructures and the community where they are located, we will observe the stations that have been transformed into railway museum nuclei and that are dependent on the Museu Nacional Ferroviário.

We will try to get answers to some questions that arise for the Portuguese territory, reflecting on good practices, namely those found in England in the scope of the tourist use they make of these structures and what strategies they use, in a sustainable perspective, capturing the interest of local communities for this process, as integral parts.

L. F. F. Rodrigues (✉)
ISPGAYA, Vila Nova de Gaia, Portugal
e-mail: lrodrigues@ispgaya.pt

© The Author(s), under exclusive license to Springer Nature Switzerland AG 2023
F. L. Almeida et al. (eds.), *Multidimensional Sustainability: Transitions and Convergences*, Springer Proceedings in Earth and Environmental Sciences,
https://doi.org/10.1007/978-3-031-24892-4_30

What we present is an embryonic approach embodied in an exploratory study that includes a brief survey of cases of sustainable use of the railroad heritage, the use of bibliography and other supports to scientifically sustain the investigation, and the execution of a semi-structured interview as a research method. The interview, conducted in a very simple way, serves the purposes of this approach, which we have established to be exploratory and in its initial phase.

The initial questions we asked were as follows: what degree of connection do the railway museum nuclei have with the territory and the surrounding community, and how do they make this relationship apparent to the tourists who seek the territory?

The criteria used to choose the railway nuclei dependent on the Museu Nacional Ferroviário are explained in the methodology. After making a brief presentation of each of the nuclei, we will present the result of the conversations that were established by telephone with the interviewees, in order to clarify the starting question and presenting the conclusions about what object of research was.

30.2 Sustainability in Tourism

Talking about sustainable development implies mentioning its beginnings, namely in the Brundtland report produced in 1987, under the title "Our Common Future", which considers sustainable development in its point 27 as: "the ability to make development sustainable to ensure that it meets the needs of the present without compromising the ability of future generations to meet their own needs" (United Nations, 1987). It is a challenge to establish a compromise between meeting the needs of the present and the availability of resources to respond to the needs of the future. Tourism sector must understand the current and future impacts on social, economic, and environmental contexts, harmonizing the needs of tourists, the environment, and local communities.

Among other development factors, network organization and the establishment of partnerships that are closely linked to local cooperation involving all actors are important (Martins & Morais, 2016). Still, with regard to community involvement and networking, it is worth taking a look at what Kayat (2014) says, as cited in Heleno (2019), who identifies five dimensions in supporting sustainable development: community involvement, empowerment and leadership; benefits to the community with the creation of infrastructure; collaboration and networking, highlighting the need for training of communities so that they can provide good service and collaborate in tourism activities; marketing and promotion; and conservation of resources.

According to Mourão (2000), the realization of the environment and heritage enhancement, simultaneously with the promotion of tourism activity, requires a firm intervention of local authorities, because it is through them that the action programs framed by strategic planning applied to areas potentially suitable for the practice of sustainable tourism are encouraged. A quality tourism experience that respects environmental and heritage resources, territory, and communities depends on the existence of adequate infrastructures. Making tourism sustainable has been envisaged for decades, particularly through the Manila Declaration of 1980. The conceptualization of sustainability refers us to the need to effectively manage the tourism resources available, having implicit economic development and, at the same time, respect for the environment and culture of each territory, two complementary and not incompatible formulas (UNWTO, 2018, p. 10).

The documents produced by the UNWTO dealing with sustainability and tourism have reflected the problem of development and the effects on the territory and communities. Of particular importance is the Charter on Sustainable Tourism produced in 1995 as part of the World Conference on Sustainable Tourism, held in Lanzarote and with reference to the Rio Declaration on Environment and Development, the Agenda 21 recommendations, the Manila Declaration, the Hague Declaration, and the Charter on Tourism Law and the Tourist Code. These state that tourism development must be sustained under sustainability criteria, must be ecologically sustainable in the long term, must be economically viable, and must be equitable to local communities from an ethical and social perspective. Thus, management strategies and tourism projects should be developed together with local communities (UNWTO, 2018, p. 121).

The same principles are advocated in the 1996 Bali Declaration on Tourism, the 2001 Osaka Millennium Declaration, the 2002 Quebec Declaration on Ecotourism, and the 2006 Tehran Declaration on Handicrafts, Tourism and Poverty Alleviation. More recently, the concerns formulated earlier continue to be discussed as demonstrated in the Beijing Declaration on Sustainable Tourism as a driver of Development and Peace of 2016.

The reflections proposed by international documents point to the need to adapt tourism to its host territories without losing sight of respect for the components that are part of the equation: the population, the available resources, whether cultural or environmental, and the tourist (Pereira, 2005). Thus, it becomes absolutely fundamental that the communities living in tourism territories actively participate in the decision-making and management processes, which may positively affect their lives. As Viegas (2013, p. 52) points out, tourism plays an important role in the economy because it ensures an improved quality of life for the population, is part of the solution to economic problems, and is a factor of dynamization of the global economy and contributes to mitigate regional imbalances.

The approaches around sustainability lead us to reflect about the increasingly important modality of community-based tourism that attributes to the community the wider exercise of power with the assumption of a comprehensive management of resources. While enclosing some dangers, "this form of alternative tourism is based on the premise of resident community involvement and community participation in the planning and management of tourism activities" (Heleno, 2019, pp. 26–27). As we have already mentioned, convening the community to play a role in the management of its tourism resources is a huge challenge for framing tourism on a course of sustainability.

30.2.1 Rail Tourism and the Role of Railway Stations

The railroad has a long history in England, that was lost in the early nineteenth century. It was synonymous with technical and scientific evolution and is one of the visible faces of the Industrial Revolution that began in the mid-eighteenth century and definitively transformed Europe and the world. In the distant year of 1825, the first public service for transporting people between Darlington and Stockton took place, using a steam locomotive.

The train arrived later in Portugal, as a result of Portugal's hesitant industrialization. Only with Fontes Pereira de Melo did the railroad become a reality. Throughout the second half of the nineteenth century, several railway lines serving a vast territory were built. In 1856, the railway became a reality with the inaugural trip from Lisbon to Carregado which opened a new page in transport in Portugal (Pereira, 2010, p. 27).

The railroad and the train became inseparable from the territory and the communities that are part of it, and the idea of progress lastingly marked the local communities that directly coexisted with the presence of railway infrastructures. In the Portuguese case, the discovery of the coastal landscape, the beaches, and the thermal environments was particularly important. It is the decisive moment of the start of the tourism industry in Portugal (Abragão, 1956; Pinheiro, 2006 as cited in Pereira, 2010, p. 29), supported by the need to provide travelers with good conditions, taking care to build "pleasant stations for tourists, (…) were the first contact that tourists had with the destination of the trip and in this sense the stations were embellished with tile panels, landscaping of the surrounding spaces and bold decorations (…)" (Matos, et al., 2009, pp. 13–14 as cited in Pereira, 2010, p. 33). In addition to the importance of the rolling stock as a sign of technological evolution, stations have a very special status. They are part of the built heritage and have received great attention throughout the history of the railroad. The station, more than any other place, is a meeting place, a place of passage and emotion. As a result of the renewed interest generated by the use of the train, there is, within the communities directly affected by the structures, an interest in safeguarding and disseminating their heritage.

England presents us with paradigmatic cases, arising from its long history related to railroads. Take, for example, the case of Community Rail Lancashire. In 2004, the British government announced the Community Rail Development Strategy, which proposed the existence of community organizations, developing inclusive, independent, and sustainable groups based on the key pillars for development: "providing a voice for the community; promoting sustainable, healthy and accessible travel; bringing communities together and supporting diversity and inclusion; and supporting social and economic development" (CRL – Community Rail Lancashire). This is an example of good practice in sustainability. According to the website, there are over sixty Community Rail Partnerships in the country and these partners are part of a federation that is overseen by the Community Rail Network.

In the case of Community Rail Lancashire, the activities developed show a very strong connection to the territory and the community, while promoting tourism activities involving the seven railway lines that are part of this group. In the events they organize, the concern to encourage the economic and cultural development of the population is clear, promoting the sale of a huge selection of local products and handicrafts, from local producers, design products, fashion, organization of exhibitions, gastronomic festivals, and organization of guided walks. Besides the events, they present some projects that are related to the community and to sustainability. This is the case of the projects "Make a Beeline Station," "Discover Amazing Women by Rail," and "Station Adoption – Friends of Stations." The latter is particularly interesting, and according to Community Rail Lancashire, Station Adoption brings considerable benefits to both the local community and the rail industry. The station becomes part of the community and is something that local people are proud of. Over time, there has been an increase in station usage, a reduction in vandalism, and an increase in external funding. Community adoption of the station adds value to what already exists and makes the station more attractive (CRL – Community Rail Lancashire, n.d.).

The focus of this chapter is the railway station for the reasons already mentioned. They are built heritage that is part of the identity of the places where they are located and are therefore subject to greater awareness to be safeguarded. They fit the classification of industrial heritage that is defined through the Nizhny Tagil Charter on Industrial Heritage (ICOMOS, 2003). This type of heritage despite its more resistant appearance is very fragile from the point of view of its conservation. Following the line of thought of Choay (2010, p. 234), it seems legitimate to consider that the reuse of railroad stations converted into museums is an effective practice at the service of the territory and the marks of its identity being an integral part of its sustainable development. Thus, many of the structures, given the urgency of their safeguarding, are reconverted and given different functions than the original ones. The case of the Musée d'Orsay in Paris, installed in the old Gare d'Orsay of 1900, is well known. The interest in stimulating industrial tourism that, in the case of railroads, has been asserting itself since the 1970s with the transformation of spaces where cultural activities are privileged, generating tourism products (Kull, 1998, cited by Rosário, 2015, p. 20), ends up assisting the task of reactivating and safeguarding these places, promoting new ways of approach in the enjoyment of heritage, and becoming identity reference points of the territory.

30.2.2 Railway Stations as Museums: The Railway Museum Nuclei in Portugal – Relationships and Roles

The conversion of a station into a museum poses very important problems due to the characteristics of the space and the collections. It has the advantage of having previously been a place of intense circulation of people and thus makes it easier to create identity and memory bonds. Museums are places of memory, and it is their goal to be mediators between the collections and the public. In this context, it is necessary to think very carefully whether the collections present in the museum fulfill this function. "In terms of the museums that have been created within Western tradition, they have been criticized for being sterile, concerned only with the past and not the present, non-emotional and non-threatening, focused on things rather than people, unaware of death, spiritually unattuned and academically distant" (Kavanagh, 2000, p. 101).

The path taken by museums has been made in the sense of getting closer and closer to the public, and this approach is more effective when it comes to museums that shelter industrial collections. The construction of memory and identity is remarkable because local communities are called to participate and because objects are made to speak beyond their form and function, reminding Kavanagh (2000, p. 105): "To think about things we have to think about people, and to think about people we have to listen, ask and learn".

Still regarding the visitors of museums and according to Edson and Dean (1996, pp. 176–177), museums make an effort to serve all publics, innovating in the way they design exhibitions in order to integrate everyone and providing unique experiences in a space where they have fun and enjoyable moments of leisure and where they feel comfortable. In fact, museums have been characterized by providing increasingly informal experiences in the way how the public approaches the museum space, the exhibition, and leisure activities.

The relationship between sustainability and museums has been a widely discussed theme and is reflected in the new concept of museums that was approved at the General Conference of ICOM, which took place on August 24 in Prague in the Czech Republic. Another aspect included in this new conceptualization is the importance of community participation. It is important to take a look at this new approach, which is in line with the study presented here. Changes have taken place in the role of museums that translate into the following: "a museum is a permanent, non-profit institution, at the service of society, which researches, collects, preserves, interprets and exhibits the tangible and intangible heritage. Open to the public, accessible and inclusive, museums promote diversity and sustainability. They act and communicate ethically, professionally and with the participation of communities, offering varied experiences of education, fruition, reflection and exchange of knowledge." (Cultura Portugal, 2022).

The National Railway Museum Foundation (FMNF) was established (Decreto-Lei, 38/2005 of 17 February) with a mission to study; to enhance and promote the historical, cultural, and technological Portuguese railway heritage, and an objective to install and manage the National Railway Museum in Entroncamento and promote and manage several museological centers in partnership with the municipalities Fundação do Museu Nacional Ferroviário (FMNF). Mayor Jorge Faria, quoted by Rosário (2015) in his inaugural speech, reminds that it is "the realization of a collective dream of a community that developed around the railroad." One of the Foundation's intentions was to create the Network of Railway Museums that "will be an organized system of museums that aims to strengthen the qualification and cooperation by promoting the enhancement of each of these railway museums (…)" Fundação Museu Nacional Ferroviário (n.d.).[1]

[1] https://www.fmnf.pt/pt/o-museu/rede-nacional-de-museus-ferroviarios/

There are seven railway museum nuclei: the Núcleo Museológico de Arco de Baúlhe; the Núcleo Museológico de Bragança; the Núcleo Museológico de Chaves; the Núcleo Museológico de Lousado; the Núcleo Museológico de Macinhata do Vouga; the Núcleo Museológico de Lagos; and Núcleo Museológico de Valença. The last two centers are closed and therefore were not included in the study.

Núcleo Museológico de Arco de Baúlhe (NMAB) This is located in the municipality of Cabeceiras de Basto. It was part of the old Tâmega line deactivated in 1990. The station was the end of the line. With emphasis on its 1940s tile panels, this nucleus includes the station building and the loading and unloading dock that is occupied with temporary exhibitions. The museum houses an important collection of carriages, locomotives, saloons, and objects that are part of the history and memory of the line, both with regard to the rolling stock, namely the vehicles in which King D. Carlos and Queen D. Amélia traveled in 1907, and the objects related to work and that have a great symbolic charge (Fundação Museu Nacional Ferroviário, n.d.).[2] The Arco de Baúlhe Station is also part of the Ecopista whose central station is Celorico de Basto, and that was built in the space previously occupied by the railway track with pedestrian or bicycle circulation (Turismo Porto e Norte de Portugal n.d., 2016, p. 34).[3] Since 2004, it is the headquarters of the Museu das Terras de Basto, also including the Núcleo de Arte Sagrada (Benedictine Monastery of Refojos), the Casa da Lã (village of Bucos), and the Casa do Pão (Abadim), thus promoting a cultural visit circuit (Município de Cabeceiras de Basto, n.d.).

Núcleo Museológico de Bragança (NMB) This is located in the center of the city in the old station. It was the terminal station of the Tua line. With the closure of the section between Mirandela and Bragança to rail traffic in 1996, the station passes to the jurisdiction of the Municipality of Bragança (Noé, 2011). According to the FMNF, the nucleus was closed from 2002 until 2019. The museum aims to tell the story of the construction and operation of the Tua line "paying tribute, very particularly, to the identity and memory of the community, whose experience would forever be linked to the railway."[4] The station also benefits from the activation of an ecopista called the Mãe de Água ecopista, which is based on the old railroad channel of the Tua Line.

Núcleo Museológico de Chaves (NMC) This occupies the old building used to park the rolling stock of the Chaves Station, which was the terminal station of the old Corgo line. Its collection tells the story of its history from 1921 to 1990, the year of its extinction (Fundação Museu Nacional Ferroviário, n.d.).[5] The nucleus is inte-

[2] https://www.fmnf.pt/pt/visita/planear-visita/nucleo-museologico-de-arco-de-baulhe

[3] http://www.portoenorte.pt/pt/o-que-fazer/ecopista-do-tamega/

[4] https://www.fmnf.pt/pt/visita/planear-visita/nucleo-museologico-de-braganca

[5] https://www.fmnf.pt/pt/visita/planear-visita/nucleo-museologico-de-chaves

grated in the Municipal Museums Network – Museu da Região Flaviense. In recent times, the municipality recovered and modified the surroundings. In the station building, the sociocultural department started working and the goods dock became an exhibition gallery. The Núcleo Museológico de Chaves occupies the old station coach-house (Câmara Municipal de Chaves, n.d.).

Núcleo Museológico de Lousado (NML) It is located in Vila Nova de Famalicão and in this case does not occupy the station as it is still active, but is next to it, in the former workshop complex of the Guimarães Railway Company. The permanent exhibition consists of rolling stock belonging to eight companies that operated on narrow-gauge lines, objects, and various equipment. The preserved workshop component is functional (Fundação Museu Nacional Ferroviário, n.d.).[6] This project is the heir of the old Secção Museológica, opened to the public in 1979, through the initiative of Armando Ginestal Machado. The Núcleo Museológico de Lousado integrates the route called "Famalicão Turismo Industrial".[7]

Núcleo Museológico de Macinhata do Vouga (NMMV) This is located in the Municipality of Águeda and occupies the old coach-house of the Macinhata do Vouga Station, which is part of the Vouga Line, the only narrow-gauge railway operating in Portugal. The permanent exhibition tells us the history of this track that until 1990 connected the towns of Viseu and Espinho, including the Aveiro branch line. In this exhibition there are a several objects and vehicles of national manufacture built between 1940 and 1947 and that makes remain the railway identity of the territory (Fundação Museu Nacional Ferroviário, n.d.).[8] One of the activities that does not happen with the other museum nuclei is the circulation of the Historical Vouga Train between Aveiro and the Macinhata do Vouga station.

30.3 Methodology

This is an exploratory study focused on the relationship between the railway museum nuclei and community belonging to their territory and on how this relationship may be perceived by tourists. Thus, in order to answer these questions, we used the information collected from both the literature and media, and developed a semi-structured interview.

In order to ensure rigor regarding methodology, the study was developed taking into account the following criteria: formulation of starting questions, bibliographic and online documentation research, data collection through the mentioned instrument, and data analysis.

[6] https://www.fmnf.pt/pt/visita/planear-visita/nucleo-museologico-de-lousado

[7] https://www.cm-vnfamalicao.pt/museu-ferroviario

[8] https://www.fmnf.pt/pt/visita/planear-visita/nucleo-museologico-de-macinhata-do-vouga

It would be unaffordable to address, in a study like this, the totality of the infrastructures and, therefore, we used the microstructures that constitute the railway museum nuclei in operation that depend on the FNMF as the only criterion. We did not include the National Railway Museum in this approach, not denying, however, its importance, but the reason is that a uniformity of analysis is necessary, since it is located in the station facilities in Entroncamento, receives a large daily flow of travelers and visitors, and has a structure whose size does not match with any of the museum nuclei already described.

The methodology of this study was a semi-structured interview made by telephone to the people in charge of each nucleus. The interviews were conducted throughout the months of July and August 2022. Two of the interviewees preferred to respond in written form to the script guide, and the interview was both sent and received by e-mail. This qualitative approach seemed to be the most appropriate methodology for this type of exploratory study.

The list of topics considered relevant in this interview followed a defined order and included previously prepared questions. The interviewee is free to answer the questions as he or she wishes and the order of the topics can be changed. The kind of information extracted from the established conversations are not possible to obtain through the questionnaire, allowing the development and clarification of the questions asked by the interviewer (Jennings, 2001, cited by Heleno 2019, p. 96).

The interview script includes the following questions, which are in line with the initial questions:

- How many tourists does the museum nucleus receive annually?
- What relationship does it have with the community: does it promote cultural activities that involve the inhabitants or cultural or economic associations, the business sector or local artisans? Are there partnerships with these entities to make the nucleus more dynamic?
- Do you promote the cultural dissemination of the community's material and immaterial heritage?
- What activities do you promote for the local population in different age groups?
- Do you sell or publicize products produced by the community on the museum facilities?
- What activities do you promote, especially for tourists visiting the nucleus?
- Do you include old railroad workers and ex-employees, collaborating in guided tours, disclosing their memories to the public?

The interviewee in a fluid conversation answered freely and shared information that was not part of the script.

30.4 Results

In a total of seven museum nuclei that are part of the network created by the Fundação Museu Nacional Ferroviário, two of them are closed (N. M. de Lagos and N. M. de Valença), and the person in charge of the N.M.C. asked for the interview script to be sent by e-mail, but did not respond in time. Thus, we successfully obtained answers from four entities (N.M.A.B.,[9] N.M.B.,[10] N.M.L.,[11] and N.M.M.V.[12]).

Regarding the first question, the four interviewees reported that during the COVID-19 pandemic, there was a significant decrease in the number of visitors, and only this year, although incomplete, they are beginning to have an increase in numbers. They found that there are more national visitors than foreign visitors. The reference year is 2019. Regarding the N.M.A.B., they had a total of 9312 visitors in 2019, of whom 667 were foreign visitors, 1840 was school public, and the rest were national visitors; in the case of the N.M.B., in 2019 (they started operating from May of that year), they recorded a total of 4920 visitors, of whom 4214 were Portuguese and 706 foreigners. The N.M.L. reported that the total number of visitors in 2019 was 11,850, of which 412 were foreigners. The N.M.M.V. presented the following data: in 2019 they had a total of 8422 visitors, of whom 1522 were foreigners.

When asked about the kind of relationship the nucleus has with the community – whether it promotes cultural activities that involve the inhabitants or cultural or economic associations involving the business sector or local artisans and whether there are partnerships with these entities for the promotion of the nucleus – the person in charge of the N.M.A.B. told us that because the nucleus is the headquarters of the Museu das Terras de Basto, a cultural relationship is automatically created between the railway museum and the other cultural spaces in the municipality. However, she did not mention links to cultural associations. Regarding the relationship with enterprises, this does not occur, but with local artisans, it is effective in the organization of temporary exhibitions that take place in the station space. The person responsible for the N.M.B. answered us that these relationships are not established and there are no partnerships with companies and associations; however, he says it is an intention in the near future. Concerning the N.M.L., it is not a promoter/organizer, but participates and integrates all local and municipal initiatives such as the Local Exhibition of Lousado, and the practice of ferromodelling of the Módulos de Comboios do Norte collective that meets regularly at the museum. This nucleus also has no partnerships with local companies and associations. The response of the person in charge of the N.M.M.V. shows a great connection with local associations and producers and claims that this relationship has been consolidated since 2017 with the signing of the protocol with the Vouga Historic Train. Thus, the museum promotes guided and staged tours with the local associations at the weekend, when the trip in the historic train takes place, and mentioned the various associations that collaborate with the nucleus: "Associação Os Descarrilados", "Amar - Associação Macinhatense de Assistência Recreio e Cultura de Macinhata do Vouga", "Grupo Folclórico e Etnográfico de Macinhata do Vouga", "Clube Macinhatense", "Atlética

[9] The interviewee was Dra. Fátima Magalhães.

[10] The interviewee was Dr. Filipe Pires.

[11] The interviewee was Dr. Pedro Costa.

[12] The interviewee was Da. Manuela Almeida.

Macinhatense", "Tema-Teatro Espontâneo de Macinhata do Vouga", "Associação Desportiva e Cultural de Jafafe". These associations will animate the trip to the museum and in the museum itself.

Regarding the question, do they promote the cultural dissemination of the tangible and intangible heritage of the community, the person responsible for the N.M.A.B. said that in addition to the permanent exhibition of the railway nucleus, temporary exhibitions are promoted that change themes every 2 years and are based on local history in order to preserve the traditions and history, as it was the case of the exhibitions they had as the theme: the linen in Cabeceiras de Basto with the collaboration of a weaver; the art of clogging with the collaboration of a clog maker; or the exhibition inaugurated in January 2022 and that will be open until May 2023, on the theme "A festa das Papas de S. Sebastião em Gondiães e Samão".[13] About this event, Fátima Magalhães explained that the tradition of the festa das papas reinforces the idea that these initiatives are promoted to establish this connection to the community. Filipe Pires from N.M.B. says that, for the reasons mentioned above, this disclosure is not yet a reality, but one of the intentions is to survey and register real testimonies around the railroad in the near future while there are still people alive who can do it. Pedro Costa of the N.M.L. says that they promote dissemination through programming and promotion of exhibitions, activities, and routes. The in-charge for the N.M.M.V. declared the importance of the animation activities, both in the museological space and in the historical train that show the cultural traditions of the territory. They also promote the collection of testimonies related to the intangible heritage.

In response to the question, what activities they promote for the local population in different age groups, all mentioned that they have activities aimed at the school public, at different levels of education, and senior citizens. Filipe Pires mentioned some activities such as, the recycling workshop, the board games; the sports vacations with the permanence in the museum for a day and the realization of a cahoot adapted to each age group. The N.M.L.'s educational service offers guided tours, visit-workshops, pied paper, workshops, and the municipal educational program "Viagens pelo Património."

Regarding the question, do they sell or promote products produced by the community on the installations of the museum nucleus, the person in charge of the N.M.A.B. said that they do not sell any products, but that tourists could find objects and handicraft products at the tourist office. In the case of the Bragança and Lousado nuclei, they do not sell local products, and at the N.M.M.V., Manuela Almeida told us that outside the museum, on weekends and taking advantage of the historic train ride, local craftsmen and farmers sell their products in a small fair.

To the question, which activities are promoted, especially aimed at tourists, Fátima Magalhães referred to the visits of the permanent and temporary exhibitions, to the experience of traveling in the self-propelled train that leaves the installations on the International Day of Museums and that makes a short trip back to its

[13] The website of Município de Cabeceiras de Basto can be checked about this exhibition: https://cabeceirasdebasto.pt/index.php?oid=29297&op=all

exhibition space, and to the activities that are related to the eco track. Filipe Pires from N.M.B. says that they are, above all, guided tours to the museum, the same happening with the N.M.L. Pedro Costa, refers, however, that the visits are made in English and the nucleus belongs to national and international routes directed to tourists, such as the "Famalicão Turismo Industrial" and the "ERIH -European Route of Industrial Heritage." Manuela Almeida mentions the animation activities in the context of the Vouga Historical Train trip and the staged visits to the museum, as well as the normal visits without this kind of animation.

Regarding the question: do they include old railroad workers and ex-employees collaborating in the guided tours, reveal their memories to the public, all were unanimous in considering this inclusion very difficult due to the disappearance and advanced age of the old workers. A negative answer came from those responsible for the N.M.A.B. and N.M.B. The N.M.L. integrates a current CP – Combóios de Portugal employee responsible for the inventory and conservation of the collection. Manuela Almeida, on the other hand, tells us that they include old employees in the visits and animation and revealed that a meeting of retired railwaymen takes place in Macinhata do Vouga to share ideas and information about the activity.

30.5 Discussion

Analyzing the answers given by the interviewees to the formulated questions, we noticed some discrepancies in the results of every museum nuclei and a non-uniformity in approach, despite being all under the purview of the FMNF, which are due to the following reasons: (a) the nature of their geographical location, (b) the fact that two of the nuclei are built in deactivated line stations and the remaining two use station spaces with railroad circulation; (c) the way COVID-19 pandemic affected the structures; (d) the diverse way in which community participation is viewed in the activities of the museum nuclei; and (e) the capacity for community mobilization.

Observing the previous premise does not mean that we are facing something negative. On the contrary, it is in the different approaches and the possibility of comparing them that their richness lies, since it implies an attitude of sharing experiences and the consequent improvement of procedures.

If we recover the initial question of this paper: what degree of connection do the railway museum nuclei have with the territory and the surrounding community, and how do they make this relationship perceived by the tourists who visit the territory? we realize through the answers given that not all nuclei have the same degree of community involvement, in the case of the N.M.B. this degree is low and at the other extreme is the N.M.M.V. with a good degree of involvement and motivation of the community in the museum activities. We believe that the involvement of communities can be a powerful element of attraction for national and international tourists to visit the places, encouraging the desire to know the culture of the local inhabitants and, therefore, alert for the preservation of their culture.

However, we needed to collect more extensive data to understand if the offers of community-related activities promoted by the nuclei for tourists, namely foreign tourists, are reflected in the number of visitors. The little involvement of local businesses and craft producers with the museum structure, except for the N.M.M.V., was evident and could also play an important role in attractiveness.

The data collected by the interview showed that all nuclei experienced a marked loss of visitors/tourists in the restriction period due to the pandemic, but year 2022 proved to be a year of recovery. From the data for 2019, and the pandemic years, it was also found that domestic tourists are unequivocally in greater numbers. We may think that the fact that N.M.L. is integrated in a railway station that is still active favors the visit of a higher number of national visitors compared to the other nuclei. It was also possible to deduct that in the case of the N.M.M.V. a larger number of foreign visitors compared to the other nuclei may be due to the circulation of the Vouga Historical Train, whose activity took place on weekends between July and September, as it happens this year. We may consider that the complement of the diversified offers, besides the museum visit, is a strong attraction motivating the tourist to consume the experience, which is also the case in the N.M.A.B. that offers thematic temporary exhibitions, that is the seat of a polynucleated museum and that benefits from an eco track. For the circumstances that have been mentioned, the N.M.B. is developing an offer of activities that involve the community and that are attractive to tourists.

We can see that the people in charge of any of the nuclei interviewed affirm the importance and the interest in relating to the local community, and N.M.A.B. and N.M.M.V. have demonstrated, in practice, this intention with the various actions carried out. This is also the case with the actions of the N.M.L. that participates and integrates local initiatives.

The promotion of local heritage, beyond the exhibition of railway collection is also a reality, especially in the organization of temporary exhibitions, calling villages, artisans, and local traditions, as is the case of the N.M.A.B. and the promotion of cultural and tourist animation by the N.M.V. or the N.M.L. All of them have guided tours and an educational service aimed at schools. Only in the N.M.M.V. is there a more active presence of animation with historical recreation, making use of a large number of local cultural and recreational associations, which becomes a very valuable instrument of integration of the local community in the museum's recreational activities, promoting the dissemination of local heritage and culture.

The promotion of local economic activities, especially agricultural production and handicrafts with the sale of products made by artisans, is only effective in the N.M.M.V., particularly when there are trips in the historic train. One of the most interesting activities in the context under analysis would be the integration of old employees in the dynamics of the museum, collecting information through oral testimonies of these people, working on a voluntary basis (or not) in the dissemination of the collection, participating actively in guided tours of the museum space. The people in charge of both the N.M.B. and the N.M.A.B. say that it is a difficult task, because many years have passed since the deactivation of the lines and, therefore, these people are either very old or have passed away. Both in the case of N.M.L. and

N.M.M.V., this integration is possible and is done, because the line is still active and therefore the old employees play an important role here, employing a CP worker in the first one and even promoting meetings of old railwaymen and active participation of others in the second one.

Concerning the initial global theme of this chapter, sustainability, and reflecting the most recent conceptualization around the museum, through all that has been exposed, we recognize that through the results of the interviews the museum centres under study, there is a real awareness and interest in including communities in the museum and, even not yet doing so continuously, have the intention to promote this relationship in the future. This dynamic is a very important step for the preservation of heritage, culture, and identity of the territories in a logic of sustainability, bringing us closer to international practices such as those mentioned above.

30.6 Conclusion

Recognizing that the train and the railroad are the areas that had and still have the greatest impact on the territories, be they inland or coastal, this exploratory study allowed us to take a first look at the relationship that some museum spaces dedicated to the railroad have with the community and what impacts they have on tourists.

Thus, we can conclude that the museum nuclei that are more interconnected with the community are the most visited; the existence of a parallel offer to the visit of the museum collection makes them more demanded; the nuclei that belong to active lines are more visited; the pandemic years may serve as a platform to rethink the offer with an even closer relationship with the communities; the museological nuclei are an added value in the dynamization of the territory and the defense of the local tangible and intangible heritage; the nuclei that are implanted in deactivated lines will have to promote a wider dissemination of their activities, using for example differentiated marketing strategies; the deactivated stations and other structures are an important landmark of memory of the territory and its populations; there are different degrees of connection between the museum nuclei and the communities, at different rhythms; the investment of the local government in the management of these spaces and their activities, which in this case is done in co-management with the Fundação Museu Nacional Ferroviário, is fundamental; and most of the observed museum nuclei have invested efforts to make sustainability a reality with the participative inclusion of local communities, which is in line with the objectives of the Plan Turismo + Sustentável – 20 23 (2021).

However, and as we have pointed out, this study is embryonic and, by its nature, presents limitations. We wanted to raise the problem and begin to lead the way to deeper research and studies in an area that is currently attracting the attention of various sectors, including tourism. It constitutes a starting point whose intention is to contribute to understanding the role of museums with these characteristics within the scope of sustainability.

We are aware that at this moment the study needs a broad perspective, inclusion of more museums as the object of study, a larger number of data, a deeper look at the scientific research that has been produced on the subject, and collection of a larger number of case studies to support an investigation like this.

Thus, based on what we have observed throughout the research, this theme has a great potential for future opportunities to produce scientific research in its most diverse aspects: heritage, museology, tourism, community intervention, leisure, events, transportation, and sustainability, besides many others that can be added.

We hope that, with this exploratory approach, we tried to answer the questions we asked and, through this particular case of tourism experience, contributed to show how, locally, sustainability solutions are sought in the area of tourism.

References

Câmara Municipal de Chaves. (n.d.). Retrieved July 2022 from https://www.chaves.pt/pages/422

Choay, F. (2010). *Alegoria do Património*. Lisboa, Ed.70.

CRL – Community Rail Lancashire. (n.d.). https://communityraillancashire.co.uk/

Cultura Portugal. (2022). https://www.culturaportugal.gov.pt/pt/saber/2022/08/nova-definicao-de-museu-icom/

Decreto-Lei 38/2005 do Ministério das obras Públicas, Transportes e Comunicações. Instituição da Fundação Museu Nacional Ferroviário Armando Ginestal Machado (2005). Diário da República n° 34/2005, Série I-A de 2005-02-17. https://dre.pt/dre/detalhe/decreto-lei/38-2005-605326

Edson, G., & Dean, D. (1996). *The handbook for museums*. Routledge.

Fundação Museu Nacional Ferroviário. (n.d.). https://www.fmnf.pt/pt

Heleno, N. (2019). *Plano de desenvolvimento de turismo de base comunitária em espaços rurais: aplicação às terras de Sicó*. Aveiro (Master Thesis Universidade de Aveiro). Repositório Institucional da Universidade de Aveiro. http://hdl.handle.net/10773/27327

ICOMOS (2003). *Carta de Nizhny Tagil sobre Património Industrial*. https://www.icomos.org/18thapril/2006/nizhny-tagil-charter-sp.pdf

Kavanagh, G. (2000). *Dream spaces. Memory and the museum*. Leicester University Press.

Martins, C., & Morais, A. (2016). O contributo do turismo para o desenvolvimento das regiões. *Gaia Scientia. Edição Especial Europa, 2*, 84–91. https://doi.org/10.21707/gs.v10.n02a09

Mourão, J. (2000). Desenvolvimento Sustentável do Turismo – princípios, fundamentos e prática. *GeoINova - Revista do Departamento de Geografia e Planeamento Regional, 2*, 87–117.

Município de Cabeceiras de Basto. (n.d.). https://cabeceirasdebasto.pt/museu-terras-de-basto

Noé, P. (2011). *Estação Ferroviária de Bragança/Estação Rodoviária de Bragança*. Sistema de Informação para o Património Arquitectónico – SIPA. http://www.monumentos.gov.pt/

Pereira, P. (2005). *A actividade Turística e a Sustentabilidade das Regiões de Destino*. Actas do X Colóquio Ibérico de Geografia.

Pereira, H. (2010). As Viagens Ferroviárias em Portugal (1845–1896). *CEM Cultura, Espaço e Memória: Revista do CITCEM, 1*, 25–40.

Rosário, D. (2015). *Turismo e Caminhos de ferro: um olhar sobre Portugal* (Master Thesis Instituto Politécnico de Tomar – Escola Superior de Gestão de Tomar). Repositório Comum. http://hdl.handle.net/10400.26/12991

Turismo de Portugal, I.P. (2021). *Plano Turismo + Sustentável – 20 23*. http://business.turismodeportugal.pt/SiteCollectionDocuments/sustentabilidade/plano-turismo-mais-sustentavel-20-23-pt-jun-2021.pdf

Turismo do Porto e Norte de Portugal. (2016). *Ciclovias, ecopistas e ecovias do Norte de Portugal*. Retrieved July 2022 from www.portoenorte.pt

Turismo do Porto e Norte de Portugal. (n.d.). http://www.portoenorte.pt/pt/porto-e-norte/

United Nations. (1987). *Our common future*; Brundtland Report 1987. Available at: https://www.are.admin.ch/are/en/home/media/publications/sustainable-development/brundtland-report.html

Viegas, M. (2013). *O uso turístico enquanto valorização patrimonial: a importância do planeamento e da gestão* (Master Thesis Universidade de Évora). Repositório da Universidade de Évora http://hdl.handle.net/10174/15844

World Tourism Organization (UNWTO). (2018). Compilation of UNWTO Declarations 1980–2018. Available at: https://www.e-unwto.org/doi/epdf/10.18111/9789284419326

Sustainability and Corporate Social Responsibility in the Portuguese Tourism Context: A Preliminary Study

31

Beatriz Rosendo, Bruno Sousa, and José Duarte Santos

31.1 Introduction

Tourism is an area that has developed over the years, and as a result, there arises a need to create strategies for organization, creation, innovation, dissemination, among others, to increase the productivity and quality of products and services provided. All services and products are different, and this implies a deeper knowledge about the needs, desires, and motivations of consumers. These factors can vary since different personalities and tastes exist. In addition to tourism development, there has also been growing concern about society's problems and the solutions that can be implemented to combat verified problems. With companies' growing concern for sustainability and social responsibility, forums dedicated to the promotion of ethical practices and models, studies on the subject, the introduction of disciplines in various courses have emerged around the world, which naturally led to the development of the area of business ethics (Santos et al., 2022). Thus, issues of social responsibility arise in the face of numerous problems, such as disability and employment, well-being and comfort of employees in the company, various ethnicities, acceptance of the LGBTI community, support for people with financial difficulties, support for victims of domestic violence or dating, support for people with health problems, combating hatred and discrimination, among others. According to UNWTO (2022), the concept of ethics and responsibility is linked to the concept of culture, because

B. Rosendo (✉)
Instituto Politécnico do Cávado e do Ave, Barcelos, Portugal
e-mail: a15069@alunos.ipca.pt

B. Sousa
Instituto Politécnico do Cávado e do Ave and CiTUR, Barcelos, Portugal

J. D. Santos
CEOS.PP and ISCAP/P.PORTO, São Mamede de Infesta, Portugal

© The Author(s), under exclusive license to Springer Nature Switzerland AG 2023

F. L. Almeida et al. (eds.), *Multidimensional Sustainability: Transitions and Convergences*, Springer Proceedings in Earth and Environmental Sciences,
https://doi.org/10.1007/978-3-031-24892-4_31

it is important to understand that companies and tourist establishments must create activities in the context of social inclusion, sustainability, and the preservation of culture and heritage. One of the objectives of creating responsibility policies is to promote ethical behavior and good practice, both on the supply side and on the demand side. Then, we feel the need to renew strategies and to create and innovate something that works on these important issues.

Then emerges the purpose of this study – to understand which strategies and solutions can be implemented in the field of tourism and to prevent disadvantageous situations and improve existing systems. Thus, this chapter first frames the concept of sustainability in tourism and second presents some considerations about the concept of corporate social responsibility (CSR) and four aspects that are assuming prominence in the tourism industry and contributing to the solidification of the CSR: disability employment in tourism, LGBTI and social responsibility, pet-friendly tourism, and eco-friendly hotels. Finally, some concluding remarks are made.

31.2 Sustainability in Tourism

According to the Tourism + Sustainable Plan 20–23, the sustainability of a territory involves the responsible use of environmental resources, respect for local culture, viable economic activities over time, information made available to stakeholders important to the company concerned, constant control of the impact of the strategies implemented, the quality of products/services, and demand-side satisfaction. The countries of Europe have already become aware of the major problem of climate change and the environmental degradation that has grown over the years. To combat these changes, those responsible for tourist destinations presented, in 2019, the European Ecological Pact that aims to create and implement strategies that make a proper use of all existing resources (Turismo de Portugal, 2021). In 2015, the United Nations launched an agenda on sustainable development aimed at reducing poverty, providing a better quality of life for all people, and protecting the planet. This agenda has 17 objectives adopted by all Member States: No Poverty; Zero Hunger; Good Health and Well-Being; Quality Education; Gender Equality; Clean Water and Sanitation; Affordable and Clean Energy; Decent Work and Economic Growth; Industry, Innovation and Infrastructure; Reduced Inequalities; Sustainable Cities and Communities; Responsible Consumption and Production; Climate Action; Life Below Water; Life on Earth; Peace, Justice and Strong Institutions; and, finally, Partnerships (UNWTO, 2022).

The world is facing a situation that has altered the entire tourism system and negatively affected the economy of all countries (Lubowiecki-Vikuk et al., 2021). This issue has caused many people and businesses to be living on the edge and with many difficulties. Although there is a need to recover the economy, tourism will not leave aside environmental and social recovery, and for this, it is necessary to develop and implement sustainable strategies. This sustainable recovery will allow the tourism sector to resume its entire activity with greater economic, social, and environmental security. Turismo de Portugal recently joined the Global Sustainable Tourism

Council and the Portuguese Pact for Plastics, which represents the responsibility of the country to intervene in improving the sustainability of the world.

According to Eger et al. (2022), sustainable development considers the needs of the present and creates strategies without compromising the future. This concern with sustainability emerged in 1980 and has evolved over time, with this concept appearing in the policies of each company, so that they develop activities that meet the preservation of the environment and the community in which the company is part. With this, sustainable objectives are constantly developed that companies intend to achieve. It is also important to evaluate the behaviors and feedback that the supply side has in relation to the sustainability policies implemented by tourist destinations. For this is the only way to improve and implement these measures.

31.3 Corporate Social Responsibility (CSR)

The term corporate social responsibility has been used for more than 50 years, but what is the responsibility of businesses and organizations is still unclear (Sousa & Soares, 2021). It should be noted that there are various synonyms for the term "corporate citizenship," "ethical business," "corporate ethics," and others. Scientists and researchers are still attempting to define social responsibility. Socially responsible marketing is a concept that emerged in 1971 after Kotler and Zalltman (1971) presented an ideology to be implemented by companies. According to these authors, each responsible person must study and understand the thoughts and desires of consumers to be able to offer them the best product, from a social responsibility perspective: "(...) the design, implementation and control of programs, calculating the influence and acceptance of social ideas, involving these considerations in the planning, price, communication, distribution and marketing research of the product" (Kotler & Zalitman, 1971, p. 5). When developing a product or service, companies should consider and analyze society's problems or ideas to subsequently include them in the strategic planning of the product or service offered. That is, the marketing managers of each company should consider the society in which the company operates and act in such a way as to include society's social responsibility standards in the offer made to their customers (Ribeiro, 2013). Therefore, cause-related marketing (CRM) has become one of the main initiatives in corporate social responsibility (CSR) (Sousa & Soares, 2021). CRM has proliferated as a marketing strategy, is being employed by numerous brands across several product categories (Galan-Ladero & Galera-Casquet, 2018), and has experienced substantial growth over the last decades, and its utilization has spread into various areas of social and public life. Therefore, its main objective is to positively influence consumer attitudes and buying behavior (Sousa & Soares, 2021).

According to Cheng et al. (2011), socially responsible marketing implies the use of marketing principles and techniques not only to influence the behavior of the target audience but also to take into consideration the benefits for society in general. In this context, marketing is the main responsible for presenting a product or a service that is aware of its actions to the side of the offer, always including the ideals and principles of this. Thus, this concept aims to respond to social problems and inequalities through inclusion and strategic development, promoting practices that must be adopted by all companies in the development of their product or service. In the twenty-first century, companies began to develop concerns about poverty and social inequality and, consequently, guided their activity in the fight against these factors. In addition to the sustainable behavior on the part of companies, responsible behavior toward society was something that began to be required of companies, to avoid any problem both on the demand side and on the supply side (Rego et al., 2007). Aware of the consequences of their actions, company managers tend to develop their role by contemplating consumer demands from the perspective of not only product or service quality but also their impact on the environment and the community at large.

31.3.1 Disabilities Employment in Tourism

In the context of socially responsible marketing, employment for people with disabilities is an equally important opportunity and should be considered in all areas and, consequently, promoted. Although there are not many companies associated with this topic and employment campaigns for people with disabilities, the opportunities should be assessed and developed to include this group of people and guarantee them the right to employment and stability (Meira et al., 2021). According to the Portuguese Association of Disabled People, well-structured and applied employability strategies entail a higher cost.

However, there will be long-term growth and improvement in the supply of jobs. This is because it is more effective to promote this practice than to bear the costs of maintaining and supporting people with disabilities and unemployed people. The National Institute of Statistics developed a study in 2002 for people with disabilities and their employability. Between 15 and 64 years of age, people with some type of disability and people without any disability were identified. About 55.7% of people with disabilities are retired people and only 7.3% are students. There are very high percentages of unemployed people resulting from the inability to integrate these people into the world of employment. This requires a strategic solution of socially responsible marketing that involves the adoption of appropriate behavior by managers and business leaders. There is a need to combat unemployment by including these people in the routine of a company (Lima & António, 2009). A more recent study, but already some time old, pointed out in 2017 that by 2020 approximately 120 million Europeans will have a disability (European Commission, 2017), representing about 16%.

31.3.2 LGBTI and Social Responsibility

A topic that also deserves to be highlighted when it comes to social responsibility in marketing concerns LGBTI tourism. According to Silva and Vareiro (2021), tourism represents a large part of the economic activity in Portugal and revenues have increased over the years. However, this sector was deeply affected by the emergence of the pandemic in 2020, which leads to a decrease in revenues and economic activity. Regardless of the economic ups or downs of tourism, it is certain that this area contributes to global and national economic data. As mentioned above, tourism relates to the social responsibility aspect and the hospitality that is necessary for visitors, as they begin to feel greater freedom to be able to travel without many complications regarding the current situation. In this context, destinations need to program a safe environment for all their visitors. Thus, the question these two authors pose is, "Does this hospitality also apply to lesbian, gay, bisexual and transgender tourists?"

Guaracino and Salvato (2017) state that not all locals, destinations, or businesses welcome the LGBTI community, and because of this, there are few destinations that welcome and develop good practices and behaviors toward the community concerned. For these authors, if a tourist destination uses the right campaign, it will create interest alongside demand and, consequently, receive LGBTI tourists. Otherwise, the destination may even have good practices and be socially responsible, but if you do not disclose your actions, tourists will not know if the place is ideal to travel. Responsible behavior will motivate your growth and prominence at the social and economic level. Currently, there is still homophobia, and the tourist industry fights this hatred by promoting hospitality and security among tourists and the receiving community. However, it is not an easy task, as there is a need for strategies appropriate to this objective, and managers develop the behavior of the receiving site through the problems that are necessary to eradicate.

According to George (2021), in 1970 the trips and tours of the LGBTI community were already recognized due to various motivations, such as the Rio carnival. These events have allowed society to move forward in cultural, social, and political terms in the face of people's thoughts toward members of the LGBTI community. These advances allowed many people to assert themselves and not deny their identity and this was possible through the different experience that was provided to them. As a young target began to travel more and more and experience new tourist destinations as they developed. Despite this progress, there are still some negative aspects, notably prejudice and discrimination.

Despite these identified problems, such as homophobia, more and more LGBTI people have gained visibility in society, which makes the community more easily accepted compared to previous times. This development occurs in terms of demand and supply, leading to more dynamic tourism. The "LGBT tourist (...) is an experienced traveller and demonstrates an above average drinking aptitude" (Silva & Vareiro, 2021, p. 305). The interest in more common destinations also increases, which makes those responsible for tourism in these places have special attention to the hospitality, comfort, safety, and well-being of people, so that they feel welcome. Thus, "LGBT Tourism must be thought and worked within the framework of the entire value chain of tourism activity to obtain quality and competitiveness in tourism" (Silva & Vareiro, 2021, p. 306). According to Ferreira (2021), from the 2000s on, society began with revolutions to support the rights of the LGBTI community, and only in 2004 Portugal considered sexual orientation as a right. Although it was too late, the country has managed to take a positive step toward the rights of the LGBTI movement.

According to Ro and Khan (2022), society does not value the issue of acceptance of the LGBTI community and does not promote as many marketing campaigns in awareness and hospitality of society. It is important that companies understand that by adopting a positive and responsible attitude when it comes to the safety of LGBTI tourists, their safety, confidence, and well-being will be promoted. For will also create market prominence, attracting more tourists who are interested in traveling interest in traveling to responsible and LGBTI friendly places. According to these authors, the key to a company is the positive attitude of all its employees toward daily situations.

31.3.3 Pet-Friendly Tourism

According to Alves e Sousa (2022), a pet has a significant impact on people as it transmits affective emotions toward them. According to these authors, people who travel with their pets end up having some benefits, namely activities together. As the world has evolved, the "traditional image" of people going on holiday and leaving animals at home is also changing. This is because there is greater openness on the part of the offer when receiving owners and their animals. Thus comes the pet friendly tourism, which allows leisure time between a person and their pet (Huang et al., 2022). With this also arise new services and products for the welfare of the animal. Despite this responsible behavior, Alves and Sousa (2022) claim that it becomes a higher cost for hotel units, due to infrastructure and some restrictions of fate or other people's little empathy for this target.

According to Machado et al. (2017), many families consider pets as a family member and, therefore, tourism managers should focus on marketing "families with pets" strategy. At case, this market segment needs places where it can practice Tourism in which their pets are also accepted. Some countries in Europe and the USA accept animals in restaurants, and some airlines allow animal travel, and this is responsible behavior by parts of these companies. Also, some hotels respect the connection between pet animals and their owners and therefore allow the pets in their space, for example, hotel "The Yeatman" located in Vila Nova de Gaia. This wine hotel is very close to Porto Wines Cellars and has a beautiful view of the Oporto (world heritage city) and the Douro River that separates the two cities.

31.3.4 Ecofriendly Hotels

Tourist destinations are crossing points for thousands of people a year, and overtourism is one of the factors that leads to the great concern of tourism managers. Hospitality and welcoming tourists are important points, but there is also a need to make good practices on both the demand side and on the supply side. Instilling responsible behavior in companies is not an easy task to manage, as this will involve higher spending (Sadiq et al., 2022). However, in the long run, it will produce positive effects on the environment and the activity of each of the companies. In this context, UNWTO has developed a sustainability policy applicable to all hotel and tourism companies, with the aim of making them sustainable and environmentally friendly (ecofriendly) companies. These policies include the proper use of all natural and environmental resources in the development of tourism activities, the conservation of natural heritage and biodiversity, respect for receiving entities and their environment, respect for local traditions and cultural values, and the implementation of "viable and long-term" economic policies that provide "socio-economic benefits stakeholders". These benefits are applied with the aim of reducing poverty and balancing economic resources through "stable opportunities for employment, income and social services for the community" (UNWTO, 2022).

According to Yadegaridehkordi et al. (2021), ecofriendly hotels are places created and developed to implement green and healthy products/services, along with responsible practice behaviors toward the community in which they are located. It is important to convey a truly ecological image when it comes to the attitudes of and other employees in carrying out all their tasks.

31.4 Conclusions

Corporate social responsibility is a concern and a trend that deserves the development of studies in the academic community and business reflection. First, some notes were elaborated on tourism and its evolution, the profile of tourists, marketing, ethics, and social responsibility. Then, some important aspects that society and tourism businesses should consider during their activity were addressed, including inclusive employment opportunities, tourism and the LGBTI community, pet-friendly hotels, and ecofriendly hotels. Addressing these aspects is extremely relevant because only in this way can society evolve in the face of the problems occurring in it and contribute to the evolution of sustainability considering its three pillars: economic, social, and environmental. In terms of conclusions, it is important to mention that tourism is an industry strongly characterized by competitiveness and (also) social concerns. In this sense, organizations try to develop their differentiating nature with socially responsible measures. In this chapter, some of these socially responsible actions (in environmental, sustainable, or community well-being domains) were presented.

Many tourist companies do not develop their activity according to the concepts discussed above. However, there is a strong need to implement responsible behaviors in terms of inclusion, safety, sustainability, quality-price of the product and service offered, comfort, among others. The world is facing a situation that has altered the entire tourism system and negatively affected the economy of all countries. This issue has caused many people and businesses to be living on the edge and with many difficulties. Although there is a need to recover the economy, tourism will not leave aside environmental and social recovery and, for this, it is necessary to develop and implement sustainable strategies. This sustainable recovery will allow the tourism sector to resume its entire activity with greater economic, social, and environmental security. This chapter represents a starting point – limited to four perspectives and not very in-depth – for future studies and research related to the theme, since it is necessary to continue to promote social responsibility and ethical behaviors in tourism. However, further research need to be carried out in this area, and the new information should be compared with the information analyzed in this document, thereby verifying the evolution, stagnation, or regression of tourism in relation to these aspects.

References

Alves, H. M., & Sousa, B. (2022). Pet-friendly tourism. In *Encyclopedia of tourism management and marketing*. Edward Elgar Publishing.

Cheng, H., Kotler, P., & Lee, N. (2011). *Social marketing for public health: Global trends and success stories*. Jones and Bartlett.

Eger, C., Munar, A. M., & Hsu, C. (2022). Gender and tourism sustainability. *Journal of Sustainable Tourism, 30*(7), 1459–1475.

European Commission. (2017). Progress report on the implementation of the European Disability Strategy (2010–2020). https://asksource.info/resources/progress-report-implementation-european-disability-strategy-2010-2020

Ferreira, E. (2021). *Crimes de Ódio e Discriminação contra a comunidade LGBTI.* Bachelor's thesis, Universidade Fernando Pessoa, Faculdade de Ciências Humanas e Sociais, Porto.

Galan-Ladero, M., & Galera-Casquet, C. (2018). Corporate social responsibility and digital tools. The cause-related marketing case. In L. Carvalho & P. Isaias (Eds.), *Handbook of research on entrepreneurship and marketing for global reach in the digital economy* (pp. 1–16). IGI Global.

George, A. A. (2021). Social and cultural influences and their implications for sustainable tourism: A case study of Trinidad & Tobago's tourism and the growth of the LGBTQ traveler segment. *Journal of Sustainable Tourism and Entrepreneurship, 2*(3), 147–141.

Guaracino, J., & Salvato, E. (2017). *Handbook of LGBT tourism and hospitality: A guide for business practice.* Columbia University Press.

Huang, R., Krier, L., Josiam, B., & Kim, H. (2022). Understanding Consumer–Pet Relationship during Travel: A Model of Empathetic Self-Regulation in Canine Companionship. *Journal of Quality Assurance in Hospitality & Tourism, 23*(4), 1088–1105.

Kotler, P., & Zalltman, G. (1971). Social marketing: An approach to planned social change. *Journal of Marketing, 35*(3), 3–12.

Lima, F., & António, J. F. (2009). O emprego de pessoas com deficiência – uma breve análise do módulo ad-hoc de 2002. *EstatINE*, 55–59.

Lubowiecki-Vikuk, A., Dercan, B., & Sousa, B. (2021). Sustainable development and leisure services: Changes and trends. In A. Lubowiecki-Vikuk, B. de Sousa, B. Dercan, & W. Filho (Eds.), *Handbook of sustainable development and leisure services, world sustainability series* (pp. 1–20). Springer.

Machado, B. L., Pereira, D. d., Oliveira, S. G., & Cunha, T. A. (2017). Turismo e o segmento Pet Friendly: Um estudo sobre o setor hoteleiro de Belo Horizonte/MG. *Revista Hospitalidade, 14*(2), 1–14.

Meira, C., Martins, I. S., & Sousa, B. B. (2021). Accessible tourism and digital platforms: A study applied to the City of Viana do Castelo. In F. Brandão, Z. Breda, R. Costa, & C. Costa (Eds.), *Handbook of research on the role of tourism in achieving sustainable development goals* (pp. 235–247). IGI Global. https://doi.org/10.4018/978-1-7998-5691-7.ch014

Rego, A., Sousa, F., Pina e Cunha, M., Correia, A., & Saur-Amaral, I. (2007). Leader self-reported emotional intelligence and perceived employee creativity: An exploratory study. *Creativity and Innovation Management, 16*(3), 250–264.

Ribeiro, M. I. (2013). *Impacto do Marketing Socialmente Responsável para as Organizações.* Master's thesis, Universidade de Évora.

Ro, H., & Khan, M. (2022). The impact of LGBT friendliness on sexual minority customers' perceptions and intentions to stay. *International Journal of Hospitality Management, 102*, 103181.

Sadiq, M., Adil, M., & Paul, J. (2022). Eco-friendly hotel stay and environmental attitude: A value-attitude-behaviour perspective. *International Journal of Hospitality Management, 100*, 103094.

Santos, V. R., Vitorino, T. F., Dias, Á. L., Martinho, D., & Sousa, B. B. (2022). Developing a commercial ethics framework for Analysing marketing campaigns. *International Journal of Service Science, Management, Engineering, and Technology (IJSSMET), 13*(1), 1–16. https://doi.org/10.4018/IJSSMET.290336

Silva, S. d., & Vareiro, L. (2021). Perceção e atitude da comunidade local do Norte de Portugal face ao mercado e turismo LGBT. *Journal of Tourism & Development, 36*(2), 303–318.

Sousa, B. B., & Soares, D. B. (2021). Digital communication as a vehicle for cause-related marketing in cancer prevention: Lactogal case study. In M. M. Galan-Ladero, C. Galera-Casquet, & H. M. Alves (Eds.), *Cause-related marketing. Cases studies from a global perspective* (pp. 247–254). Springer. https://doi.org/10.1007/978-3-030-65455-9_20

UNWTO. (2022). *Tourism definitions.* CEDRO, Centro Español de Derechos.

Yadegaridehkordi, E., Nilashi, M., Nasir, M. H. N. B. M., Momtazi, S., Samad, S., Supriyanto, E., & Ghabban, F. (2021). Customers segmentation in eco-friendly hotels using multi-criteria and machine learning techniques. *Technology in Society, 65*, 101528.

32 The Impact of COVID-19 on Active Population Perception, Motivation, and Decision to Travel in the Northern Zone of Portugal

Óscar Silva, Catarina Lucas, and José Oliveira

32.1 Introduction

The emergence of the COVID-19 pandemic started around January 2020 in the Chinese city of Wuhan, Hubei Province. This is a respiratory disease caused by a virus called SARS-CoV-2 from the coronavirus family. It is a highly transmissible disease among humans causing respiratory failure or severe acute respiratory syndrome. It has proved to be a serious public health problem due to the direct impact on the lives of hundreds of thousands of people. Epidemics and pandemics generate severe impacts on the economy of countries (WHO, 2020; Feyisa, 2020).

In April 2020, all destinations worldwide had some type of travel restrictions related to COVID-19, forcing countries and states for the first time in recent history to adopt extreme restrictions on international travel (UNWTO, 2020). The most pessimistic loss forecasts were felt on airlines, hotel companies, restaurants, event companies, and even major entertainment centers, such as the Disney Complex (Tsionas, 2021; Segal & Gerstel, 2020). In the first 6 months of 2020 alone, the losses verified by the airlines exceeded US$ 84 billion (IATA, 2022). The imposed lockdown took over our daily lives in the form of closed borders, restrictions, a ban on the movement of people, and interrupted transport (Baum & Hai, 2020).

Ó. Silva (✉)
Centro de Investigação Techn&Art – Instituto Politécnico de Tomar,
Vila Nova de Gaia, Portugal
e-mail: osilva@ispgaya.pt

C. Lucas
ISPGAYA, Vila Nova de Gaia, Portugal

J. Oliveira
REMIT – Research on Economics, Management, and Information Technologies,
Vila Nova de Gaia, Portugal

© The Author(s), under exclusive license to Springer Nature Switzerland AG 2023

F. L. Almeida et al. (eds.), *Multidimensional Sustainability: Transitions and Convergences*, Springer Proceedings in Earth and Environmental Sciences,
https://doi.org/10.1007/978-3-031-24892-4_32

Consequently, social isolation is a strategy found to reduce and counteract the contagion curve, directly impacting the different economic activities, and with greater expressions in non-essential sectors such as tourism. According to UNWTO (2020), the pandemic could put at risk 100–120 million direct jobs in tourism, without considering indirect ones. The effects of the pandemic caused by COVID-19 caused an impact on the development of tourism activities worldwide (Gössling et al., 2020). Although many travel and tourism companies are used to including risk management and assessment model in their business planning (Ural, 2015; Ritchie & Jiang, 2019), the nature of this pandemic placed these models under strong pressure as their impacts were manifested on a global scale. According to Horta (2020), Portugal is one of the European countries where a strong reduction in tourism activity is expected, with a forecast of more than 40% in the number of visitors. The economic dependencies of multisectoral business organizations, regardless of the degree of magnitude, will survive, due to the combination of government aid and private funding.

Presently, it is important to think and discuss about "post-Covid tourism" (Baba et al., 2020; Chang et al., 2020; Haywood, 2020). When looking to the possible manisfestions of tourism that were in this pandemic context, virtual spaces, or imaginative mobility (Elliott & Urry, 2010; Urry, 2000), thus, gaining greater prominence, in counterpoint to the impossibility of conventional tourist flows. In this context, the different communication/promotion policies on the diverse tourist destinations have disseminated images in unrecognizable manners: alerting to the risk of traveling, not recommending, and advising to travel to a particular destination or staying at home. All these messages are usually accompanied by images of the destination(s), attraction(s), culture, and traditions, which, in a contradictory way, continue to feed a brand image dimension and imagination of tourist mobility. Assuming mobility as a category of analysis (Cresswell, 2010; Freire-Medeiros et al., 2018; Hannam et al., 2006; Kaufman, 2010; Sheller, 2014; Sheller & Urry, 2006), its complexity is recognized, leading to immobility.

The aim of this work consists of investigating how the COVID-19 pandemic influenced the tourist's motivation to travel and destination choice in the North of Portugal. This area is composed of 86 municipalities.

This chapter is organized as follows: It starts with an introduction to the problem under study. In Sect. 32.2, an extensive literature review is performed on the consequences of COVID-19 and the post-pandemic on the tourism sector. In Sect. 32.3, the methodology and data collection used for this research are explained. In Sect. 32.4, the obtained results and their discussion are presented. Finally, in Sect. 32.5, general conclusions, contributions, limitations, and clues for future investigations are presented.

32.2 Literature Review

In recent years, tourism has been the engine of economic growth in Portugal. 2019 was a very positive year for the tourism sector, with approximately 1.5 billion arrivals of international tourists (UNWTO – World Tourism Organization, 2020), with an effective growth of 4% in global tourism activity compared to the homologous period recorded in the previous year. In this same report, the OMT reaffirms the continued growth of the sector over the last decade, reaffirming a growth of 3–4% in 2020, with the advent of a series of impactful events such as the Tokyo Olympics and the cultural Expo 2020 in Dubai, impacted positively and reinforced the sector (Brito, 2020). In 2018, tourism in Portugal provided 1047 million jobs, which is equivalent to 21.8% of total employment (PRESSTUR, 2019). Furthermore, it represents circa 14.6% of the national GDP (TravelBI, 2019).

According to the World Travel & Tourism Council (WTCC, 2021), it is the fourth country with the highest rate of wealth generation by tourism (WEF, 2018). On the same period, the country followed the worldwide growth on arrivals and tourist revenues, having been awarded for such in 2017, and for the first time, considered the best tourist destination in Europe and the world, a prize obtained consecutively until the year 2020 (Seabra et al., 2020). The year 2020 could have been the best year ever for the Portuguese tourism industry (Turismo de Portugal, 2021). The months of January and February 2020, shortly before the registration of the first case of COVID-19 in Portugal, indicated that the number of overnight stays was increasing by 11.41% compared to 2019 (INE – National Institute of Statistics, 2020). According to Turismo de Portugal (2021), in 2020 there was a drastic decrease in the demand for tourist accommodation by 63% compared to the same equivalent period of 2019, a figure seen only in 1994. Citing the INE report (2020), tourist accommodation decreased by 60.4% and 61.1%, respectively. Additionally, the data released by Banco de Portugal in 2021 mentioned a decrease of 62.2% after the emergence of the pandemic, compared to the balance of travel and tourism, and also recorded a decrease in tourist exports, which in percentage terms means a decrease of 57.6% when compared to 2019, concerning the expenses and debts of tourist imports, there is a reduction of 46.1% when compared to 2019 (National Institute of Statistics, 2021).

According to WTTC (2021), in 2020 due to the pandemic situation, it is estimated that around 62 million jobs have been lost, corresponding to a decrease of 18.5% globally. Consequently, the tourism sector represents only 8.9% of all jobs. Also in 2020, the tourism sector affected by national and international mobility restrictions suffered a global loss of US$4.7 trillion and the contribution to gross domestic product decreased from 10.4% in 2019 to 5.5% in 2020. The expectations for the second decade of the twenty-first century were quite optimistic, riding on the exponential growth that was dominant; it was thus predicted that the number of arrivals of international tourists increased to 1.8 billion arrivals in 2030 (UNWTO – World Tourism Organization, 2011). However, this pandemic caused by this new coronavirus (SARS-CoV-2) imposed a sharp reduction and burst in tourist flows, with the adoption of containment measures implemented to try to control the spread of the virus (Gössling et al., 2020).

According to studies by Roehl and Fesenmaier (1992) on leisure and travel, several types of risk were identified with three dimensions of travel: leisure, physical equipment, and travel destination (Seabra et al., 2020). The risk factors associated with travel destinations have motivated the attention and interest of researchers: (i) health risk: physical danger, injury, or illness (Park & Reisinger, 2010); (ii) risk of natural disasters (Becken & Hughey, 2013); (iii) crime, political instability, and violence (Brunt et al., 2000; Fletcher & Morakabati, 2008; Saha & Yap, 2014); and (iv) risk of terrorism (Seabra et al., 2014). The health risk, due to the increasing frequency of pandemics, epidemics, and outbreaks in recent decades (Abrantes et al., 2021), has been increasingly explored in the literature on tourism. Thus, the risks introduced by changes in population mobility, business travel, tourism, migration, and emigration, in a much more globalized world, have caused this virus to spread differently (Shi et al., 2020). In this sense, the COVID-19 and the closing of the original celebration (festivities, leisure experiences, and dream fulfillment) generated a weakening of social ties and hospitality (Korstanje, 2020). Sensorial elements such as smells, sounds, taste, touch, and visual aspects can stimulate the emotions and memory of travel and travelers. Despite the thought, memories are dynamic and subject to change over time, through daily routine processes (Park & Santos, 2017). Coelho et al. (2018) point to the need to expand the classification of the study of the emotions of tourist experiences, whether positive or negative, and the scales and studies present are manifestly insufficient to describe how such emotions can be better understood and/or managed.

The state of emergency created by COVID-19 restricts individual rights to free mobility, a situation that is similar to terrorism and demonstrates material asymmetries between social classes. When it comes to post-pandemic tourism, the views are divided between an expectation of recovery of the levels of tourism of previous years, the implementation of protocols and medical-sanitary measures, and the search for reform, an opportunity to rethink its principles and ethics (Gössling et al., 2020; Hall et al., 2020; Higgins-Desbiolles, 2020a, b; Jamal & Budke, 2020). Furthermore, the expectations with the so-called "proximity tourism" (Ioannides & Gyimóthy, 2020) and valuing traditional practices in a fairer mobility regime (Sheller, 2021). Some studies investigated the relationship between communications and tourism during the pandemic, dealing with tourist perceptions and racial discrimination (Yu et al., 2021), as well as its effects on the mental health of Chinese tourists' victims of racist attacks (Zheng et al., 2020). If they notice that certain destinations are at risk, the probability of the tourist choosing them is virtually nil (Kozak et al., 2007).

Tourists, when traveling, may be exposed to risks associated with their health, especially when traveling to foreign and unknown destinations (Chua et al., 2020). COVID-19 created a level of fear and panic in individuals never before felt (Slanina et al., 2021; Zheng et al., 2022). The world is facing an unprecedented pandemic (Chebli & Said, 2020), a pandemic which persists and causes numerous uncertainties, notably in the real negative impacts it will have on the tourism industry (Neuburger & Egger, 2021). Roughly speaking in post-pandemic tourism, the views are divided between an expectation of recovery of previous tourism levels, as soon as medical and sanitary protocols are implemented, and others looking for retirement, and an opportunity to rethink its principles and ethics (Gössling et al., 2020; Hall et al., 2020; Higgins-Desbiolles, 2020a, b; Jamal & Budke, 2020).

The Strategic Vision for the Economic Recovery Plan of Portugal 2020–2030 points to options and priorities, which should guide the recovery of the harmful economic effects caused by the pandemic on tourism. This document states that Portugal should rely on its geographical and landscape diversity as a way to attract the reference markets again, betting on a diversified, competitive, and quality offer, through a good harmonization between conventional tourism and nature tourism, health tourism, cultural tourism, and recreational nautical. Many are the predictions of behavior change, and habits of tourists, in the face of the pandemic context. Demand for unmastified destinations is expected to increase, as they can offer unique and quality experiences to travelers. Places full of nature and without large crowds of people, where travelers can find small hotel units, can become the most sought-after places by tourists in the post-COVID-19 era (Almeida & Silva, 2020).

32.3 Methodology

This research investigated the touristic behavior during the COVID-19 pandemic of certain locations in Portugal. The methodology employed in this research was of qualitative nature, i.e., a case study, whereby questionnaires were applied. A case study should be performed when investigating contemporary events (Lopes et al., 2018; Whitehead, 2003). A case study is the most suitable method for this research as it allows to gather interpretations through observations (Ferreira et al. 2021; Lopes et al., 2022).

32.3.1 Data Collection Procedure

This chapter presents preliminary research results on how the COVID-19 pandemic influenced the tourist choice and preferences of the destinations located in Porto and Northern Portugal.

To achieve this goal, a questionnaire was distributed by e-mail to a sample of active population in the north of Portugal between 10 November2021 and 30 April 2022 (after the COVID19 pandemic).

A voluntary non-random (non-probabilistic) sampling was adopted. More than 3,000 individuals from the business community were contacted by email and invited to answer the questionnaire via a link. The sample contained 70 individuals, which represents a response rate of approximately 2%. The online survey was designed and created on Google Forms, and disseminated by companies, universities, and institutes.

It contained 51 questions that generated data related to the motivations that led participants to travel (e.g., leisure, gastronomic, and cultural); the activities they usually practice in the Porto and North of Portugal region; the criteria of choice that led them to travel to this region (e.g., climate, safety, and proximity); the attributes of the region that exceeded their expectations; the recommendations they would like to implement to improve the Porto and North of Portugal region as a safe and preferred destination in a post-COVID-19 scenario (e.g., greater mobility and access and greater hotel supply); municipalities visited in this region; and municipalities of choice.

The following control variables were used: age, gender, qualification level, job conditions, and profession. The database was cleaned, categories were grouped, variables were recorded, and negative items were reversed. All participants were informed about the study and the anonymity of the research process.

Of the 70 individuals who answered the questionnaire, 56% (N = 39) were female, with age described as approximately 37.91 ± 13.69 years, 67% had concluded university studies, 76% reported as being employed (21% self-employed and 54% employee), 16% were students, and 9% unemployed. The participants had professions mostly related to management, administration, and technical positions.

32.3.2 Measures

A 5-point Likert scale was used to study and measure the degree of perception of tourists regarding the safety of the region, given the current context of generalized deconfinement (1 = "Very unsafe" to 5 = "Very safe"); additional safety measures that could be accepted (1 = "Definitely not" to 5 = "Definitely yes"); and implication and relationship of COVID-19 with the choice of tourist destinations (1 = "Definitely not" to 5 = "Definitely yes").

32.3.3 Data Analysis

In terms of statistical software, Microsoft Excel, R version 4.1.2 (2021-11-01), and JASP were used. Given some gaps in the characteristics of the sample (the sampling technique and sample size), at this point, inferential statistics were not used, and only descriptive statistics of the different variables were used.

To characterize the tourists' motivations to travel, or their recommendations to turn Porto and the North of Portugal region into a safe and preferred destination, a Pareto analysis was implemented.

This methodology is based on the observation that operational results and economic wealth are not distributed evenly and that some inputs contribute more than others. It is referred to as the 80/20 rule, a nomenclature that has popularized a complex economic concept introduced by Vilfredo Pareto, a nineteenth-century Italian economist. The underlying concept is that the majority of problems (roughly 80%) are often caused by a small number of sources (roughly 20%). The 80/20 rule implies that most efforts are not efficient and should be reduced.

The strategic objective would be to leverage and maximize the efforts that produce most of the results. In strategic management, Pareto analysis is linked to the analysis of an organization's internal environment. It is particularly useful to identify internal strengths and weaknesses through the evaluation of an organization's internal resources and capabilities, which are the source of its core competencies and which in turn create competitive advantage (Powell & Sammut-Bonnici, 2015).

For example, when tourism entities have a complete picture of tourists' needs or the target audience's motivations to travel, action can be focused on the most relevant and sustainable measures. The use of this methodology, in this case, will consist of segmenting tourist motivations to travel or their recommendations to turn Porto and the North of Portugal region into a safe and preferred destination. In other words, list the motivations in the order they are most cited by tourists and, from there, create actions targeted at these motivations to improve the number of sales in the tourism sector in a more sustainable manner.

32.4 Results and Discussion

In the Pareto Diagram shown in Fig. 32.1, it can be seen that in order to meet the needs of the participants in this study and satisfy 83.3% of the factors that motivate them to travel, companies in the tourism sector must act essentially to promote programmes involving leisure, holidays, culture, nature and gastronomy in order to have a greater impact on increasing demand. When analyzing the activities effectively carried out by tourists and specifically in the Porto and North of Portugal region, it can be observed that 81% of the activities performed in the region corresponded to leisure, cultural, and sports activities (Fig. 32.2). Also, 81% of the criteria used to decide to travel to Porto and the Northern Portugal region corresponded to proximity, hospitality, monumental heritage attractions, price/quality, safety, weather, and tourism offer (Fig. 32.3). An inverse analysis suggests that the

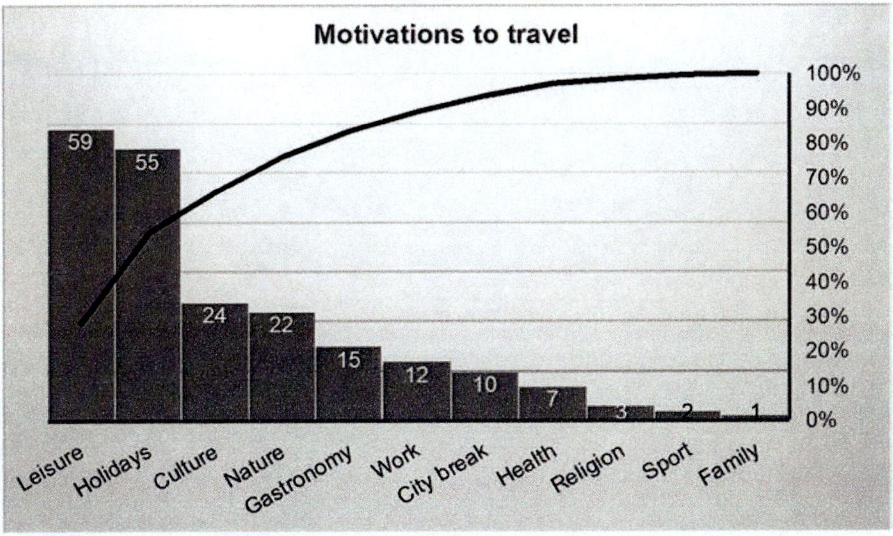

Fig. 32.1 Motivations to travel

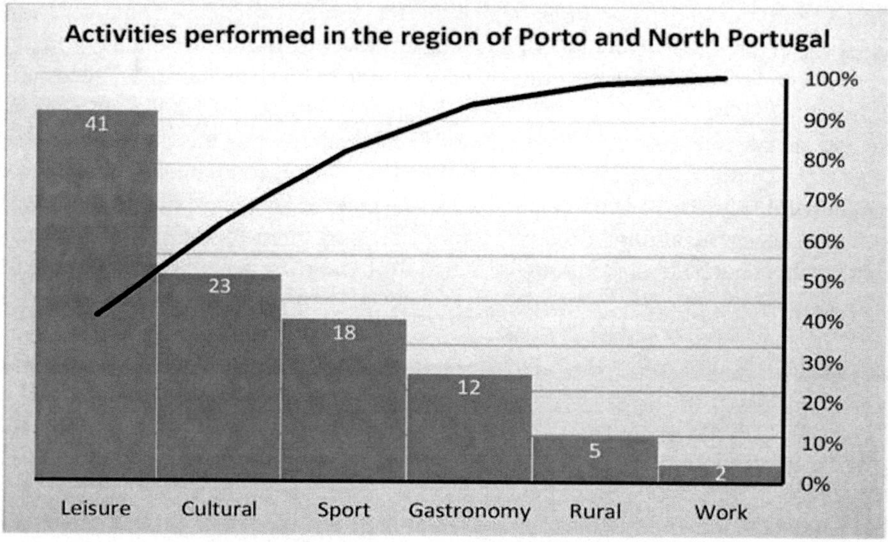

Fig. 32.2 Activities performed in Porto and North of Portugal region

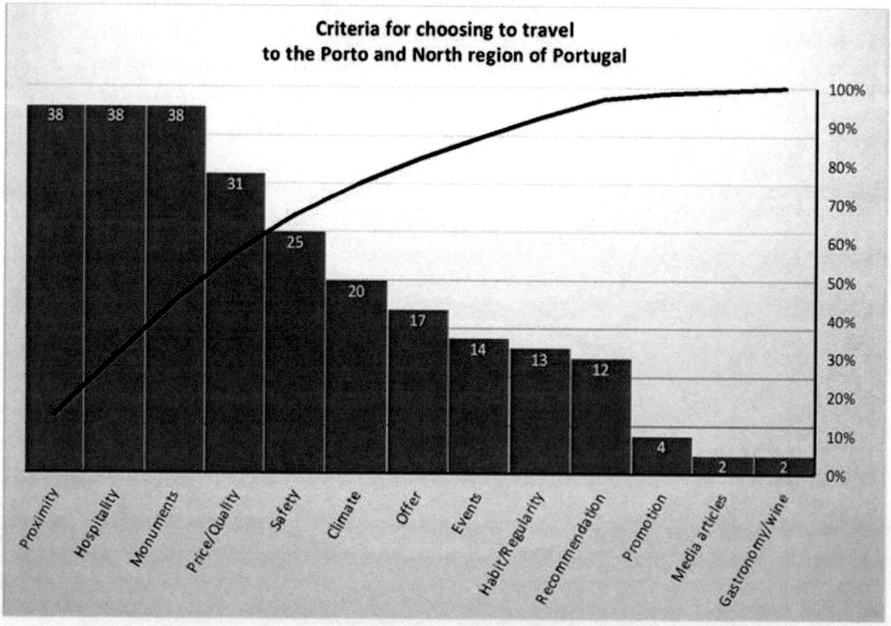

Fig. 32.3 Criteria for choosing to travel to the Porto and North of Portugal region

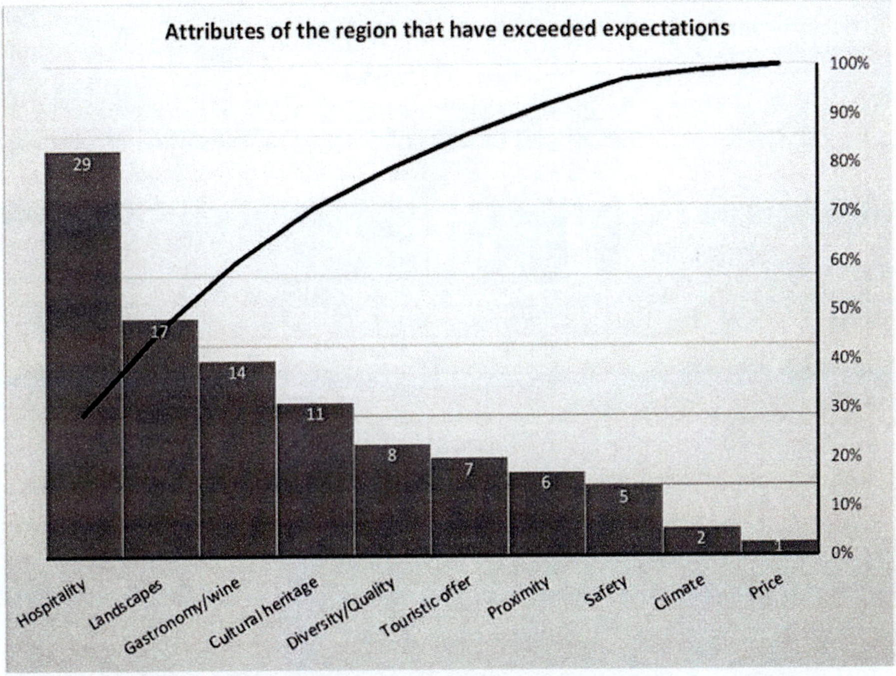

Fig. 32.4 Attributes of the region that have exceeded expectations

attributes with lower or null frequencies such as price, safety, and tourist offer, among others, which were not even mentioned as attributes by the participants, should be improved (Fig. 32.4). Thus, by investing in these characteristics, they may be considered attributes of the region by tourists in the future.

Furthermore, when analyzing the total number of motivations, it can be observed that, on average, each individual presents three motivations to travel with a standard deviation of 1.51. By looking into the total number of activities practiced in the Porto and North region of Portugal, it can be observed that on average each individual has 1.44 activities practiced with a standard deviation of 0.77. In terms of sports activities, hiking, canoeing, stand up paddle (SUP), and surfing are the most frequent. Considering the total number of decision criteria, selected by the participants, it is observed that, on average, each individual presents 3.63 criteria in a total of 12 options or more (min 0 and max 10) to travel to this region with a standard deviation of 1.94. Considering the total number of attributes of the region that have exceeded the expectations of the participants, it is observed that, on average, each individual presents 1.43 attributes to the region in an unlimited total (because it is an open-ended question) (min 0 and max 9) with a standard deviation of 1.17.

The following graphs show the results obtained with the analysis of the three dimensions (described by several items on a 5-point Likert scale) to study and assess the degree of tourists' perception regarding the security of the region, given the current context of widespread deconfinement (Fig. 32.5); additional security measures

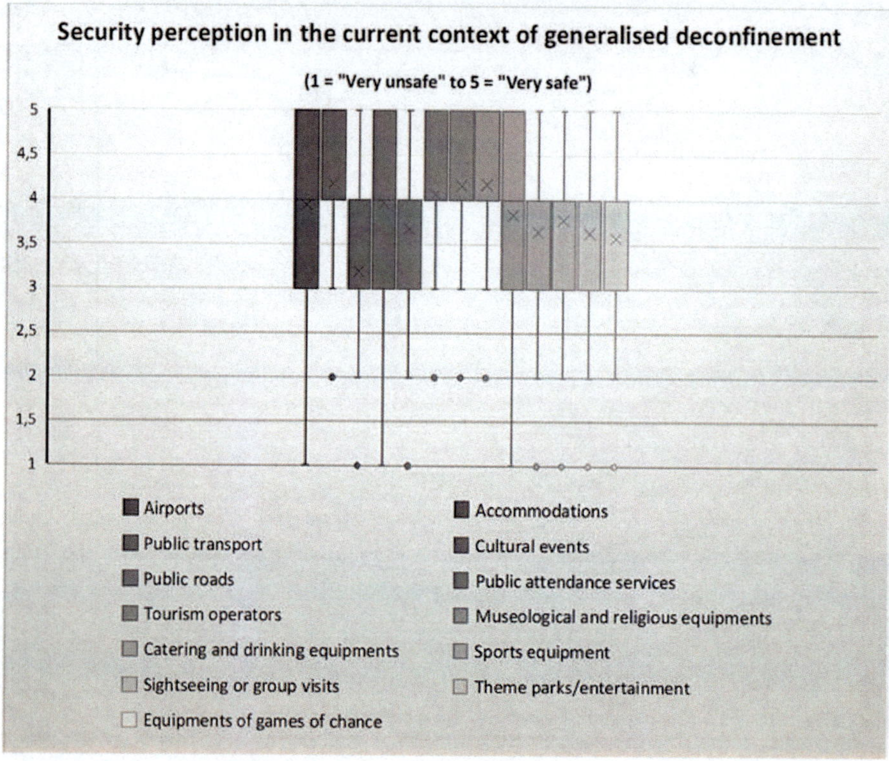

Fig. 32.5 Tourists' perception on the security of the region, given the current context of widespread deconfinement

that could be accepted (Fig. 32.6); and the implication and relationship of COVID-19 with the choice of tourist destinations (Fig. 32.7).

Considering the current context of generalized deconfinement, the degree of respondents' perception of security in relation to different measures implemented in the tourism sector presents always median above 3 (scale: 1 = "Very unsafe" to 5 = "Very safe"). Thus, the results indicate that tourists have a high degree of perception of safety in relation to different measures implemented in the tourism sector, considering the wild deconfinement.

In terms of extra security measures, there are large differences in terms of response agreements for certain items. Some of them present a more homogeneous distribution of answers, while other items do not so much.

Regarding the influence of COVID19 on the choice of tourist destinations, all indicator items presented responses with medians greater than or equal to 2.5 (1 = "Definitely not" to 5 = "Definitely yes").

To understand the data a little more and to go further in the analyses, the variables average safety, average of "extra measures," and average of "effect of COVID19 on choices" were created. The overall scores were obtained as an average

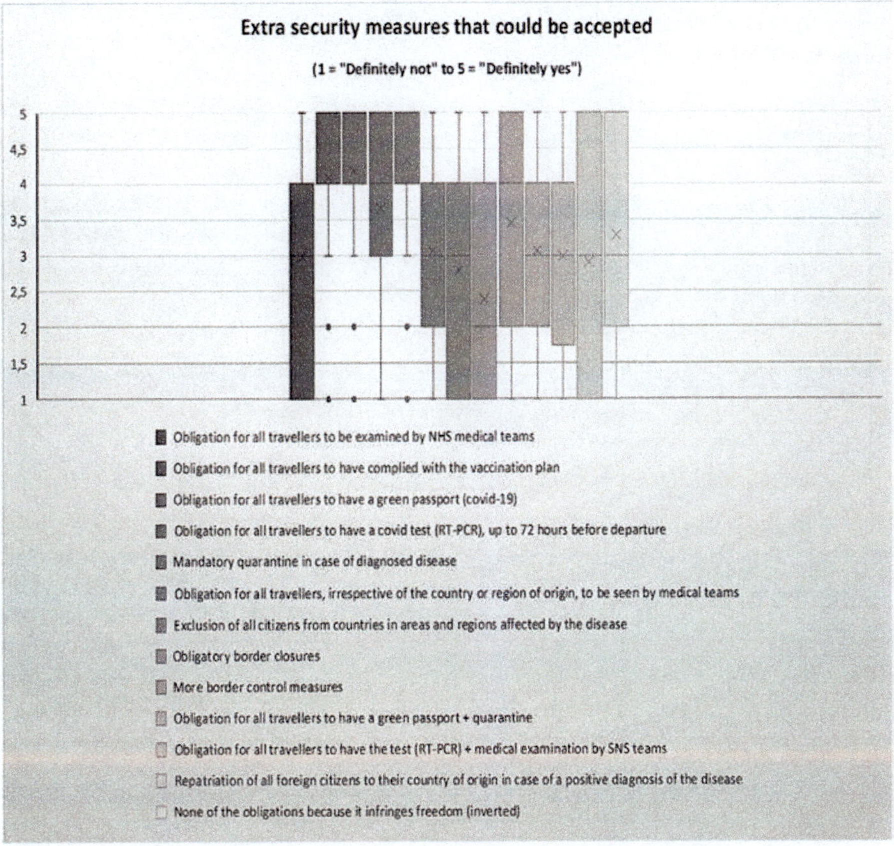

Fig. 32.6 Tourists' perception of the additional security measures that could be accepted

of the responses to all items regarding each dimension of perception. Table 32.1 shows the descriptive statistics for the variables of interest in this study.

Regarding the municipality of choice, and considering only the first answer, Oporto is the municipality with the highest frequency, followed by Viana do Castelo. More than 50% of the participants recommend an increase in the offer of transport and frequency, more investment, and a greater concern with the environment and heritage preservation for Porto and the North of Portugal to be considered a safe destination of choice in a post-Covid scenario. On the other hand, in terms of hospitality and hotel offers, there are not many recommendations (11% and 13%, respectively). According to Almeida and Silva (2020), these results refer that demand for unmastified destinations is subject to increase since they can no longer offer unique and quality experiences for travelers, one particular example that can be referred is the region of Porto and North of Portugal. It was predicted that after the COVID-19 era, small hotels in regions where nature is abundant and with fewer crowds would be more sought by travellers (Almeida & Silva, 2020).

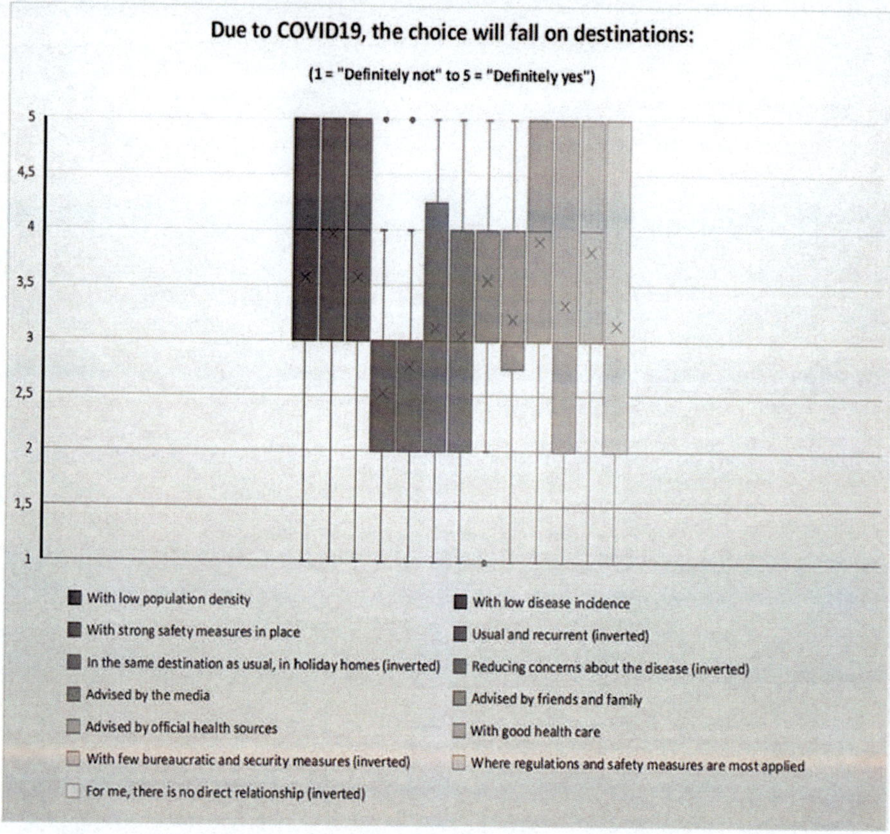

Fig. 32.7 Tourists' perception of the implication of COVID-19 on tourist destinations

32.5 Conclusions

32.5.1 General Conclusions

The conclusions denote a trend in the touristic behavior in Porto and the Northern Region during the COVID-19 Pandemic. However, although only preliminary results of a broader study of the perception of the security of the region, additional security measures that could be accepted, and the implication and relationship of COVID-19 with the choice of tourist destinations are presented here, an attempt was also made to formulate some hypothesis to verify the existence/absence of a dependency relationship between some variables.

As mentioned earlier and according to Almeida and Silva (2020), these results show that demand for unmastified destinations is subject to increase since they can no longer offer unique and quality experiences for travelers, one particular example

Table 32.1 Descriptive statistics of some variables of interest

Variable of interest	Mean (SD)	Scale
Perceptions' dimensions		
Overall security score	3.83 (0.67)	*1 = "Very unsafe" to 5 = "Very safe"*
Overall extra measures score	3.32 (0.92)	*1 = "Definitely not" to 5 = "Definitely yes"*
Overall effect of COVID19 on choices score	3.34 (0.54)	*1 = "Definitely not" to 5 = "Definitely yes"*
Some specific strong variables		
None of the obligations because it infringes freedom (not inverted)	2.71 (1.43)	*1 = "Definitely not" to 5 = "Definitely yes"*
For me, there is no direct relationship between COVID19 and the choice of destinations (not inverted)	2.86 (1.39)	*1 = "Definitely not" to 5 = "Definitely yes"*
Some new variables		
Number of criteria to travel to Porto and North of PT	3.63 (1.94)	*Out of 12 criteria options*
Number of suggestions for improvement so that the Porto and North of PT become a safe destination of choice post-COVID19	5.61 (2.60)	*Out of 16 suggestions options*
Number of attributes of the region of Oporto and North of PT that exceeded expectations	1.43 (1.17)	*With no limit of options (open answer)*
Number of municipalities visited in Porto and North of PT	36.01 (23.65)	*Out of 86 existing municipalities in the region of Porto and North of Portugal*

is the region of Porto and North of Portugal. It was predicted that after the COVID-19 era, small hotels in regions where nature is abundant and with fewer crowds would be more sought by travellers (Almeida & Silva, 2020).

Nevertheless, the results indicate that tourists have a high degree of security perception in relation to the different measures implemented in the tourism sector; furthermore, a large portion of the sample will not accept any additional security measures, unless mandatory by law. The respondents consider that COVID-19 had an impact on the choice of tourist destinations and recommend a greater frequency of transport and added investment so that the Porto and North region of Portugal are considered as a post-COVID-19 destination of choice. Proximity and hospitality stand out as the main decision-making criteria for traveling to the Porto and Northern Portugal region. However, when analyzing the activities being practiced in Porto and the northern region of Portugal, it is observed that most of the activities practiced in these regions correspond to leisure, cultural, and sports activities. As managerial advice, the authors suggest that companies in the tourism sector should act essentially in the promotion of programs that involve leisure, holidays, culture, nature, and gastronomy to obtain a greater impact, thus increasing the demand.

32.5.2 Limitations

Due to some limitations of the sample related to its small size and the sampling technique that was applied, inferential statistics were not used in this study, but only descriptive statistics to compare the proportions (%) of responses for the different categories. In particular, the fact that a non-random voluntary sampling was used, in an online survey disseminated through a link sent by e-mail, may lead to a non-representative sample of the population and consequent bias in the results (as only those who made themselves available responded to the questionnaire). To overcome this situation, future research intends to use a stratified sampling technique that allows obtaining a representative sample of the population.

Moreover, there was no data on travel frequencies (neither before, during nor after the pandemic), and in the questionnaire, there was no question about how often participants traveled to North Portugal: 1× per week, per month, per year. So, it is unknown who travels more or less. Another limitation of the study is that by using the variable describing "the number of municipalities already visited in the North of Portugal", it may be analysing individuals who travelled a lot before COVID and stopped travelling during and after COVID. Furthermore, this measure is relative: on the one hand, knowing a few municipalities does not imply travelling little because individuals can travel to the same municipality every weekend and get around more than others; on the other hand, knowing many municipalities does not imply travelling a lot, for example, individuals who have only visited the Caminhos de Santiago once, or a single trip through the North of Portugal, can be considered well-travelled using this single measure.

32.5.3 Clues for Future Investigations

Thus, in future work, with a more adjusted questionnaire and using a more adequate sampling technique, it is intended to test the following hypotheses: (i) Did the government measures have an impact on the decision to travel? (ii) Did the most "insecure people" (in terms of perception of security) travel less? (iii) Did individuals who had already circulated/or started to circulate through more municipalities correspond to the people with an average perception of higher safety in the pandemic period?

References

Abrantes, M., Abrantes, J., Silva, C., Reis, P., & Seabra, C. (2021). Health risks, pandemics and epidemics affecting tourism: understanding COVID-19 pandemic. In C. Seabra, C. Silva, O. Paiva, & J. Abrantes (Eds.), *Pandemics and travel: COVID-19 impacts in the tourism industry* (pp. 7–28).

Almeida, F., & Silva, O. (2020). The impact of COVID-19 on tourism sustainability: Evidence from Portugal. *Advances in Hospitality and Tourism Research (AHTR), 8*(2), 440–446. https://doi.org/10.30519/ahtr.775340

Baba, C., Stăncioiu, A.-F., Gabor, M. R., Alexe, F.-A., Oltean, F. D., & Dinu, A. C. (2020). Considerations regarding the effects of COVID-19 on the tourism market. *Theoretical and Applied Economics, 27*(3), 271–284.

Baum, T., & Hai, N. T. T. (2020). Hospitality, tourism, human rights and the impact of COVID-19. *International Journal of Contemporary Hospitality Management, 32*(7), 2397–2407. Recovery from: https://doi.org/10.1108/IJCHM-03-2020-0242.

Becken, S., & Hughey, K. (2013). Linking tourism into emergency management structures to enhance disaster risk reduction. *Tourism Management, 36*, 77–85. https://doi.org/10.1016/j.tourman.2012.11.006

Brito, J. (2020). *Turismo mundial sobe e registra 1,5 bi de chegadas de turistas internacionais em 2019, diz OMT*. Portal Brasileiro do Turismo. https://www.mercadoeeventos.com.br/_destaque_/turismo-mundialsobe-e-registra-15-bi-de-chegadas-de-turistas-internacionais-em-2019-dizomt/

Brunt, P., Mawby, R., & Hambly, Z. (2000). Tourist victimization and the fear of crime on holiday. *Tourism Management, 21*(6), 417–424. https://doi.org/10.1016/S0261-5177(99)00084-9

Chang, C.-L., McAleer, M., & Ramos, V. (2020). A charter for sustainable tourism after COVID-19. *Sustainability, 12*(9), 1–4. https://doi.org/10.3390/su12093671

Chebli, A., & Said, F. B., (2020). The impact of Covid-19 on tourist consumption behaviour: A perspective article. *Journal of Tourism Management Research, Conscientia Beam, 7*(2), 196–207. Recovered: https://ideas.repec.org/a/pkp/jotmre/v7y2020i2p196-207id2876.html

Chua, B. L., Al-Ansi, A., Lee, M. J., & Han, H. (2020). Impact of health risk perception on avoidance of international travel in the wake of a pandemic. *Current Issues in Tourism, 24*(7), 985–1002. https://doi.org/10.1080/13683500.2020.1829570

Coelho, M. d. F., Gosling, M. d. S., & Almeida, A. S. A. d. (2018). Tourism experiences: Core processes of memorable trips. *Journal of Hospitality and Tourism Management, 37*, 11–22. https://doi.org/10.1016/j.jhtm.2018.08.004

Cresswell, T. (2010). Towards a politics of mobility. *Environment and Planning D: Society and Space*, 28. https://doi.org/10.1068/d11407

Elliott, A., & Urry, J. (2010). *Mobile lives*. Routledge.

Feyisa, H. L. (2020). The world economy at COVID-19 quarantine: Contemporary review. *International Journal of Economics, Finance and Management Sciences, 8*(2), 63–74. https://doi.org/10.11648/j.ijefm.20200802.11

Ferreira, C., Cardoso, C., Travassos, M., Paiva, M., Pestana, M., Lopes, J. M., & Oliveira, M. (2021). Disorders, vulnerabilities and resilience in the supply chain in pandemic times. *Logistics, 5*, 48. https://doi.org/10.3390/logistics5030048

Fletcher, J., & Morakabati, Y. (2008). Tourism activity, terrorism and political instability within the Commonwealth: The cases of Fiji and Kenya. *International Journal of Tourism Research, 10*(6), 537–556. https://doi.org/10.1002/jtr.699

Freire-Medeiros, B., Telles, V., & Allis, T. (2018). Por uma teoria social on the move. *Tempo Social, 30*(2). https://doi.org/10.11606/0103-2070.ts.2018.142654

Gössling, S., Scott, D., & Hall, C. M. (2020). Pandemics, tourism and global change: a rapid assessment of COVID-19. *Journal of Sustainable Tourism, 29*(1), 1–20. https://doi.org/10.1080/09669582.2020.1758708

Horta, A. (2020). *Tourism in Portugal expected to fall dramatically*. Recovery from: https://www.theportugalnews.com/news/tourism-in-portugalexpected-to-fall-dramatically/53926

Hall, M., Scott, D., & Gössling, S. (2020). Pandemics, transformations and tourism: be careful what you wish for. *Tourism Geographies, 22*(3), 577–598. https://doi.org/10.1080/14616688.2020.1759131

Hannam, K., Sheller, M., & Urry, J. (2006). Mobilities, immobilities and moorings. *Mobilities, 1*(1), 1–22. https://doi.org/10.1080/17450100500489189

Haywood, K. M. (2020). A post COVID-19 future – Tourism re-imagined and re-enabled. *Tourism Geographies, 22*(3), 599–609. https://doi.org/10.1080/14616688.2020.1762120

Higgins-Desbiolles, F. (2020a). Socialising tourism for social and ecological justice after COVID-19. *Tourism Geographies, 22*(3), 610–623. https://doi.org/10.1080/14616688.2020.1757748

Higgins-Desbiolles, F. (2020b). The "war over tourism": challenges to sustainable tourism in the tourism academy after COVID-19. *Journal of Sustainable Tourism, 29*(4), 551–569. https://doi.org/10.1080/09669582.2020.1803334

Instituto Nacional de Estatística. (2020). *Estatísticas do Turismo, 2021*, 79.

International Air Transport Association (IATA). (2022, junho 09). *Prejuízo do setor aéreo deve atingir US$ 84 bilhões em 2020*. https://www.iata.org/contentassets/060a388cfde24a83b7f38fda101a2239/2020-06-09-01-pt.pdf

Ioannides, D., & Gyimóthy, S. (2020). The COVID-19 crisis as an opportunity for escaping the unsustainable global tourism path. *Tourism Geographies, 22*(3), 624–632. https://doi.org/10.1080/14616688.2020.1763445

Jamal, T., & Budke, C. (2020). Tourism in a world with pandemics: Local-global responsibility and action. *Journal of Tourism Futures, 6*(2), 181–188. https://doi.org/10.1108/JTF-02-2020-0014

Lopes, J. M., Gomes, S., & Mané, L. (2022). Developing knowledge of supply chain resilience in less-developed countries in the pandemic age. *Logistics, 6*(1), 3. https://doi.org/10.3390/logistics6010003

Lopes, J. N., Farinha, L. M., Ferreira, J. J., & Ferreira, F. A. (2018). Peeking beyond the wall: Analysing university technology transfer and commercialisation processes. *International Journal of Technology Management, 78*(1–2), 107–132. https://doi.org/10.1504/IJTM.2018.093936

Kaufman, V. (2010). *Re-Thinking Mobility: Contemporary Sociology*. Ashgate.

Korstanje, M. E. (2020). Passage from the tourist gaze to the wicked gaze: A case study on COVID-19 with special reference to Argentina. In *International case studies in the management of disasters* (pp. 197–211). Emerald Publishing Limited. https://doi.org/10.1108/978-1-83982-186-820201012

Kozak, M., Crotts, J. C., & Law, R. (2007). The impact of the perception of risk on international travellers. *International Journal of Tourism Research, 9*(4), 233–242. https://doi.org/10.1002/jtr.607

Neuburger, L., & Egger, R. (2021). Travel risk perception and travel behaviour during the COVID-19 pandemic 2020: A case study of the DACH region. *Current Issues in Tourism, 24*(7), 1003–1016. https://doi.org/10.1080/13683500.2020.1803807

Park, S., & Santos, C. A. (2017). Exploring the tourist experience: A sequential approach. *Journal of Travel Research, 56*(1), 16–27. https://doi.org/10.1177/0047287515624017

Park, K., & Reisinger, Y. (2010). Differences in the perceived influence of natural disasters and travel risk on international travel. *Tourism Geographies, 12*(1), 1–24. https://doi.org/10.1080/14616680903493621

Powell, T., & Sammut-Bonnici, T. (2015). Pareto analysis. In C. L. Cooper, J. McGee, & T. Sammut-Bonnici (Eds.), *Wiley encyclopedia of management*. https://doi.org/10.1002/9781118785317.weom120202

PRESSTUR. (2019). *Portugal é o 4° país onde o turismo tem mais 'peso' no emprego total*. http://www.presstur.com/mercados/portugal/portugal-e-o-4%2D%2Dpais-onde-oturismo-tem-mais-peso-no-emprego-total/

Ritchie, B., & Jiang, Y. (2019). A review of research on tourism risk, crisis and disaster management: Launching the annals of tourism research curated collection on tourism risk, crisis and disaster management. *Annals of Tourism Research, 79*, 1–15. https://doi.org/10.1016/j.annals.2019.102812

Roehl, W. S., & Fesenmaier, D. R. (1992). Risk perceptions and pleasure travel: An exploratory analysis. *Journal of Travel Research, 30*(4), 17–26. https://doi.org/10.1177/004728759203000004

Saha, S., & Yap, G. (2014). The moderation effects of political instability and terrorism on tourism development: A cross-country panel analysis. *Journal of Travel Research, 53*(4), 509–521. https://doi.org/10.1177/00472875134964

Seabra, C., Abrantes, J. L., & Kastenholz, E. (2014). The influence of terrorism risk perception on purchase involvement and safety concern of international travellers. *Journal of Marketing Management, 30*(9–10), 874–903. https://doi.org/10.1080/0267257X.2014.934904

Seabra, C., Pereira, A., Silva, C., Abrantes, J. L., Reis, M., & Paiva, O. (2020). Destination image perceived by domestic tourists. *European Journal of Tourism Research, 25*, 2506. http://hdl.handle.net/10362/99909

Segal, S. & Gerstel, D. (2020, março). *The global economic impacts of Covid-19*. Center for Strategic & International Studies. https://www.csis.org/analysis/global-economic-impacts-covid19

Sheller, M. (2014). Sociology after the mobilities turn. In P. Adey, D. Bissel, K. Hannam, & M. Sheller (Eds.), *The Routledge handbook of mobilities*. Routledge.

Sheller, M. (2021). Reconstructing tourism in the Caribbean: connecting pandemic recovery, climate resilience and sustainable tourism through mobility justice. *Journal of Sustainable Tourism, 29*(9), 1436–1449. https://doi.org/10.1080/09669582.2020.1791141

Sheller, M., & Urry, J. (2006). The new mobilities' paradigm. *Environment and Planning A: Economy and Space, 38*(2), 207. https://doi.org/10.1068/a37268

Shi, Q., Dorling, D., Cao, G., & Liu, T. (2020). Changes in population movement make COVID-19 spread differently from SARS. *Social Science & Medicine, 255*, 113036. https://doi.org/10.1016/j.socscimed.2020.113036

Slanina, H., Madhugiri, R., Bylapudi, G., Schultheiß, K., Karl, N., Gulyaeva, A., et al. (2021). Coronavirus replication–transcription complex: Vital and selective NMPylation of a conserved site in nsp9 by the NiRAN-RdRp subunit. *Proceedings of the National Academy of Sciences, 118*(6), e2022310118. https://doi.org/10.1073/pnas.20223101

TravelBI. (2019). *Consumo Turístico representa 14,6% do PIB em 2018 | Conta Satélite do Turismo 2018*. https://travelbi.turismodeportugal.pt/pt-pt/Paginas/conta-satelite-do-turismo2018.aspx

Tsionas, M. G. (2021). COVID-19 and gradual adjustment in the tourism, hospitality, and related industries. *Tourism Economics, 27*(8), 1828–1832. https://doi.org/10.1177/13548166209330

UNTWO – World Tourism Organization. (28 de Abril de 2020). UNTWO – World Tourism.

UNWTO – World Tourism Organization. (2011). *Tourism towards 2030/Global overview*. UNWTO – World Tourism Organization.

Ural, M. (2015). Importance of risk management for the sustainability of tourism. *Balıkesir University the Journal of Social Sciences Institute, 18*(33).

Urry, J. (2000). *Sociology beyond societies: Mobilities for the twenty-first century*. Routledge.

WEF. (2018). *The travel & tourism competitiveness report 2018*. World Economic Forum.

Whitehead, D. (2003). Case study research design and methods 3rd edition. *Journal of Advanced Nursing, 44*, 108. https://doi.org/10.1046/j.1365-2648.2003.02790_1.x

World Health Organization (WHO). (2020, March 11). *WHO Director-General's opening remarks at the media briefing on COVID-19 – 2020*. https://www.who.int/dg/speeches/detail/who-directorgeneral-s-opening-remarks-at-the-media-briefing-on-covid-19%2D%2D-11-march-2020

World Tourism Organization [UNWTO]. (2020, April 28). *COVID-19 related travel restrictions a global review for tourism* (Second report). https://webunwto.s3.eu-west-1.amazonaws.com/s3fs-public/2020-04/TravelRestrictions%20-%2028%20April.pdf

World Travel & Tourism Council. (2021). Economic impact 2021: Global economic impact & trends 2021. *WTTC, 27*.

Yu, M., Li, Z., Yu, Z., He, J., & Zhou, J. (2021). Communication related health crisis on social media: A case of COVID-19 outbreak. *Current Issues in Tourism, 24*(19), 2699–2705. https://doi.org/10.1080/13683500.2020.1752632

Zheng, D., Luo, Q., & Ritchie, B. W. (2022). The role of trust in mitigating perceived threat, fear, and travel avoidance after a pandemic outbreak: A multigroup analysis. *Journal of Travel Research, 61*(3), 581–596. https://doi.org/10.1177/0047287521995

Zheng, Y., Goh, E., & Wen, J. (2020). The effects of misleading media reports about COVID-19 on Chinese tourists' mental health: A perspective article. *Anatolia, 31*(2), 337–340. https://doi.org/10.1080/13032917.2020.1747208

Index

© The Editor(s) (if applicable) and The Author(s), under exclusive license to Springer Nature Switzerland AG 2023

F. L. Almeida et al. (eds.), *Multidimensional Sustainability: Transitions and Convergences*, Springer Proceedings in Earth and Environmental Sciences,
https://doi.org/10.1007/978-3-031-24892-4